Second Edition

RESISTANCE WELDING

Fundamentals and Applications

Second Edition

RESISTANCE WELDING

Fundamentals and Applications

Hongyan Zhang and Jacek Senkara

CRC Press is an imprint of the
Taylor & Francis Group, an **informa** business

CRC Press
Taylor & Francis Group
6000 Broken Sound Parkway NW, Suite 300
Boca Raton, FL 33487-2742

First issued in paperback 2017

Version Date: 20110812

ISBN 13: 978-1-4398-5371-9 (hbk)
ISBN 13: 978-1-138-07524-5 (pbk)

Library of Congress Cataloging-in-Publication Data

Zhang, Hongyan, 1962-
Resistance welding : fundamentals and applications / Hongyan Zhang and Jacek Senkara. -- 2nd ed.
p. cm.
Includes bibliographical references and index.
ISBN 978-1-4398-5371-9 (hardcover : alk. paper)
1. Electric welding. I. Senkara, Jacek. II. Title.

TK4660.Z43 2012
671.5'213--dc23 2011029702

Visit the Taylor & Francis Web site at
http://www.taylorandfrancis.com

and the CRC Press Web site at
http://www.crcpress.com

To

Kevin, Jackie, and Qun

Ela, Agatha, Anna, and Christina

Contents

Preface

Resistance welding, largely represented by and referred to as resistance spot welding, is a complex, yet exciting subject for both research and engineering practice. The multi-process nature of resistance welding and its wide range of industrial applications have attracted sufficient attention from researchers around the world, and a huge number of publications have been devoted to this topic. However, because of the complexity of resistance welding processes, research papers and books generally take two drastically different approaches: they either focus on a specific topic, or they cover most of the multiple facets of a welding process in a handbook style. In order to obtain a general knowledge of resistance welding of necessary depth, one must rake through a large quantity of research works on resistance welding as well as many related essential subjects, such as statistical analysis. This book is intended to provide the reader a systematic view of the fundamentals and applications in resistance welding, which may benefit both students and researchers in academia and welding practitioners in the sheet metal industry.

Resistance welding is quite different from other well-cultivated subjects, such as plasticity or dynamics, in which one may start with a number of reasonable assumptions, derive a set of equations based on these assumptions, and solve them. In welding, however, the engineering sense, which is largely qualitative rather than quantitative, is more important in understanding the process, solving problems, and drawing valid conclusions. For instance, the electrical current, an important process parameter for resistance welding, is not determined through solving a set of equations, although some commercial software packages claim that they can do so through numerical calculation alone. The actual value of an electric current is almost always determined experimentally, often with the guidance from handbooks or computer programs. The difficulties in both teaching and research in welding lie in dealing with the complicated multi-physical processes involved in welding, especially resistance spot welding. It is very hard, if not impossible, to extract *"a"* dominant process even for a particular aspect of resistance welding. Two or more coupled simultaneous processes, such as electrical and metallurgical processes, in welding make quantitative analyses impractical, as evidenced by the fact that there are no commercially available programs that are capable of treating the main physical processes in a fully coupled manner.

As a result, resistance welding is often treated as an "art" rather than a "science." Based on the state-of-the-art research results, this book presents the fundamental aspects of the important processes in resistance welding and discusses their implications on real-world welding applications in a systematic manner. As educators and researchers in welding, the authors have extensive interaction with the sheet metal manufacturing industry. The experiences they gained over the years and the desire to have suitable textbooks for their teaching provoked the inception of writing this book.

Welding metallurgy is presented first because of its obviously important role in resistance welding. Phase transformations in resistance spot welding, including melting, solidification, and solid-state phase transformations in a heat-affected zone, determine the quality of a weldment. They are introduced in Chapter 1 and emphasized in other chapters whenever an application requires such a consideration. Another largely metallurgy-dependent process also presented in the chapter is cracking in either the nugget or the heat-affected zone. This subject deserves more attention, and a more in-depth discussion can be found in Chapter 3. Significant changes and additions are made in the current edition to reflect advances in the past few years in adopting magnesium alloys in the automobile industry, whereas the previous (first) edition of the book dealt with steels and aluminum alloys only. In Chapter 2, the basics of welding schedule selection based on the fundamental thermo-electrical consideration are presented. The electrode life is discussed in the chapter, considering both the thermo-electrical and metallurgical effects. Although the quality of a weld refers to a number of performance characteristics of the weld and is measured in various ways, mechanical testing is the most common means to test its quality because of its simplicity and consistency. The commonly conducted mechanical tests are presented in Chapter 4, and a more general definition of weld quality together with its measures, either obtained in a destructive or non-destructive manner, is covered in Chapter 6. The monitoring and control of a welding process are essential in ensuring weld quality, and they are presented in Chapter 5.

The mechanisms of expulsion, an important process in resistance welding largely responsible for defect formation and other unwanted features, are thoroughly analyzed in Chapter 7. This edition also presents a different type of expulsion observed in welding magnesium AZ91D that is not covered by previous expulsion models. The metallurgical, electrical, and thermo-mechanical influences are discussed, and the methods to predict and suppress expulsion are proposed and experimentally verified. The influence of the mechanical aspects of welding machines, which is often overlooked in resistance welding-related research, is presented in Chapter 8. Chapter 9 presents the procedure for numerically simulating a resistance welding process. Although it is not as mature as people have hoped, simulation proves a significant help in understanding the welding process, as it provides an insight to the process that is impossible to obtain otherwise. The last chapter of the book has been devoted to statistical design and analysis that is a method especially suitable for welding-related applications. This chapter explains the procedure of statistical analysis using welding research as an example. The research results by the authors have shown that a statistical analysis can indeed provide useful information of a welding process. This chapter provides an accurate yet convenient step-by-step procedure to apply statistical approaches to welding research. Overall, this book places its emphasis on establishing a relationship among welding parameters, characteristics of welded joints, and their performance. We have tried our best to provide, based on the available resources, a thorough review of the state-of-the-art results in resistance welding research and a solid foundation for solving practical problems in a scientific and systematic manner to the reader.

This book would not be possible without the encouragement and collaboration of our colleagues, friends, families, and the staff at CRC/Taylor & Francis over a span of more than 14 years. Although it is impossible to list everybody who has helped at various stages in the course of preparing this book, the authors feel obliged to acknowledge those based on their best recollections. While omission cannot be avoided, it is certainly not intentional as the planning and writing of this book have stretched over such a long period of time. The authors are extremely grateful for Mr. John C. Bohr and Prof. C.-L. Tsai whose encouragement was the determinant factor for the authors to undertake this exciting task. Many of our colleagues in Advanced Technology Program, Intelligent Resistance Welding (IRW), have provided various support in a span of 5 years, which formed the basis of this book. Drs. S. J. Hu, X. Wu, W. Li, H. Peng, M. Zhou, W.-K. Hou, and H. Tang, Mr. J. Grasse, Drs. M. V. Li, X. Sun, and Z. Feng, Mr. M. Kimchi, Mr. M. Fleming, Mr. D. Androvich, Mr. D. Boomer, Mr. J. W. Dolfi, Mr. T. Mackie, Mr. T. Morrissett, Mr. A. M. Turley, and Dr. W. Trojanowski, all contributed in some way to the first edition of this book. Special thanks go to Dr. S.-W. Cheng who helped in writing the chapter on statistical analysis, Mr. P. Deshpande who helped in writing the chapter on numerical simulation, and Dr. S. Babu who generously allowed the authors to use many of his illustrations in this book. Dr. J. Jakubowski helped in preparing several photos used in the book. The help from the current and former students of the authors, Dr. A. Shayan, Ms. H. Zheng, Mr. N. Ari, Mr. G. Karve, Mr. S. Agashe, Mr. V. Vaddadi, Ms. X. Su, and Mr. K. Yadav, is highly appreciated. The authors have benefited tremendously from their professional interaction with welding practitioners in the US automobile industry. In particular, the biannual Sheet Metal Welding Conferences chaired by J. Bohr, M. Karagoulis, M. Palko, and M. Gugel have inspired the authors in many ways.

Revising this book was proposed by our editor at CRC Press/Taylor & Francis, Allison Shatkin, senior editor of materials science and chemical engineering, and the planning was critiqued and perfected by our peers at the universities and the sheet metal industry, including Drs. W. Li, M. D. Tumuluru, X. Wu, D. L. Chen, and S. Ramasamy.

A high standard of professionalism has been demonstrated in the course of publishing the first edition by CRC/Taylor & Francis Group, LLC, through its staff members including S. Kronzek and C. R. Carelli (acquisitions editors), T. Delforn (project coordinator), and K. L. Nazzaro (project editor). The significant improvement of the new edition would not have been possible without the detailed guidance, constant encouragement, and extraordinary patience of A. Shatkin (senior editor), K. A. Budyk (senior project coordinator), E. Curtis (project editor), A. Dale (editorial assistant), and A. Nanas (project manager). It has been a rewarding process working with such a group of knowledgeable, friendly, and enthusiastic individuals.

Finally, the authors express their gratitude toward their respective families for their unconditional support, love, understanding, and belief in their husbands and fathers in the course of writing this book.

Hongyan Zhang
The University of Toledo, Toledo, Ohio

Jacek Senkara
Warsaw University of Technology, Warsaw, Poland

Authors

Dr. Hongyan Zhang is an Associate Professor at the Department of Mechanical, Industrial, and Manufacturing Engineering (MIME), University of Toledo. He holds a BS in Applied Mathematics, an MS in Metal Physics, and a PhD in Materials Science. Among many of his research and teaching interests are materials, forming, welding, and mechanical fastening; manufacturing process monitoring and control; failure analysis; and hybrid propulsion systems. He has published over 70 peer-reviewed journal and conference papers and contributed to a number of American Welding Society Standards. Dr. Zhang has served as a principal reviewer for *Welding Journal* and as a reviewer for several other journals.

Dr. Zhang has extensive experience working with the automotive industry. He has been active in a number of AWS/SAE technical committees, and he has served as an organizer for several conferences such as Sheet Metal Welding Conferences and ASME annual meetings.

Dr. Jacek Senkara is a Professor of the Production Engineering Faculty at Warsaw University of Technology in Poland. He holds an MS in Metallurgy, a PhD in Materials Science, and a DSc in Welding Engineering. His research interests include fundamental research in welding such as materials aspects of welding and welding-related processes, along with surface modification of materials.

Dr. Senkara is the author of about 100 scientific and technical papers published in professional journals and conference proceedings. He has served as a principal investigator for a number of government, industry, and university-supported research projects. He has been teaching courses of Fundamentals of Welding and Welding Metallurgy for undergraduate, MS, and PhD students. From 1995 to 1999, while on leave from his parent university, he worked as a visiting scientist at the University of Michigan at Ann Arbor.

Dr. Senkara's service to professional organizations includes active involvement in the State Committee for Scientific Research, Polish Welding Society, and Polish Vacuum Society.

1

Welding Metallurgy

Welding is a metallurgical process—all aspects of a welding process can be, more or less, related to the metallurgy of the materials involved in welding, either the base metal or the electrodes. There are a number of books dedicated specifically to welding metallurgy.[1–7] Although most of them are on fusion welding, the general metallurgical principles are applicable to resistance spot welding (RSW). In this chapter, the metallurgical principles governing the various aspects of RSW are discussed. They are critical in understanding the formation of the structures of an RSW-welded joint, the mechanisms of defect formation, and their impact on a weld's strength. This chapter is categorized according to the materials most relevant to RSW as workpieces and electrodes. The metallurgical characteristics of steels, aluminum alloys, and magnesium alloys that affect welding processes and weld quality are discussed. In addition to "conventional" materials used in RSW such as steels, magnesium alloys are also included because of their increasing presence in automobile construction for significant weight reduction. The impact of electrode material on resistance welding has been widely recognized by the resistance welding community, yet little can be found from the public domain that directly aids the understanding and control of the RSW process. In fact, many processes in RSW are electrode dependent. For instance, resistance heating at the electrode–workpiece interface introduces unwanted changes such as alloying and others, affecting the life and performance characteristics of the electrodes, and the integrity of the weld. Therefore, Cu is included in this chapter as it is the most common material for electrodes. Finally, the metallurgical aspect of cracking is presented. For additional information regarding the metallurgy in resistance spot welding these materials, the reader is referred to the recommended reading listed at the end of this chapter.

1.1 Solidification in Resistance Spot Welding

The cast structure of ingots in the sheet materials used in RSW, such as steels, is deliberately modified by hot or cold working, such as rolling and heat treatment operations. In the process, grains are refined through cold working and recovery/recrystallization, and structures are homogenized through solution annealing or quenching and tempering. However, such operations are difficult to perform in welding, especially in RSW, as melting and solidification occur between two sheets in a short period. Welding parameters, such as hold time and post-heating, may alter the microstructure to a certain extent. However, because of the steep temperature gradient in a weldment, the extremely high cooling rate, and the very short time elapsed in welding, such a treatment is not comparable to the controlled heat treatment processes of the parent sheets. Therefore, the microstructures and properties of a weldment are generally not as optimized as in the base metal.

During welding, solidification of a liquid nugget is similar to that in a metal casting. It consists of two steps: nucleation of solid phases and subsequent crystal growth, same as solidification in an ingot mold. The crystallization process is controlled by the heat dissipation into the base metal and the electrodes. The direction and rate of cooling, in addition to the alloy's composition, decisively affect the type, size, and orientation of the crystals formed. During solidification of a liquid nugget, a change in alloy composition takes place in the crystals being precipitated, compared to the original composition of the alloy. In the case of a very rapid cooling of a spot weldment, the insufficient diffusion in the precipitated solid crystals and the remaining liquid, and the difference in solubility of certain elements in solid and liquid, produce a sharp gradient in the composition distribution through microsegregation. The difference in composition between the core and outer layer of a crystal increases with increasing distance between the liquidus and solidus lines in a phase diagram, and decreases with increasing diffusion rate and the time span for solidification. In addition to microsegregation, which occurs in the scale of crystals, segregation also takes place as the solid–liquid interface advances into the liquid, as solidification proceeds, and results in enrichment in concentration in the remaining melt of alloying elements. Some of the elements form eutectics of lower melting temperature that exist in the liquid state, mainly around the central portion of a nugget after it is cooled to a temperature below the solidus of the alloy but above the eutectic temperature. Examples of such eutectics are Al–Cu, Al–Mg, and Al–Mg–Si in aluminum alloys, and certain compounds such as sulfur and phosphorous eutectics in steels. Because of their lower melting temperatures, they are the last bits of liquid to solidify, mainly at grain boundaries, as they are rejected from the solidified crystals due to reduced solubility during cooling. Grains surrounded by such liquid at the boundaries can be torn apart as the liquid has no strength when they are stretched, either by external loading or thermal stresses in the same way as in the case of fusion welding. However, such cracking rarely occurs in RSW as it may be suppressed by the pressure from the electrodes during cooling if proper electrodes and welding schedule are used. After solidification, solid-phase transformation may occur and it may alter a weld's microstructure, which may be drastically different from the just solidified structure. For instance, martensitic transformation may occur in certain steel welds which may result in a significantly more complex structure than the austenite solidified from the liquid.

The formation of various crystals, such as dendrites, globular, and cellular crystals, is controlled by the composition and heat transfer through the liquid–solid interface. Solidification occurs when the liquid nugget reaches the liquidus temperature of the alloy and there is a net heat loss in the liquid; that is, the heat dissipated from the liquid is greater than that into the liquid. Under proper welding conditions, the water-cooled electrodes may act as a large heat sink during welding. The parent sheet metal also absorbs heat from the periphery of the liquid nugget. A possible scenario of solidification during RSW can be constructed based on understanding the metallurgical and thermal changes that may occur in welding. Solid grains in the partially molten or mushy zone at the nugget–HAZ (heat-affected zone) borders may serve as nuclei for crystal growth, and solidification starts in this region. Further cooling results in columnar grains in directions approximately normal to the fusion line, and the solid–liquid interface advances toward the center of the nugget. The remaining molten metal in the central portion of the nugget solidifies last and forms equiaxed grains when the liquid volume is small after much of its surrounding is solidified. Shrinkage cracks or voids, if created, tend to be located in the nugget center that is last solidified. The actual structures formed in a weld nugget depend strongly on welding schedules and other conditions. A carefully created spot weld on a TRIP (TRansformation Induced Plasticity) steel is shown in Figure 1.1, with a clearly defined HAZ and columnar structure in the weld nugget.

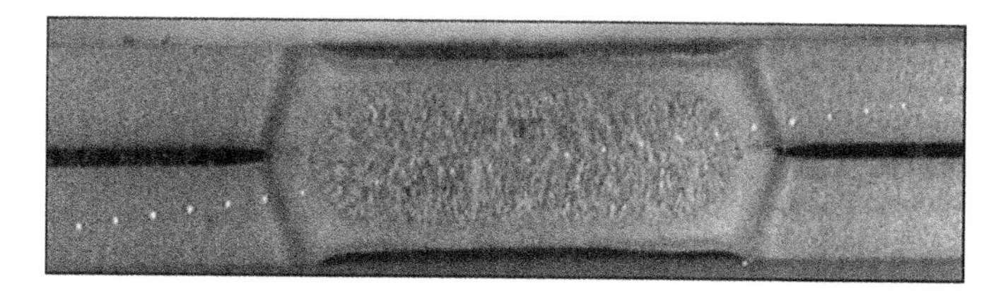

FIGURE 1.1
A spot weld made on 780-MPa TRIP steel.[8] Equally spaced white dots are indentation marks formed during microhardness testing.

During solidification of the last bit of liquid, usually at a location close to the original faying surface of the sheets, a deficit of volume can easily create cracks or voids. In general, a volume deficit of liquid metal during solidification may result from insufficient pressure exerted onto the weldment, insufficient molten metal volume, and excessive cooling rate. A large electrode force can effectively compensate the volume shrinkage of a weldment during cooling, and can suppress the formation of voids or cracks. Insufficient heating, such as that generated by low welding current and/or short welding time, can result in a small volume of molten metal and a high cooling rate. Under a small electrode force, such insufficient melting can easily form voids and cracks. One such example is shown in Figure 1.2, where the fracture surface of a weldment failed in interfacial fracture mode, revealing a macroscale shrinkage void with a clear evidence of freely solidified surface.[9] The fracture surface along the original faying interface is shown in Figure 1.2a. There is a clear evidence of fusion and fracture of fused metals in the nugget area. The central part of the nugget,

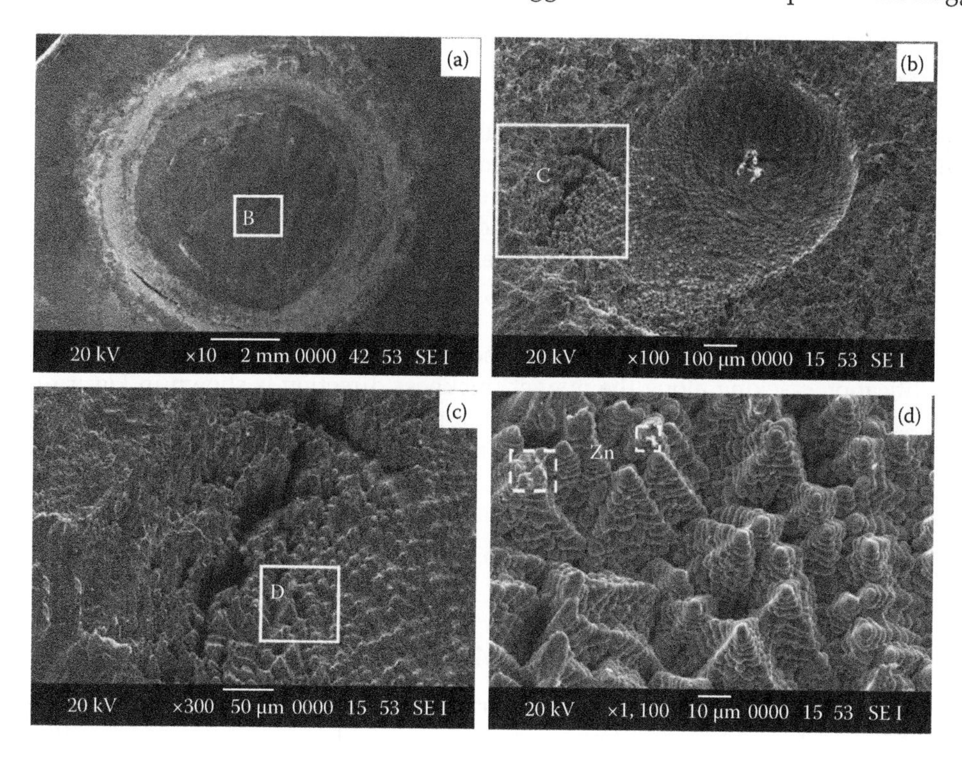

FIGURE 1.2
Microstructure of interfacial fracture surface in DP600 steel weld.[9]

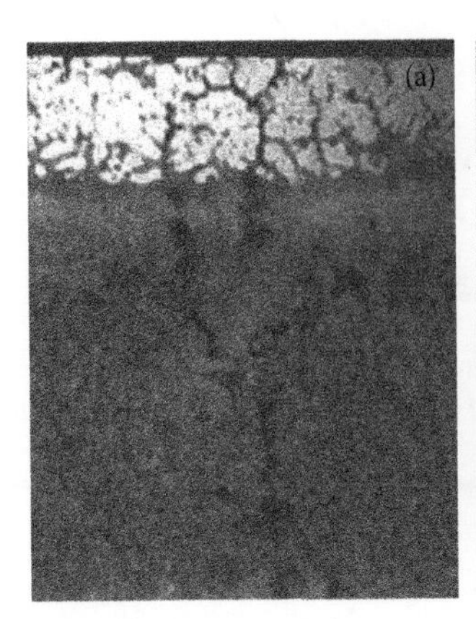

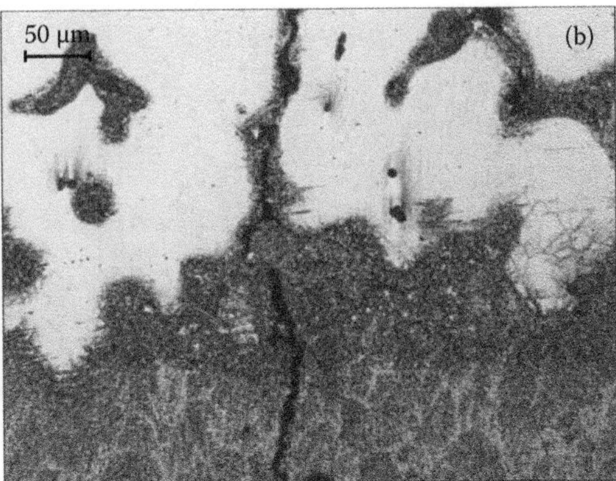

FIGURE 1.3
Morphologies of cracked sections of AZ91D weld: (a) shrinkage cracks extended from nugget, through HAZ, to surface; (b) a closer look at cracks near fusion line.[10]

marked as Box B, is enlarged in Figure 1.2b. There is a void of about 600 μm in diameter. The dendrites observed on the surface of the large void consumed the last liquid during cooling and remained intact. The opening in Figure 1.2c which corresponds to Box C in Figure 1.2b, near the border of the void could result from mechanical loading the cracks created due to volume deficit during solidification of the weld. A closer look of the structure marked by Box D in the figure is shown in Figure 1.2d. It possesses the characteristics of free solidification structure. The white boxes in the figure show the dendrites enriched in zinc, from the hot-dipped zinc coating of the DP steel. This is an evidence of insufficient melting of the nugget.

When cooling from electrodes is impeded, for instance, when the actual electrode–sheet contact area is small due to electrode misalignment or electrode wear, most of the heat is conducted out through the sheet metal. Therefore, the last bit of liquid solidifies around the center of the nugget in the thickness direction. Because of the small volume of such a liquid and the often accompanied volume deficit, cracks and porosity are often formed around the center of the nugget along the electrode direction. As these discontinuities are far from the HAZ at the faying interfaces, they should have a small effect on strength. However, such cracks very often propagate from the center to the edges of the nugget in the form of branching out. This is discussed in more detail in Chapter 3. An example of solidification cracking along the nugget thickness direction is shown in Figure 1.3. In welding a magnesium alloy AZ91D, it was found that cracks were formed around the center of a spot weld, extending from the faying surface, across the fusion line, to the electrode–sheet interface.[10]

1.2 Metallurgical Characteristics of Metals

The welding-related metallurgical characteristics of commonly used structural materials, such as steels, aluminum alloys, and magnesium alloys, are presented in this section. Copper alloys are also discussed since they are the most common material for electrodes.

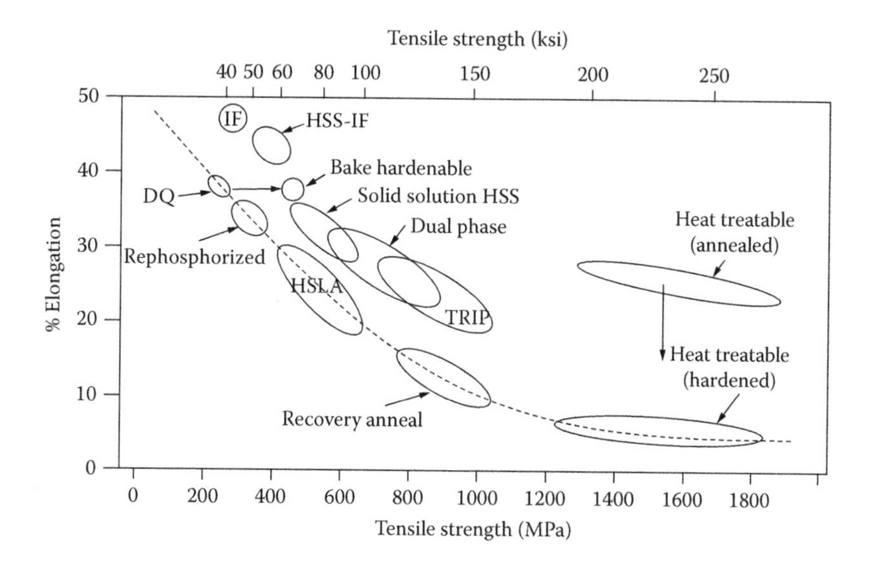

FIGURE 1.4
Mechanical property diagram of various steels. (Courtesy of Auto-Steel Partnership.)

1.2.1 Steels

Properties of the parent sheets and those of the weld metals are determined by both the chemical composition of the alloys and the fabrication conditions, such as heat treatment and hot and cold working. The property map of various steels, shown in Figure 1.4, illustrates the influence of chemistry and processing. In general, low-carbon steels have low tensile strength and high ductility, whereas ductility diminishes as strength rises. The figure shows that by altering the chemical composition and controlling phase transformations, desirable properties of an alloy can be achieved. However, for a weld nugget and the heat-affected zone in RSW, there is only a limited control on transformations and processing. Therefore, the sheet strength obtained through sophisticated metallurgical and mechanical processes during fabrication may disappear in a weld metal.

1.2.1.1 Solid Transformations in Steels

The upper-left corner of the equilibrium iron–carbon phase diagram is shown in Figure 1.5. Consider a steel with a carbon content lower than the eutectoid composition (0.77 wt.% C) cooled from a temperature above the A3 temperature, such as in the case of cooling a solidified nugget or the HAZ. Face-centered cubic austenite is the stable phase at this temperature. When it is slowly cooled to below A3 temperature, the body-centered cubic (BCC) ferrite phase is produced, containing a smaller amount of dissolved carbon. The volume fraction of austenite grains decreases, yet they are progressively enriched in carbon. At the eutectoid temperature (727°C), the residual austenite transforms into a laminated eutectoid mixture of ferrite and cementite (Fe_3C), called pearlite. Therefore, the resultant steel has a structure of ferrite and pearlite mixture. Cementite is not *stable*; rather, it is termed *metastable*, as it decomposes to iron and graphite if held at an elevated temperature for a long period. Same phase transformations occur when cooling at a higher rate,

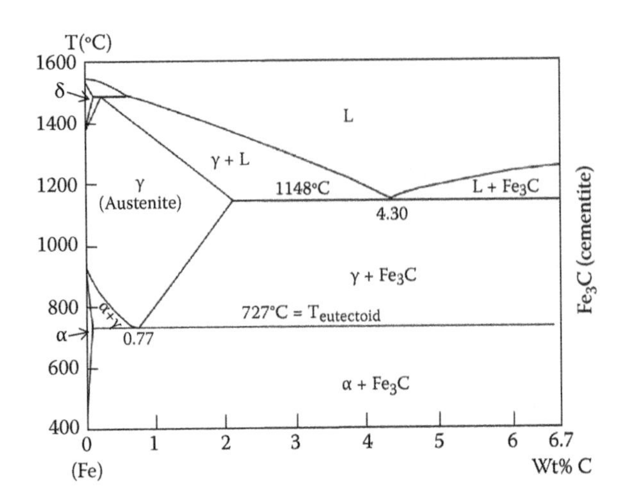

FIGURE 1.5
Fe–C phase diagram. (Adapted from Callister, W.D., Jr., *Materials Science and Engineering: An Introduction*, 6th edition, John Wiley & Sons, Inc., New York, 2003.)

but usually at temperatures lower than those marked on the equilibrium phase diagram. Although a mixture of soft ferrite and hard cementite is the typical structure for low-carbon steels, the morphology of the phases is a strong function of cooling rate, and the mixture can be either pearlite or bainite depending on the cooling rate.

Isothermal phase transformation, or sometimes called time–temperature transformation (TTT) diagram, is an important tool in understanding the microstructures that may occur upon cooling. A TTT diagram is developed isothermally by quenching samples into molten salt baths of fixed temperatures and keeping them for predetermined periods, then quenching quickly in an ice–salt brine. These diagrams show how metals transform with time at given temperatures. Figure 1.6 is a TTT diagram for an iron–carbon alloy. A typical TTT diagram of a plain carbon steel shows the starts and completions of pearlite formation, bainite formation, and martensite formation.

Such diagrams are generated under equilibrium conditions that are rarely met in practice. Especially in an RSW process, the heating and cooling rates are extremely high and transformations are far from equilibrium. Because most industrial heat treatment processes use controlled cooling rather than isothermal transformation, continuous-cooling transformation (CCT) diagrams are more representative of actual transformations than TTT diagrams. Cooling of a weldment of RSW is also far from isothermal; therefore, CCT diagrams are more applicable to understanding the microstructures of a weldment. CCT diagrams are similar to TTT diagrams except that in CCT diagrams, transformations occur over a range of temperatures. A typical CCT diagram of a mild steel is shown in Figure 1.7. A continuous cooling with a slow cooling rate results in a mixture of ferrite and pearlite; an intermediate cooling tends to produce a mixture of ferrite, bainite, and martensite; and a rapid cooling (above the critical cooling rate) creates a structure of all martensite. Although some techniques such as CCT diagrams take into account the dynamic nature or kinetics of phase transformations, they are usually material dependent, and there is a serious lack of information on transformations occurring at such a high cooling rate as in RSW. Therefore, most phase diagrams are not adequate when used in a quantitative manner. Nevertheless, information of possible transformations and reactions during welding can be obtained from the phase diagrams.

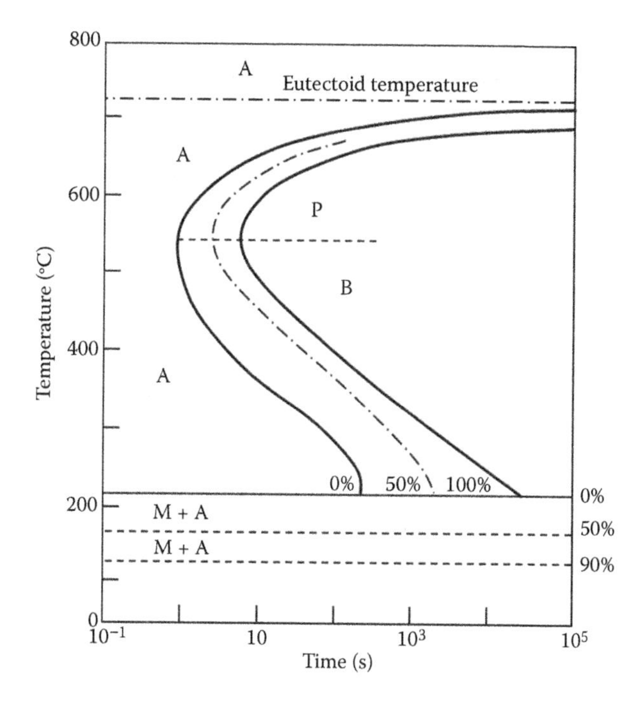

FIGURE 1.6
A TTT diagram for an iron–carbon alloy of eutectoid composition: A, austenite; B, bainite; M, martensite; P, pearlite. (Adapted from Callister, W.D., Jr., *Materials Science and Engineering: An Introduction*, 6th edition, John Wiley & Sons, Inc., New York, 2003.)

Under certain conditions, such as when the carbon (or carbon equivalence) content is sufficiently high, a very high cooling rate, as what often occurs during RSW, may result in martensitic transformation. The rapid cooling makes equilibrium phase transformations impossible, and it tends to depress the transformation temperatures. At low temperature, the nucleation rate is high, whereas the growth rate is low. The resultant structure (ferrite

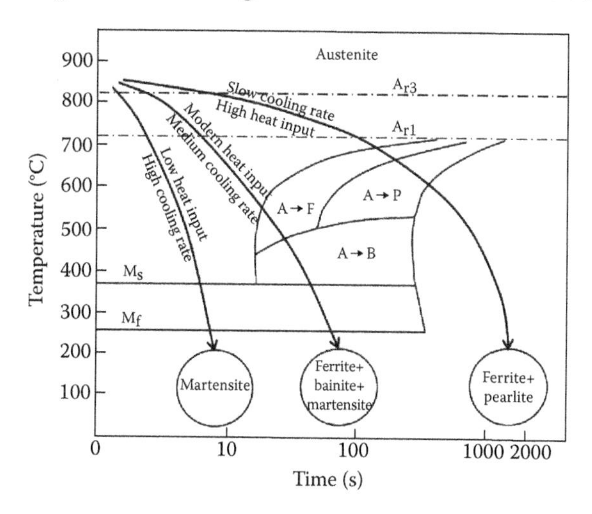

FIGURE 1.7
Typical CCT diagram of a mild steel: A, austenite; F, ferrite; P, pearlite; B, bainite; M, martensite.

+ cementite) appears in the form of fine needles rather than thick laminar plates. Further depression of the transformation by a higher cooling rate may result in the transformation of austenite to martensite. In general, a higher cooling rate results in a lower transformation temperature and a harder structure. Alloying elements are commonly added to control the phases produced in steel making, and they have significant effects on the temperatures as well as the shape and location of the C-shaped curves in the phase diagrams. Elements such as titanium, molybdenum, chromium, and tungsten lower the eutectoid carbon content and raise the transformation temperature and, therefore, they are called ferrite stabilizers. The existence of such elements raises the pearlite nose and moves it to the right side. Other elements such as nickel and manganese lower the eutectoid carbon content and lower the transformation temperature, so they are austenite stabilizers. Their effect is demonstrated on the TTT and CCT diagrams by lowering the pearlite nose and moving it to the right side. In fact, all metals except cobalt increase the hardenability of steels, a measure of how rapid a quench is necessary to form martensite. That is, they move the nose of the pearlite curve to the right, allowing martensite to form with less rapid quenching.

Martensite is responsible for the high strength of most steels. It has a distorted BCC lattice structure. The amount of distortion and, therefore, the properties of martensite are a strong function of carbon content. For low-carbon steels (less than 0.2 wt.%), the lattice structure of the martensite is very close to BCC and it is a little brittle. On the other hand, for higher-carbon steels, martensite is body-centered tetragonal (BCT) and is brittle. High carbon content promotes the formation of martensite and increases its hardness. Because the martensitic transformation is diffusionless and instantaneous, the start and finish of this transformation are represented by horizontal lines. More horizontal lines as in Figure 1.6 are used to indicate the percentage of completion of the austenite-to-martensite transformation. The influence of alloying elements on the effectiveness of carbon in martensitic formation is measured by the so-called "carbon equivalent," and it is discussed in Section 1.2.1.3.

1.2.1.2 Transformations in HAZ of a Steel Weld

The heat-affected zone of a resistance spot weld experiences thermal cycles and its microstructure is determined accordingly. Upon heating a steel through its upper critical temperature, the stable austenite forms and grows. The austenite grain growth is very sensitive to temperature, and aluminum and other elements are added to steel in order to produce fine grains by impeding the growth of austenite grains during various thermal cycles. Lancaster[4] divided an HAZ into three zones from a metallurgical viewpoint: supercritical, intercritical, and subcritical:

- The supercritical region is divided into two parts: grain growth region and grain refinement region. A thermal cycle during welding above the grain-coarsening temperature promotes grain growth, and it refines the grain structure below that temperature. This region is located near the fusion line, next to the weld nugget. Different steels contain different grain growth inhibitors, and they have different grain-coarsening temperatures.
- In the intercritical region, the peak temperature is lower than that in the supercritical region and, therefore, partial phase transformation is experienced in this region. New phases that do not exist in the original base metal may form in this region, and such transformation depends on the duration of the metal exposed to the peak temperature and on the cooling rate.

- The subcritical region does not normally undergo any observable microstructural changes as the temperature is generally low. It is usually difficult to distinguish this region from the base metal. In some cases, very fine precipitates may appear in the region.

Nonmetallic inclusions such as sulfides and oxides may have an effect on the hardenability of the HAZ. They produce a lower hardness by nucleating ferrite within the transforming austenite grains and reducing the amount of austenite for transforming to martensite or bainite. In some cases, a low hardenability is preferred in the HAZ in order to minimize the risk of stress corrosion cracking.

The microstructure of a nugget is determined by the composition of the base alloy and the thermal history, and it can be predicted using the relevant phase and transformation diagrams. Therefore, it is critical to obtain the temperature distribution as a function of welding time in a weldment. However, it is difficult to obtain the temperature profile of a weldment during resistance welding, as directly and accurately measuring temperature is impossible. Using sensors such as thermal couples may interfere with the welding process and result in invalid temperature readings. Numerical simulation such as finite element modeling can provide an approximation, yet its lack of ability to fully couple the electrical–thermal–mechanical effects and a lack of temperature-dependent material property data make accurate prediction impossible. Nevertheless, the temperature profile can be estimated based on the structures and sizes of various zones in a weldment revealed by metallographic examination, and the temperature ranges of the structures for the alloy determined on a phase diagram.

Figure 1.8 shows an approximated relation between the phase diagram and microstructures linked by the possible temperature distribution in a steel weldment at the peak of heating. The regions of various structures indicate the possible phase transformations experienced at such locations upon heating and cooling. These structural changes are closely associated with the heights of the phase regions in the phase diagram, which outline the temperature limits for phase transformations. The peak temperature in the melt can be approximated as a few hundred degrees above the liquidus, and its value does not drastically affect the temperature distribution. By drawing such lines from the phase diagram and the cross section of a weldment, the possible temperature distribution can be established by the intersections of the lines. In Figure 1.8, an HSLA steel weld shows regions of various structures. These structures are different from the base metal as they were modified during welding by the heating and cooling cycles. In the nugget region, a clear casting structure indicates melting and solidification, and, therefore, the peak temperature in this region has to be over the melting point of the alloy. Next to this region is the partial melting zone, as it is filled partially by columnar grains. This region corresponds to the temperature limits between the liquidus and solidus in the phase diagram. Beyond this region, no melting occurred during welding, but changes in structures, such as grain shape and size, can be clearly observed in the solid structures. They have experienced solid-phase transformations. The temperature range in the Fe–C phase diagram for this region is fairly wide, yet the region is narrow, resulting in a large temperature drop in the temperature distribution. When determining the possible temperature ranges from the phase diagram, it is reasonable to assume that the temperature distribution is continuous in the weldment at any instance during welding. From the figure, it can be seen that the temperature gradient in the molten nugget is not large, and it increases dramatically in the HAZ. It again drops near the base metal. In the region next to the partial melting zone, a supercritical region exists. In the overheated zone, as marked in the figure, grain growth is evident; therefore,

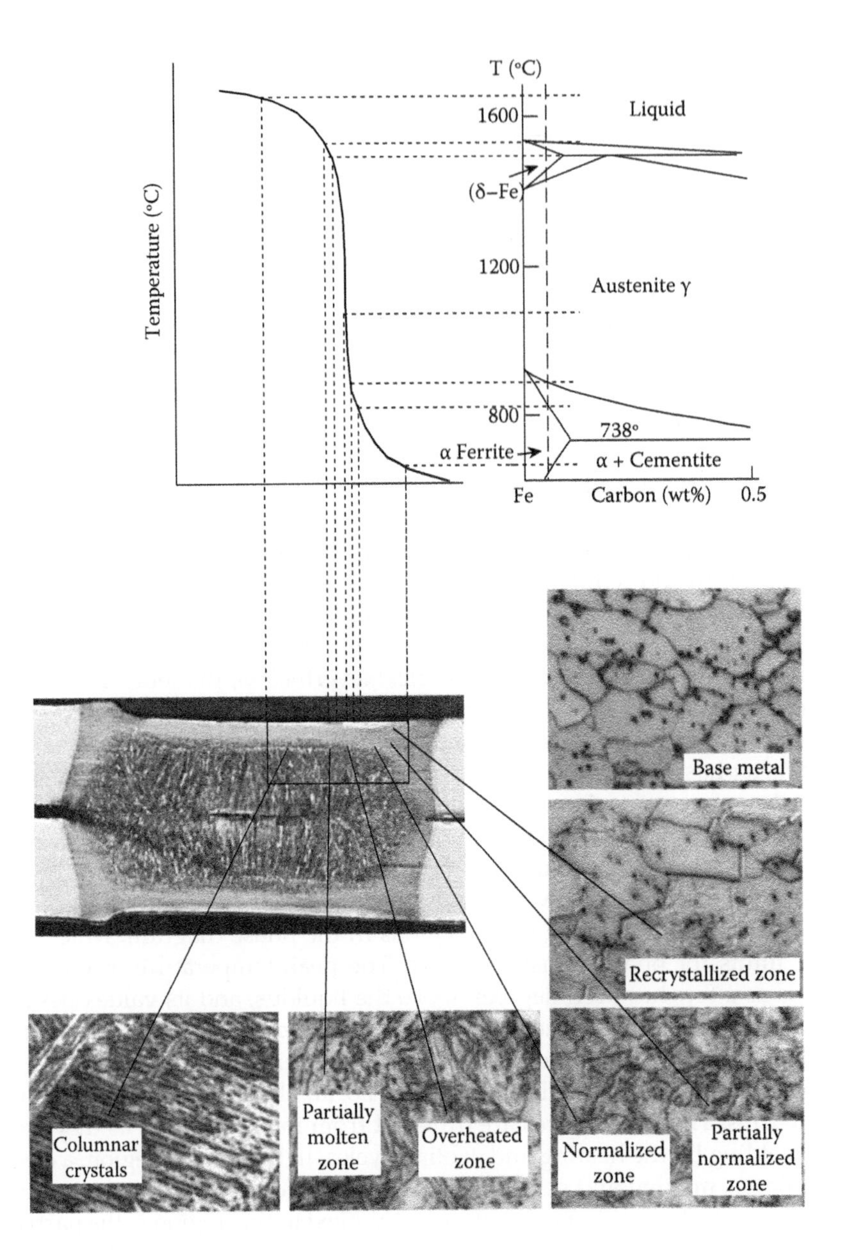

FIGURE 1.8
Structure of high-strength, low-alloy steel (HSLA) steel weldment against temperature gradient and basic Fe–C phase diagram.[12]

the peak temperature of this region exceeded the grain-coarsening temperature. Such an overheated region may be embrittled and have a coarse intergranular fracture when impact loaded. The embrittlement is mainly due to the solution of inclusions such as sulfides and aluminum nitride at high temperature, and their reprecipitation at grain boundaries on cooling. Next to the grain-coarsening region is the grain-refined region of the supercritical zone. The refinement is mainly due to a process similar to that of normalization.

Next to this region, there is a recrystallized structure, which is the so-called intercritical region. This region basically retains the original structure with parts of the grains showing a slight sign of recrystallization and grain growth. The figure does not show a subcritical region, as its difference from the base metal may be invisible under the magnification.

The microstructure of a weldment, including the base metal, the HAZ, and the fusion zone are dictated by the chemical composition of the weld metal as well as the welding process. The microstructure shown in Figure 1.9 for a DP600 steel weld[9] is significantly different from that of the HSLA steel weld in Figure 1.8. In Figure 1.9, a significant change in microstructure is observed from the original, unaltered base metal to the molten and then solidified weld center, through the HAZ. Each of the regions consists of the two typical phases, α-ferrite (BCC) and martensite (BCT), but with different amount and morphology. In the base metal, the martensite is evenly distributed in a matrix of ferrite. More martensite is observed in the HAZ, and the volume fraction of martensite increases rapidly from the base metal to the fusion line. Finer constituents of martensite and ferrite are observed in the HAZ. According to the authors, the fine structure in the HAZ results from the repeated rapid heating–cooling cycles and the restricted grain growth in the region. Because of the rapid heating and then cooling, austenitizating is not complete and austenite grain growth is interrupted by the rapid cooling that results in martensite transformation. Other phases, such as retained austenite and lower bainite, are also observed in the region. It is filled with martensite due to the presence of high carbon and manganese contents in DP600, and the rapid cooling during RSW, which is estimated at a few thousands of degrees Celsius per second.

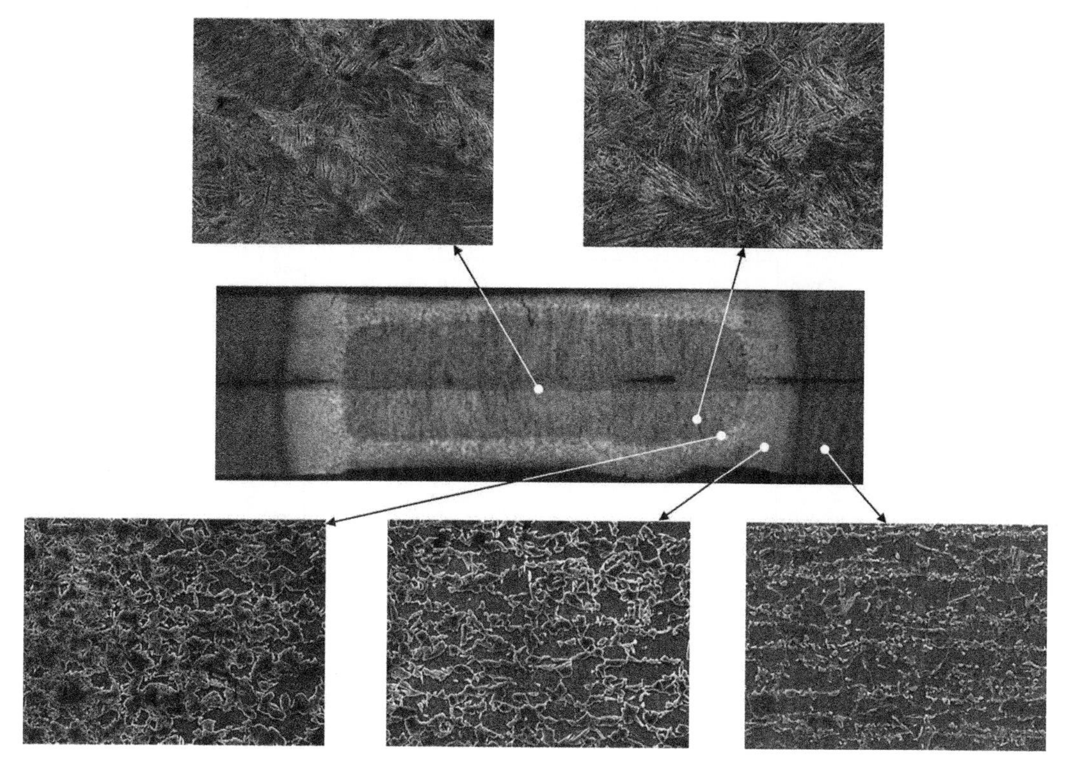

FIGURE 1.9
Microstructure of various zones in DP600 steel weld.[9]

1.2.1.3 Effect of Carbon Content

The ability to form hard (often brittle and hydrogen cracking-prone) metallurgical constituents, such as martensite, bainite, carbide, and other hard phases in a ferrous alloy, either steel or cast iron, depends on the content of carbon and other alloying elements, and the cooling rate. In RSW, it directly affects the integrity as well as the strength of a weldment.

Although carbon and many other alloying elements such as manganese, chromium, silicon, molybdenum, vanadium, copper, and nickel can all raise the hardness of a steel, their mechanisms and contributions are different. Carbon content in the parent austenite phase is directly responsible for, and has the largest effect on the formation of martensite. For instance, the plate martensite formed with a high carbon content has a higher hardness than the lath martensite formed with a low carbon content. As summarized by Krauss[13] on the hardness data from the literature for Fe–C alloys and steels in Figure 1.10, the hardness increases monotonically with carbon content in most of the carbon range of steels.

The effects of other alloying elements, however, are different from that of carbon. They raise the hardness of steels through alteration of the metallurgical process, micro-alloying, and precipitation of hard particles. Cr, Ni, Mo, Si, and Mn may help in retaining austenite by retarding the eutectoid transformation $\gamma \rightarrow \alpha + Fe_3C$ and, therefore, promote martensite formation. The addition of small amount of vanadium can significantly increase the strength of steels through refinement of the ferrite grain size and the formation of hard vanadium carbides. In addition, alloying elements such as vanadium, niobium, and titanium may react preferentially with carbon and/or nitrogen, to form fine dispersion of precipitated hard particles in the steel matrix. All carbide formers are also nitride formers, and the tendency of several alloying elements to form hard nitrides and, therefore, to increase the hardness of a steel by precipitation hardening is shown in Figure 1.11. In general, carbon is the most significant alloying element affecting the hardness of steel and, therefore, the influence of other alloying elements is accounted for in the form of equivalent carbon content (CE), or carbon equivalence.

Equivalent carbon content, or carbon equivalence [or carbon equivalent (CE)], is an empirical value in weight percent, relating the combined hardening effect of different

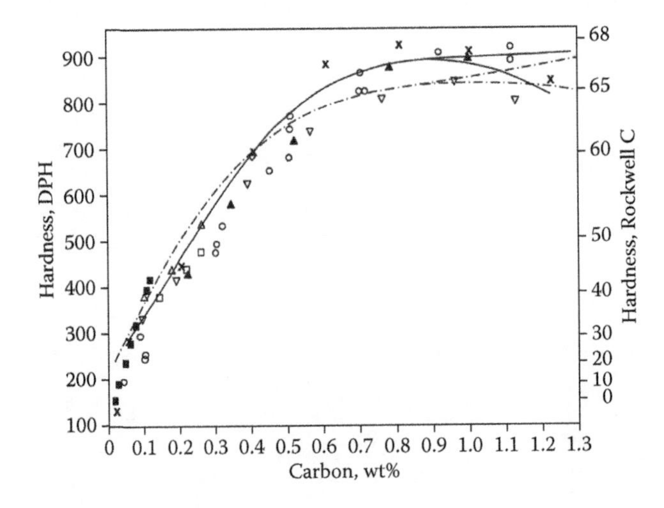

FIGURE 1.10
Summary of hardness data from literature for Fe–C alloys and steels by Krauss.[13] Symbols on the curves indicate the sources of the data.

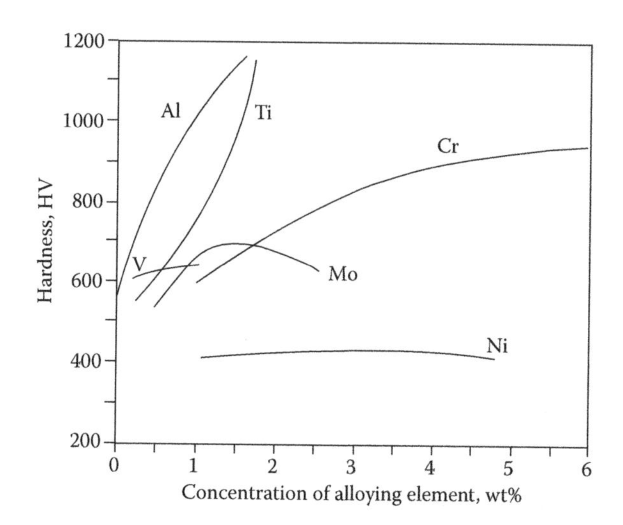

FIGURE 1.11
Effect of alloying element additions on hardness after nitriding. Base composition: 0.25% C, 0.30% Si, 0.70% Mn.[14]

alloying elements used in the making of carbon steels to an equivalent amount of carbon. It is usually expressed in the form of a simple mathematical summation of weighted contributions of various alloying elements, and the weighting factors are obtained through a systematic experimentation. A first-order model containing only the contents of the alloying elements is usually used for simplicity. As a simplified "equivalency" of alloying elements to carbon may not account for the nonlinear effect of alloying elements on hardness, as seen in Figure 1.11, and the interactions among the alloying elements and with material processing, etc., carbon equivalence should be used with discernment. The application of such a model should be limited to the specific class of steels on which the model is developed. There are a number of commonly used CE formulas for particular material systems. For instance, a well-known carbon equivalent formula developed by Dearden and O'Neill[15] based on carbon–manganese steels and later modified by the International Institute of Welding, works well for high-carbon, low-alloy steels:

$$CE = C + \frac{Mn}{6} + \frac{Cr + Mo + V}{5} + \frac{Cu + Ni}{15} \tag{1.1}$$

A different formula was developed for low-carbon steels or micro-alloy steels:

$$CE = C + \frac{Si}{25} + \frac{Mn + Cr}{16} + \frac{Cr + Ni + Mo}{20} + \frac{V}{15} \tag{1.2}$$

Another formula, the Ito–Bessho carbon equivalent, is often used for low-carbon (with C between 0.07% and 0.22%) and micro-alloyed steels[16]:

$$CE = C + \frac{Si}{30} + \frac{Mn + Cu + Cr}{20} + \frac{Ni}{60} + \frac{Mo}{15} + \frac{V}{10} + 5B \tag{1.3}$$

For steels with a carbon content between 0.02% and 0.26%, the Yurioka[17] formula can be used to calculate the CE:

$$\mathrm{CE} = C + A(C) * \left\{ 5B + \frac{\mathrm{Si}}{24} + \frac{\mathrm{Mn}}{6} + \frac{\mathrm{Cu}}{15} + \frac{\mathrm{Ni}}{20} + \frac{\mathrm{Cr}}{5} + \frac{\mathrm{Mo}}{5} + \frac{\mathrm{Nb}}{5} + \frac{\mathrm{V}}{5} \right\} \tag{1.4}$$

where $A(C) = 0.75 + 0.25 * \tanh[20(\mathrm{C} - 0.12)]$.

If the contents of some alloying elements are not available, the following formula is sometimes used:

$$\mathrm{CE} = \mathrm{C} + \frac{\mathrm{Mn}}{6} + 0.05 \tag{1.5}$$

In addition to being used as an indicator of the hardness of steel, CE is more applied directly to describing the processing and predicting performance of steels. By varying the amount of carbon and other alloying elements, the desired strength levels can be achieved through proper heat treatment. Other properties such as weldability and low-temperature toughness can also be controlled or predicted by altering CE. The American Welding Society states that for an equivalent carbon content above 0.40%, calculated based on Equation 1.1, there is a potential for cracking in the HAZ on flame-cut edges and welds.[18] The following carbon equivalent formula was used to determine the weldability of spot welding high-strength, low-alloy steels with excessive hardenability[19]:

$$\mathrm{CE} = \mathrm{C} + \frac{\mathrm{Mn}}{30} + \frac{\mathrm{Cr} + \mathrm{Mo} + \mathrm{Zr}}{10} + \frac{\mathrm{Ti}}{2} + \frac{\mathrm{Cb}}{3} + \frac{\mathrm{V}}{7} + \frac{\mathrm{UTS}}{900} + \frac{h}{20} \tag{1.6}$$

where UTS is the ultimate tensile strength in ksi and h is the sheet thickness in inches. Note that the formula contains the mechanical strength and sheet thickness, in addition to the alloying element contents. Therefore, it can be used to describe the failure mode of a spot weld.

In a study on a DP600 steel with a nominal chemical composition of 0.08 C, 1.91 Mn, 0.04 Si, 0.018 P, 0.006 S, 0.035 Al, and 0.005 N (in wt.%), Ma et al.[9] studied the effect of alloying elements on the hardness profile in a spot weld, and established a relationship between the CE value and fracture mode. Because of the high content of the alloying elements in steel and high cooling rate during welding, a significant increase in hardness is observed in a typical DP600 steel weld, from the base metal to the center of the weld due to the formation of martensite (Figure 1.12).

A large proportion of the welds tested exhibit interfacial fracture failure mode in their study. Using the Nippon Steel CE formula and a variation of Equation 1.5 as in the following,[20,21]

$$\mathrm{CE} = \mathrm{C} + \frac{\mathrm{Si}}{30} + \frac{\mathrm{Mn}}{20} + 2\mathrm{P} + 4\mathrm{S} \tag{1.7}$$

$$\mathrm{CE} = \mathrm{C} + \frac{\mathrm{Mn}}{6} \tag{1.8}$$

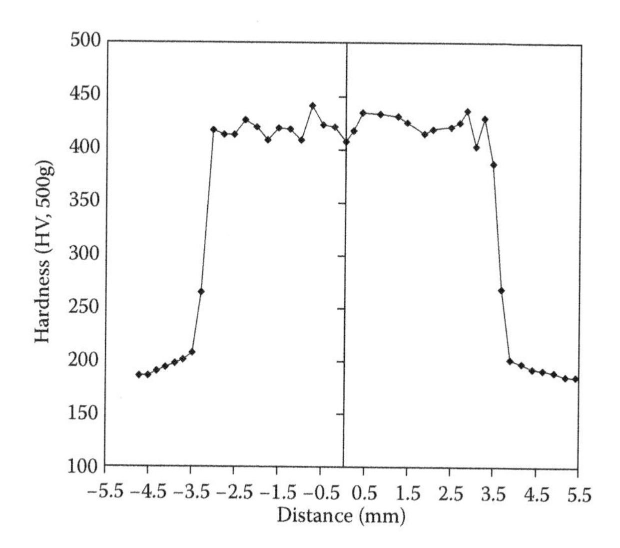

FIGURE. 1.12
Microhardness profile of a typical DP600 steel spot weld. Welding current = 8.26 kA, electrode force = 3.34 kN, and welding time = 300 ms. (From Ma, C. et al., *Mater. Sci. Eng. A*, 485, 334–346, 2008. With permission.)

CE values of 0.242 and 0.367, respectively, were obtained. Both formulas yield a CE value larger than 0.24, which is considered as the threshold of interfacial failure of spot welds.[20]

In another study by Khan et al.,[22] experiments were conducted on welding dissimilar material combination of HSLA350/DP600 steels. The fusion zone of such a spot weld was found to be predominantly martensitic with some bainite. The hardness of such a dissimilar material spot weld is different from that of the similar ones in each of the HSLA350 and DP600 steels, and tests revealed that the DP600 weld properties played a dominating role in the microstructure and tensile properties of the dissimilar material spot welds.

In their study, spot welds of 7.5 mm diameter were made using various material combinations: DP600/DP600, HSLA350/HSLA350, and HSLA350/DP600. Their microhardness profiles are shown in Figure 1.13. The DP600 weld has the highest hardness, whereas that of HSLA is the lowest. It is interesting to see that the hardness of the dissimilar material weld lies between those of the same material combinations. Using the same material combination (HSLA350/DP600), the 5.5-mm weld has a higher hardness than that of the 7.5-mm nugget, possibly due to the more martensitic microstructure resulting from the faster cooling it experienced because of a smaller weld diameter. The CE values were calculated according to the formula in Equation 1.3, and 0.208 and 0.141 were obtained for the DP600 steel and the HSLA350 steel, respectively, using the data listed in Table 1.1. The higher CE of the DP600 steel corresponds to a more martensitic and harder microstructure as compared to the HSLA350 steel. Combining these two steels produces a CE in the weld nugget lying between those of these two materials, as well as a hardness as seen in Figure 1.13. The difference in the peak hardness values of the HAZ of the HSLA350/DP600 nugget is an indication of the difference in hardenability between the base metals, similar to the observations by Marya and Gayden[20] for the TRIP/HSLA combination.

1.2.2 Aluminum Alloys

As an important structural material, aluminum alloys are often the material of choice because of their balanced overall quality of strength, weight, corrosion resistance,

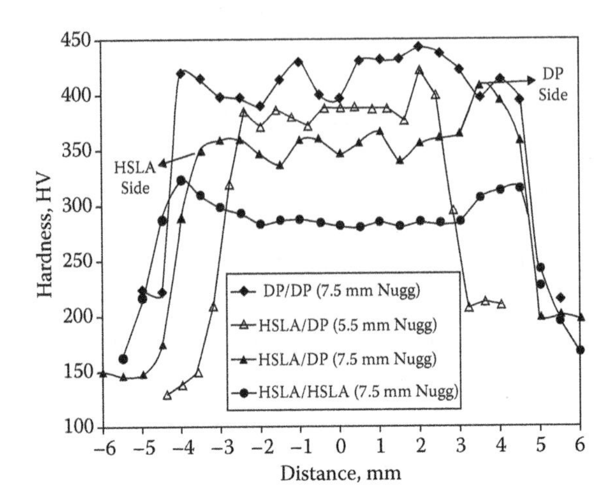

FIGURE 1.13
Microhardness profiles of DP/DP, HSLA/DP, and HSLA/HSLA spot welds. (From Khan, M.S. et al., *Sci. Technol. Welding Joining*, 14(7), 616–625, 2009. With permission.)

manufacturability, and cost. Pure aluminum is rarely used as engineering material, and various alloying elements are used in order to achieve a wide range of engineering properties. Aluminum alloys are made either as casting alloys or wrought alloys. Because of Al alloys' light weight and high specific strength, they are widely used in the aerospace industry. Over the past two decades, they have been introduced to the automobile industry mainly for weight reduction. Significant knowledge has been accumulated in manufacturing such as welding Al alloys.

1.2.2.1 Classifications and Properties

Aluminum alloys are classified as casting alloys and wrought alloys, and both are further divided as heat-treatable and non–heat-treatable alloys. More specific classifications are made by a number system by the American National Standards Institute, or by names indicating their main alloying elements (as by the German Institute for Standardization and International Organization for Standardization), and others. Wrought alloys are identified with a four-digit number indicating the alloying elements. Casting alloys use a four- to five-digit number with a decimal point, which is used for the form of casting (cast shape or ingot). Table 1.2 lists the major alloying elements in the aluminum alloy series, as well as the main hardening mechanisms of these alloys. As resistance welding is usually performed on sheet materials, only wrought aluminum alloys are discussed in this section.

TABLE 1.1
Chemical Composition of Test Materials (wt.%)[22]

C	Mn	Si	Ni	Cr	V	Mo	Ti	P	S	Cu	Nb
Hot-dip galvannealed HSLA350 steel											
0.05	0.6	0.05	0.01	0.04	0.003	0.004	0.001	0.03	0.004	0.043	0.01
Hot-dip galvannealed DP600 steel											
0.10	1.5	0.19	0.01	0.18	0.005	0.24	0.02	0.009	0.002	0.02	0.007

TABLE 1.2

Designation and Hardening Mechanisms for Alloyed Wrought Aluminum Alloys

Major Alloying Element	Designation	Work-Hardened	Solution Heat-Treated/ Age-Hardened
None (99% + Aluminum)	1XXX	Yes	
Copper	2XXX		Yes
Manganese	3XXX	Yes	
Silicon	4XXX		Yes
Magnesium	5XXX	Yes	
Magnesium + Silicon	6XXX		Yes
Zinc	7XXX		Yes
Lithium	8XXX		Yes

The typical alloying elements for aluminum alloys are copper, magnesium, manganese, silicon, zinc, and lithium. Small quantities of chromium, titanium, zirconium, lead, bismuth, and nickel are also added, and the presence of iron in small quantities is invariable. The taxonomy of aluminum alloys reflects the composition, hardening mechanism, processing, and strength of the alloys. In addition to the digits for the major alloying elements, designations are also made to specify the work-hardening or heat-treatment conditions.

The 1000, 3000, and 5000 series aluminum alloys are work hardened by cold work such as cold rolling. Their properties are determined by the degree of cold work and heat treatment following the cold work. The nomenclature used to describe these conditions is presented in Table 1.3. On the other hand, the designation, as listed in Table 1.4, is different for the solution heat-treated and age-hardened aluminum alloys, that is, the 2000, 4000, 6000, 7000, and 8000 series alloys. The large number of possible combinations of composition, solution heat treatment temperature and duration, quench rate, shaping, and other factors make it possible to tailor the alloys' properties in a wide range.

TABLE 1.3

Standard Nomenclature for the Heat Treatment of Work-Hardened Aluminum Alloys

Symbol	Description
O	Annealed, full soft
F	As fabricated
H12	Work hardened, without thermal treatment, 1/4 hard
H14	Work hardened, without thermal treatment, 1/2 hard
H16	Work hardened, without thermal treatment, 3/4 hard
H18	Work hardened, without thermal treatment, fully hard
H22	Work hardened, partially annealed, 1/4 hard
H24	Work hardened, partially annealed, 1/2 hard
H26	Work hardened, partially annealed, 3/4 hard
H28	Work hardened, partially annealed, fully hard
H32	Work hardened, stabilized, 1/4 hard
H34	Work hardened, stabilized, 1/2 hard
H36	Work hardened, stabilized, 3/4 hard
H38	Work hardened, stabilized, fully hard

TABLE 1.4
Heat Treatment Designation for Aluminum Alloys

Symbol	Description
T1	Cooled from hot working and naturally aged (at room temperature)
T2	Cooled from hot working, cold worked, and naturally aged
T3	Solution heat-treated, cold worked, and naturally aged
T4	Solution heat treated and naturally aged
T5	Cooled from hot working and artificially aged (at elevated temperature)
T6	Solution heat treated and then artificially aged
T7	Solution heat treated and overaged/stabilized
T8	Solution heat treated, cold worked, and artificially aged
T9	Solution heat treated, artificially aged, and cold worked
T10	Cooled from hot working, cold worked, and artificially aged
W	Solution heat treated only

When welding an aluminum alloy, it is important to consider the influence of its metallurgical characteristics on welding process and weld quality. For instance, the structure of the HAZ in an alloy 6111 may be quite different from that of a 5754 alloy, because the former is heat treatable. Another metallurgical factor, the high affinity of aluminum for oxygen, makes aluminum easily oxidized when exposed to air. Therefore, there is always a clear, protective layer of aluminum oxide on the surface of aluminum alloys. Such an oxide layer has a significant implication in the electrical contact resistance and, therefore, affects a welding process.

1.2.2.2 Resistance Welding Aluminum Alloys

Welding aluminum alloys is significantly different from welding other metals such as steels due largely to their unique metallurgical properties. For instance, aluminum welding is more prone to expulsion and cracking. The wide solidus–liquidus gap in the Al–Mg phase diagram (Figure 1.14) indicates the existence of a partial melting zone for a relatively long period of time during heating and cooling an alloy with Mg as the major alloying element. The presence of low-melting-point eutectics and impurities also weakens the HAZ in a weldment. A typical microstructure of the region in the HAZ close to the nugget is presented in Figure 1.15 for aluminum alloy 5754. Precipitates inside the grains and at grain boundaries (intergranular precipitates), where they form chains or even continuous layers, are visible. Energy-dispersive x-ray (EDX) and wavelength dispersive x-ray analyses revealed an increased amount of Mg in such regions. This is most probably due to an Al_3Mg_2 secondary phase (the presence of which should be about 6% in the Al–Mg 3.5% alloy, according to the Al–Mg equilibrium phase diagram), which exists in the alloy before welding and serves as the source of liquid at elevated temperature. This was confirmed by an x-ray diffraction examination. More discussion of the effect of low-melting eutectics on cracking of aluminum alloys can be found in Chapter 3. In addition to cracking, resistance welding aluminum alloys has other unique characteristics associated with their metallurgical properties, such as massive shrinkage voids/porosity, rapid electrode wear. They will be discussed in detail in other chapters specifically dealing with these subjects.

The microstructures of an aluminum weldment can be linked to the metallurgical characteristics of aluminum alloys through a possible thermal history during resistance welding. Such a relation for an aluminum AA5754 weldment is shown in Figure 1.16. The

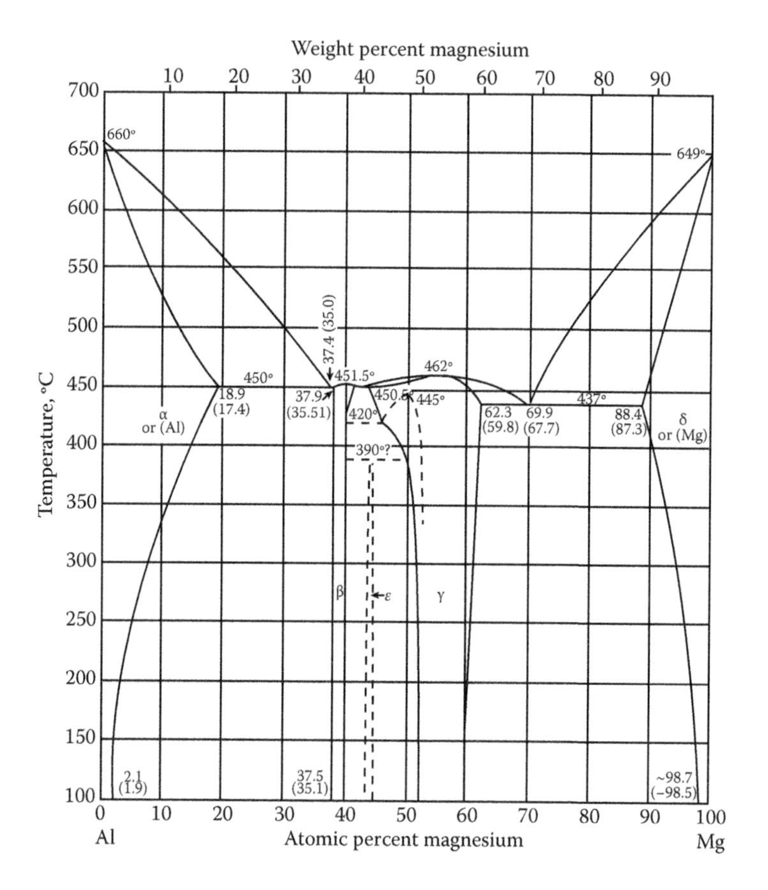

FIGURE 1.14
Al–Mg binary phase diagram. (From Hansen, M., Anderko, K., *Constitution of Binary Alloys*, 1958, 106. With permission.)

FIGURE 1.15
Precipitation zone in HAZ.

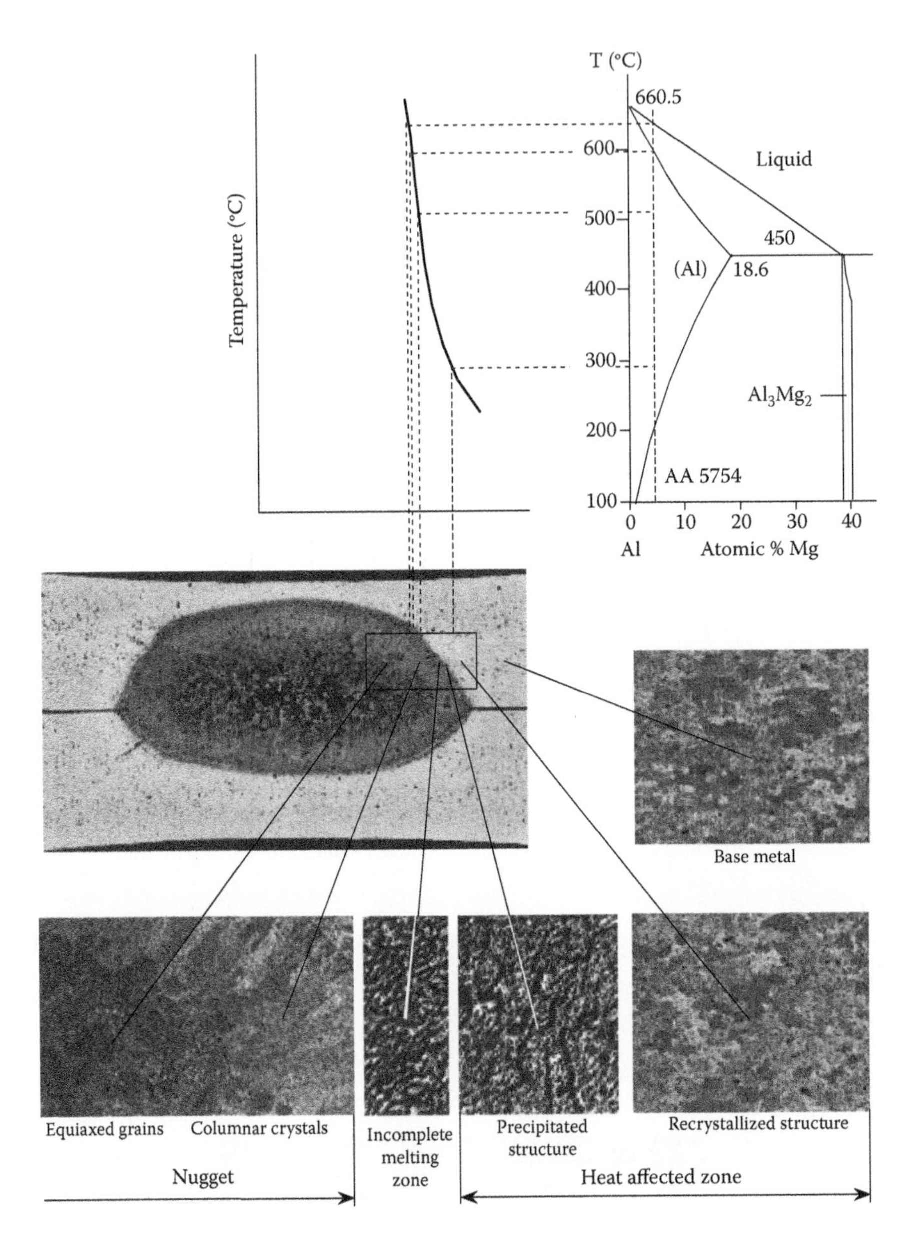

FIGURE 1.16
Weldment structure of a 5754 aluminum alloy against temperature gradient and basic Al–Mg phase diagram.[12]

structures of aluminum welds are usually not as clearly distinguishable as in steel welds, and the HAZ is significantly narrower for an aluminum weld. These make the identification of various zones difficult. The Al–Mg phase diagram used for this alloy indicates precipitation at temperatures just below the solidus, and under that there is a recrystallization temperature range. The temperature gradient is very high in the HAZ, indicating a possible large thermal stress development in the region.

1.2.3 Magnesium Alloys

Magnesium alloys are an attractive alternative to steels and even Al alloys for auto components due to their low density, high specific strength, excellent recyclability, and many other advantages. In order to achieve substantial vehicle weight reduction and improving fuel efficiency, a vehicle's average magnesium content is expected to increase to as much as 350 lb by the year 2020, replacing heavier components, as stated in *Magnesium Vision 2020*.[24] A crucial issue in large-scale application of Mg alloys is an industrial joining means of Mg components, such as RSW for joining sheet materials. Because of the relatively small volume of Mg used in sheet metal industry, limited knowledge is accumulated in joining Mg alloys, and the knowledge on resistance welding steels and Al alloys cannot be directly applied to welding Mg.

1.2.3.1 Properties and Applications of Mg Alloys

Among the lightest of all the metals, magnesium is principally used as an alloying element for aluminum, lead, zinc, and other nonferrous alloys. When magnesium is used as a structural material, however, it is usually used in the form of alloys, as pure magnesium has too low a strength for any engineering application. The strength of magnesium alloys containing small amounts of aluminum, manganese, zinc, zirconium, etc., is similar to that of mild steels. Commercially available magnesium alloys can be classified into three groups:

1. Mg–Mn alloys. Because of their good weldability, they are often used for sheet metal fabrications.
2. Mg–Al–Zn alloys. These alloys can be heat treated by solution treatment and precipitation hardening, and can be processed through die casting, sand casting, extrusion, and forging.
3. Mg alloyed with rare earths. This group also includes zirconium as alloying element. They are used in both cast and wrought form.

Magnesium alloys are coded differently by different organizations. ASTM and SAE use two letters, indicating the two major alloying elements, followed by two digits for the nominal percentage contents of these two elements. The letters usually used are A for aluminum, B for bismuth, C for copper, D for cadmium, E for rare earth, F for iron, H for thorium, K for zirconium, M for manganese, and Z for zinc. There are other coding systems such as the Unified Numbering System and the British Standard for magnesium alloys.

Magnesium alloys are an ideal material for weight reduction mainly because of their high specific strength. They have a wide range of other mechanical properties, depending on the composition, condition (whether cast or wrought), fabrication process, heat treatment, and other factors. In addition, magnesium alloys have higher damping capacities than cast iron (up to about 3 times) and aluminum (about 30 times), which make them ideal for automobile applications. Their high electrical and thermal conductivities make them preferred for certain applications such as heat sink for heat dissipation. Because of their high chemical reactivity, magnesium alloys usually have a protective film on their surface to protect against corrosion. However, it is not as dense as the oxide layer on an aluminum alloy, and it is easily corroded by chlorides, sulfates, and other chemicals. Therefore, magnesium is often anodized for corrosion protection.

Both casting and wrought magnesium alloys are used in automobile applications. In sheet metal fabrication, however, wrought alloys are prevalent. As the mechanical properties of such alloys are determined by the fabrication and they are usually anisotropic, attention is needed in identifying the direction dependence of the material properties. The inference of such anisotropy in properties on welding can be a subject of study. For rolled sheets, the longitudinal properties are normally a little lower than the transverse properties, and the tensile properties of sheets are usually determined on specimens cut parallel to the direction of rolling. The yield strengths of wrought magnesium alloys normally vary with the direction of metal flow. The modulus of elasticity of commercial magnesium alloys in tension and compression is about 45 GPa, the shear modulus is about 16 GPa, and Poisson's ratio is 0.35. The modulus of elasticity decreases with temperature. It was found that cyclic cold loading beyond the yield strength in both tension and compression reduces the modulus of elasticity of magnesium alloys. A number of low-melting eutectics between Mg and other elements may form during material processing, and they play an important role in the behavior of Mg alloys. For instance, wetting of grain boundaries by low-melting eutectics was considered responsible for the super-plastic deformation of these alloys.[25] They may affect the weldability by promoting hot cracking during welding.

Magnesium alloys have found a wide range of applications in automobile, electronics, aerospace, and other industries because of their advantages over other structural materials, and the abundance of Mg resources. In automobile construction, for instance, magnesium alloys have been used as structural components, chassis system support, and interior components. Their unique contributions to vehicle weight reduction, fuel economy, emission reduction, noise reduction, safety, recyclability, and many others make them a focus in automobile material and processing research. Because of the limitations in fabricating rolled products and in material joining techniques, a large portion of magnesium products are in the form of castings.

1.2.3.2 Welding Mg Alloys

Welding magnesium alloys is not as robust as welding steels or aluminum alloys. The difficulties in welding Mg alloys arise from their intrinsic physical properties. Nevertheless, extensive research work has been carried out in fusion welding Mg alloys, and a number of important findings have been made. Some of them are summarized in the following[26,27]:

1. Excessive grain growth in the fusion zone. A high heat input rate is necessary in welding Mg alloys because of their high thermal conductivity. In addition, the melting temperatures of the alloys and, therefore, the recrystallization temperatures are fairly low. As a result, coarse grains are usually observed in the fused and solidified area, accompanied by large segregation of alloying elements. These severely affect the strength of a welded joint.
2. Excessive thermal stresses and distortion. The large coefficient of thermal expansion (CTE) and rapid heat input during welding Mg alloys produce significant deformation, distortion, and thermal stress in the weldment during welding.
3. Cracking. Mg may form a number of eutectics with other alloying elements having much lower melting temperatures than the Mg matrix. Therefore, there is a partial melting zone in which the melting of eutectics weakens the material and makes it prone to cracking, with the aid of thermal stresses.
4. Void formation. Hydrogen from various sources such as moisture and coating compound may dissolve in the molten metal. As the solubility of hydrogen in Mg drops drastically during cooling, gas bubbles may form during solidification.

The aluminum content in the Mg–Al–Zn alloys, such as AZ31B, AZ80A, AZ91, and AZ92A, generally promotes weldability, as it helps refine the grain structure. Zinc of more than 1% makes the material prone to hot cracking. Therefore, Mg alloys of high zinc content (ZH62A, ZK51A, ZK60A, and ZK61A) have poor weldability. The weldability of these alloys can be improved by adding a small amount of thorium.

Because of the difficulties in welding Mg alloys, alternative means of joining such as mechanical fastening and adhesive bonding have been explored. However, the automobile industry still prefers RSW because of its robustness proven in the practice of joining other metals over a few decades, and the high comfort level of the operators.

1.2.3.3 Resistance Welding Mg Alloys

There is very limited information available on resistance spot welding Mg alloys in the public domain. In a study on welding AZ31B alloys,[28,29] it was found that the center of a weld nugget consists of fine equiaxed grains, decorated with β-$Mg_{17}Al_{12}$ precipitated from α-Mg. Cracking was speculated to initiate in the welds during solidification, and electrode deterioration and expulsion were found to be the most common defects in welding Mg alloys. In another work of welding AZ91D and AM50, low-melting phases at the grain boundaries in the base metal, possibly formed by segregation, melted during heating (below the melting temperature of the alloy), and solidified during cooling.[30]

Because of the similarities between Al and Mg in electrical, thermal, and metallurgical properties, knowledge of welding Al alloys can be utilized as guidance for welding Mg alloys. In welding both alloys, care must be taken to avoid expulsion, cracking, and premature electrode failure. However, one has to recognize that welding Mg alloys is different from welding Al alloys in several aspects. For instance, experiments have found that Mg alloys are more prone to both surface and interfacial expulsion, as discussed in more detail in Chapter 7. In general, resistance welding Mg alloys has the following characteristics:

1. High electric current. The low electrical resistivity of Mg and its alloys warrants a high electric current to be applied, in order to generate sufficient heat through the Joule process.
2. Short welding time. The resistance heating has to be done in a short period of time, as the high thermal conductivity of the alloys makes the heat dissipate rapidly.
3. Large electrode force. This serves two purposes: reducing the contact resistance by breaking up the oxide layer and creating sufficient electric contact at the electrode–sheet interface, and confining the expansion of the weldment, which is critical for containing expulsion and reducing defect formation. These alloys have a fairly large CTE and, therefore, a large electrode force is needed.
4. High expulsion tendency. Both surface and interfacial expulsion are prevalent in welding Mg alloys, and the expulsion mechanisms are different in welding different alloys.
5. Short electrode life. The electrode life can be significantly shorter than in welding Al alloys, as a result of surface expulsion and alloying between the sheet and electrode.
6. Defect formation. Both cracks and voids are common in Mg welds, due to the large expansion of the alloys, and volume deficit created by expulsion.

In a study on the characteristics of resistance welding Mg alloys AZ31B and AZ91D, Luo et al.[31] studied cracking and expulsion during resistance welding. Some of the phases in the base metal and the weldment can be found from a binary Mg–Al phase diagram in Figure 1.17. As seen from a typical microstructure of the AZ91D casting shown in Figure 1.18a, in addition to small voids possibly formed during casting, a large amount of β phase ($Mg_{17}Al_{12}$) exists at the grain boundaries. The structures and properties of different phases are drastically different, and the difference exists even after the material is heat-treated for tempering and homogenization. The rolled AZ31B sheets have a significantly different morphology of microstructure with much finer grain boundaries, possibly due to the smaller amount of alloying elements and a different fabrication process (Figure 1.18b).

Due to the different metallurgical characteristics of AZ31B and AZ91D, these two alloys behaved quite differently in welding. Expulsion is observed in welding AZ31B, at both surface and interface. Expulsion is also prevalent in welding AZ91D. However, the liquid metal is ejected both from the faying interface and through a liquid network consisting of low-melting eutectics.

Unlike in welding most metals such as steels, columnar dendritic grains are not common in the fusion zone of Mg welds. In welding AZ31, it was found that the edge of a weld tends to have a cellular dendritic structure, whereas the nugget center has an equiaxed dendritic structure (Figure 1.19).[32] These structures are similar to those revealed from the opened crack surfaces in the same work, as shown in Figure 1.6a and b, respectively. The epitaxial growth of cellular dendritic crystals is believed to be driven by the supercooling created by the rapid heat dissipation during cooling, which is possible through the solid portion of the weldment as a result of the high thermal conductivity of Mg alloys. The temperature gradient decreases as the solid–liquid interface advances toward the weld center; and the Al and Zn concentrations increase due to segregation, which creates a large constitutional supercooling. Both conditions favor the equiaxed dendritic crystal growth

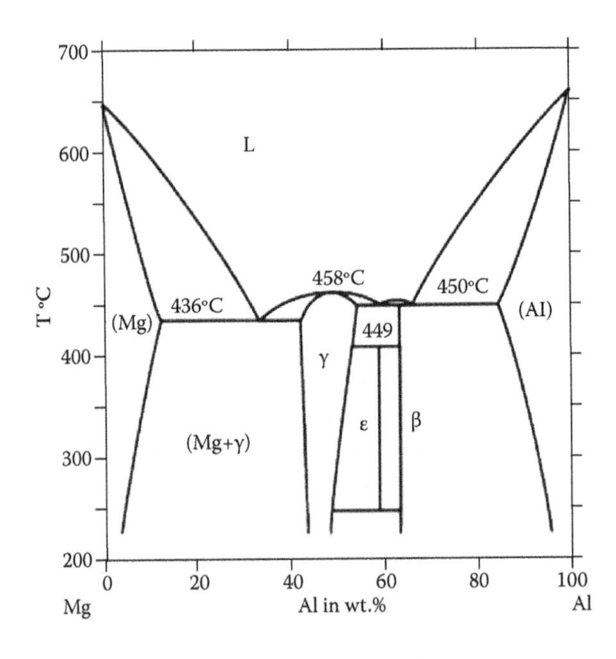

FIGURE 1.17
Mg–Al phase diagram.

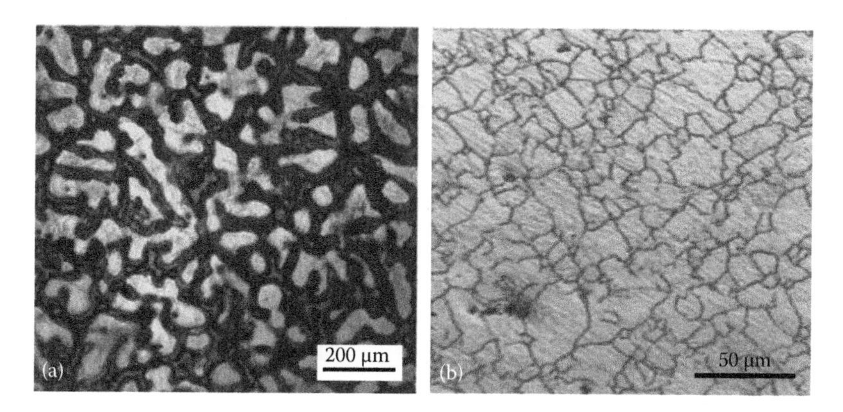

FIGURE 1.18
Microstructures of (a) AZ91D and (b) AZ31B sheets.

at the weld center. Another study on AZ31 confirmed the equiaxed dendritic structure in the weld nugget.[29]

The microstructure of a Mg weld also depends on the heating/cooling process, or welding parameters. In a study on welding AZ31B,[10] it was found that the welds usually consist of cellular dendritic structure in the nugget near the fusion line, and equiaxed dendrites away from the fusion line (Figure 1.20a and b), same as observed by other researchers. Cellular dendrites may also form at the interior of a weld, as shown in Figure 1.20c under certain heating/cooling conditions. Similar structures are observed in an AZ91D weld (Figure 1.21).[10]

The existence of low-melting eutectics in Mg alloys has a significant effect on the structure and defect formation in Mg welds. Because these eutectics are generally rich in alloying element, they are concentrated around grain boundaries, largely resulting from segregation. They may be created during fabrication of the material, as shown in Figure 1.18a in the AZ91D casting, or they may be formed during solidification in the fusion zone. Grain boundary melting of such eutectics in the HAZ may weaken the structure, and cracking may occur under tensile loading. Figure 1.22a shows a crack in the HAZ near the base metal. Its intergranular nature is an indirect evidence of grain boundary melting. Similar cracking is observed inside of a weld, as seen in Figure 1.22b, which shows

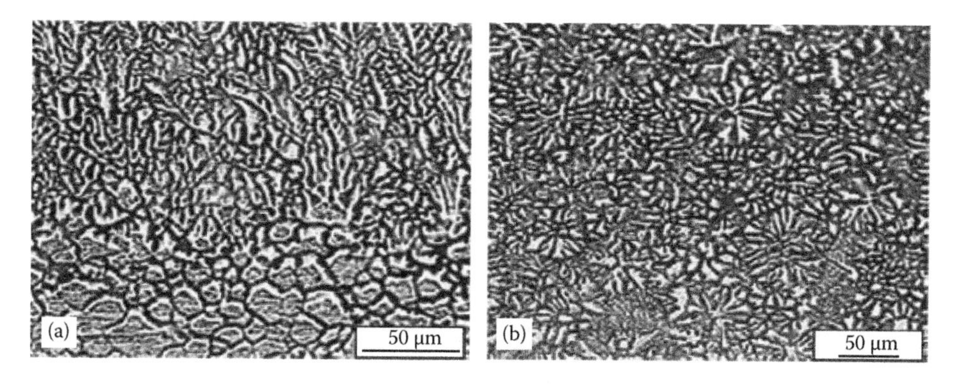

FIGURE 1.19
Microstructure of AZ31 weld nugget: (a) cellular dendritic structure near fusion line, and (b) equiaxed dendritic structure at center. (From Sun, D.Q. et al., *Mater. Sci. Eng. A*, 460, 461, 494–498, 2007. With permission.)

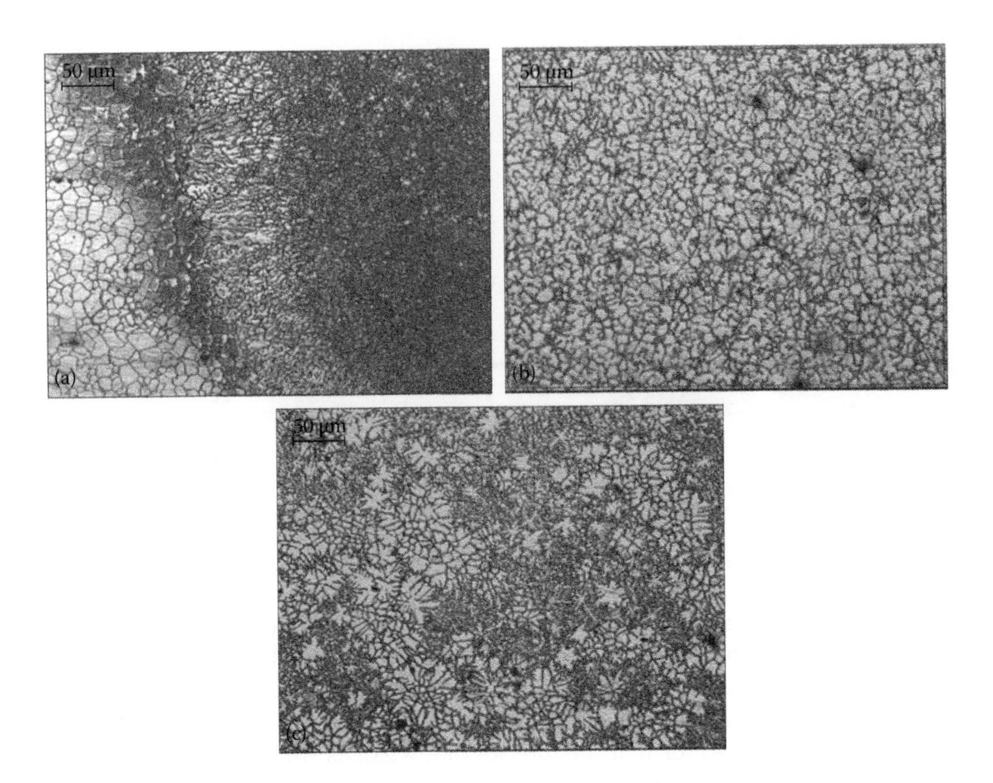

FIGURE 1.20
Microstructures of AZ31B welds. Those of the HAZ and nugget near the fusion line (a), and at a distance from the fusion line (b). The interior of an AZ31B weld created under a different welding condition is shown in (c). (From Luo, H., New joining techniques for magnesium alloy sheets, MS thesis, Institute of Metal Research, Chinese Academy of Sciences, May 2008. With permission.)

both cracks and voids. The formation of these defects is driven by the liquid eutectics that remain as liquid until all their surroundings are solidified, the volume deficit created by the irreversible thermal expansion and expulsion, and the thermal stress developed during cooling. The structures shown in Figure 1.23 from crack surfaces are the result of free solidification that occurred in such a process.

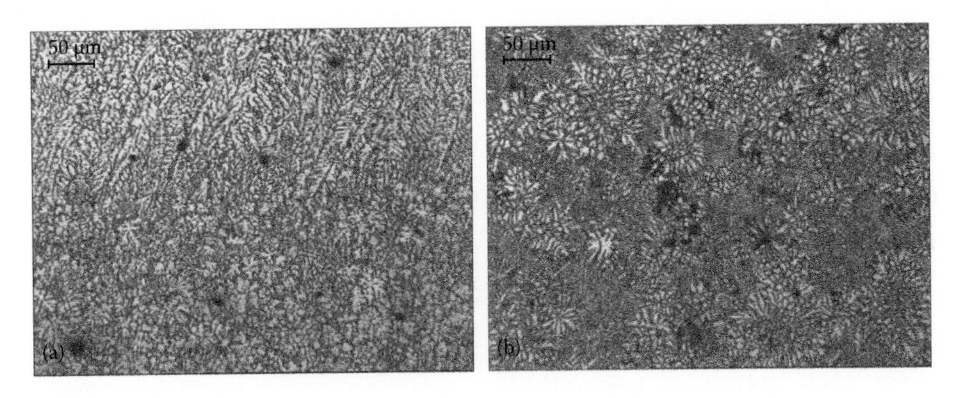

FIGURE 1.21
Weld microstructure of AZ91D weld. (From Luo, H., New joining techniques for magnesium alloy sheets, MS thesis, Institute of Metal Research, Chinese Academy of Sciences, May 2008. With permission.)

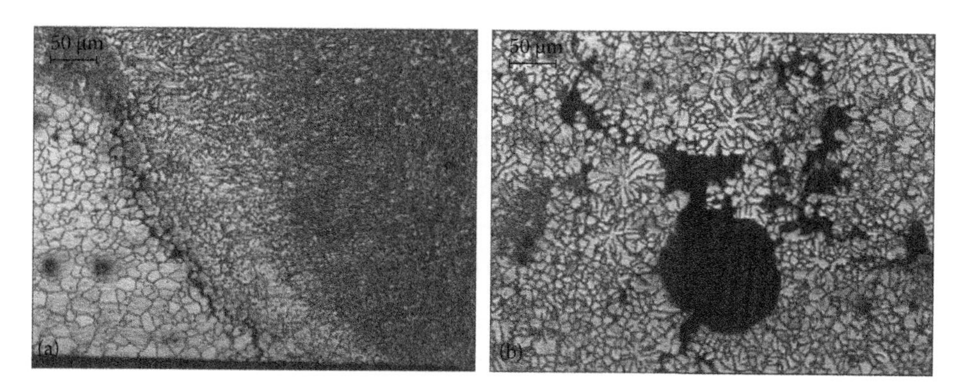

FIGURE 1.22
Microstructure near HAZ (a) and inside (b) of an AZ31B weld. (From Luo, H., New joining techniques for magnesium alloy sheets, MS thesis, Institute of Metal Research, Chinese Academy of Sciences, May 2008. With permission.)

1.2.4 Copper Alloys

Copper is the base metal normally used for resistance welding tongs and tips. The major functions of the electrodes are to conduct the welding current, to apply pressure to the weld joint, and to transfer heat from the workpiece. Copper is an industrial metal that is widely used in both pure and alloyed forms, mainly because of its unique characteristics of high electrical conductivity, high thermal conductivity, high corrosion resistance, good ductility and malleability, and reasonable mechanical strength. Pure copper is the ideal material for electric current conductors, but the electrodes for resistance welding require a strength that exceeds the level attainable with pure copper. Therefore, the use of copper alloys becomes necessary. The most common Cu alloys used as electrode materials are:

1. Cu–Cr
2. Cu–Cr–Zr
3. Cu–Zr
4. Dispersion-strengthened copper (DSC)

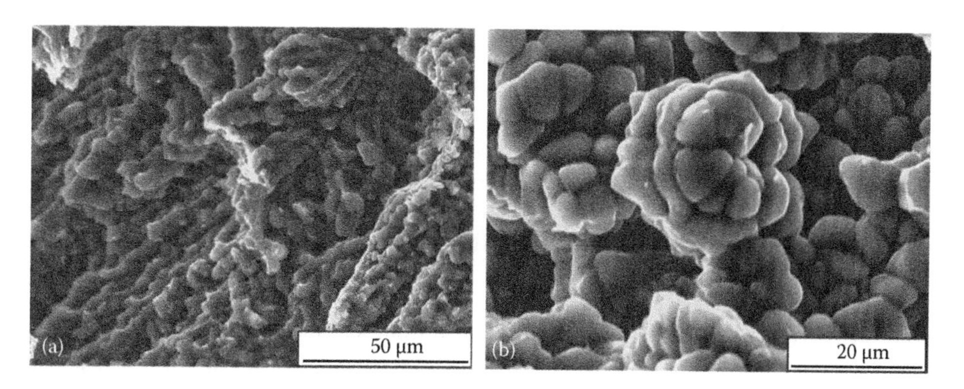

FIGURE 1.23
Morphology of opened crack surface: (a) cellular dendrites, and (b) equiaxed dendrites. (From Sun, D.Q. et al., *Mater. Sci. Eng. A*, 460, 461, 494–498, 2007. With permission.)

Cu–Cr and Cu–Cr–Zr are the original materials used to weld bare steels. Cu–Zr and DSC were introduced to deal with coated steels when sticking became a major production issue.[33] In general, copper alloys used as electrode materials must exhibit high electrical and thermal conductivities, combined with high strength, at elevated temperatures.

1.2.4.1 Strengthening of Cu Alloys

The difficulty in developing a suitable material for electrodes lies in achieving a balance between mechanical strength and electrical conductivity, and resistance to softening at elevated temperatures. Significant research efforts have been put forward in order to raise the mechanical strength while maintaining the high electrical and thermal conductivities of copper. Over the years, a large variety of high-copper alloys with reasonable strength has been developed through solid solution hardening and precipitation hardening, and often aided by cold work.

Copper lattice is able to dissolve a certain amount of atoms of other metals to form a solid solution. In such a solid solution, the lattice in the vicinity of the impurity atoms is distorted because of the difference in size of the impurities from that of copper. Such distortion creates local stress fields that impede dislocation motion and, therefore, strengthens the metal. This is called solid solution strengthening. Solid solution strengthening is often accompanied by a certain degree of cold work to further increase the strength. However, a tradeoff for the increase in strength is the decrease in electrical conductivity, resulting from the lattice distortion caused by alloying atoms. In general, all dissolved additions to copper reduce electrical conductivity, and the extent of this effect varies widely from element to element. Cadmium additions, for example, affect conductivity the least, whereas others, such as phosphorus, tin, and zinc, are more detrimental. In order to minimize the drop in electrical conductivity, solid solution-hardened high copper alloys with low content of alloyed elements have been developed for applications such as resistance welding electrodes. Among them, Cu–Cr alloys have a favorable combination of mechanical properties and electrical conductivity, and are commonly used as electrode materials for resistance welding of low-carbon steels.[34]

There is a limit, called solubility limit, as to how much a particular impurity can be dissolved in a metal. It in general increases with temperature. A different phase may start to form if a solid solution is cooled down and the solubility limit is undershot. The second phase, usually an intermetallic, is in the form of fine (often lower than 100 nm in size) and hard particles, and it strengthens the material. This is precipitation strengthening. In addition to the strength, this process also affects the electrical conductivity of the metal. As the impurity atoms leave the lattice, the lattice distortion is undone and the electric conductivity of the material is restored to a certain extent. The solubility of Cr in Cu is fairly low: no more than 0.7 wt.% Cr could be dissolved into solid solution at a safe solution treatment temperature. An addition of a very small amount of Zr (less than 0.1 wt.%) improves hardness and electrical conductivity of the Cu–Cr alloy due to the very low solubility of Zr in Cu matrix at room temperature. Zr in the alloys adjusts the orientation relationship between Cr and the matrix, and tends to increase the conductivity of aged Cu–Cr–Zr alloys after deformation. In the Cu–Cr–Zr alloys, the precipitated phases are characterized as Cr, $Cu_{51}Zr_{14}$, and Cu_5Zr,[35] as well as the Hesuler phase $CrCu_2Zr$.[36] The plastic deformation during fabrication of electrodes involves cold work, which increases the hardness but reduces the conductivity. Such cold work after the solution annealing but before the age annealing promotes the formation of fine and homogeneously distributed precipitates. An important characteristic of precipitation-hardened alloys is their high relaxation resistance. This is critical for resistance welding electrodes as they are exposed to fairly high temperature and pressure during welding. Solid solution-hardened alloys generally have insufficient relaxation resistance at elevated temperatures, and precipitation strengthening becomes essential.

The International Annealed Copper Standard (IACS; a high purity copper with a resistivity of 0.0000017 Ω cm) is sometimes used as an electrical conductivity standard for metals, and the effect of alloying on electrical conductivity is often expressed by the resultant electrical conductivity value of the alloy as a percentage of IACS. For instance, it was found that in Cu–0.4%Cr–0.08%Zr, the maximum of electrical conductivity, which is between 89% and 92% IACS, appears at 480°C.[34]

Sticking of the electrodes to the workpiece is a major problem when welding galvanized steels. The reaction of the galvanized Zn and Cu can produce a bond between the electrode and the steel sheet and cause the electrode to stick to the sheet. This may result in the alloyed layer of brass on the electrode to be peeled from the electrode, or even the electrode pulled from its holder.[37] Using Cu–Zr electrodes can dramatically reduce sticking, and another type of material, DSC, has better sticking resistance.[33] Both Cu–Zr alloys and DSC have excellent high-temperature stability and, therefore, prolonged electrode life. Because dispersed oxides such as alumina are often used in making DSC, they are often referred to as oxide dispersion-strengthened copper (ODSC or ODS copper). The high thermal stability of ODSC comes from the fact that the aluminum oxide particles can retain their original size and spacing, and retard the copper matrix from recrystallization even after prolonged heating.[38]

There are several means of preparing ODS coppers. One is powder metallurgy, in which a mixture of very fine powdered copper and oxides are compacted and sintered to form a solid metal. A more commonly used method nowadays is internal oxidation, which forms oxides of a reactive alloy constituent *in situ*. ODS coppers are commercially available in three grades: C15715, C15725, and C15760, with 0.3, 0.5, and 1.1 wt.% Al_2O_3, respectively. For instance, Glidcop® AL-15, a C15715 material, is a low-alumina content grade of DSC with 0.15 wt.% aluminum as Al_2O_3.[39] It is strengthened by an ultrafine dispersion of Al_2O_3 particles through *in situ* internal oxidation of the aluminum in a dilute solid solution copper–aluminum alloy powder. Along with superior strength retention, an ODS copper's thermal and electrical conductivities are higher than conventional copper alloys. Its electrical conductivity is rated at 92% IACS at 20°C.

1.2.4.2 Classifications of Electrodes

Electrode tips for resistance welding are made of copper alloys and other materials. The Resistance Welder Manufacturers' Association has classified electrode tips into two groups according to the chemical compositions of the materials[40]—Group A: copper-base alloys, and Group B: refractory metals. These groups are further classified according to the chemical compositions and performance characteristics of the materials.

Group A, made of copper alloys, is further divided into five classes (I, II, III, IV, and V) with Class I electrodes the closest in composition to pure copper. As the class number goes up, the hardness and annealing temperature increase, whereas the thermal and electrical conductivities decrease. In Group B, Classes 10, 11, 12, 13, and 14 are the refractory alloys, made of sintered mixtures of copper and tungsten, etc., designed for wear resistance and compressive strength at high temperatures. Refractory metals, including the tungsten–copper composites (Classes 10–12), are used in specialty applications in which the high heat, long weld time, inadequate cooling, and high pressure involved may cause rapid deterioration of DSC-base alloys (Group A).[38] The main metallurgical aspects of these classes of electrodes are presented in the following.

GROUP A: Copper-Base Alloys

Class 1. Cadmium–copper is suitable for welding aluminum and magnesium alloys, coated materials, and brass and bronze. It is not heat treatable. This alloy is superior to pure copper as an electrode material because of its high electrical conductivity and reasonable strength at elevated temperatures.

Class 2. Chromium–copper is suitable for welding cold- and hot-rolled steels, stainless steel, and low-conductive brasses and bronzes. It is heat treatable. It has good strength, and when hardened, has a conductivity of 80% that of pure copper. It is recommended for high-production operations. A special heat-treated alloy that meets the minimum electrical conductivity and hardness specifications of Class 2 alloy is Zr–Cr–Cu. It is suited to welding galvanized steel and other metallic-coated steel.

Class 3. Nickel–copper and beryllium–nickel–Copper are suitable for welding steels having high electrical resistance, such as stainless steels. They are heat treatable. These alloys have higher strength than the previous classes and, therefore, they are often used as electrode shanks and electrode holders.

Class 4. Beryllium–copper has lower conductivity than Class 3 alloys but has exceptional strength and hardness, in some cases approaching the levels attained in heat-treated steels. It is available in the annealed condition that is more readily machined and then heat treated. It is often used in the form of inserts, die facings, and seam welder bushings.

GROUP B: Refractory Metals

Class 10. Tungsten55%–copper45%, is recommended where (relatively) high electrical conductivity and some degree of malleability are desirable. It is suitable for facings and inserts for projection welding electrodes and flash and butt welding dies.

Class 11. Tungsten75%–copper25%, is harder than Class 10 alloys, and suited to similar applications as Class 10 for facing on electrode forming dies, and for projection welding electrodes.

Class 12. Tungsten80%–copper20%, is suitable for electro-forming and electro-forging die facings, and for electrode facings used as upset studs and rivets.

Class 13 (tungsten) and **Class 14** (molybdenum). These two classes of materials are used primarily for welding or electro-brazing nonferrous metals having relatively high electrical conductivities. They are suitable for cross-wire welding of copper and brass, and for welding copper wire braid to brass or bronze terminals. Special setups and procedures are required.

1.2.4.3 Copper Electrode and Coating/Sheet Interaction

The interaction between Cu electrodes and coating/sheets is a complicated process consisting of mechanical, electrical, and thermal, in addition to metallurgical factors. When Cu-base alloy electrodes are used for welding coated metals, certain metallurgical reactions may happen between Cu and the coating, and such interaction may significantly affect the welding process and electrode life. For instance, Cu may form a brittle low-melting intermetallic with Zn. The alloyed electrode surface has a lower electrical conductivity, and during welding, the temperature at the electrode–workpiece interface increases as a result. This may promote sticking of electrodes to the coated workpieces when welding galvanized steels. Such alloying may happen between Cu electrodes and a number of coating and substrate materials such as Ni, Sn, Al, and Mg. The alloying of electrode material with the coating/substrate is the primary reason for electrode deterioration. Examples illustrating the interaction between Cu electrodes and Zn-coated steels, and that between Cu and Al sheets are shown in the following.

GALVANIZED STEEL WELDING

In a study on the electrode wear mechanisms, Gugel et al.[41] systematically investigated the evolution of electrode face in terms of surface chemistry through secondary electron imaging and EDX analysis. The chemical change of the Cu–Cr electrode face clearly shows the severe interaction between the electrode and the Zn coating, as seen in Figure 1.24. The first weld already picked up substantial amount of zinc, and by 10 welds, zinc overtook Cu on the electrode surface. The concentration of zinc steadily increases to nearly 80 wt.% by 1000 welds, and the concentration of iron remains around 10%. Since the Cu–Zn alloy has higher electrical resistivity than Cu–Cr, a large amount of heat was generated at the electrode–sheet interface. This was evidenced by the substantial sticking. Similar trends were observed when welding using Cu–Zr electrodes (Figure 1.25), and a Glidcop AL60 grade DSC electrode (Figure 1.26). Although there are similarities among these three types of electrodes in reacting with the Zn coating, the DSC electrodes had shorter electrode life than the other two, possibly because the higher relative hardness of the DSC electrodes may hinder the plastic deformation of the Cu, and are unable to heal small surface imperfections.[42] It was also observed that oxides existed in the depressions on the electrode faces; the effect of which on electrode deterioration is not clear.

The interaction between Cu electrodes and Zn coating of the steel sheets is also affected by the Al content in the coating. Matsuda et al.[43] studied the effect of Al in the coating on electrode life by intentionally varying the Al content in the range from 0.22 to 0.87 wt.% in hot-dip galvanized steel sheets, and from 0.19 to 0.78 wt.% in the galvannealed steel sheets. They found that when the aluminum content in the coating is sufficiently high (more than 0.4 wt.%), a thin layer of intermetallic $Fe_2Al_{5-x}Zn_x$ is present at the interface between the coating and substrate in hot-dip galvanized steel sheets. This layer inhibits Fe diffusion into the coating, and a low Fe content in the coating makes the coating low in melting temperature with high Zn concentration, promotes the Cu–Zn alloying, and accelerates electrode wear. On the other hand, there is no such intermetallic layer between the coating and substrate in galvannealed steel sheets, possibly resulting from the post-heating process of galvannealing. The aluminum content in the coating, therefore, has no effect on electrode

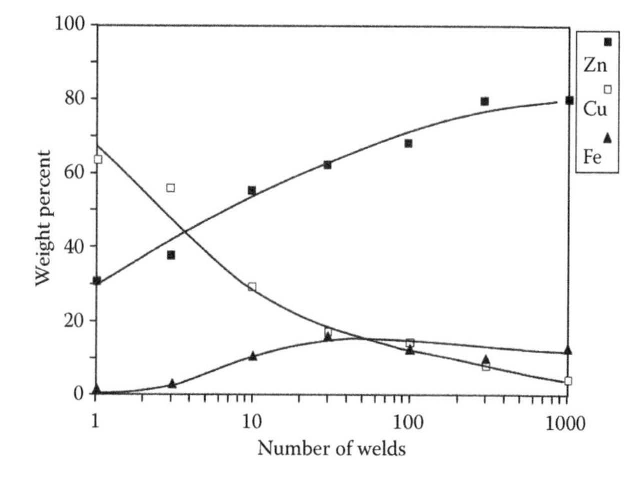

FIGURE 1.24
Evolution of element concentrations on Cu–Cr electrode face. (From Gugel, M.D. et al., Progression of electrode wear during RSW of EG steel, SMWC V, Paper A4, 1992. With permission.)

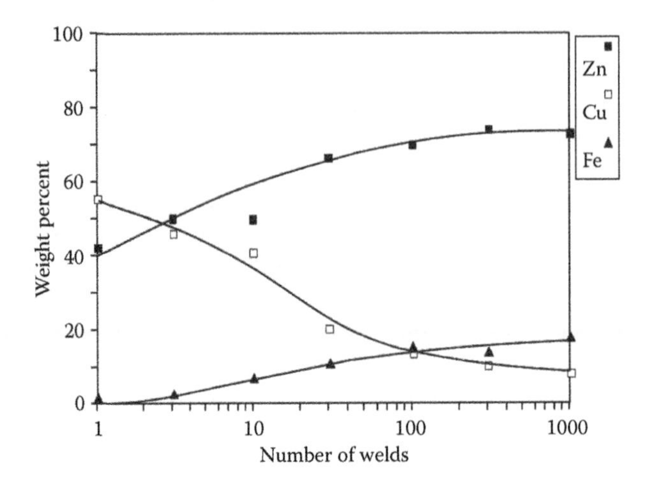

FIGURE 1.25
Evolution of element concentrations on Cu–Zr electrode face. (From Gugel, M.D. et al., Progression of electrode wear during RSW of EG steel, SMWC V, Paper A4, 1992. With permission.)

life in this welding material. The coating of galvannealed steel has a high concentration of the Fe–Zn intermetallic compounds formed during galvannealing, which have higher melting temperatures and hardness than mostly free Zn in the hot-dip galvanized steel coating. Therefore, in general, the electrode life is significantly longer when welding galvannealed steels than welding galvanized steels.

As revealed by Matsuda et al.,[43] the amount of free Zn in the coating directly affects the reaction between the electrode and the coating. Figure 1.27 shows the differences in geometry and composition of the depressions on the electrode faces. In the two types of galvannealed steels with different Al contents in the coating, there is a significant amount of Fe in the material picked up by the electrodes from the coatings, which largely consist of Fe–Zn

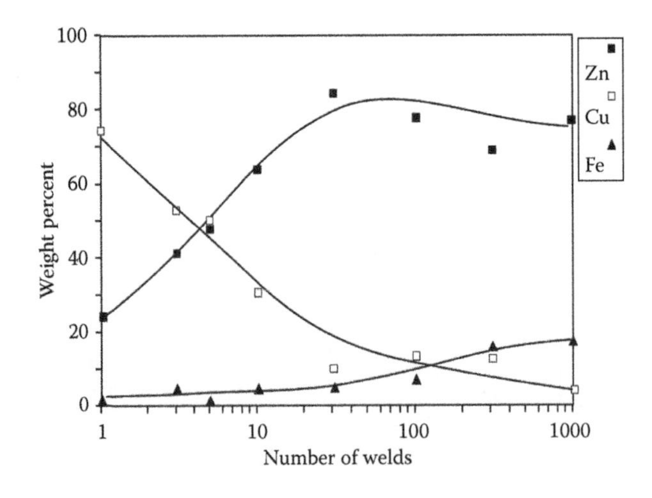

FIGURE 1.26
Evolution of element concentrations on DSC electrode face. (From Gugel, M.D. et al., Progression of electrode wear during RSW of EG steel, SMWC V, Paper A4, 1992. With permission.)

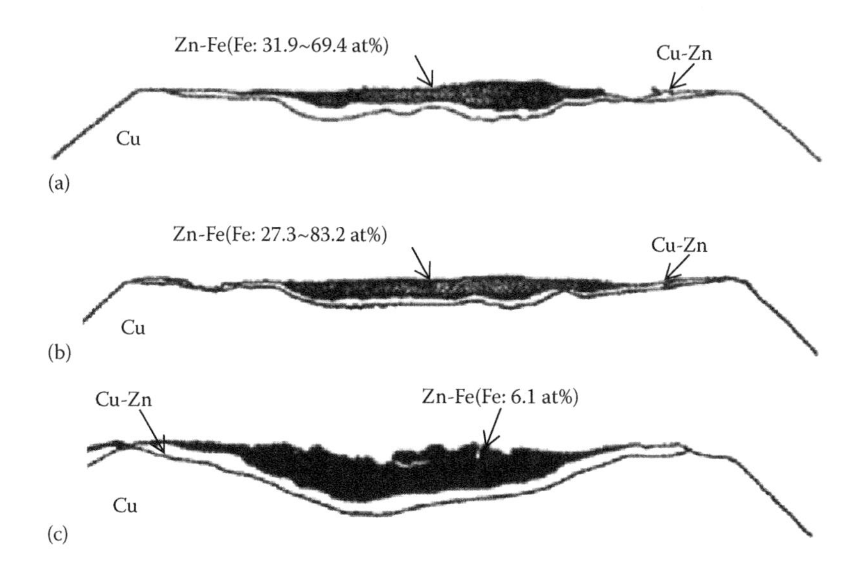

FIGURE 1.27
Cross-sectional structures of electrode tip faces after 1500 welds on galvannealed (GA) and hot-dip galvanized (HDG) steels.[43] (a) GA-2 (Al: 0.78 mass%); (b) GA-6 (Al: 0.23 mass%); (c) HDG-6 (Al: 0.22 mass%).

intermetallic compounds. Al content has no impact on the depression on the electrodes. The electrode used for welding hot-dip galvanized steel, however, has a much deeper depression on the surface, filled with metals picked from the coating of mostly free zinc.

The low-melting intermetallic formed between Cu and Zn on the surface of a galvanized steel may attack the grain boundaries and create cracks. This effect is called liquid metal embrittlement (LME) cracking, and it has significant impact on the integrity of certain Zn-coated steels. More detail can be found in Section 1.4. When the alloying process completely covers the contacting electrode tip surface, weld quality degradation slows down considerably but still continues because mechanical wear creates newly exposed copper surfaces. Thus, pre-seasoning (conditioning) the electrodes is necessary not for prolonged electrode life, but for stabilization of the welding process.[44]

AL WELDING

A study on the effects of sheet surface conditions on electrode life revealed that electrode wear is the result of localized heating at the interface, which creates the condition for metallurgical reaction between the elements in the electrode and sheet/surface.[45] 5A02 aluminum sheets (2 mm in thickness), and a number of surface condition and welding schedule combinations were used in the study.

The effects of electrode force and welding time on electrode alloying are clearly seen in Figure 1.28. The line scanning of the chemical composition along the center of an electrode shows the percentage of Cu, Al, and Mg at each location on a line through the electrode center. When the electrode force is small (4.5 kN), long welding time (180 ms) generates more heat at the electrode–sheet interface and, therefore, a larger amount of alloying with Al and Mg (Figure 1.28b) than shorter welding time (Figure 1.28a). The effect of welding time is similar when electrode force is higher (9.0 kN, as in Figure 1.28c and d), but the severity of alloying is significantly lessened with high electrode force, as can be seen by comparing Figure 1.28a with 1.28c and Figure 1.28b with 1.28d. This is because a large

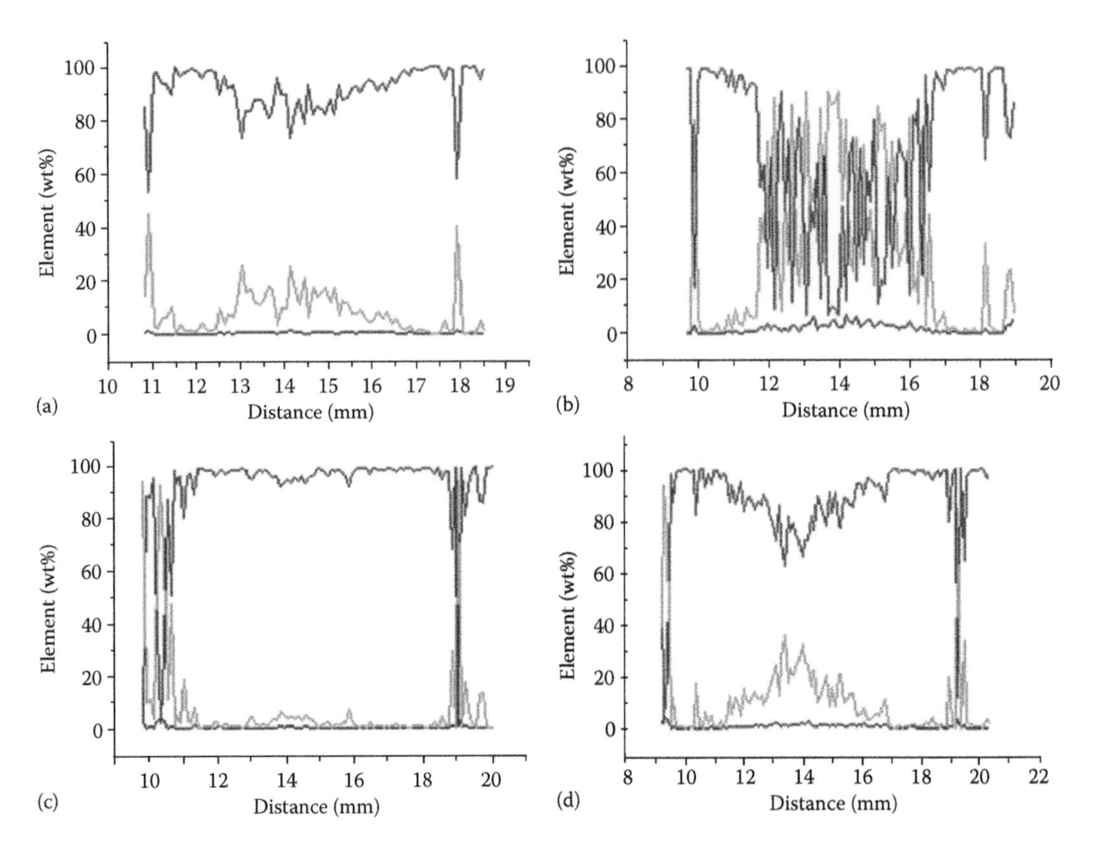

FIGURE 1.28
(See color insert.) Composition profiles of electrode surfaces after 60 welds using schedules of (a) $F = 4.5$ kN, $\tau = 60$ ms; (b) $F = 4.5$ kN, $\tau = 180$ ms; (c) $F = 9.0$ kN, $\tau = 60$ ms; and (d) $F = 9.0$ kN, $\tau = 180$ ms. Red line is for Cu, green is for Al, and blue is for Mg.

electrode force results in a low contact resistance and, therefore, less heat generation and alloying at the electrode–sheet interface.

1.3 Embrittlement of Weldment

Embrittlement of metals refers to the loss of ductility resulting from metallurgical reactions or other processes. In general, its occurrence in certain material systems requires specific conditions. There are several types of embrittlement in metals and the most common ones are described below:

1. Liquid metal embrittlement. It is also known as liquid metal-assisted cracking, or liquid metal-induced cracking. Deep liquid grooves may form at the grain boundaries when a specific liquid metal is in contact with a polycrystalline solid in certain material systems. The most common LME-causing material is mercury (melting point –38.8°C), which is liquid at room temperature. It presents a significant danger for airplanes as the aircraft material Al–Zn–Mg–Cu alloy DTD 5050B

is especially susceptible to LME by Hg.[46] Another commonly cited example is the Al–Ga system in which liquid Ga quickly penetrates into grain boundaries in Al, leading to intergranular fracture even under very small stresses. As elevated temperature is needed in order for most metals to melt, LME generally occurs only above a certain temperature. A tensile stress is usually needed for LME to occur, which is readily available in most metals, and only modest levels of stresses are needed to break a polycrystalline structure in which the grains are only "glued" by the liquid metal. Such stresses may come from externally applied loading, or from residual stresses created during fabrication of the material such as cold working. Although this phenomenon is relatively common, the mechanisms of LME are not well understood.

2. Hydrogen embrittlement (HE). The presence of atomic hydrogen within the crystal lattice structure of a metal may significantly impair its mechanical strength, and may result in failure under a tensile loading. Hydrogen can be introduced into a metal at various stages such as in service or during material processing. Common causes of HE are pickling, electroplating, and welding. However, HE is not limited to these processes. High-strength steels, such as quenched and tempered steels or precipitation-hardened steels, are particularly susceptible to HE. The susceptibility to HE of steels is directly related to their strength. Steels with a tensile strength in the order of 1000 MPa or higher are generally subject to HE, whereas it is usually not a concern for steels of lower strength. Currently, the majority of hot-dip galvanized steels are generally in the range of 200–450 MPa and they are generally not so susceptible to HE. It can be avoided if necessary precautions are taken, for instance, using mechanical cleaning instead of acid pickling for preparing the surfaces before hot-dip galvanization of high-strength steels.
3. Strain age embrittlement. Strain aging is associated with the strains resulting from plastic deformation, usually generated from cold working. The residual stresses that resulted from plastic deformation are the driving force for the changes in structures and properties of metals. In steels, carbon atoms in the iron crystals tend to move to dislocations under the residual stress field produced by plastic deformation. The resultant high concentration of carbon atoms at dislocations impedes their mobility and, therefore, a reduction in ductility results in the steel. Because it depends on diffusion of carbon atoms, strain aging is a strong function of temperature, in addition to stress level.
4. Intermetallic-compound embrittlement. Formation of brittle intermetallic compounds, usually at the grain boundaries of a crystalline solid, may lower the strength and ductility of the material. Such intermetallic compounds may form during fabrication or during service, either due to segregation of alloying elements to grain boundaries, or diffusion of the elements from coating or environment. For instance, an improperly controlled casting process may create a brittle grain boundary network resulting from segregation of alloying elements. A long exposure of galvanized steel to temperature slightly below the melting point of zinc (420°C) causes zinc diffusion into the steel, resulting in the formation of a brittle iron–zinc intermetallic compound in the grain boundaries. For the same reason, it is difficult to weld Al directly to steels.

All of these four types of embrittlement may occur during resistance welding. For instance, if the surface of a steel sheet is not properly cleaned, hydrogen from decomposition of grease or surface treatment agents at the faying interface may be trapped in the molten nugget and cause HE. A large sheet distortion/separation is an indication of large residual stresses, which may embrittle the weldment through strain aging. Certain confinement of sheets during welding, such as applying additional constraints to prevent cracking as discussed in Chapter 3, can effectively reduce the distortion and, therefore, the risk of strain age embrittlement. In general, these two types of embrittlement can be avoided if proper precaution is taken. On the other hand, it could be difficult to avoid LME and intermetallic-compound embrittlement in certain material systems as they are more material related than process related. Many structural alloys can be embrittled by low-melting-point metals. For instance, aluminum is embrittled by mercury, indium, tin, and zinc; steel by tin, cadmium, zinc, lead, copper, and lithium; stainless steels by cadmium, aluminum, lead, and copper; titanium by cadmium and mercury; and nickel by zinc, cadmium, and mercury.[47] Significant efforts have been made in understanding the mechanisms of LME. Brittle intermetallic compounds may form during welding, especially in the HAZ where stress concentration often occurs and fracture initiates. HE can be avoided if proper precaution is taken, and it is not common in resistance spot welds except under certain special conditions, such as when the welded structure is subject to corrosive environment with hydrogen charging. As the strain age embrittlement involves complex interaction between the deformation process and material structure and composition, it is case dependent and difficult to be singled out. The other three types of embrittlement are discussed in more detail in this section.

1.3.1 Liquid Metal Embrittlement

LME is basically a metallurgical process. The tendency and severity of LME strongly depend on the material systems. LME is significantly affected by alloying of the solid metals: some alloying elements may increase the severity, whereas others prevent LME. The alloying elements affect segregation to grain boundaries of the metal and, therefore, alter the grain boundary properties. Maximum LME is observed in cases where alloying elements have saturated the grain boundaries of the solid metal.[48] It is a strong function of the mutual solubility of a solid–liquid metal combination.[49] Excessive solubility makes sharp crack propagation difficult, yet zero solubility prevents wetting of the solid surfaces by the liquid metal and, therefore, prevents LME. Presence of an oxide layer on the solid metal surface also prevents good contact between the two metals and deters LME. Figure 1.29 shows the effects of solute elements on mechanical strength of a polycrystalline Al. Addition of third elements to the liquid metal may increase or decrease the embrittlement and also alters the temperature region over which embrittlement occurs. Metal combinations that form intermetallic compounds do not cause LME. The susceptibility to LME of a solid metal is also affected by its hardness and grain size. Harder metals are more severely embrittled, and solids with larger grains are more susceptible. The microstructure of LME cracks is shown in Figure 1.30.[50] The grains are separated along grain boundaries, that is, it is an intergranular fracture.

A complete review of LME mechanisms can be found from a huge amount of literature in the public domain on this subject, such as the work by Joseph[48] and Glickman,[51] and some of the theories are listed here[46]:

1. The dissolution–diffusion model[52,53]: Adsorption of the liquid metal on the solid metal induces dissolution and inward diffusion, leading to crack nucleation and propagation.
2. The brittle fracture theory[54,55]: Adsorption of the liquid metal atoms at the crack tip weakens inter-atomic bonds and promotes cracking.

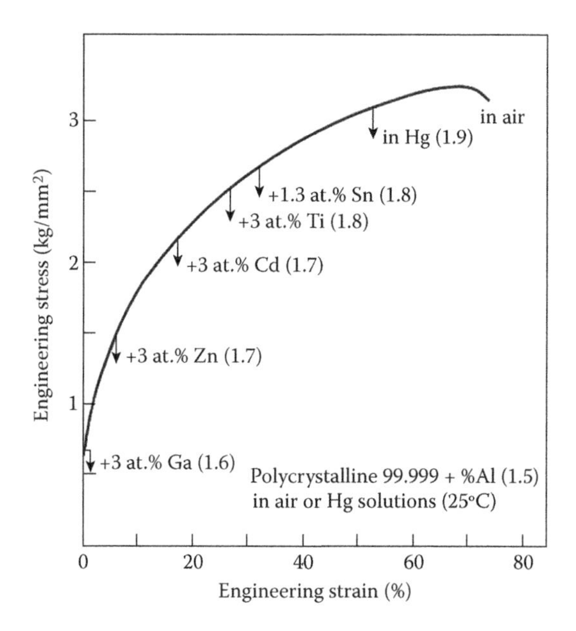

FIGURE 1.29
Polycrystalline pure Al embrittled by various Hg solutions. Electronegativities of solute elements are listed in parentheses. (From Westwood, A.R.C. et al., in Liebowitz, H. (ed.) *Fracture: An Advanced Treatise,* Academic Press, New York, 589–644, 1971. With permission.)

3. The diffusion–penetration model[56]: Crack nucleation and propagation are promoted through a diffusion–penetration process of liquid metal atoms into the grain boundaries.
4. The ductile failure model[57]: Nucleation and migration of dislocations under stress are promoted by adsorption of the liquid metal that weakens the atomic bonds. The dislocations pile up and work-harden the solid. In addition, such dissolution may also help the nucleation of voids that grow under stress and cause ductile failure.

FIGURE 1.30
Microstructure of an LME fracture. (From AET_Service_Capability.pdf. Available online at http://www.aet-int.com/capability/AET_Service_Capability.pdf. Accessed in Nov. 2010. With permission.)

All of these theories are based on the fundamental assumption that the adsorption of liquid metal lowers the surface energy of the solid metal, which leads to LME. This mechanism of LEM can be easily understood through fracture mechanics consideration. The strength of a material to resist cracking is associated with the energy needed to create new surfaces, or the surface energy density, γ_s. Or, if such strength is expressed in terms of fracture stress, σ_f[58]

$$\sigma_f = \sqrt{\frac{2Ew_f}{\pi a}} \tag{1.9}$$

where E is Young's modulus and a represents the crack length. w_f is a unified fracture energy that could include elastic, plastic, viscoelastic, and other effects. During an LME, the adsorption of liquid at the crack tip (usually along a grain boundary) reduces w_f and, therefore, σ_f. Efforts have been spent on characterizing the dependence of w_f on alloying elements of a solid, solution composition, temperature stress, etc. Because of the large number of factors involved and the difficulties in measuring w_f, limited qualitative, and not quantitative, data have been obtained on certain material systems.

LME could occur in two separate but related processes that may have impact on the quality of a resistance spot weld. One is associated with hot-dip galvanizing, which is one of the most efficient ways of providing durable corrosion protection for steel components. When hot-dip galvanizing steel, the molten zinc interacts with the steel, and a layered coating is formed consisting of zinc–iron intermetallics as reaction products, covered by a layer of solid zinc, as shown in Figure 1.31. The coating thickness is normally in the order of 100 μm. During a galvanizing process, the free zinc (often with other additives) in its liquid state may attack the grain boundaries of the sheets, resulting in cracking and rendering the sheets unusable. In general, cracking in galvanized steel structures may occur at various stages of fabrication. For certain steels, especially some stainless steels or heavily cold-worked products such as rolled sheets, the condition for LME may exist when they are galvanized, that is, when they are in contact with molten zinc. For instance, a steel structure used in a sign bridge that spans the freeway and carries informational sign boards consisted of a triangulated space frame fabricated from four long parallel

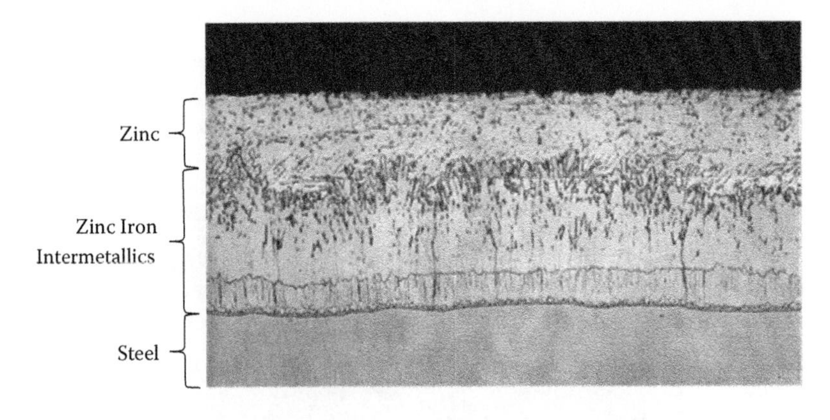

FIGURE 1.31
Microstructure of a galvanizing layer. (From Kinstler, T.J., Current knowledge of the cracking of steels during galvanizing—A synthesis of the available technical literature and collective experience for the American Institute of Steel Construction, GalvaScience LLC. Available online at http://www.aisc.org/uploadedFiles/Research/Files/Final5906.pdf. Accessed in Nov. 2010. With permission.)

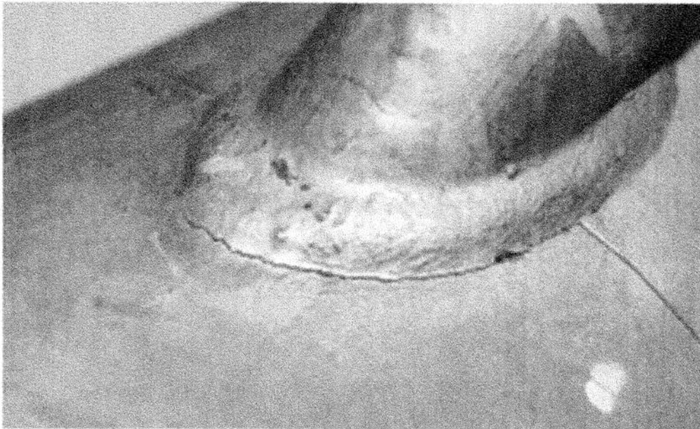

FIGURE 1.32
Cut-out section of cracked area on a sign bridge (left), and cracks at weld joining a diagonal tube to the chord (right). Note crack growth into chord at right. (From Website of Metallurgical Associates, Inc., http://www.metassoc.com/pdf/MAI_Minutes-6_04.pdf. Accessed in Nov. 2010. With permission.)

large diameter tubes, called chords, cross braced by smaller diameter perpendicular and diagonal tubes. After welding, the structure was hot-dip galvanized by submerging it in a bath of molten zinc to provide corrosion protection. While the welding was appropriately performed, cracking appeared after the structure was galvanized, as seen in Figure 1.32.[59] Metallurgical examination revealed that there was no sign of hot cracking, a defect associated with improper welding process, and all the cracks, near the weld or away from it, were filled with zinc. When galvanized steels are used in fabrication involving elevated temperature operation, LME may occur as well. Figure 1.33 shows a crack in a welded structure using galvanized steels, which initiates at the termination of the weld (in the HAZ) and propagates into the base metal.[60]

LME may happen during RSW. As shown in Figure 1.34, LME cracking is apparent near the surface of an HSLA steel weld that is in contact with the Cu electrode during welding.[50] The electron dispersive spectroscopy maps of elements clearly show that the cracks are filled with Cu and Zn (Figure 1.34b and c), which means that Zn in the coating and Cu

FIGURE 1.33
Crack at weld termination. (From Kinstler, T.J., Current knowledge of the cracking of steels during galvanizing—A synthesis of the available technical literature and collective experience for the American Institute of Steel Construction, GalvaScience LLC. Available online at http://www.aisc.org/uploadedFiles/Research/Files/Final5906.pdf. Accessed in Nov. 2010. With permission.)

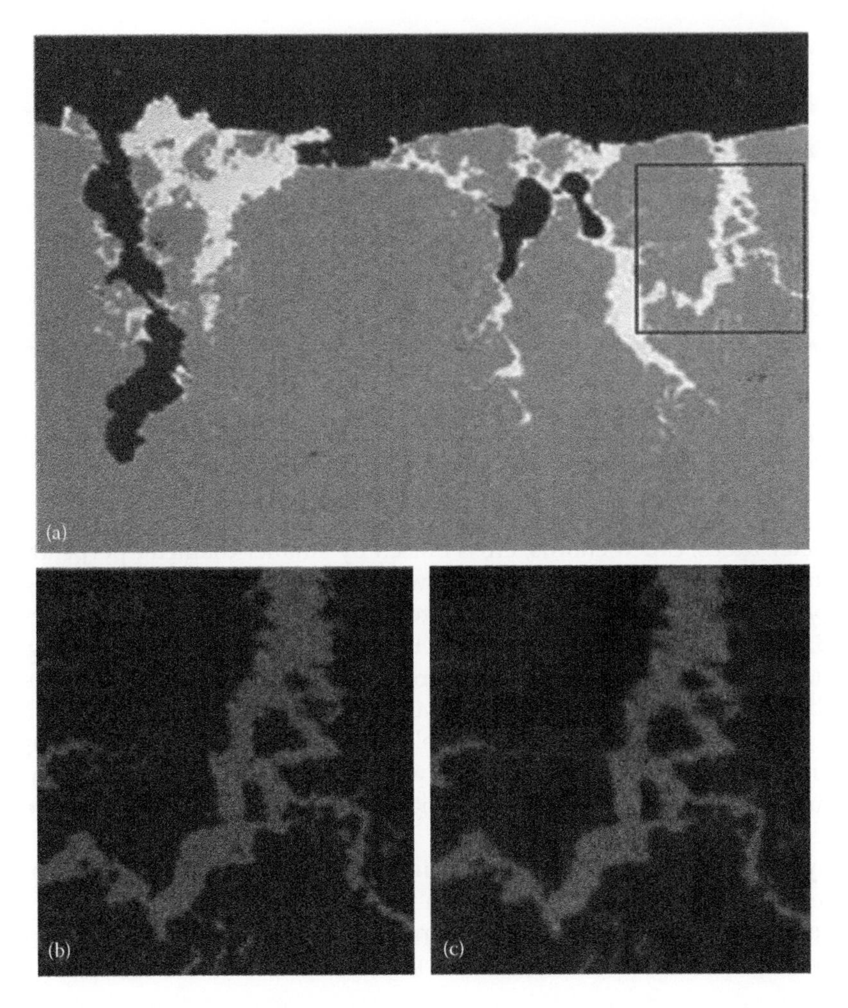

FIGURE 1.34
(See color insert.) Liquid metal embrittlement cracking in a Zn-coated HSLA steel spot weldment (a), and x-ray maps of Cu (b) and Zn (c) of area outlined in (a). (From AET_Service_Capability.pdf. Available online at http://www.aet-int.com/capability/AET_Service_Capability.pdf. Accessed in Nov. 2010. With permission.)

in the electrode were in liquid state during welding, possibly due to an improper welding schedule. LME-related cracking in RSW is further discussed in Section 3.1.1 of Chapter 3.

1.3.2 Hydrogen Embrittlement

Atomic hydrogen has to be introduced to the steel structure for HE to occur. Because only some of the steels are susceptible to it in service under certain conditions, there is very little research on HE in the context of RSW. Interstitial-free (IF) steels have been used in a number of studies on HE in RSW, because the high diffusivity and permeability of hydrogen in their ferritic microstructures make IF steels susceptible to HE. IF steels are commonly used in automobile body-in-white construction because of their good formability. In the work of Mukhopadhyay et al.,[62] the effect of HE was investigated on an IF steel of a tensile strength of 295 MPa, and interesting results were obtained. In their study, the RSW joints were submerged in an aqueous solution of 3.5% sodium chloride and loaded quasistatically under *in situ* cathodic hydrogen charging in the solution. The effect of hydrogen charging on mechanical strength was measured.

There is a complex interaction between process variables in affecting the behavior of a spot weldment charged with hydrogen. The susceptibility of an IF steel to hydrogen is shown in Figure 1.35. The effect of HE is measured by the failure load, or load bearing capacity (LBC), and by the maximum displacement, or extension up to maximum loading (EML) of a weldment when tensile tested. The variables are hydrogen charging and pre-straining, and LBC and EML are used as responses. The specimens were either as-received or pre-strained in a certain range, and they were tested either with (submerged in a solution) or without (in air) hydrogen charging.

The hydrogen charging has a negligible effect on the spot weldments made on as-received sheets on the failure load, and it slightly increases the ductility, which is possibly statistically insignificant. Therefore, it can be concluded that HE has no effect on the performance of welds made on unstretched IF steel sheets. When the sheets are pre-strained, however, hydrogen charging clearly weakens the welds made on such sheets, and the weakening severs with the amount of pre-strain. In general, the trend in both peak (failure) load and ductility (or EML) is consistent with what has been observed in metallic materials, that is, cold work strengthens a metal but makes it less ductile. However, the level of influence of pre-straining is clearly different with and without HE. The increase of failure load with pre-straining level is clear in specimens tested in air, and at a very moderate level with hydrogen charging. The loss of ductility is more significant in hydrogen-charged specimens than in those tested in air. This is consistent with the findings of other researchers.[63] The effect of HE requires a critical level of hydrogen in the lattice. It is believed that cold work or plastic deformation helps hydrogen entrapment in the steel by generating dislocations and vacancies in the lattice that serve as the sites for trapping hydrogen atoms.

The pre-straining was found to affect the properties of the base metal, not the weld including the HAZ. The melting and solidification in a weld nugget, and recrystallization in the HAZ essentially erase the work hardening induced by cold work. As a result, the hardness of the HAZ was found by the researchers[62] to be unaltered by pre-straining. The hardness decreases from the base metal toward the HAZ in a pre-strained specimen, and the damage caused by hydrogen charging is limited in the base metal, up to its border

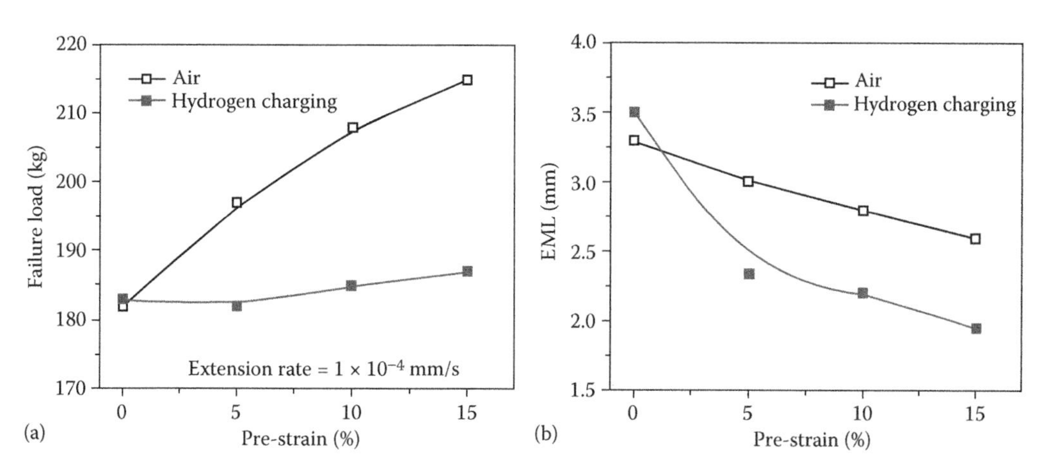

FIGURE 1.35
Peak load (a) and ductility (b) as a function of pre-strain level tested on spot welds with and without hydrogen charging. (From Mukhopadhyay, G. et al., *Mater. Corros.*, 61(5), 398–406, 2010. With permission.)

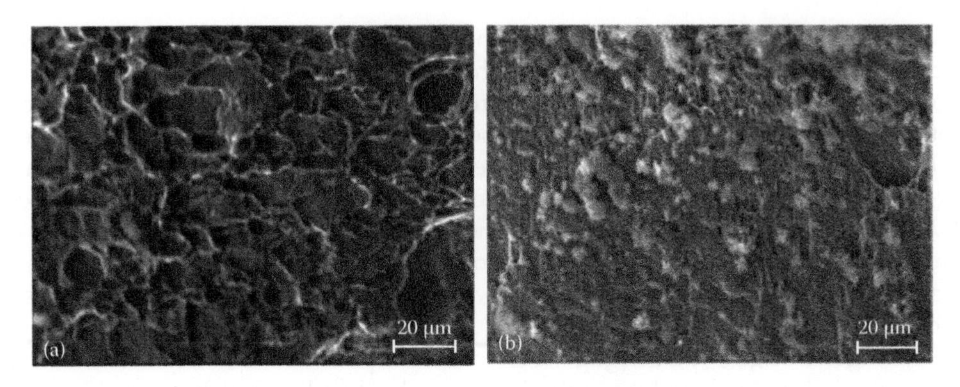

FIGURE 1.36
Fracture surfaces of specimens tested without (a) and with (b) hydrogen charging. (From Mukhopadhyay, G. et al., *Mater. Corros.*, 61(5), 398–406, 2010. With permission.)

with the HAZ. It is common to observe that fracture of a weldment starts from its HAZ, due to the stress concentration associated with its geometric location in the weld and the change of material properties in the region. The fluctuation in material properties was found to increase with the level of pre-straining and, therefore, large pre-straining generates a sharp change in mechanical strength. The fracture of the weld specimens was found to initiate in the region of the HAZ–base metal junctures, and the fracture surfaces clearly illustrate the effect of HE as shown in Figure 1.36. The specimen without hydrogen charging in Figure 1.36a has dimples on the fracture surface, whereas the one tested after immersion is fairly smooth, with small voids and certain corrosion products (Figure 1.36b).

1.3.3 Intermetallic-Compound Embrittlement

Most of the intermetallic compounds formed in structural materials are hard and brittle compared with their constituent metals, with the exception of several rare earth intermetallic compounds that are ductile at room temperature, discovered by researchers at the U.S. Department of Energy's Ames Laboratory at Iowa State University.[64] In controlled material fabrication processes, hard and brittle compounds may be used to strengthen the material, such as in dispersion–strengthening processes. In a study, particles of intermetallic-compounds Al_3Ni, Al_3Ti, Al_3Zr, Al_7Cr, and $Al_{12}Mo$ were produced by reactions when these metal powders were added to molten aluminum, and reinforced aluminum matrix composites were created as a result.[65]

Intermetallic compounds often deteriorate materials in unintended ways. In a study on casting AA6082 aluminum alloy, it was found that cracks initiate at the interface between the intermetallic β-Al_5FeSi or α-Al(FeMn)Si particles and the matrix through nucleation of voids.[66] Although such particles are generated during fabrication and mostly intended to strengthen the material, large hard and brittle intermetallic particles may break under loading, as shown in Figure 1.37, and they reduce the overall ductility of the alloy. Materials containing intermetallic compounds, intentionally and unintentionally created, may be affected in the following ways:

1. Brittle fracture of the compounds
2. Debonding at the interface
3. Loss of strength with low melting compounds

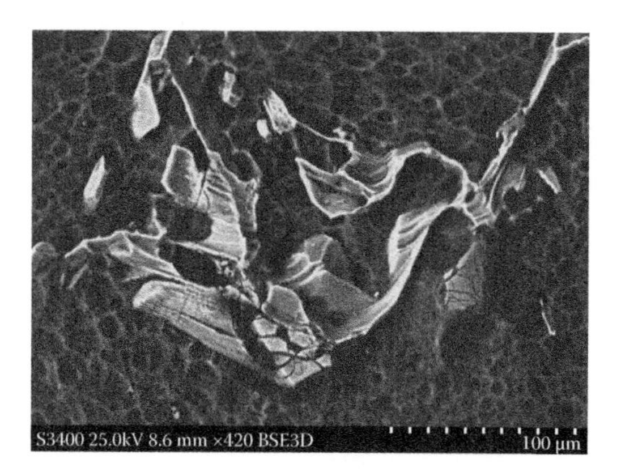

FIGURE 1.37
Broken particles at fracture surface in AA6082 alloy. (From Mrówka-Nowotnik, G., *J. Achievements Mater. Manuf. Eng.*, 30(1), 35–42, September 2008. With permission.)

In many cases, intermetallic compounds are unintentionally created and their shape, size, and distribution are not controlled. In Figure 1.38, the $Al_{12}Mg_{17}$ precipitates at the grain boundaries of a magnesium alloy AZ80 serving as the sites of stress concentration when the material is loaded. In addition, the coarse intermetallic compound particles embrittle the metal by fracture in the compounds, and debonding the interface between the compound particles and the matrix. As a result, the toughness of the material is adversely affected. There are numerous examples of such embrittlement in the literature. In Al–Cu joints for electric connection, intermetallic compounds Al_x–Cu_y can form after a certain period of service because of electric resistance heating, as shown in Figure 1.39.[67] A similar phenomenon may be observed in welding aluminum alloys using Cu electrodes. These intermetallics raise the electrical contact resistance between the electrode and sheet and promote localized heating at the interface. They tend to break off from the electrode and affect the integrity of the electrode face, and result in rapid electrode deterioration. On the other hand, such hard and brittle intermetallic compounds can be utilized to slow down the electrode deterioration process in the case of welding Zn-coated steel sheets, in which a layer of Cu–Zn intermetallic is formed after a number of welds (usually between

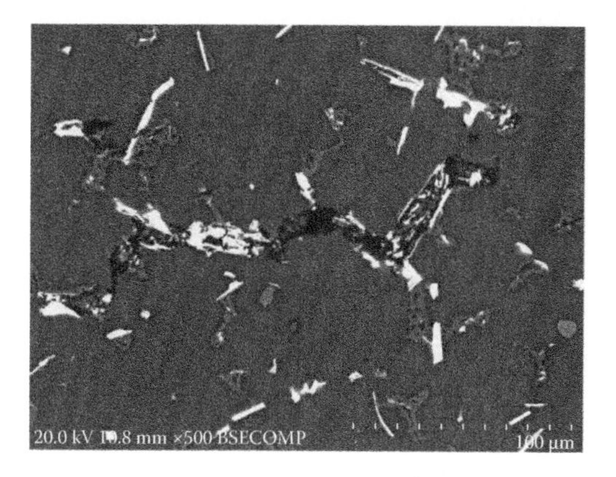

FIGURE 1.38
Precipitation of $Al_{12}Mg_{17}$ at grain boundaries of an AZ80 alloy. (Courtesy of B. Wang and J. Wang.)

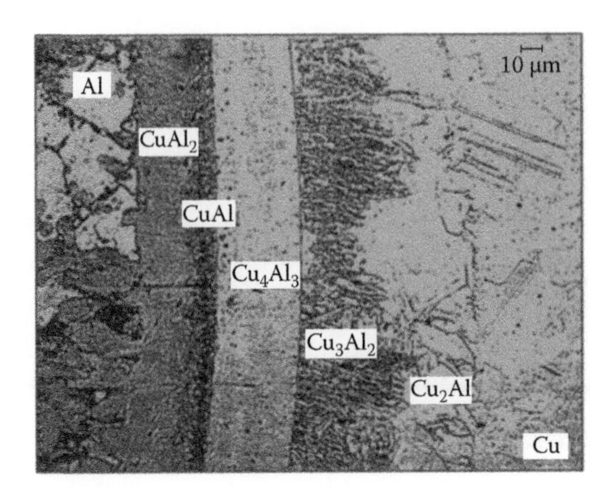

FIGURE 1.39
Various intermetallic compounds formed at an Al–Cu joint. (From Slade, P.G. (ed.), *Electrical Contacts: Principles and Applications*, Marcel Dekker, Inc., New York, 1999. With permission.)

20 and 50) are made using new electrodes. This is called pre-seasoning, or conditioning of electrodes. Welding schedules for production should be selected based on pre-seasoned/conditioned, not new electrodes.

Many of the intermetallic compounds have low melting temperature, and they may melt before the rest of the structure. This may have a serious implication in welding, especially when the volume of the compounds is large and their particles are interconnected in a structure. During RSW, the HAZ is usually heated to a temperature, although not high enough to melt the material, sufficient to melt certain low-melting intermetallic compounds such as eutectics in it. When the HAZ is stretched during welding, the liquid phase effectively provides no resistance to fracture. Even discretely distributed low-melting particles may weaken the structure as the molten volume of a precipitate may act as a crack or pore under loading. Fracture associated with the low-melting phases during welding, or liquation cracking, is presented in Section 1.4. In addition to the embrittlement such intermetallic compounds bring to the weldment during heating/cooling, a connected

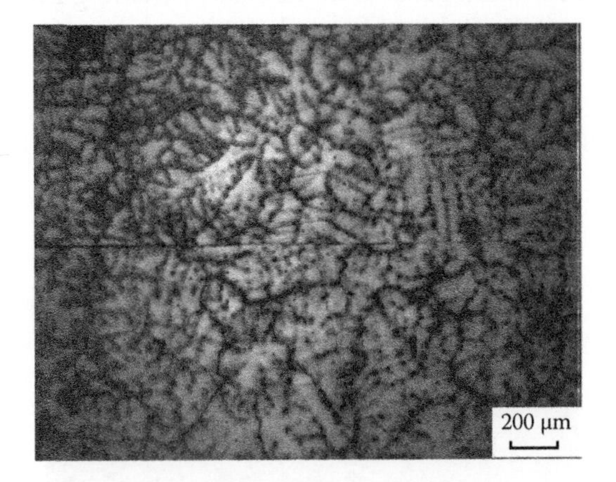

FIGURE 1.40
Microstructure of partial melting zone of AZ91D weld. (From Munitz, A. et al., Resistance spot welding of Mg–AM50 and Mg–AZ91D alloys. Magnesium Technology, TMS [The Minerals, Metals & Materials Society], 2002. With permission.)

liquid network in the HAZ also provides a pathway for the molten metal in the nugget to eject from the nugget to the faying interface. Such an expulsion phenomenon has been observed in Al and Mg welding and is discussed in Chapter 7.

In a study by Munitz et al.,[30] it was observed that in AZ91D sheets, a large amount of intermetallic β phase ($Al_{12}Mg_{17}$) exists in the form of a network at grain boundaries, in the partial melting zone of a spot weldment (Figure 1.40). In three-dimensions, the β phase forms a continuous "foam" made of grain boundaries. A portion of this network adjacent to a weld nugget during RSW may stay as liquid for a significant proportion of the welding time, which embrittles the material, as the melting temperature of the eutectic is only 437°C, far below that of Mg matrix (650°C for pure Mg).

1.4 Cracking

Cracking in RSW occurs in a similar manner as in other fusion welding processes. A weldment expands under Joule heating, and at the same time, it is distorted due to electrode squeezing and confinement from the surrounding metals. Such a distortion is irreversible and the weldment undergoes a restrained contraction during cooling. It induces tensile stresses in the joint that are directly responsible for cracking. Cracking may occur at all locations of a weldment: the nugget, HAZ, and parent metal. If it occurs during solidification of the liquid nugget, it is called *solidification cracking* or *hot cracking*. The tensile stresses may also induce cracking in the solid phases, that is, the HAZ or even the parent metal. Fracturing the low-melting components in the HAZ due to liquation is called *liquation cracking*, and cracking the weld or HAZ in high-carbon steels at low temperatures (below solidus temperatures) or even after welding is called *cold cracking*. In general, cracking requires two concurrent conditions: a weakened structure, such as one embrittled due to melting of low-melting eutectics or corrosion, and a tensile stress field. This section discusses the metallurgical aspect of cracking. Examples of cracking and its suppression can be found in Chapter 3.

1.4.1 Solidification Cracking

The stress level on the just solidified nugget, together with the mechanical strength of the material at elevated temperatures, determines the occurrence of solidification cracking. The stresses induced by the restrained contraction are proportional to the cooling, and they are released if the material is plastically deformed under such stresses. Cracking occurs when the material is relatively weak, yet the stresses reach a certain level, usually in the *brittle temperature range*. It is between the temperature at which ductility sharply drops and the liquidus temperature.

Figure 1.41 shows that the ductility of AA5754 increases rapidly right after the solidus temperature is reached during cooling, whereas the ultimate strength grows at a slower pace. In the figure, the solidus temperatures for two AA5754 alloys with 2.6 and 3.6 wt.% Mg, respectively, are also plotted. The brittle temperature range can be in this case approximated by the difference between solidus and liquidus temperatures. By comparing the liquidus and solidus in Figure 1.14, it can be seen that the brittle temperature range for the alloy with 3.6 wt.% Mg is larger than that for the alloy with 2.6 wt.% Mg. Therefore, a high concentration of Mg in AA5754 makes it more prone to cracking. Therefore, solidification

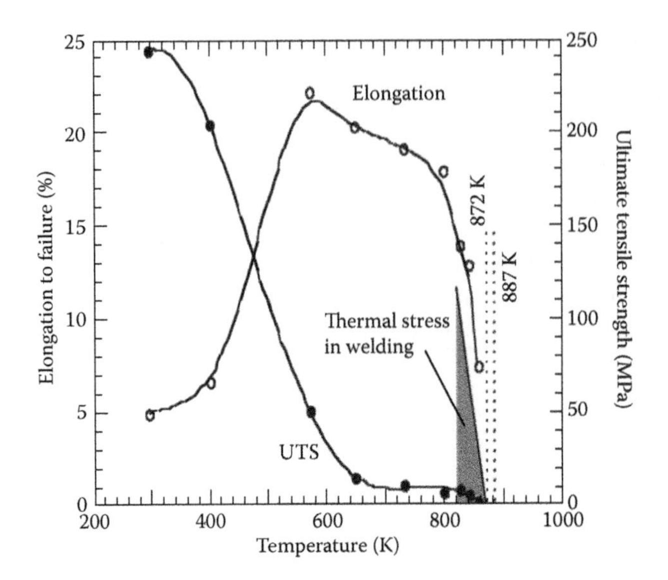

FIGURE 1.41
Dependence of ductility (elongation) and UTS on temperature. Dashed lines mark the solidus temperatures (887 and 872 K) for AA5754, with Mg content ranging from 2.6 to 3.6 wt.%. (From Zhang, H. et al., *Trans. ASME—J. Manuf. Sci. Eng.*, 124, 79, 2002. With permission.)

cracking of aluminum alloys is associated with the intentionally added alloying elements, and the presence of low-melting impurities.

Another type of cracking in the nuggets that is closely related to solidification is associated with the shrinkage of a nugget. The volume deficit created in a nugget by large expansion and deformation during heating under the electrode force, and by restrained contraction, may not be made up during cooling. Therefore, the last part of the nugget to solidify has an insufficient amount of liquid to form a coherence. Free solidification occurs as a result. A crack formed by this mechanism has a clear trace of free solidification on the crack surfaces, in the form of visible columnar or equiaxed grains. Higher strength heat-treatable aluminum alloys are prone to solidification cracking. Examples of solidification cracking in a weld are shown in Figures 1.2 and 1.3. Solidification cracking was also observed by Ma et al.[9] during welding of a DP600 steel, as seen in Figure 1.42. In addition to cracking, voids are also visible in the nugget, possibly formed because of expulsion as seen from the metal remnants on both sides of the weld at the interfaces.

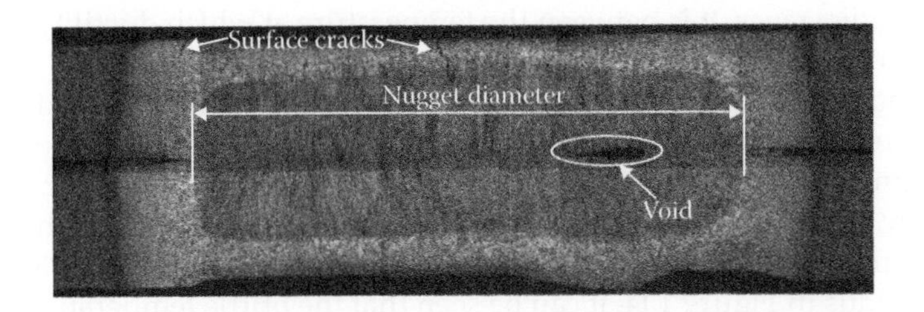

FIGURE 1.42
Microstructure of a DP600 steel weld. (From Ma, C. et al., *Mater. Sci. Eng. A*, 485, 334–346, 2008. With permission.)

1.4.2 Liquation Cracking

At certain moments during welding, the temperature at certain locations in a weldment may be lower than the solidus of the alloy, but higher than the melting temperatures of some low-melting components, such as eutectics or impurities. This may happen both at the HAZ near the nugget and in the solidified part of a nugget after it has cooled from the peak temperature. As such, components are usually solvent rich; they tend to have a higher concentration at grain boundaries than in the grains due to segregation. Therefore, continuous intergranular liquid films may present at elevated temperatures, and they have no strength to resist thermal stresses. As a result, cracking may occur. Cracks formed due to liquation have a clear intergranular characteristic. The amount of low-melting eutectics, such as sulfur and phosphorus eutectics in steel and Al–Mg eutectics in aluminum and magnesium alloys, depends on the solubility of the element at the eutectic's melting point. Only the excess of these elements over their solubility limits forms the respective eutectics and contributes to liquation cracking. Therefore, it is important to know the type, amount, and solubility of elements, as well as the melting temperatures of their eutectics, in order to determine the possibility of liquation cracking.

Cracking during welding and solidification cracking during casting have similar characteristics, and the knowledge of cracking in casting is helpful for understanding the cracking formation in welding. According to the classical works by Pellini and Flemings, hot tearing in casting alloys occurs at the last stage of crystallization, during which solid grains are surrounded by the liquid; such a structure has very low strength. Tensile stresses and strains, resulting from nonuniform temperature distribution and cooling, may cause material failure. Hot cracking tendency in casting increases with grain size, solidus–liquidus gap, and solidification shrinkage, which is especially high for Al as well as Mg alloys. The presence of impurities and grain boundary segregation also promotes cracking. The mechanism of hot cracking in welding, similar to that in casting, can be understood from the theory developed by Borland and Prokhorov. The occurrence of cracking in the coherence temperature range (Borland's definition) depends on both critical strain and critical strain rate. Comparisons of various Al alloys in casting and arc welding revealed that the Al–Mg system is second to the Al–Cu system in crack susceptibility among aluminum alloys, in spite of only a small amount of eutectics formed during solidification.

1.4.3 Corrosion Cracking

The spot-welded structures are usually protected against corrosion before they are put into service. Therefore, the corrosion resistance of spot welds is usually not an issue, and corrosion cracking of spot welds is rarely investigated. However, improper manufacturing and usage may expose resistance welds to corrosive environment. For instance, welding of zinc-coated steel sheets destroys the protective coating and exposes the indented weld area. Residual stresses along the periphery of the indentation marks may promote cracking if the weld surface is not properly protected. A study by Mukhopadhyay et al.[62] has found that the strength and ductility of an IF steel weld are reduced if the weldments are hydrogen charged and pre-strained, as discussed in Section 1.3.2, through a mechanism called hydrogen embrittlement. In the experiment, the welded specimens were cathodically charged with hydrogen for certain days in a 3.5% sodium chloride solution. The specimens were then tensile–shear tested with *in situ* hydrogen charging. Figure 1.43 shows the comparison between hydrogen-charged and original welded specimens. After 40 days of immersion in the sodium chloride solution, the weldment lost 11% of its thickness as

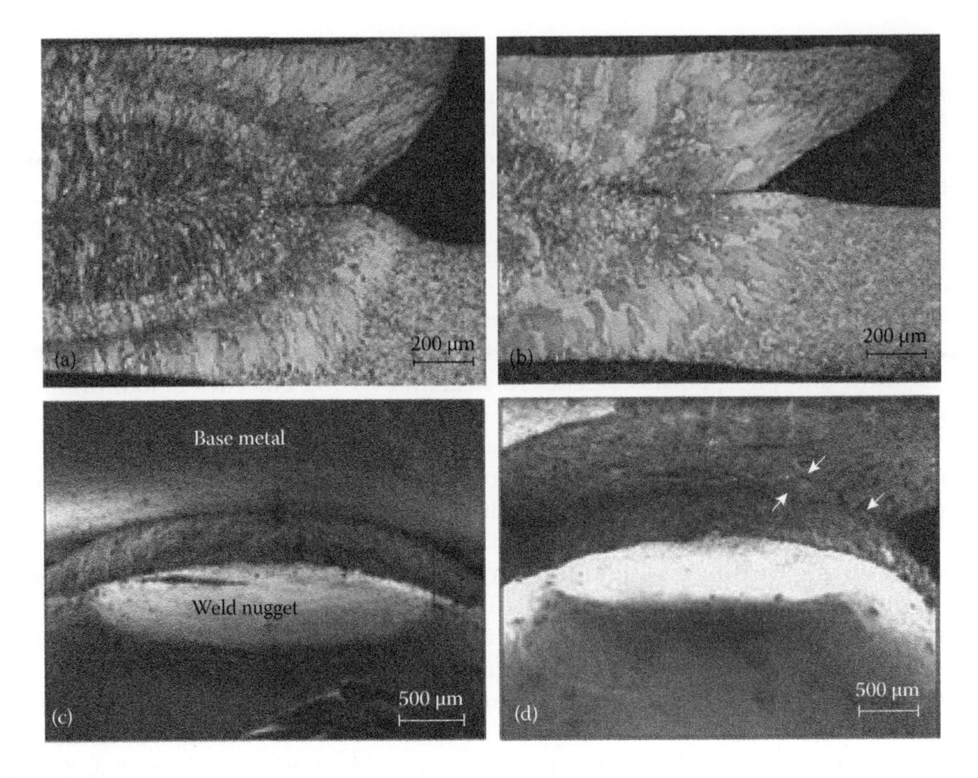

FIGURE 1.43
Effect of corrosion on welds and cracking.[62] Cross-sections of torn spot-welds without (a) and with (b) immersion, and pulled weld buttons without (c) and with (d) immersion in a corrosive environment. Arrows indicate secondary cracks near the fracture surface.

shown in Figure 1.43b, compared with the original one (Figure 1.43a). In addition, a large number of secondary cracks, indicated by the arrows in Figure 1.43d, are observed in the tested specimen, in the periphery of a hydrogen-charged weld nugget. The one without immersion in the solution, shown in Figure 1.43c, does not have any secondary cracks. Clearly, the corrosive solution weakens the weldment by corroding the faying interface, and creating cracks at or near the HAZ. In general, the HAZ of a weld is particularly vulnerable to corrosion cracking because of the residual stresses resulting from welding, and a modified microstructure in the region that often loses certain corrosion resistance intentionally created in fabricating the base metal.

References

1. Easterling, K. E., *Introduction to the Physical Metallurgy of Welding*, 2nd edition, Butterworth-Heinemann, Cambridge, UK, 1992.
2. Granjon, H., *Fundamentals of Welding Metallurgy*, Woodhead Publishing, Abington, UK, 1991.
3. Kou, S., *Welding Metallurgy*, 2nd edition, J. Wiley & Sons, Inc., Hoboken, NJ, 2003.
4. Lancaster, J. F., *Metallurgy of Welding*, 6th edition, Abington Publishing, Abington, UK, 1999.

5. Linnert, G. E., *Welding Metallurgy: Carbon and Alloy Steels, Volume 2: Technology*, 3rd edition, American Welding Society, New York, 1967.
6. Linnert, G. E., *Welding Metallurgy: Carbon and Alloy Steels, Volume 1: Fundamentals*, 4th edition, American Welding Society, Miami, FL, 1994.
7. Lippold, J. C. and Kotecki, D. J., *Welding Metallurgy and Weldability of Stainless Steels*, John Wiley & Sons, Inc., New York, 2005.
8. Tumuluru, M. D., Effect of post-weld baking on the behavior of resistance spot welds in a 780-MPa TRIP Steel, SMWC XI, Paper 6-2, 2004.
9. Ma, C., Chen, D. L., Bhole, S. D., Boudreau, G., Lee, A., and Biro, E., Microstructure and fracture characteristics of spot-welded DP600 steel, *Materials Science and Engineering A*, 485, 334–346, 2008.
10. Luo, H., New joining techniques for magnesium alloy sheets, MS thesis, Institute of Metal Research, Chinese Academy of Sciences, May 2008.
11. Callister, W. D., Jr., *Materials Science and Engineering: An Introduction*, 6th edition, John Wiley & Sons, Inc., New York, 2003.
12. Hu, S. J., Senkara, J., and Zhang, H., Quality definition of resistance spot welds: A structural point of view, in *Proceedings of International Body Engineering Conference IBEC'96*, Body and Engineering Section, Detroit, MI, 1996, 91.
13. Krauss, G., Martensitic transformation, structure and properties in hardenable steels, *Hardenability Concepts with Applications to Steel*, edited by D. V. Doane and J. S. Kirkaldy, AIME, Warrendale, PA, 1978, 229–248.
14. http://www.leonghuat.com/articles/articles1.htm. Accessed Nov. 2010.
15. Dearden, J. and O'Neill, H., A guide to the selection and welding of low alloy structural steels, *Institute of Welding Transactions*, 3, 203–214, 1940.
16. Ito, Y. and Bessho, K., *Journal of Japan Welding Society*, 37 (9), 983, 1968.
17. Yurioka, N. et al., *Welding Journal*, 62 (6), 147s, 1983.
18. American Welding Society, *Structural Welding Code*, AWS D1.1, 2004.
19. Ginzburg, V. B. and Ballas, R., *Flat Rolling Fundamentals*, Marcel Dekker, New York, 2000.
20. Marya, M. and Gayden, X. Q., Development of requirements for resistance spot welding dual-phase (DP600) steels. Part 1—the causes of interfacial fracture, *Welding Journal*, 84 (11), 172s–182s, 2005.
21. Kuntz, M. L. and Bohr, J. C., Modeling projection welding of fasteners to AHSS sheet using finite-element method, Sheet Metal Welding Conference XII, Paper 8-6, 2006.
22. Khan, M. S., Bhole, S. D., Chen, D. L., Biro, E., Boudreau, G., and Deventer, J. V., Welding behaviour, microstructure and mechanical properties of dissimilar resistance spot welds between galvannealed HSLA350 and DP600 steels, *Science and Technology of Welding and Joining*, 14 (7), 616–625, 2009.
23. Hansen, M. and Anderko, K., *Constitution of Binary Alloys*, McGraw-Hill, New York, 106, 1958.
24. *Magnesium Vision 2020*, A North American Automotive Strategic Vision for Magnesium, USAMP, United States Automotive Materials Partnership, 2006.
25. Straumal, B., Lopez, G. A., Mittemeijer, E. J., Gust, W., and Zhiyaev, A. P., Grain boundary phase transitions in the Al–Mg system and their influence on high-strain rate superplasticity, *Defect and Diffusion Forum*, 216–217, 307–321, 2003.
26. Munitz, A., Cotler, C., Shaham, H., and Kohn, G., Electron beam welding of magnesium AZ91D plates, *Welding Journal*, 79, 202s–208s, 2000.
27. Munitz, A., Cotler, C., Stern, A., and Kohn, G., Mechanical properties and microstructure of gas tungsten arc welded magnesium AZ91D plates, *Materials Science and Engineering A* 302, 68–73, 2001.
28. Wang, Y., Feng, J., and Zhang, Z., Influence of surface condition on expulsion in spot welding AZ31B magnesium alloy, *Journal of Materials Science and Technology*, 21 (5), 749–752, 2005.
29. Wang, Y., Zhang, Z., and Feng, J., Effect of welding current on strength and microstructure in resistance spot welding of AZ31 Mg alloy, *Chinese Welding Journal*, 16 (4), 37–41, 2007.
30. Munitz, A., Kohn, G., and Cotler, C., Resistance spot welding of Mg–AM50 and Mg–AZ91D alloys. Magnesium Technology, TMS (The Minerals, Metals & Materials Society), 2002.

31. Luo, H., Hao, C., Zhang, J., Gan, Z., Chen, H., and Zhang, H., Characteristics of resistance welding magnesium alloys AZ31 and AZ91, *Welding Journal*, in print, July 2011.
32. Sun, D. Q., Lang, B., Sun, D. X., and Li, J. B., Microstructure and mechanical properties of resistance spot welded magnesium alloy joints, *Materials Science and Engineering A*, 460, 461, 494–498, 2007.
33. Nippert, R. A., Composite resistance welding electrode, AWS Sheet Metal Welding Conference IX, Paper 3-4, 2000.
34. Jovanovic, M. T. and Rajkovic, V., High electrical conductivity Cu-based alloys, Association of Metallurgical Engineers of Serbia, *Metalurgija, Journal of Metallurgy*, 15 (2), 125–133, 2009.
35. Li, H., Xie, S., Wu, P., and Mi, X., Study on improvement of conductivity of Cu–Cr–Zr alloys, *Rare Metals*, 26 (2), 124–130, 2004.
36. Holzwarth, U. and Stamm, H., *Journal of Nuclear Materials*, 279, 31, 2000.
37. Kimchi, M., Gould, J. E., Helenius, A., Keippi, K., and Nippert, R. A., The evaluation of various electrode materials for resistance spot welding galvanized steel, SMWC IV, Paper 7, 1990.
38. ASM Specialty Handbook, *Copper and Copper Alloys*, edited by J. R. Davis, ASM International, 2001.
39. http://www.spotweldingconsultants.com/GlidCop_AL_15.pdf. Accessed in Nov. 2010.
40. Resistance Welder Manufacturers' Association Bulletins #16—Resistance Welding Equipment Standards, 1996.
41. Gugel, M. D., White, C. L., and Wist, J. A., Progression of electrode wear during RSW of EG steel, SMWC V, Paper A4, 1992.
42. Wist, J. A. and White, C. L., Metallurgical aspects of electrode wear during resistance spot welding of zinc-coated steels, SMWC IV, Paper 7, 1990.
43. Matsuda, H., Matsuda, Y., Nagae, M., and Kabasawa, M., Effect of aluminum on spot weldability of hot-dipped galvanized and galvannealed steel sheets, SMWC VIII, Paper 1-5, 1998.
44. Steinmeier, D., Resistance welding—Electrode seasoning-1, microJoining Solutions—microTips, Available online at http://www.microjoining.com/microTip_Library/microTip_Resistance_Electrode_Seasoning-1.pdf. Accessed in Nov. 2010.
45. Li, Z., Hao, C., Zhang, J., and Zhang, H., Effects of sheet surface conditions on electrode life in aluminum welding, *Welding Journal*, 86 (4), 34s–39s, 2007.
46. http://en.wikipedia.org/wiki/Liquid_metal_embrittlement. Accessed in Nov. 2010.
47. Lai, G. Y., *High Temperature Corrosion and Materials Applications*, ASM International, Materials Park, OH 44073, 2007.
48. Joseph, B., Picat, M., and Barbier, F., Liquid metal embrittlement: A state-of-the-art appraisal, *European Physical Journal Applied Physics*, 5, 19–31, 1999.
49. Liquid metal assisted cracking of galvanized steel work, Topic Paper, SC/T/04/02, Standing Committee on Structural Safety, London, U.K., June 2004. Available online at www.scoss.org.uk. Accessed in Nov. 2010.
50. AET_Service_Capability.pdf. Available online at http://www.aet-int.com/capability/AET_Service_Capability.pdf. Accessed in Nov. 2010.
51. Glickman, E. E., Mechanism of liquid metal embrittlement by simple experiments: From atomics to life-time, *Multiscale Phenomena in Plasticity*, Kluwer Academic Publisher, Ouranopolis, Greece, 2000.
52. Robertson, W. M., Propagation of a crack filled with liquid metal, *Transactions of the Metallurgical Society of AIME*, 236, 1478–1482, 1966.
53. Glickman, E. E. et al., Round table discussion: II. Grain boundary wetting and liquid grooving, *Defect and Diffusion Forum*, 156, 265–272, 1998.
54. Stoloff, N. S. and Johnston, T. L., Crack propagation in a liquid metal environment, *Acta Materialia*, 11 (4), 251–256, 1963.
55. Westwood, A. R. and Kamdar, M. H., Concerning liquid metal embrittlement, particularly of zinc monocrystals by mercury, *Philosophical Magazine*, 8 (89), 787–804, 1963.
56. Gordon, P., Metal-induced-embrittlement of metals—An evaluation of embrittler transport mechanisms, *Metallurgical Transactions A*, 9A, 267–273, 1978.

57. Lynch, S. P., Environmentally assisted cracking: Overview of evidence for an adsorption-induced localised-slip process, *Acta Metallurgica*, 36 (10), 2639–2661, 1988.
58. Anderson, T. L., *Fracture Mechanics: Fundamentals and Applications*, 2nd edition, CRC Press, Boca Raton, FL, 1995.
59. Website of Metallurgical Associates, Inc., http://www.metassoc.com/pdf/MAI_Minutes-6_04.pdf. Accessed in Nov. 2010.
60. Kinstler, T. J., Current knowledge of the cracking of steels during galvanizing—A synthesis of the available technical literature and collective experience for the American Institute of Steel Construction, GalvaScience LLC. Available online at http://www.aisc.org/uploadedFiles/Research/Files/Final5906.pdf. Accessed in Nov. 2010.
61. Westwood, A. R. C., Preece, C. M., and Kamdar, M. H., Adsorption-induced brittle fracture in liquid–metal environments, *Fracture: An Advanced Treatise*, edited by H. Liebowitz, Academic Press, New York, 1971, 589–644.
62. Mukhopadhyay, G., Bhattacharya, S., and Ray, K. K., Strength of spot-welded steel sheets in corrosive environment, *Materials and Corrosion*, 61 (5), 398–406, 2010.
63. Liu, P. W. and Wu, J. K., *Materials Letters*, 57, 1224, 2003.
64. Ductile intermetallic compounds discovered, Public release date: 15-Sep-2003, DOE/Ames Laboratory. Available online at http://www.eurekalert.org/pub_releases/2003-09/dl-dic091503.php.
65. Mitsumaki, M. and Tadashi, K., In-situ fabrication of intermetallic compound-dispersed Al matrix composites by addition of metal powders, *Osaka Furitsu Sangyo Gijutsu Sogo Kenkyujo Hokoku*, 20, 63–67, 2006 (in Japanese).
66. Mrówka-Nowotnik, G., The effect of intermetallics on the fracture mechanism in AlSi1MgMn alloy, *Journal of Achievements in Materials and Manufacturing Engineering*, 30 (1), 35–42, September 2008.
67. Slade, P. G. (ed.) *Electrical Contacts: Principles and Applications*, Dekker, Inc., New York, 1999.
68. Zhang, H., Senkara, J., and Wu, X., Suppressing cracking in RSW AA5754 aluminum alloy by mechanical means, *Transactions of the ASME—Journal of Manufacturing Science and Engineering*, 124, 79, 2002.

2

Electrothermal Processes of Welding

In resistance welding, the heat needed to create the coherence is generated by applying an electric current through the stack-up of sheets between the electrodes. Therefore, the formation of a welded joint, including the nugget and the heat-affected zone (HAZ), strongly depends on the electrical and thermal properties of the sheet and coating materials. With the knowledge of phases and their transformations, as discussed in Chapter 1, a weld's formation can be linked to the electrical and thermal processes of welding. Controlling the parameters of these processes is a common practice in resistance welding. Governed by the principle of Joule heating, the general expression of heat generated in an electric circuit can be expressed as

$$Q = I^2 R \tau \tag{2.1}$$

where Q denotes heat, I is the current, R is the electrical resistance of the circuit, and τ is the time the current is allowed to flow in the circuit. When the current or resistance is not constant, integrating the above expression will result in the heat generated in a time interval τ. For resistance welding, the heat generation at all locations in a weldment, rather than the total heat generated, is more relevant, as heating is not and should not be uniform in the weldment. In addition, the heating rate is more important than the amount of heat, as how fast the heat is applied during welding determines the temperature history and, in turn, the microstructure. This can be easily understood by considering an aluminum welding. If the welding current is low, melting may not be possible no matter how long the heating is, because of the low electrical resistivity of aluminum, and the fact that the heat generated is conducted out quickly through the water-cooled electrodes and the sheets due to the high thermal conductivity of aluminum. In general, the electrical and thermal processes should be considered together in welding, and such a consideration is essential in understanding resistance welding process and choosing correct process parameters.

2.1 Electrical Characteristics of Resistance Welding

The total electrical resistance of a sheet stack-up can be attributed to the contributions of the contact resistance at the electrode–sheet interfaces (R1 and R5 in Figure 2.1), that at the sheet faying interface (R3), and the bulk resistance (R2 and R4). These quantities are usually not constant—contact resistance is a strong function of both temperature and pressure, and bulk resistance is sensitive to temperature, not pressure.

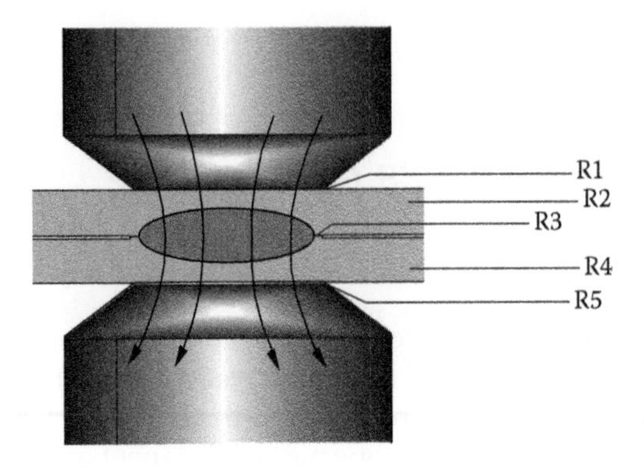

FIGURE 2.1
Electrical resistance in a sheet stack-up during RSW.

2.1.1 Bulk Resistance

Figure 2.2, created based on data presented in Refs. 1–6, shows the dependence of bulk resistivity on temperature for the metals commonly used in resistance welding. They all increase with temperature, although at different rates. The bulk resistivity of iron (for steels) is very sensitive to temperature, and its value is significantly larger than that of the pure copper. Although copper alloys, such as Cu–Cr–Zr alloys rather than pure copper, are used as electrode materials, the resistivity of pure copper provides an important indication of the relative value of resistivity of the copper alloys compared to the workpieces. The resistivity of copper is significantly lower than that of iron, even at elevated temperatures. Therefore, when an electric current is applied, more heat is generated in the steel sheet stack-up than in the electrodes. This is even more the case as welding time elapses when the sheets are heated, resulting in higher electrical resistivity in the sheets, as the electrodes are usually water cooled.

Compared with steel, aluminum's electrical resistivity is very low, or its electric conductivity is very high. In fact, it is very close to that of copper before it melts, as shown in

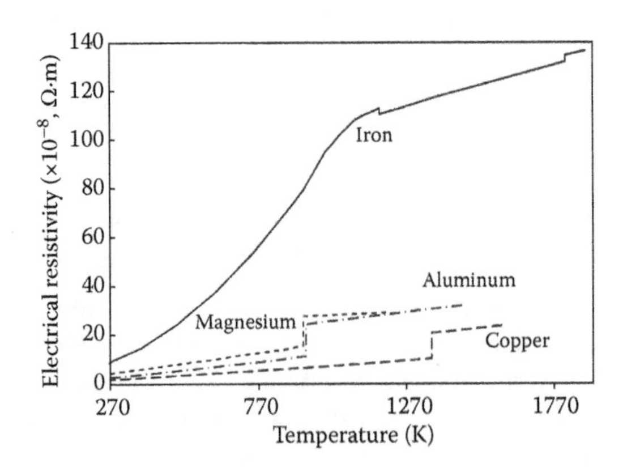

FIGURE 2.2
Bulk resistance vs. temperature for Fe, Mg, Al, and Cu.

Figure 2.2. Therefore, it is difficult to weld pure aluminum even when using pure copper as electrodes. However, pure aluminum is rarely used in practice. Its various alloys, such as Al–Mg, and Al–Cu are more commonly encountered, and they have significantly higher electrical resistivity than pure aluminum, which makes welding aluminum (alloys) possible. Another important factor in welding aluminum is the usually fairly high surface resistivity from aluminum oxides. In addition, liquid aluminum has significantly higher resistivity than solid aluminum. Therefore, welding aluminum sheets is possible if the chemistry of the electrodes is properly chosen (so their resistance is sufficiently lower than that of the aluminum sheets) and the water cooling of electrodes is tightly controlled.

As seen in Figure 2.2, pure Mg behaves in a similar manner as Al. In fact, welding Al and Mg has very similar electrical and thermal characteristics. Welding Mg alloys, however, is more difficult because of their more volatile metallurgical nature compared with Al.

2.1.2 Contact Resistance

Although the bulk resistivity for most metals can be considered independent of pressure, contact resistance is usually very sensitive to pressure distribution, in addition to the surface conditions at the contact interfaces. The apparent contact area at the faying interface is slightly larger than that at the electrode–sheet interface, with a radius ratio of about 1.2 as estimated by Eager and Kim.[7] In general, only a small portion of the apparent contact area is in actual contact, which is formed by irregularities in the form of crests and troughs between the contacting surfaces. During resistance spot welding (RSW), the pressure at the interfaces created by electrode squeezing smashes the irregularities and causes a decrease in the contact resistance. A small electrode force may not be able to create sufficient electrical contact at the interfaces and may produce concentrated heating and possibly localized melting or even vaporization.

Contact resistance is affected by the surface condition of the sheets. The presence of oil, dirt, oxides, scales, paints, and any other foreign content causes a change in the resistance. For bare steels, the surface is usually contaminated by oil/greases, possibly rusts, etc. Their effects on contact resistance diminish quickly after an electrode force is applied, especially after the interface is heated by electric current application. Therefore, the contact resistance is usually not a concern when welding bare steels.

Coatings deliberately introduced for corrosion protection and other purposes, on the other hand, may significantly affect the contact resistance. For instance, hot-dip galvanized steel sheets require significantly higher welding current, due to the reduced contact resistance from the zinc layer, than those for uncoated or bare steel. When welding galvanized steel sheets, the zinc layer at the electrode–sheet interface is melted first (due to the low melting point of zinc) and most of the molten zinc is squeezed out during the first few cycles of welding. The contact resistance is significantly reduced at the interface, even after a significant portion of the zinc is pushed out of the compressed area under the electrode face. Only a small amount of zinc from the coating, trapped at the sheet–sheet interface, is needed to fill the gap between the interfaces that are not in direct contact. This, in addition to the low bulk resistance of Zn, contributes to the reduction of the overall resistance. Therefore, welding of zinc-coated steels requires a large electric current. Galvannealed steel sheets have a surface coating in the form of an Fe–Zn compound, not free zinc. This compound has a significantly higher bulk resistivity than free zinc. Therefore, the influence of the coating on contact resistance is not as dramatic as in the galvanized coating.

An Al_2O_3 layer, which is inherent to aluminum sheets, plays an important role in affecting the contact resistance at both the electrode–sheet and sheet–sheet interfaces. The layer at the as-fabricated state is usually not uniform or may be broken under an electrode force during welding. As a ceramic, Al_2O_3 is highly insulating with a high melting temperature. A nonuniform or broken Al_2O_3 layer on a sheet surface results in uneven distribution of electric current, with very high electric current density at low resistance locations, and produces significantly localized heating or even melting on the surfaces.[8] According to the German standard DVS 2929, a stable welding process with uniform weld nuggets can be achieved if the sheet–sheet contact resistance is controlled between 20 and 50 μΩ. Such contact resistance can only be achieved if the sheet surface is properly treated.[9] The contact resistance for aluminum welding can be measured by following the German standard DVS 2929 using a resistance spot welder, as illustrated in Figure 2.3.

Li et al.[10] investigated the contact resistance following the German standard using an electrode force of 7.5 kN. Twenty randomized measurements were taken for four types of surface conditions. The electrodes were cleaned after each measurement using same-grit sandpaper in order to create a consistent electrode surface condition. The four sheet surface conditions consist of the original, or as-fabricated surface, and three types of cleaned surfaces. The cleaning methods are degreasing, chemical cleaning, and electric arc cleaning, as described in the following:

1. Degreasing. Al sheets are soaked in a water solution of a metal-degreasing detergent for 5 min, wiped using cotton, and then water rinsed three times. The sheets are air-dried afterward.
2. Chemical cleaning. Sheets are cleaned first following the degreasing procedure as described in 1. Then they are soaked in a water solution of 5% NaOH at 60°C for 4 min. After being water rinsed three times, they are soaked in 30% HNO_3 for 2 min at room temperature, then water rinsed three times before being air dried.
3. Electric arc cleaning. The sheets are cleaned using an arc welder on both sides. The electric current for cleaning is kept at an appropriate level for a short period, in order to avoid melting the Al sheet surface.

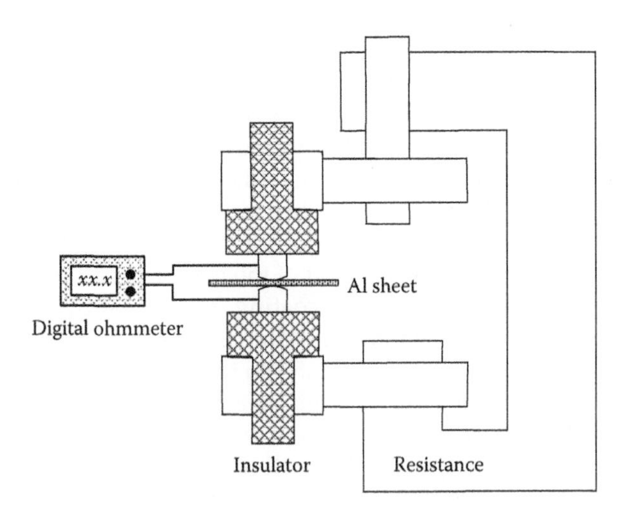

FIGURE 2.3
Setup for contact resistance measurement.[9]

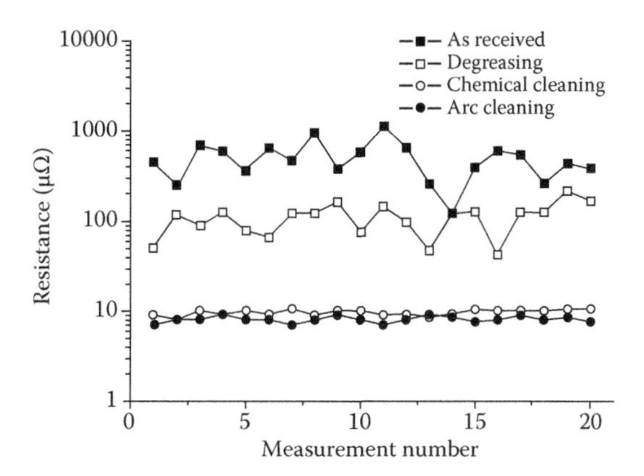

FIGURE 2.4
Resistance measurement of various surface conditions.

Using 2-mm 5A02 aluminum alloy sheets, the contact resistances of sheets with various surface conditions were measured using the setup shown in Figure 2.3 with a digital micro-ohmmeter. The measured resistances are plotted in Figure 2.4. As shown in the figure, there is a significant difference in contact resistance among the sheets of different surface conditions. Electric arc cleaning results in the lowest contact resistance, possibly because the layer of grease and oxides on the surface was burned off under the intensive heat of electric arc. The base metal was exposed and little oxidation occurred after cleaning, as it was made under the protection of Ar gas. The time elapsed between cleaning and measurement was a few hours, in which only a thin layer of Al_2O_3 was expected to form. Softening of the base metal in the cleaned surface area might also occur under the electric arc heating, making the contact area between the electrode and sheet larger than it would be for untreated sheets or those treated by other means.

The resistance of chemical cleaning is fairly uniform, with a magnitude slightly higher than that of electric arc cleaning. Degreased surfaces have a higher contact resistance that is still significantly lower than the untreated surfaces. The distribution of contact resistance for degreased and untreated surfaces is not as uniform as that for chemically or electric arc-cleaned surfaces. The clear differences in both the level and distribution of contact resistance can be largely attributed to whether the Al_2O_3 is removed. The resistance values for chemically cleaned, degreased, and untreated surface conditions are consistent with those reported by other researchers.

2.1.3 Total Resistance

In RSW, the total heat is determined by the total electrical resistance of the sheet stack-up between the electrodes, which is the sum of individual resistances (contact and bulk resistances) at various locations. Therefore, a change in the total resistance reflects the changes in individual resistance values that are induced by the underlining physical processes during welding. Therefore, an examination of the electrical resistance value during welding provides an opportunity to understand the welding process. Figure 2.5 shows a comparison of total resistance changes during welding for a steel and an aluminum alloy. For aluminum, the presence of Al_2O_3 on the surfaces makes the total resistance very high at the beginning. There is a steep drop once welding current is applied, implying that

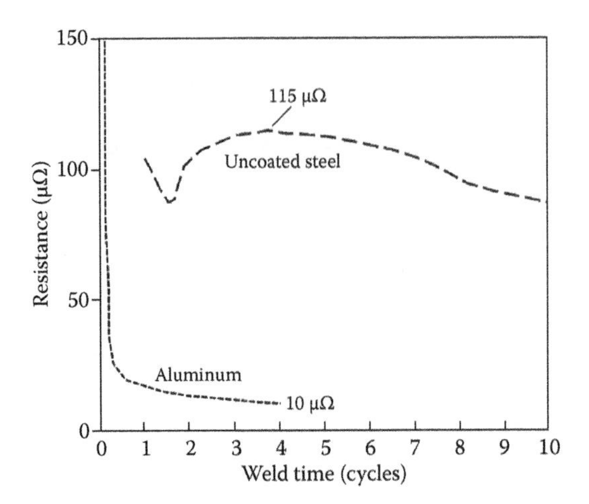

FIGURE 2.5
Dynamic resistance for welding a steel and an aluminum alloy.[11]

the alumina layer is broken under the electrode force and heat. After the initial decrease, the resistance continues to decline, but at a much slower pace. The bulk resistance of aluminum dominates in this period. The bulk resistivity of aluminum increases with temperature in both solid and liquid states and during melting, so the total resistance should increase. However, aluminum is softened or even melted under heating, which results in an increase in contact area and a decrease in the thickness of the stack-up. After initial breakdown of the alumina layer at the faying interface, Al_2O_3 particles are gradually mixed into the softened as well as the molten bulk metal, and their contribution to the total resistance slowly diminishes. Therefore, the total resistance slightly decreases with temperature, which means that the decrease in resistivity due to softening/melting outpaces the increase in resistivity resulting from heating.

Because of the differences in electrical and mechanical properties, the electrical resistance characteristics of welding are quite different for uncoated steel and aluminum. There is also a drop in total resistance during steel welding, as in welding aluminum, but at a much slower pace, at the beginning of welding. This decrease can be attributed to the changes in contact resistance through burning surface agents such as grease. The surface layer on the steel has a much lower resistance than the alumina on aluminum. Then an increase in resistance value is observed when welding continues. This corresponds to the rise of bulk resistivity when steel is heated. As seen in Figure 2.2, the magnitude of increase in resistivity with time is significantly higher in steel than in aluminum. In this stage, although the steel is softened and its yield stress is lowered, the steel in the solid state still has sufficient strength to resist large deformation of the stack-up. Therefore, a net increase in total resistance results. Further heating induces melting and significant lowering of the yield stress of the solid and produces a decreasing resistance. From the figure, it can be seen that the resistance in welding steel is significantly higher than that in welding aluminum. Because aluminum has higher electrical and thermal conductivities than steels, high electric current and short welding time have to be used in welding aluminum alloys. For example, a current of about 10,000 A may be needed when welding a 2-mm to 2-mm steel sheet combination, but more than 40,000 A is usually required to make similar combinations of aluminum sheets.

The electrical and thermal characteristics of electrodes significantly affect the welding process, especially electrode life. The existence of impurities and all alloying elements, except for silver, will decrease the electrical and thermal conductivities of copper, and the electrical conductivity decreases as the amount of the impurity or alloying element increases. Cadmium has the least effect on copper's electrical conductivity, followed by increasing effects from zinc, tin, nickel, aluminum, manganese, and silicon, with the largest effect from phosphorus. Zinc has a very minor effect on the thermal conductivity of copper. When choosing electrode materials, the effects of alloying elements on the electrical and thermal properties, as well as mechanical properties, should be considered, and an optimal design can be reached to achieve quality welds and long electrode life.

2.1.4 Shunting

Previously made welds may affect the subsequent welding if the welds are spaced close to each other due to electric current shunting, as in Figure 2.6. The welding current may be diverted from the intended path by the previously made welds. As a result, the current or current density may not be sufficient to produce a quality weld. The shunting effect is a strong function of the bulk resistivity of the sheet material. A highly conductive metal, such as aluminum, requires a large space between the welds. This should be taken into account when welds are designed into structures, as putting too many welds close to each other may not provide the intended strength. A good practice is to follow the recommendations of International Institute of Welding (IIW),[12] which suggests a minimum weld spacing of 16 times the sheet thickness, or 3 times the electrode face diameter for the given sheet thickness.

In the work by Howe[13] and a group of researchers, several types of coated and bare steel sheets of various thicknesses were tested, and results were analyzed through analysis of variance. Most of the weld spacing in the test was larger than that recommended by IIW and, consequently, it was found that the weld spacing was not significant. The exception is a bare thick sheet material. That study concluded that the minimum weld spacing should be determined based on a number of factors: the sheet thickness (because of low resistance through the cross-sectional area), the conductivity of the substrate, and the coating.

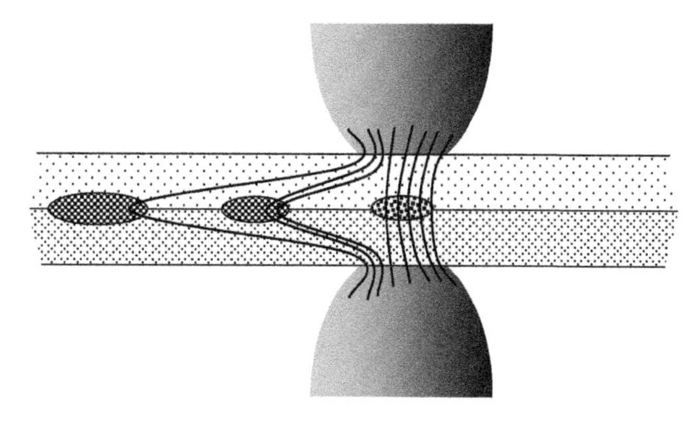

FIGURE 2.6
Shunting in resistance spot welding.

2.2 Thermal Characteristics of Resistance Welding

Heat dissipation does not start after the electric current is shut off—at any instant, the heat generated in the system is conducted through either the sheets or the electrodes. Maintaining a low temperature in the electrodes is vital to weld quality and electrode life. The thermal conductivity of Cu, as shown in Figure 2.7, is significantly higher than that of iron or aluminum. Therefore, heat generated in or conducted to the electrodes can quickly dissipate if they are properly cooled. Compared with steel, pure aluminum has a higher thermal conductivity. The heat generated in an aluminum sheet stack-up dissipates rapidly during welding. This means that a more concentrated heating, in the form of high electric current in a short period of welding, is needed for welding aluminum, as a significant concurrent heat loss makes welding impossible if the heating rate is low. Because aluminum has a high chemical affinity for copper to form a brittle alloy (bronze) with lower electrical and thermal conductivities than copper, a high, concentrated heating promotes the bronze formation and reduces electrode life. Mg is similar to Al in all aspects, especially in electrode life.

Both solid and liquid phases of a metal expand upon heating. However, the amount of expansion is different for different materials. Figure 7.16 presents calculated specific volume changes of an aluminum alloy and pure iron in heating. A similar observation can be made from Figure 2.8 on the coefficients of thermal expansion (CTE) for pure aluminum, magnesium, copper, and iron. A drastic difference is visible for these metals. During RSW, the sheets between the electrodes are not allowed to expand freely, but are constrained by the electrodes. Such constraining is necessary to maintain electrical and thermal contact at the interfaces, and to contain the liquid metal from expulsion, as discussed in Chapter 7. Calculations revealed that the pressure in the liquid aluminum nugget is twice as high as in iron, and that of Mg is located between the other two. A large expansion needs a large electrode force to suppress, and often results in a large/deep electrode indentation. Thermal stress levels in an aluminum or magnesium weldment can also be high, which may lead to cracking and formation of shrinkage porosity. Because aluminum expands more than either copper or steel, a significantly larger electrode force is usually needed for welding aluminum than for welding steel. Because of this, special equipment, such as stiffer welder arms and larger pneumatic cylinders, may be needed for welding aluminum

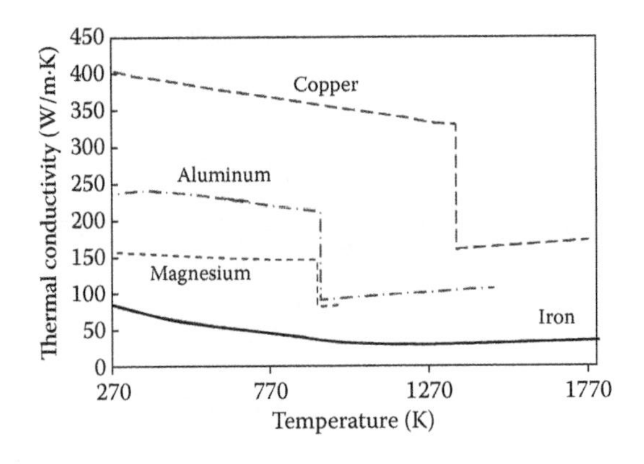

FIGURE 2.7
Thermal conductivity of various metals.[1–5]

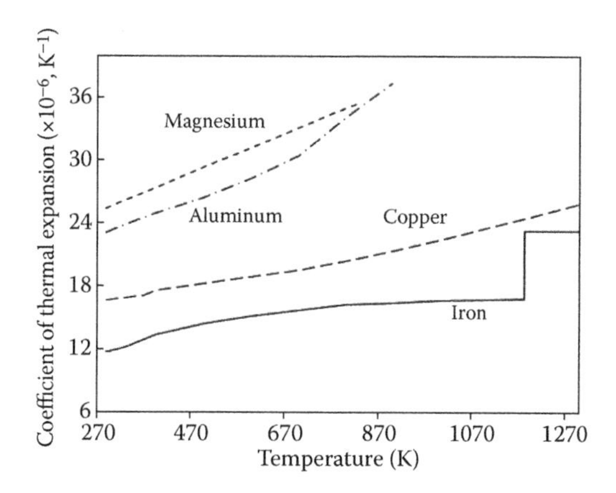

FIGURE 2.8
CTE of various metals (plotted using data from Refs. 1–4, 14, and 15).

and magnesium alloys, in addition to larger transformers, as discussed in the previous section.

The thermal process of a welding is directly reflected in the electrode displacement. A net increase in heat, when the heat generated due to Joule heating surpasses the heat loss due to dissipation through the electrodes and the sheets, causes the sheet stack-up to tend to push the electrodes apart. When there is a net heat loss, the stack-up shrinks. Therefore, the electrode displacement serves as a good indication of the possible thermal and even metallurgical processes of welding. Figure 2.9 shows the electrode displacement during a steel welding. The corresponding electric current profile is used as a reference. After electric current is applied, the solid is heated, as seen in region I. After about three cycles of heating, the electrode displacement shows a sudden increase in both magnitude

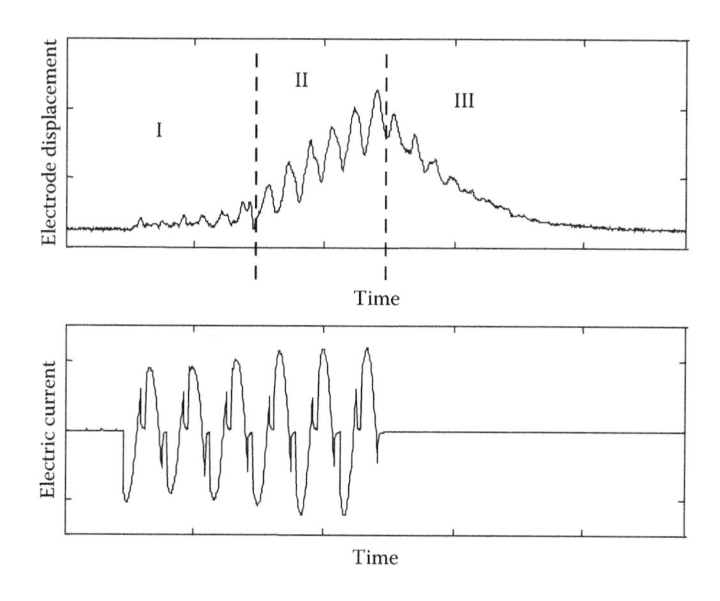

FIGURE 2.9
Electrode displacement when welding a steel.

and fluctuation. This possibly corresponds to the melting of part of the sheets, as the volume increase due to melting is significantly higher than that during solid-state heating (Figure 7.16). After the electric current is shut off, the sheet stack-up is cooled and shrinks, as in region III. Because an alternating current (AC) is used, the amount of heat generated increases when the current value goes from zero to the peak, and decreases when it returns to zero. Assuming that the cooling rate does not change much during welding, the electrode displacement mimics the heating/cooling cycles induced by the AC. Because the heat generation is insensitive to electrical polarity, the fluctuation of the electrode displacement signal has a frequency twice that of the electric current.

2.3 Electrode Life

The metallurgical interaction between electrodes and sheets is possible because of Joule heating during welding. Therefore, the electrical and thermal processes determine the electrode wear and, ultimately, electrode life. In this section, the effects of electrical, thermal, and metallurgical processes and their interactions on electrode life are discussed on both galvanized steels and aluminum alloys.

2.3.1 Welding Galvanized Steels

An important factor in electrode life is contact resistance, which is a strong function of the surface condition, and it determines the heat generation and metallurgical reactions at the electrode–sheet interface. When welding galvanized steel sheets, the low contact resistance due to the high conductivity of zinc warrants a significantly higher welding current than welding bare steel. A more profound influence of free zinc at the electrode–sheet interface is on electrode wear. The alloying of copper with zinc to form brass increases the resistivity of the electrode face. This in turn raises the electrode temperature during welding. The face of an electrode is deformed through repeated heating and mechanical impacting at the interface, and the brass formed on the electrode face is often picked up by the sheets, leaving a golden-colored ring of the indentation mark on the sheets. Therefore, the zinc coating promotes electrode wear. Figure 2.10 shows a pair of electrodes after a large number of welds on zinc-coated steel sheets. The electrode surfaces have a clear sign of oxidation and brass formation. The flat electrodes are damaged, with significantly enlarged face areas. Welding using a current value as originally designated when the electrodes were new will result in substandard welds due to insufficient current density.

FIGURE 2.10
(See color insert.) Electrode wear in welding steel.

In an electrode life study, Kimchi et al.[16] systematically studied Cu–Cr, Cu–Zr, and Al_2O_3 dispersion-strengthened Cu (78% IACS) electrodes on both hot-dip galvanized and electrogalvanized steels. Two types of electrode life testing were performed: a conventional test using a current just below the expulsion limit and a current producing nominal-sized welds; and an oscillating weldabilty test. The end of electrode life was detected when three out of five peel tests produced no weld buttons. The Cu–Zr, Cu–Cr, and dispersion-strengthened copper (DSC) electrodes had electrode lives of 3400, 2400, and 1700 welds, respectively, when tested at a current level just below the expulsion limit; and 2600, 1100, and 500 welds when tested at the nominal current. In the oscillating weldability tests, the DSC electrode performed exceptionally well, with the longest life and the least amount of sticking compared with the other two electrodes. The different behaviors of the electrodes were attributed to the different electrode wear mechanisms. In Cu–Zr and Cu–Cr, few large pits were observed, whereas a large number of small pits were uniformly distributed on the surface of the DSC electrodes.

Significant efforts have been made by White et al.[17–21] in a series of publications on the electrode wear mechanisms. The deterioration of electrodes generally involves two processes: mushrooming of the electrode face due to plastic deformation at elevated temperatures, and pitting that results from material depletion from the electrode surface largely due to alloying, that is, the formation of low-melting phases with low electrical conductivity.

Harder materials such as dispersion-strengthened materials are more resistant to mushrooming than softer materials. Therefore, harder materials tend to have longer electrode life. However, in an electrode life comparison between AL60 electrodes and Resistance Welder Manufacturers' Association (RWMA) Class II Cu–Cr and Cu–Zr alloy electrodes, it was found that the harder AL60 generally have a less life than the latter in a constant current service. The main reason for this is that pitting evolves differently in these two types of materials. The precipitation-hardened Cu–Cr and Cu–Zr electrodes tend to form a single visible pit, and it grows throughout the life of the electrode. However, a number of small pits are observed in the dispersion-strengthened AL60 electrodes, which also grow over the electrode life. The different effects of plastic deformation on the growth of pits in these two types of materials determine that they have different growth rates and, therefore, different electrode lives. It was observed in experiments[17] that the pits in the surfaces of Cu–Cr and Cu–Zr electrodes can heal by localized deformation of the electrode material, whereas such deformation was not possible in the AL60 electrodes. Therefore, pits in the AL60 electrodes tend to grow larger at a faster pace, which results in a shorter life than the softer materials. In the same work, dispersion-strengthened copper Glidcop AL25 electrodes were tested, and a 34% longer electrode life was obtained compared with AL60 grade DSC electrodes. The failure mechanism of AL25 is similar to that of Cu–Cr and Cu–Zr electrodes.

The evolution of craters on electrode face during welding galvanized steel using Cu–Cr–Zr electrodes was also investigated by Kusano.[22] Under high pressure and temperature, a layer of Cu_5Zn_8 alloy may form at the electrode surface. This intermetallic compound has higher electrical resistivity than the electrode, and more heat is generated in this layer due to Joule heating. This accelerates the alloying process, and more of this phase is formed as welding proceeds. When the alloy buildup reaches a certain level, it may spall off of the electrode face upon electrode retraction. Craters may form on the electrode surface because of the material deficit created by this process. Figure 2.11 shows the evolution of craters on the Cu electrode face as a result of repeated cycles of alloying of the Cu–Zn compound, buildup of the alloy with the number of welds, and depletion of the alloy layer from the electrode. Electrode wear due to alloying between the electrode and coating of

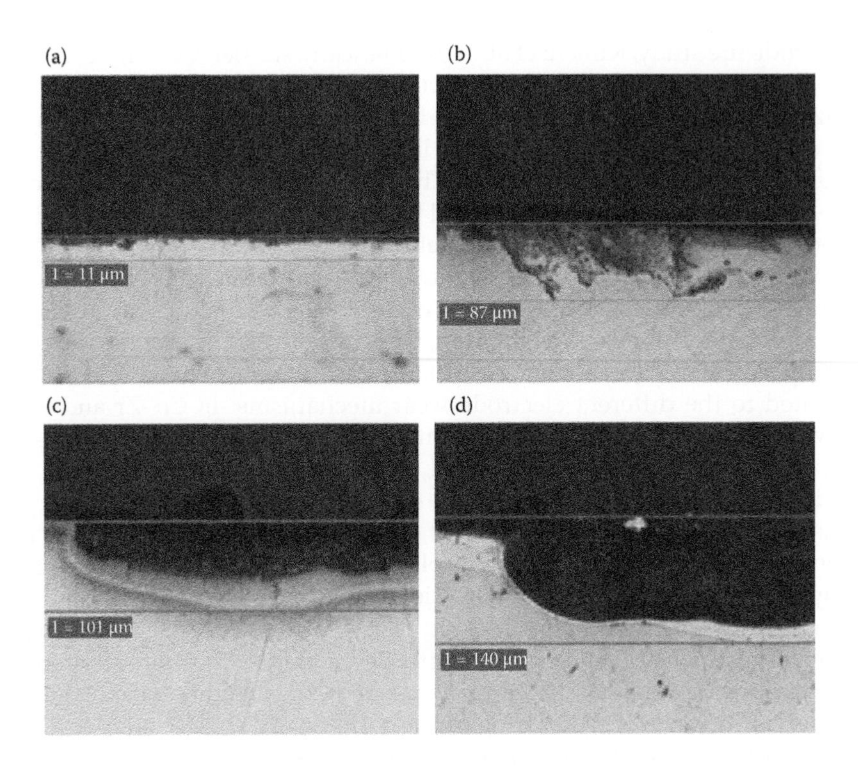

FIGURE 2.11
(See color insert.) Crater formation and growth on electrode surface as a function of number of welds.[22] (a) Alloy layer after 100 welds (11 μm); (b) crater depth after 300 welds (87 μm); (c) crater depth after 400 welds (101 μm); (d) crater depth after 500 welds (140 μm).

the workpieces has been widely reported such as in the work on welding ZnNi-UC EG sheet steels by Howe.[23]

2.3.2 Welding Aluminum Alloys

The mechanism of electrode wear in aluminum welding is similar to that in welding galvanized steels. As seen in Figure 2.12, domed electrode surfaces used for welding an Al alloy are worn with significant Al pickup and alloying.

The rapid deterioration of electrodes is the collective consequence of high pressure, high temperature, and a rapid metallurgical (alloying) process. Due to the presence of

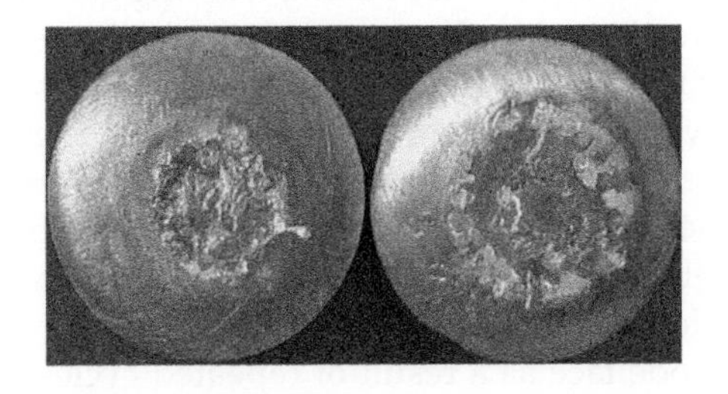

FIGURE 2.12
(See color insert.) Electrode wear in aluminum welding.

a nonuniform Al_2O_3 layer on aluminum sheets, localized heating and even melting at the start of welding may occur. Therefore, the deterioration of electrodes due to alloying between Cu and Al is largely affected by the contact resistance at the electrode–sheet interfaces. The electrode face deteriorates rapidly due to alloying and material depletion under high pressure (electrode force) and high temperature. In a continuous welding process, a repeated and accelerated (due to accumulative alloying and material depletion) deterioration of the electrode surfaces makes electrode life so short that such electrodes and sheets cannot directly be used in automated large-volume automotive production.

As the heating of the interface between a copper electrode and an aluminum sheet determines the electrode life, a surface without an Al_2O_3 layer is preferred for a long electrode life. Alternatively, if that is difficult to achieve, a thin, uniform layer can be tolerated, as it will result in uniform heating. This statement is on electrode life only; it does not take into consideration the fact that welding Al would be very difficult without an oxide layer on the surface because its bulk electrical resistivity is very close to that of Cu electrodes (Figure 2.2). Welding Mg alloys is very similar to welding Al alloys because of the similarity in thermal, mechanical, and metallurgical characteristics between Mg and Al. An experimental study by Thornton et al.[24] revealed that an electrode life of up to 1000 welds can be achieved if the sheets are properly degreased or chemically cleaned when welding a 2-mm sheet. A similar electrode life was obtained using aluminum sheets covered by a specially designed thin film.[25] A recent study by Li et al.[10] systematically investigated the effects of sheet surface conditions on electrode life. Using 2-mm 5A02 aluminum sheets, a schedule that appeared most preferable concerning the electrode life was used for testing the effects of sheet surface conditions. An electrode life of more than 2000 welds without dressing has been reported. The following sections show the details of the study.

2.3.2.1 Experiment

The chemical composition of 5A02 aluminum alloy is listed in Table 2.1. Dome-shaped Cu–Cr–Zr electrodes of face radius 100 and 20 mm in diameter were used for welding, using a 300-kVA, three-phase direct current (DC) pedestal welder. Four types of surface conditions—untreated, chemically cleaned, degreased, and electric arc cleaned—were compared.

2.3.2.2 Rapid Electrode Life Determination

Under the belief that electrode life can be estimated through examining the features of electrode surfaces, a set of electrodes was used to make 60 welds under each of the four surface conditions. The electrodes were then compared with those tested for life. Because the electrode life is closely related to the welding parameters, a set of experiments was conducted first to understand the influence of welding schedules. Welds were made to determine appropriate welding schedules, as listed in Table 2.2, with the minimum weld size of $5\sqrt{t}$ (t is Al sheet thickness in millimeters) maintained for each surface condition during the 60 welds.

TABLE 2.1

Chemical Composition of 5A02 Aluminum Alloy (wt.%)

Si	Fe	Cu	Mn	Mg	Ti	Al
0.40	0.40	0.10	0.25	2.5	0.15	Balance

TABLE 2.2
Welding Schedules for Rapid Electrode Life Determination

Surface Condition	Untreated	Degreased	Electric Arc Cleaned	Chemically Cleaned
Welding time	80 ms	80 ms	120 ms	100 ms
Welding current	27.2 kA	29.1 kA	34.3 kA	32.7 kA

Note: Electrode force was 9 kN for all welding tests.

The electrode faces after 60 welds using the schedules listed in Table 2.2 on four different surface conditions are shown in Figure 2.13. As the difference between upper electrodes and lower electrodes was small after welding 60 welds, only the upper electrodes are shown in the figure. A silver-colored ring of aluminum pickup is visible in all electrodes. However, such a ring appears different for the four electrodes. It is thin and clear in that of chemical cleaning, and wider yet still clear in that of degreasing. The ring becomes blurry and wider for that of electric arc cleaning, and very fuzzy and wide for that of the untreated condition. A wide and blurry ring of Al pickup may represent an unstable electrode–sheet contact during a continuous welding process, resulting from alloying between Cu and Al and removal of the bronze from the electrode surface at many locations. The degree of oxidation of Cu is also different on the electrode faces. Figure 2.13a and b appears less oxidized than Figure 2.13c and d. The surface of the electrodes shows a significant roughening (in the form of small craters) on those untreated and electric arc-cleaned sheets, whereas the electrodes used on chemically cleaned and degreased sheets are much smoother.

Figure 2.13 clearly shows that welding using sheets of different surface conditions creates distinctively different appearances of electrode faces. Recognizing such differences will help in predicting electrode life by making only a small number of welds. This is possible by linking the features of these electrodes to their respective electrode lives produced in electrode life tests.

2.3.2.3 *Electrode Life Test*

Using the same schedules as in Table 2.2, electrode life tests were conducted on sheets with the four surface conditions. All welds were peel tested to measure the weld size. A failure was defined as no weld button or when it is smaller than $3.5\sqrt{t}$. When 5% or more of 100 welds failed, the end of an electrode life was reached. For specimens with chemical, electric arc-cleaned, and degreased surfaces, 1 of every 100 welds was tensile–shear tested, and 1 of every 50 welds was tested on specimens with untreated surfaces.

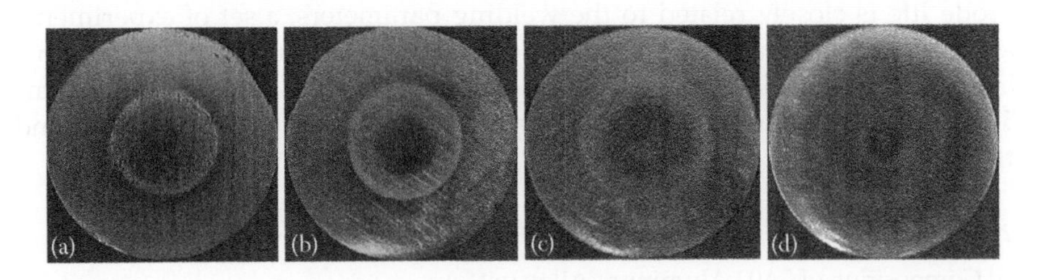

FIGURE 2.13
(See color insert.) Electrode surfaces after making 60 welds on sheets of different surface conditions: (a) chemically cleaned, (b) degreased, (c) electric arc cleaned, and (d) untreated.

The electrode lives determined by welding using sheets of the four different surface conditions are shown in Figure 2.14. In welding chemically cleaned sheets, more than 2300 quality welds were produced, with the electrodes slightly worn and far from the end of their lives, as shown in Figure 2.14a. Electrodes used to weld degreased sheets have a life of more than 2000 welds (Figure 2.14b). The variation in weld diameters grows large at the end of electrode life, but significantly smaller than those for electric arc-cleaned and untreated sheets (Figure 2.14c and d). As shown in Figure 2.14c, the electrode life was about 1700 welds when sheets were electric arc cleaned. When untreated sheets were used, the electrode life was about 200 welds, which is significantly shorter than any of those for treated sheets. Therefore, the surface condition of sheets plays a key role in determining electrode life in welding Al.

The faces of electrodes after electrode life tests using various surface conditions are shown in Figure 2.15, arranged in the same order as their electrode lives (Figure 2.14). The electrodes used in welding untreated sheets (Figure 2.15d) appear less worn than others. However, they were used to make only about 200 welds, whereas the others made 1700 welds (electric arc cleaning, Figure 2.15c) or more than 2000 welds (chemical cleaning, Figure 2.15a; degreasing, Figure 2.15b).

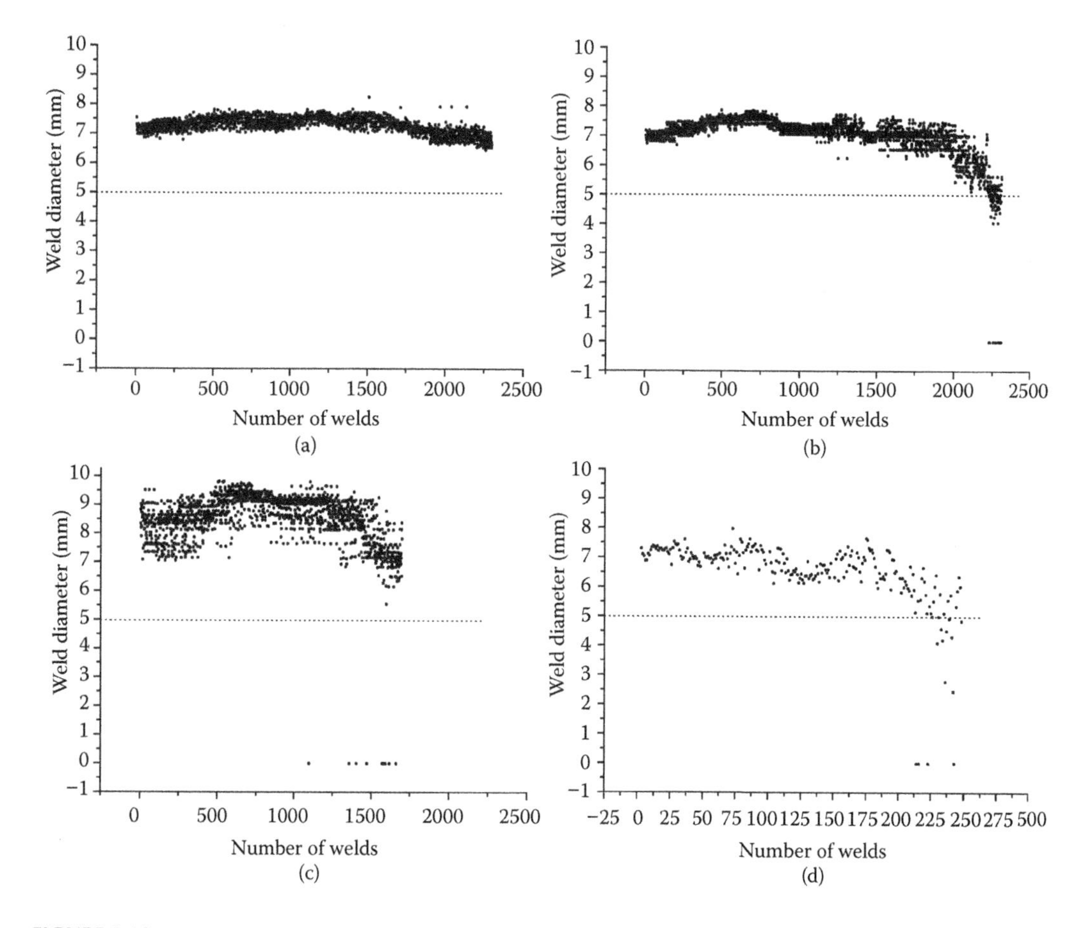

FIGURE 2.14
Electrode life testing results using (a) chemically cleaned, (b) degreased, (c) electric arc-cleaned, and (d) untreated surfaces. Dashed line represents minimum weld diameter ($3.5\sqrt{t}$) for sheets.

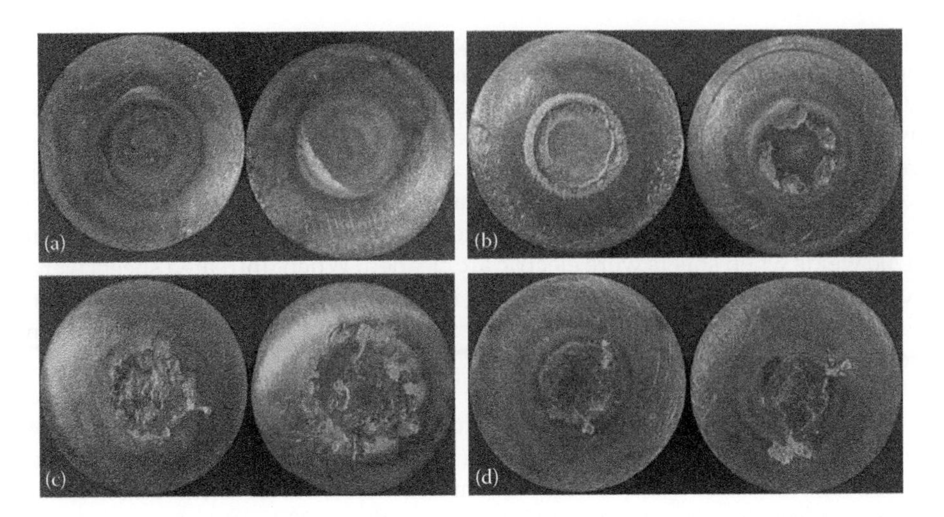

FIGURE 2.15
(See color insert.) Electrode surface morphology after life tests using (a) chemical cleaning, (b) degreasing, (c) electric arc cleaning methods, and (d) untreated aluminum sheets. Electrodes on left side are from the lower arm of the welder (negative), and those on right side are from the upper arm (positive).

The effects of surface conditions on electrode life can be characterized using the visually estimated magnitude and uniformity of contact resistance across the contact area between the electrode and sheet. As seen from Figure 2.15, chemically cleaned surfaces had the lowest contact resistance, and they produced the longest electrode life. On the other hand, untreated sheets exhibited the highest contact resistance and yielded the shortest electrode life. Low contact resistance benefits the electrode life primarily due to less oxidation and alloying; both contribute to less heating at the contact interface and long electrode life. The electrodes used to weld chemically cleaned sheets have slight alloying and oxidation on the surface after the life test, and those for degreased sheets have craters due to the depletion of bronze, and a large area of Cu–Al alloying. The electric arc-cleaned sheets deteriorate the electrodes the most, as evidenced by the large number of craters and Al pickup/alloying on the electrode surfaces.

The consistency and uniformity of electrode–sheet contact also play an important role in affecting electrode life. As discussed in previous sections, nonuniformly distributed contact resistivity on a sheet surface induces uneven, localized heating between the sheet and Cu electrode during welding. Severe oxidation and alloying may occur at these locations, resulting in a nonuniform distribution of surface resistivity on the Cu electrode face. When making a subsequent weld, the resistivity at the contact interface is not uniform due to the uneven resistivity distribution produced by the preceding welding on the electrode, and due to the nonuniform surface resistivity of the sheet, which changes from weld to weld. New oxidation and alloying will occur on the electrode surface during welding. An accumulative effect of such a process is the continuous deterioration of the electrode surface. As a result, the current distribution changes from weld to weld and produces inconsistent welds if the electrode surface is damaged severely enough. Although Figure 2.15 shows that electric arc-cleaned sheets have a lower and more consistent contact resistance than degreased sheets, in the experiment, it was observed that it was not trivial to make electric arc cleaning consistent, especially when melting of the surface had to be avoided.

This explains that electric arc cleaning produces lower contact resistance but shorter electrode life than degreasing. Another possible reason is that the thin aluminum oxide layer remaining on the sheets after degreasing would serve as a protection layer, which prohibits the interdiffusion between Cu and Al, whereas it would not produce much localized heating in the contact area.

Another factor responsible for the relatively short electrode life is the modified surface properties of the electric arc-cleaned sheets. The surface of such a treated sheet is softened by the electric arc heating, which results in a large contact area between the electrode and the sheet under an electrode force of 9 kN. Therefore, the welding of such sheets requires a high electric current in order to achieve a minimum current density for making a weld. As shown in Table 2.2, welding electric arc-cleaned sheets requires the highest welding current and time among all surface conditions. The surfaces of the electrodes used to weld electric arc-cleaned sheets in Figure 2.15 have significantly more damage than those using other cleaning methods. There are many large and deep craters, a large area of Al deposit, and the contact area appears significantly larger than others. When intensive alloying and alloy depletion from an electrode surface occur, the effective contact area between the electrode and a sheet surface becomes unstable—it can be small at one weld and result in a large current density, and it can be large at the next weld and result in a very low electric current density, producing low weld penetration or an undersized weld. Such change in contact area is random and produces large variations in the welds created.

The effects of electrode force and welding time on electrode alloying are discussed in Chapter 1. In general, long welding time generates more heat at the electrode–sheet interface and, therefore, a larger amount of alloying with Al and Mg than shorter welding time. A similar trend can be observed with higher electrode force, but the severity of alloying is significantly lessened with high electrode force, as a large electrode force creates low contact resistance and, therefore, less heat generation and alloying at the electrode–sheet interface. Thus, large electrode force is preferred for electrode life.

Electrical polarity appears to have some effects on electrode deterioration. In Figure 2.15, for each pair of electrodes, the one on the left side was taken from the lower or negative electrode arm. These electrodes appear less damaged than those on the right side, which were taken from the upper, or positive, electrode arm. This phenomenon might be explained considering the micro-morphology of the contact interface and the dynamics of resistance heating, and it deserves a dedicated study.

The variability of weld quality in a welding process is an important index in production. It may also provide a useful indicator for electrode life, as a large variability indicates that the welding process becomes unstable, and it may be close to the end of electrode life. Average weld diameters and standard deviations are plotted in Figure 2.16. These quantities were calculated on every 50 welds in the electrode life tests. Welding chemically cleaned sheets produces fairly consistent welds and a low, almost constant standard deviation (Figure 2.16a). It can be seen, when welding under other surface conditions, that accompanying a drop in the average weld diameter when an electrode life approaches its end, the standard deviation increases dramatically. From the plots in the figure, an increase of about 300% in standard deviation is observed for all surface conditions when the electrode life is reached. The standard deviation before the sudden increase is about 0.4 mm, and it jumps to about 1.4 mm or more, accompanied by a visible drop in weld diameters, when it is close to an electrode life. In the case of electric arc cleaning, the first such increase does not correspond to an average weld diameter falling below the desired value. However, it is fairly close to the end of the electrode life. Therefore, the change in standard deviation of weld diameters during welding serves as a useful index for electrode life.

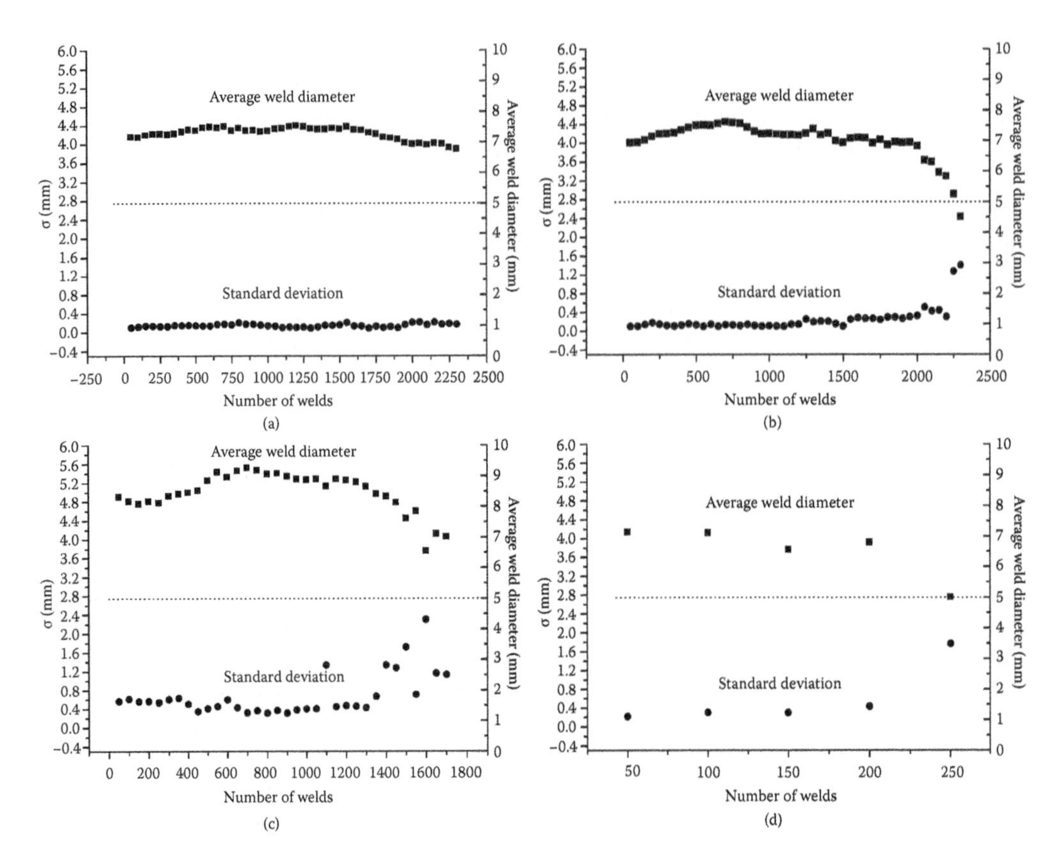

FIGURE 2.16
Average diameters and standard deviations of welds using (a) chemical cleaning, (b) degreasing, (c) electric arc cleaning methods, and (d) untreated sheets.

The tensile strength of welds tested during electrode life tests shows a trend similar to that of the weld diameter. Tensile–shear tests on the welds of various surface conditions are shown in Figure 2.17. One of every 100 welds was tested during electrode life tests, except for untreated sheets, where 1 of every 50 welds was tested. Chemical cleaning again produces the highest strength with the least variability. Degreasing has lower strength and larger variability, and electric arc cleaning is quite unstable, similar to those observed in Figure 2.16 for measured weld sizes. Such differences can be attributed to the magnitude and distribution of contact resistivity of the sheets cleaned using different methods.

2.3.2.4 Relation between 60-Weld Electrodes and Electrode Life

As shown in Figures 2.13 and 2.15, the electrodes used to weld sheets of different surface conditions have distinctively different characteristics. The subsequent electrode life tests proved that the electrode lives are different. Therefore, it is possible to predict the electrode life for particular stack-up of sheets and welding schedule, only after a small number of welds, such as 60 welds, as in this study. By analyzing the features shown in Figure 2.13 and linking them to the corresponding electrode lives, the following observations can be made:

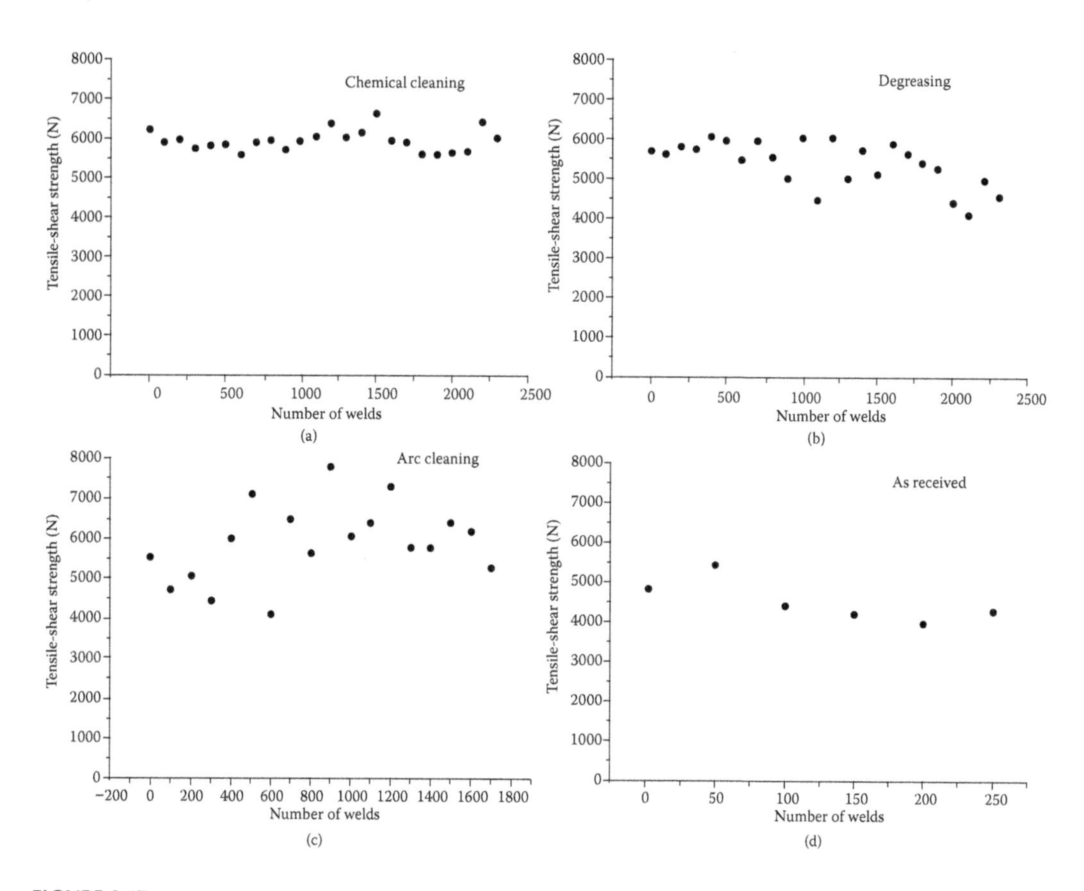

FIGURE 2.17
Tensile–shear strengths of welds made using (a) chemically cleaned, (b) degreased, (c) electric arc-cleaned, and (d) untreated sheets, taken during electrode life tests.

1. Silver-colored band on the electrode face. An electrode of long life tends to have a small, thin, yet clear silver band on the surface after a few welds. On the other hand, a large, thick, and fuzzy silver band may indicate a short electrode life, as in the cases of electric arc cleaning and untreated sheets.
2. Black oxidation (burning) marks at the center of an electrode face. Inside the silver band, there is usually an area of oxidation that is directly related to the cleanness of the electrode–sheet interface. Greases and other organic compounds at the interface may be burned under the intense heating during welding. Because such a reaction is due to the existence of low conductive, or even insulating substances at the interface, it directly reflects the contact resistivity or resistance. Therefore, it affects the deterioration of electrodes and electrode life. By comparing Figures 2.13 and 2.14, it can be seen that small and light burning marks on the electrode face after 60 welds indicate a long electrode life, and large and dark burning marks correspond to a short electrode life.

Therefore, it is possible to predict the electrode life of a combination of sheets, electrodes, and welding parameters by conducting a small number of welds that produce visible characteristics on the electrode faces, as observed in this study.

2.4 Heat Balance

A good knowledge of the thermal and electrical characteristics significantly helps in understanding a welding process and developing optimal welding schedules. In RSW, a primary concern for a practitioner is to select correct welding parameters, including welding current, welding time, electrode force, and electrode face diameter, which would produce a weld with desired features, such as certain geometric dimensions and weld strength. The practices of the RWMA[26] as well as those of many other organizations and corporations are available for this purpose. Most of the welding schedules are empirically developed. Although they are very useful in finding good welding parameters for even-thickness welding, schedules for welding uneven-thickness sheets are generally developed by and practiced within individual manufacturers. Because even-thickness combinations are rarely used in practice, there is clearly a practical need of welding schedules for uneven-thickness combinations. Welding schedules are commonly developed using theoretical, empirical, or combined theoretical–empirical methods, considering the thermal–electrical characteristics of welding. The common techniques used in this respect are summarized in the following sections.

2.4.1 Law of Thermal Similarity

The law of thermal similarity (LOTS) has been commonly used in the Japanese automotive industry to develop resistance welding schedules.[27] It is based on a heat flow analysis that attempts to make the temperature distributions in various weld thicknesses similar. The LOTS has been used to develop schedules to obtain desirable temperature profiles based on the known data, that is, to extrapolate the results obtained from known standard specimens for predicting the temperature profile for a different combination of sheet stack-up.[28] The LOTS gives a relationship between the distance and time that makes the temperature distributions similar for different thickness stack-ups, assuming that a temperature distribution reflects the corresponding welding process as well as weld formed. It has been mostly used as a guideline for choosing welding schedules for thick sheets based on those verified for thin sheets.

The LOTS states that similar temperature profiles will be produced if the weld time is proportional to the square of the sheet thickness,[28] or

$$t \propto h^2 \tag{2.2}$$

that is, if the welding time is t_1 for a sheet of thickness h_1, then $n^2 \times t_1$ is needed to weld a sheet of thickness $n \times h_1$. The total weld time is determined by the total thickness of the stack-up, and the thinnest outer sheet determines the maximum duration of any weld pulse. Other welding parameters can be derived similarly.

In general, when the plate thickness and diameter of the electrodes are magnified n times, the welding time should be increased to n^2 times and the current density decreased to n times in order to have the new temperature distribution be similar to the original one.[27,29]

Let h_1 (h_2), d_{e1} (d_{e2}), δ_1 (δ_2), and t_1 (t_2) be the sheet thickness, electrode diameter, current density, and welding time, respectively, of the original sheet stack-up (the new sheet stack-up). Then, the temperature distributions for the two stack-ups are similar if[27]

$$h_2 = n \times h_1 \tag{2.3}$$

$$d_{e2} = n \times d_{e1} \tag{2.4}$$

$$\delta_2 = (1/n) \times \delta_1 \tag{2.5}$$

$$t_2 = n^2 \times t_1 \tag{2.6}$$

Although the LOTS may theoretically yield similar temperature profiles for different stack-ups, it was found that many welds obtained using the LOTS schedules were either undersized or with heavy expulsion. The LOTS better serves as a tool for phenomenally understanding the RSW process, and its use for predicting schedules for actual welding practice is limited. The law does not hold well in welding sheets of dissimilar thicknesses, as it only considers the total thickness of a stack-up, and does not account for the individual thicknesses of the sheets. The LOTS does not consider the effect of the actual heat input to make a weld. Table 2.3 shows a comparison on welding an uncoated mild carbon steel of three thicknesses using schedules suggested by the *Welding Handbook*[30] and those predicted by the LOTS.

The table shows that using different schedules based on the *Welding Handbook* or the LOTS yields significantly different welds, even for even-thickness combination welding. A schedule based on the LOTS for welding one thickness is derived from that proven for welding another thickness (selected using the schedules from the *Welding Handbook*). For instance, a schedule for welding 0.75 mm steel can be directly obtained from the *Welding Handbook* (the first row of Table 2.3), or it can be derived using LOTS based on the proven schedule (provided by the *Welding Handbook*) for 1.21 mm steel or 1.89 mm steel (second and third rows of the table). A good welding schedule is defined as one that yields a large weld without expulsion. It can be clearly seen that there is a large difference between the weld schedules suggested by the *Welding Handbook* and those predicted by the LOTS; the former are usually more realistic and yield significantly better welds than the latter. Due to the limitations mentioned above, the LOTS cannot be directly used for practical welding, as it was not intended for such a purpose in the first place.

Based on the consideration of the heat needed to form a weldment, a new theory was proposed by Agashe and Zhang,[31] which overcomes some of the limitations of LOTS. The weld schedules predicted as per the new theory are based on heat balance equations, and are therefore closer than the LOTS to the actual physical welding conditions. Besides providing more accurate welding schedules, the theory can accommodate different sheet thicknesses in a single stack-up and is thus closer to the practical welding scenario.

2.4.2 Heat Balance

In RSW, heat balance can be defined as a condition in which the fusion zones in both pieces being joined undergo approximately the same degree of heating and pressure application.[30] It describes the ideal situation when a symmetric weld (with equal depth of nugget penetration) is made. Heat balance is influenced by the relative thermal and electrical conductivities of the materials to be joined, the geometry of the weldment, and the geometry of the electrodes.

A heat balance can be achieved if two identical sheets are welded together with electrodes of equal mass and contour, and heat is generated in both the pieces uniformly,

TABLE 2.3

Welds Made with Various Schedules

Sheet Thickness (mm)	Source of Weld Schedule	LOTS Factor n	Electrode Diameter (mm)	Current (A)	Time (ms/cycles)	Force (kg/lb)	Weld Diameter (mm)	Expulsion Occurrence	Surface Appearance
0.75	*Welding Handbook*	1.00	6.35	10,500	150/9.00	227/500	6.10	No	Good
	LOTS from 1.21	0.62	4.41	8,367	77/4.60	136/299	3.20	No	Good
	LOTS from 1.89	0.40	3.15	6,547	45/2.68	93/204	2.89	No	Good
1.21	*Welding Handbook*	1.00	7.11	13,500	200/12.00	354/780	7.07	No	Good
	LOTS from 0.75	1.61	10.24	16,940	390/23.43	590/1301	7.76	Very heavy	Damaged
	LOTS from 1.89	0.64	5.08	10,563	116/6.97	242/532	3.4 0	No	Good
1.89	*Welding Handbook*	1.00	7.94	16,500	283/17.00	590/1300	8.01	No	Good
	LOTS from 0.75	2.52	16.00	26,460	953/57.15	1,440/3175	11.30	Very heavy	Damaged
	LOTS from 1.21	1.56	11.11	21,086	488/29.28	863/1903	10.80	Very heavy	Damaged

with an oval-shaped weld cross section. However, if one of the pieces has a higher electrical resistivity than the other, heat will be generated more rapidly in this piece, resulting in a less-than-perfect weld, depending upon the amount of heat imbalance. In the case of dissimilar metals, such as when welding plain carbon steel to stainless steel, this dissimilarity can be compensated for by increasing the electrode contact area on the high-resistivity stainless steel side, or by using an electrode material of higher resistance on the low-resistivity carbon steel side. In the case of similar metals of unequal thickness, proper heat balance can be achieved by using a smaller contacting electrode area on the thinner sheet, with short times and high current densities.[30,32]

2.4.3 Modified Heat Balance Theory

The basic idea of the theory is that the total heat needed to create a weldment can be partitioned into that needed for the fusion zone, the HAZ, and the indented area, where a significant amount of heat is required. Instead of considering the overall thickness of the stack-up only (as in the LOTS), the heat input into each of the zones of the stack-up is accounted for. Therefore, rather than treating it as one entity, a weldment is split into different zones for heat calculation. Such a division is necessary when the sheets have different thicknesses, material properties, etc. Thus, for a two-sheet stack-up, there are two nugget zones near the center, surrounded on either side by a HAZ and an outer indentation zone, as shown in Figure 2.18.

The heat for heating a solid or a liquid is

$$Q = m \times C_p \times \Delta T \tag{2.7}$$

where m is the mass, C_p is the specific heat of the material, and ΔT is the change in temperature due to heating. Each zone is idealized as a short cylinder for simplicity. For instance, assuming that the HAZ has the same diameter as the electrode face, the mass of the HAZ can be expressed as

$$m = \frac{\pi}{4} \times d_e^2 \times h \times \rho \tag{2.8}$$

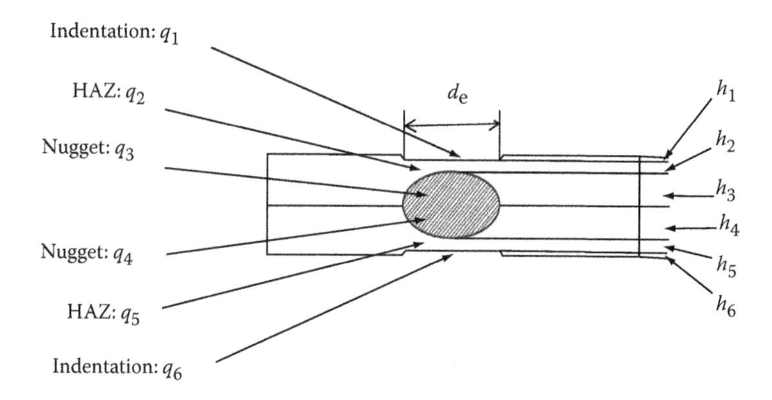

FIGURE 2.18
Partition of zones in a weldment for heat calculation. q = heat (Joules), h = thickness (mm).

where d_e is the electrode diameter, h is the height of the HAZ, and ρ is the density of the sheet. Then the heat in the HAZ can be calculated using Equation 2.7.

The heat components can be calculated once the mass, thermal properties, and the possible maximum temperature increases are known. The volumes of various zones in a weldment are not equal in practice. However, an assumption that the zones are (short) cylinders of the same diameter but different heights can be made for simplicity. Actually, their diameters are not much different in a well-controlled weld. The electrode diameter d_e can be used as the diameter of all the zones (cylinders). A nominal value is used for d_e, as it is usually not a constant during welding, and it is desirable to have the size of a weld close to that of the electrodes.

For the indentation, heat is accounted for by assuming that a (empty) cylinder of indentation experiences heating from room temperature up to (but below) the melting temperature. Indentations from both sides of a weldment contribute to the total heat:

$$q_1 = \frac{\pi}{4} d_{e_1}^2 \rho_1 C_{p_1} h_1 \Delta T_1 \tag{2.9}$$

$$q_6 = \frac{\pi}{4} d_{e_2}^2 \rho_2 C_{p_2} h_6 \Delta T_6 \tag{2.10}$$

where $\Delta T_1 = \Delta T_6 = T_{melt} - T_{amb}$ is the difference between the melting temperature and room temperature, d_{e1} and d_{e2} are the face diameters of the electrodes on both sides, C_{p1} and C_{p6} are the specific heats of the respective materials, and h_1 and h_6 are the depths of indentations for the upper and lower sheets, respectively.

Similarly, for the HAZ, the heat inputs on both sides are

$$q_2 = \frac{\pi}{4} d_{e_1}^2 \rho_1 C_{p_1} h_2 \Delta T_2 \tag{2.11}$$

$$q_5 = \frac{\pi}{4} d_{e_2}^2 \rho_2 C_{p_2} h_5 \Delta T_5 \tag{2.12}$$

where $\Delta T_2 = \Delta T_5 = T_{melt} - T_{amb}$ as an approximation.

The solid–liquid phase transformation, or melting, takes place in the nugget zone. The heat of the nugget includes that needed for heating the metal from room temperature to the melting point, the latent heat for melting, and the heat needed to raise the temperature beyond the melting point of the metal. The density and specific heat are usually different in different stages. However, the specific heat does not change much in the stages, and hence can be assumed constant.

Let

$$q_3 = \frac{\pi}{4} d_{e_1}^2 h_3 \left[\rho_1 C_{p_1} \left(T_{melt} - T_{amb} \right) + \rho_1' L_{f_1} + \rho_1'' C_{p_1}' \left(T_{max} - T_{melt} \right) \right] \tag{2.13}$$

$$q_4 = \frac{\pi}{4} d_{e_2}^2 h_4 \left[\rho_2 C_{p_2} \left(T_{melt} - T_{amb} \right) + \rho_2' L_{f_2} + \rho_2'' C_{p_2}' \left(T_{max} - T_{melt} \right) \right] \tag{2.14}$$

be the heat input to the two halves of the nugget, where h_3 and h_4 are the heights of the fusion zone, ρ_1' and ρ_2' are the liquid densities at melting temperature, ρ_1'' and ρ_2'' are the average densities, C_{p_1}' and C_{p_2}' are the average specific heats of the liquid between T_{max} and T_{melt}, and L_{f_1} and L_{f_2} are the latent heats of fusion.

The total heat supplied for making the weldment is $q = q_1 + q_2 + q_3 + q_4 + q_5 + q_6$, or

$$\begin{aligned} q = {} & \frac{\pi}{4} d_{e_1}^2 \rho_1 C_{p_1} h_1 \Delta T_1 + \frac{\pi}{4} d_{e_1}^2 \rho_1 C_{p_1} h_2 \Delta T_2 \\ & + \frac{\pi}{4} d_{e_1}^2 h_3 \left[\rho_1 C_{p_1} \left(T_{melt} - T_{amb} \right) + \rho_1' L_{f_1} + \rho_1'' C_{p_1}' \left(T_{max} - T_{melt} \right) \right] \\ & + \frac{\pi}{4} d_{e_2}^2 h_4 \left[\rho_2 C_{p_2} \left(T_{melt} - T_{amb} \right) + \rho_2' L_{f_2} + \rho_2'' C_{p_2}' \left(T_{max} - T_{melt} \right) \right] \\ & + \frac{\pi}{4} d_{e_2}^2 \rho_2 C_{p_2} h_5 \Delta T_5 + \frac{\pi}{4} d_{e_2}^2 \rho_2 C_{p_2} h_6 \Delta T_6 \end{aligned} \tag{2.15}$$

Based on the heat components calculated, a characteristic dimension H can be defined as

$$H = h_1 \frac{q_1}{q} + h_2 \frac{q_2}{q} + h_3 \frac{q_3}{q} + h_4 \frac{q_4}{q} + h_5 \frac{q_5}{q} + h_6 \frac{q_6}{q} \tag{2.16}$$

This characteristic dimension is used instead of the actual thickness of the entire stack-up (as in the LOTS), as it differentiates the contributions of various regions in a heating process. Although they are closely related in a welding process, h_i values ($i = 1, \ldots, 6$) are independently defined and can be independently altered to obtain the desired features of a weldment.

This theory was verified in the case of developing welding schedules for uneven-thickness sheet stack-ups. The first step is to develop good schedules for welding even-thickness sheet stack-ups. One can use proven welding schedules for equal-thickness sheets, such as those listed in the *Welding Handbook*.[33] The schedules for welding uneven thicknesses can then be developed based on the parameters for welding even-thickness sheets and this theory.

For an even-thickness stack-up, the weld time, welding current, electrode force, and electrode diameters can be chosen from established sources such as the *Welding Handbook*. The schedules for welding various-thickness uncoated low-carbon steel sheets are listed in Table 2.4.

For a sheet stack-up, the heat input needed for making a weldment is proportional to the square of welding current, weld time, and the resistance of the sheet material:

$$q \propto I^2 \times R \times \tau \tag{2.17}$$

TABLE 2.4

Welding Handbook Schedules for Uncoated Low-Carbon Steel Sheets[30,33]

Sheet Thickness (mm)	Welding Current (A)	Weld Time (ms/ cycle)	Electrode Force (kg/lb)	Electrode Diameter (mm)
0.508	8,500	117/7	181/400	4.78
0.635	9,500	133/8	204/450	4.78
0.762	10,500	150/9	227/500	6.35
0.889	11,500	150/9	272/600	6.35
1.016	12,500	167/10	317/700	6.35
1.143	13,000	183/11	340/750	6.35
1.270	13,500	200/12	363/800	7.92
1.397	14,000	217/13	408/900	7.92
1.524	15,000	233/14	454/1,000	7.92
1.778	16,000	267/16	544/1,200	7.92
2.032	17,000	300/18	635/1,400	7.92
2.286	18,000	333/20	726/1,600	9.53
2.667	19,500	383/23	816/1,800	9.53
3.048	21,000	467/28	952/2,100	9.53

The resistance can be assumed proportional to the characteristic dimension of the stack-up, and inversely proportional to the square of the nugget diameter[16]:

$$R \propto \frac{H}{d_e^2} \tag{2.18}$$

Therefore,

$$q \propto I^2 \times \frac{H}{d_e^2} \times \tau \tag{2.19}$$

The derivation of these equations does not consider the heat loss through the electrodes and sheets (a variable during welding). Therefore, the heat calculated is not the total input heat, as the heat loss can take a large portion of the total heat generated during welding. Only the heat needed to create various dimensions of a weldment is taken into consideration by this method.

Consider a case of two-sheet welding. Let I_1, H_1, τ_1, d_{e1}, and F_1 be the current, characteristic dimension, time, nugget diameter, and electrode force, respectively, for one stack-up, and I_2, H_2, τ_2, d_{e2}, and F_2 be the current, characteristic dimension, time, nugget diameter, and electrode force, respectively, for another stack-up. Based on the assumption that the amount of heat needed to make the uneven-thickness welding is the sum of one half of that for the thin even-thickness welding and one half of that for the thick even-thickness welding, the parameters for the uneven-thickness welding can be approximated as

$$H_3 = \frac{\dfrac{I_1^2 \times H_1 \times \tau_1}{d_{e_1}^2} + \dfrac{I_2^2 \times H_2 \times \tau_2}{d_{e_2}^2}}{\dfrac{I_1^2 \times \tau_1}{d_{e_1}^2} + \dfrac{I_2^2 \times \tau_2}{d_{e_2}^2}} \tag{2.20}$$

$$\tau_3 = \frac{\dfrac{I_1^2 \times H_1 \times \tau_1}{d_{e_1}^2} + \dfrac{I_2^2 \times H_2 \times \tau_2}{d_{e_2}^2}}{H_3 \times \left(\dfrac{I_1^2}{d_{e_1}^2} + \dfrac{I_2^2}{d_{e_2}^2} \right)} \tag{2.21}$$

$$I_3^2 = \frac{\dfrac{I_1^2 \times H_1 \times \tau_1}{d_{e_1}^2} + \dfrac{I_2^2 \times H_2 \times \tau_2}{d_{e_2}^2}}{H_3 \times \tau_3 \times \left(\dfrac{1}{d_{e_1}^2} + \dfrac{1}{d_{e_2}^2} \right)} \tag{2.22}$$

The electrode force is assumed proportional to the square of the electrode diameter to keep a constant pressure[32]:

$$F \propto d_e^2 \tag{2.23}$$

Therefore,

$$F_3 = \frac{\dfrac{F_1}{d_{e_1}^2} + \dfrac{F_2}{d_{e_2}^2}}{\dfrac{1}{d_{e_1}^2} + \dfrac{1}{d_{e_2}^2}} \tag{2.24}$$

The temperature in the weldment is assumed proportional to the heat generated, and inversely proportional to the characteristic thickness and the square of the electrode diameter[34] when the welds formed are similar:

$$T \propto \frac{q}{H \times d_e^2} \tag{2.25}$$

Because the zones in a weldment are assumed similar to their counterparts in the even-thickness welds, Equation 2.25 can be used to approximate the temperature of a weldment. Let q_1 and q_2 be the heats of the two even-thickness stacks; then for the combination stack-up, the heat content q_3 can be derived from Equation 2.25, assuming an average temperature is reached in the new stack, similar to those of the even-thickness stacks:

$$q_3 = \frac{\dfrac{q_1}{H_1 \times d_{e_1}^2} + \dfrac{q_2}{H_2 \times d_{e_2}^2}}{\dfrac{1}{H_3} \times \left(\dfrac{1}{d_{e_1}^2} + \dfrac{1}{d_{e_2}^2}\right)} \tag{2.26}$$

2.4.4 Experimental Verification

Experiments were carried out to verify the theory and have proven it suitable for use as a guideline for selecting welding schedules that are not available. The experiments were conducted on an RSW machine equipped with a programmable weld control unit. A 35-kVA transformer was used along with a C-type gun. The raw material used was bare mild carbon steel sheets of 14 (0.75 mm), 18 (1.21 mm), and 22 (1.89 mm) gauge of ASTM A569 and ASTM A366 grade.

The parameters in the experiments are: ambient temperature = 27°C, melting point for mild steel = 1535°C, maximum temperature reached was assumed to be 1735°C (with 200°C overheating), specific heat of mild steel = 502 J/kg K, latent heat of fusion = 275,000 J/kg, the average density of mild steel between room temperature (27°C) and melting temperature (1535°C) is 7470 kg/m^3, (liquid) density at 1535°C = 7190 kg/m^3, and density at 1735°C = 6991 kg/m^3.[5,14,15] Several sets of welds were made with the calculated schedules based on the schedules for even-thickness sheet welding listed in the *Welding Handbook*.[33] The welds were peel tested and the weld diameter was measured. Samples were prepared for metallographic examination and measuring various dimensions. With the help of these measured dimensions, the welding parameters for other sheets were predicted using the equations of the proposed theory. Then using these predicted weld parameters, new sets of welds were made, and the procedure was repeated. Finally, the weld parameters obtained from welding schedules predicted by the proposed theory were compared with those obtained in actual welding.

Even-thickness sheets of 0.75- and 1.21-mm thickness were welded first, with schedules very close to the ones given by the *Welding Handbook*.[15] The weld diameter was measured, and microscopic observations revealed the heights of various zones for calculating the characteristic dimensions. From this data, weld schedules for a stack of 0.75 + 1.21-mm sheets were predicted using the equations of the modified heat balance method. Welding using these schedules yielded the expected weld sizes without expulsion. The results are tabulated in Table 2.5.

In Table 2.5, the welding current in experiment 3 was searched, based on the predicted value, to obtain a similar characteristic height as that predicted. This practice was to show that a characteristic height (and therefore a weldment) can be created in welding using schedules derived by the theory. The table reveals that the experimental results obtained for uneven-thickness combinations are in good agreement with those predicted by the theory. The welding schedules predicted yielded good welds in terms of size and surface quality. Several additional tests were carried out to further verify the results of the

TABLE 2.5

Experimental Results

Experiment No.	Thickness (mm)	Electrode Diameter (mm)	Weld Current (A)	Weld Time (ms/cycle)	Electrode Force (kg/lb)	Heat (J)	Characteristic Thickness (mm)	Minimum Weld Diameter (mm)[a]	Average Weld Diameter (mm)
1	0.75 + 0.75	6.35	9,750	150/9	227/500	414	0.492	3.43	5.302
2	1.21 + 1.21	7.14	13,500	200/12	354/780	714	0.779	4.58	7
Prediction	0.75 + 1.21	6.35/7.14	11,557	180/10.81	283/624	598	0.684	3.43	–
3	0.75 + 1.21	6.35/7.14	10,500	183/11	286/630	561	0.699	3.43	5.9

Source: *Welding Handbook, Vol. 2, Welding Processes*, 9th edition, American Welding Society, Miami, FL, 2004. With permission.

[a] Minimum weld size required as listed in Welding Handbook.[33]

proposed theory and to build a confidence interval on its ability to predict correct welding schedules.

1. Using the same welding schedule, several welds were made and measured to establish a variance on the weld diameter. The variance on the mean diameter of the welds was found to be very small ($\mu_d = 4.93$, $\sigma^2 = 0.0514$).
2. With all other welding parameters kept constant, the weld time was varied over a range. It was revealed that a weld by the selected schedule had the largest size without expulsion. Shorter weld times resulted in undersized welds, whereas longer weld times led to expulsion.
3. With all other weld parameters kept constant, the weld current was varied over a range. It also proved that the weld current predicted by the theory gave the largest nugget size without expulsion. Again, lower-than-selected currents produced a smaller weld button, whereas expulsion resulted from higher currents.

Experiments have shown that the predicted welding parameters used are optimized to have the largest weld diameter without expulsion. The parameters can be predicted with 98% confidence. Thus, the theory can be used to predict welding schedules for uneven-thickness sheets with good accuracy and easiness for practical use.

2.5 Electric Current Waveform

Electric current is the foremost important factor in RSW, as it is directly responsible for generating the heat needed to melt the metal. Most spot welders in the industry use AC, which is usually characterized by the root mean square (RMS) value, an integration that over time has been used as an indicator of the total heat generated. However, experiments have shown that welding is affected by both the total heat input and the heat input rate. The latter is directly associated with the change in current value (magnitude) with time, or the waveform of electric current. Different current profiles, for example, AC and DC, with similar RMS values, have produced significantly different spot welds in both aluminum alloys and steels.[35–37] In addition to single-phase AC, which is the most popular current waveform used in resistance welding, there are a number of alternatives such as single-phase DC, three-phase DC, and a very promising form of current, medium-frequency direct current (MFDC).

The electric current used in RSW is converted from the power supply. Such a conversion produces the desirable waveform of the current for welding. Roth[38] explained in detail the principles of power conversion in resistance welding, of single-phase transformers, frequency convertor transformers for AC to AC conversion, and rectifier equipment for AC to DC conversion. A schematic of the conversion of single-phase AC to AC, and three-phase AC to three-phase DC in Figure 2.19, shows the basic steps of conversion and rectification in resistance welding transformers.

Roth and Hofman[40] presented the differences between welding using AC and DC from an electrical point of view. A resistance welding circuit can be considered a series circuit consisting of a resistance and an inductive reactance when powered with an AC, or a resistance and an inductance when powered with a DC. These electrical components determine the current waveforms. For instance, the impedance, which is a function of the resistance and reactance in an AC circuit, determines the power factor, and the

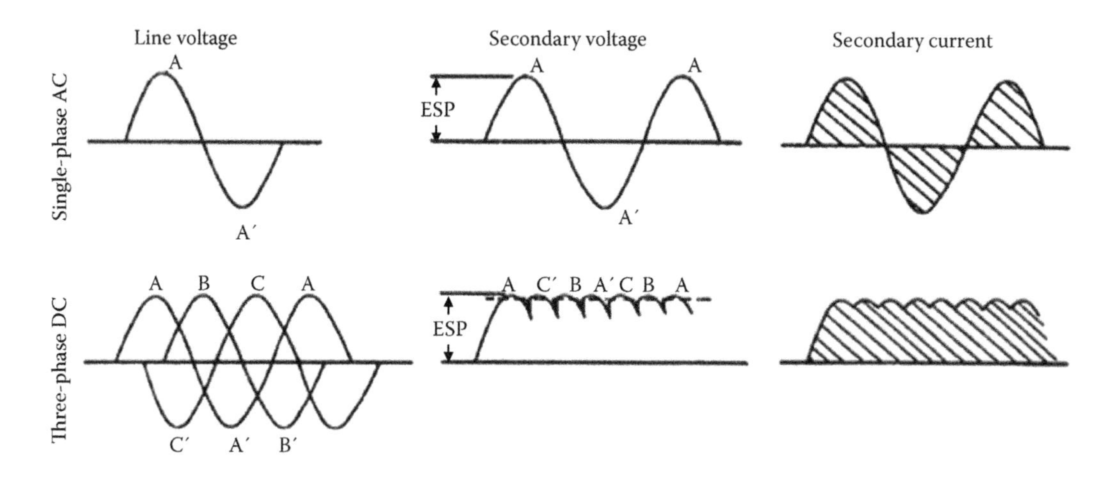

FIGURE 2.19
Wave forms of primary and secondary voltages, and secondary currents of single-phase AC and three-phase DC power supplies. (Adapted from Moss, L.E., Bates, F.E., Three phase D.C. flash welding, SMWC I, Paper 05, 1984.)

displacement between the voltage and current waves. When a DC circuit is energized, the current does not jump to its designated value immediately. Rather, it goes up exponentially and reaches the steady-state current level after a certain period of time, as shown in Figure 2.20, affected by the electrical inductance of the circuit. The time elapsed is often called the time constant of the circuit, and it can be calculated by the ratio L/R, where L is the inductance and R is the resistance of the circuit. As L and R depend on both the electric circuit of the welder and the workpieces, the time constant of a circuit varies with welding setup. Such an effect may be significant and should be considered when transferring welding schedules derived in a laboratory setup to a production environment. When the electric current is shut off, that is, the DC circuit is de-energized, the current value decays exponentially to zero, which is also determined by the time constant. The rise and decay of DC current should be factored in when selecting welding time and hold time, respectively.

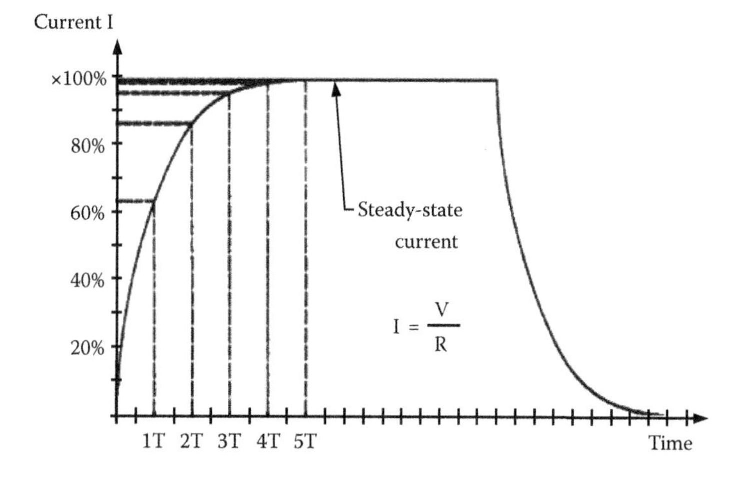

FIGURE 2.20
Rise of DC current to its steady-state value and its decay after power is shut off. (From Roth, D.K., Hofman, K.A., Alternating current versus direct current in resistance welding, SMWC IV, Paper D19, 1990. With permission.)

They are especially important when short welding time is used, as in the case of Al or Mg welding, in order to accurately determine the amount of heat input.

2.5.1 Single-Phase AC

Single-phase AC is extensively used in welding because it is reliable and relatively simple to control, and the equipment cost is low. The basic control of a single-phase AC welder is shown in Figure 2.21a.

The main advantage of single-phase AC convertors is their simplicity. However, their inherent electromagnetic characteristics need to be accounted for in accurately controlling a welding process. Comparing the measured single-phase AC in Figure 2.21b with the idealized AC waveform in Figure 2.19, it can be seen that they are different in several aspects. Deviated from the perfect sinusoidal waves, the measured current has non-constant peak values and there are periods during which the current ceases flowing. In addition, the peak current of the first cycle in Figure 2.21 is about 20% lower than the rest. These are the result of electric impedance of the circuit. The shortcomings of an AC welder also include unbalanced line loading, higher primary current draw, lower power factor, and large inductive losses in the throat of the welding gun.[41]

An effect associated with AC, but often overlooked by many welding practitioners is the skin effect.[42] Because of the self-inductive properties of a conductor, an AC tends to flow on the outer part of the conductor, making the electric current distribution nonuniform in the cross section. The skin effect is characterized by the "skin depth," which is defined as the distance below the surface where the current density falls to $1/e$, or 37% of its value at the surface, and it is linked to the applied current and material properties of the conductor in the form[43]

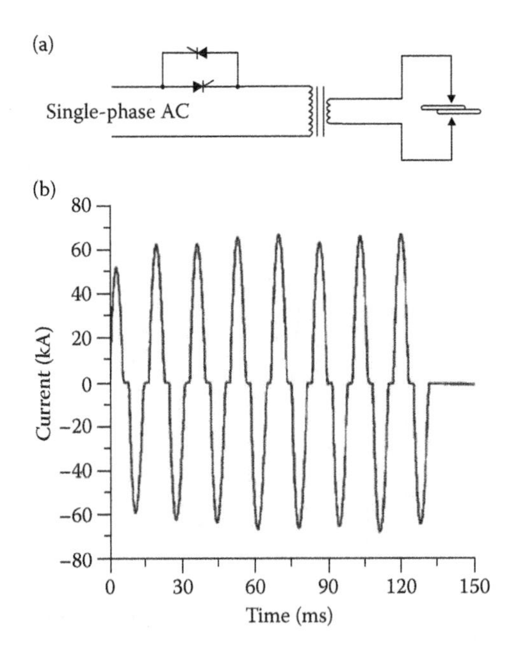

FIGURE 2.21
A single-phase AC weld control (a) and a typical measured AC current profile (b). (From Michaud, E.J., Renaud, S.T., A comparison of AC and mid-frequency DC resistance spot weld quality, SMWC VII, Paper A1, 1996. With permission.)

$$\delta \approx 503\sqrt{\frac{\rho}{\mu_r f}} \quad (2.27)$$

in meters. f is the frequency of the AC pulsation (Hz), ρ the electrical resistivity of the medium (Ω·m), and μ_r is the relative permeability of the medium. The current density decreases exponentially from the surface, and a large portion, or 67% of the total current is located between the surface and a distance of δ from the surface. Although the formula of skin depth is derived on an infinitely thick plane conductor, it provides an approximate to the current distribution in a conductor with a circular cross section, such as the contact at the faying interface in spot welding. Considering a cast iron with a resistivity of 10.0×10^{-8} Ω·m at room temperature, a relative permeability of 100,[44] then a 60-Hz AC current has a skin depth of

$$\delta = 503\sqrt{\frac{10.0 \times 10^{-8}}{100 \times 60}} \approx 2.1 \text{ mm} \quad (2.28)$$

For pure Al and Mg, the skin depths are 10.7 and 13.7 mm, respectively, calculated using the resistivity values at room temperature for these two alloys as 2.724×10^{-8} and 4.45×10^{-8} Ω·m, respectively. The permeability is taken as 1.0 for Al and Mg.[44] During spot welding, the contact area created by electrode squeezing at the faying interface depends on the electrode size, electrode force level, and the sheet thickness and mechanical properties. The diameter of the contact area can be assumed in the range from 6 to 10 mm. Therefore, when welding Al or Mg, there is no skin effect because of their low permeability. When welding ferrous alloys, however, the skin effect may be significant in current distribution, heat generation, and weld formation. As most of the AC is concentrated in the periphery of the contact area between the sheets, more heat is generated in the vicinity of the edges of the contact than near the center. As a result, the initial melting may start near the periphery, and the weld formed with insufficient heating may have an annular shape. However, this cannot be easily verified, and it is difficult to de-couple the skin effect from the others as the heat generation is also affected by the contact resistivity distribution at the faying interface, which is a strong function of pressure and temperature. As discussed in Chapter 8, the pressure distribution at the faying interface is not even, with the maximum near the edges of the contact area. As contact resistivity is inversely proportional to pressure, electric current density at the edge may be higher than the center where resistivity is high due to low pressure. The higher amount of heat created near the periphery could come from either the higher current density due to the skin effect or the lower resistance resulting from the pressure distribution, or both. It is not uncommon to observe more heating and higher degree of melting near the edges of a weld nugget than in the center, from the cross sections of steel welds. The skin effect diminishes once a sizable portion of the contact area is melted. An observation by Dupuy and Fardoux,[42] possibly due to skin effect, is shown in Figure 2.22 when comparing the weld growth processes in welding hot-dip galvanized interstitial-free steels using an AC and an MFDC, respectively. The welds made using the AC current have more melting on the edges than the center after both 4 and 6 cycles, possibly because of higher current density at the edges of the contact between the sheets. The DC welds do not possess such features.

An operation window, often defined as the difference between the electric current for the minimum sized welds and expulsion current, is usually determined when a material

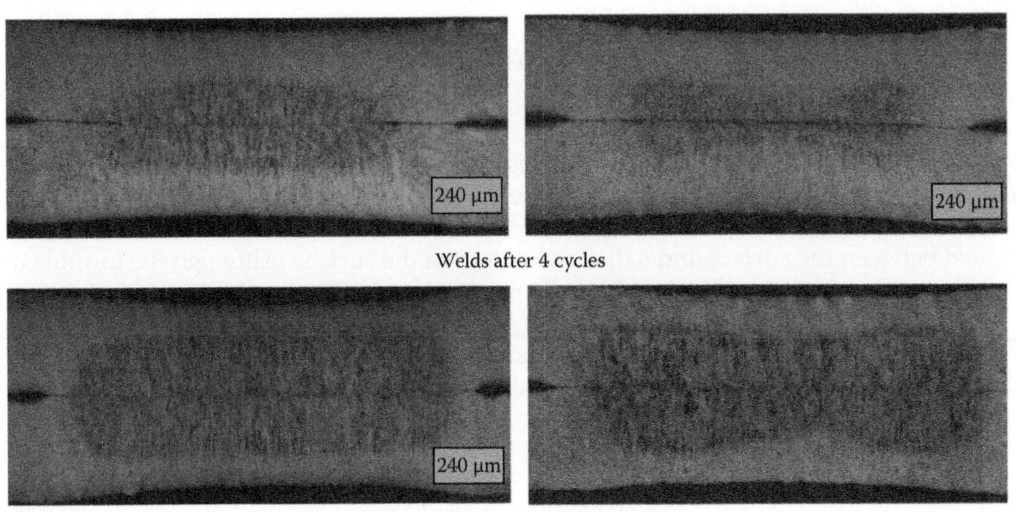

FIGURE 2.22
Cross sections of welds made using MFDC (left) and AC (right) currents. (From Dupuy, T., Fardoux, D., Spot welding zinc-coated steels with medium-frequency direct current, SMWC IX, Paper 1-2, 2000. With permission.)

is chosen for spot welding. In general, a large operation window is beneficial to practical welding, as a small deviation from the designated current level, which is quite common due to random effects in welding, will not significantly affect the weld size or the occurrence of expulsion. Various efforts have been made to enlarge the operation windows of single-phase AC welders, based on the belief that weld formation and expulsion depend not only on the amount of heat input, but also on how the heat is input.[45–49] Tawade et al.[45] have explored modified current pulse in zinc-coated advanced high-strength steels. Enhanced lobe width was achieved by using two current pulses one after the other, with the second pulse stepped down in magnitude from the first pulse. A study by Yadav,[50] shown in the following paragraphs, compared the effects of welding current patterns on operation windows, and showed that modified current patterns or profiles tend to raise the expulsion limits and enlarge the operation windows.

Raising the expulsion limit is an effective means of enlarging operation windows. Expulsion is generally observed in the final stage of welding steels when enough heat is accumulated. This corresponds to the pressure change in a liquid nugget, as discussed in Chapter 7, which goes up with temperature, or the amount of heat input. Therefore, the additional heat input at the end of welding may play a decisive role in expulsion. Reducing the heat input at the end may reduce the expulsion occurrence, and therefore raise the expulsion limit and enlarge the operation window. The descent of electric current magnitude can be achieved in many ways. For instance, the current could drop linearly from the maximum to a desired minimum or zero. This, however, may cause a rapid decrease in the current magnitude and a large heat loss, which may have a negative impact on the weld quality. A gradual decrease in magnitude toward the end of welding seems more logical in achieving sizable welds and containing expulsion.

In the study by Yadav,[50] profiles resembling a sine curve were used to simulate a variable heat input during a welding cycle. Three current profiles—the constant, sinusoidal, and half-sine current profiles—were studied. Their effects on delaying expulsion limits or enlarging windows of operation were investigated. The following sections show the procedure of calculating current values for various profiles, and the experimental results using the current profiles.

2.5.1.1 Constant Current

In the study, constant current refers to the conventional constant-magnitude (peak value) electric current pattern. An example is shown in Figure 2.23. The RMS values, instead of the actual current values, are shown in the figure.

2.5.1.2 Half-Sine Current Profile

A declining heat input is assumed following a half-sinusoidal curve. The current magnitude follows the sine function from $\pi/2$ (90°) to π (180°), as shown by the dashed line in Figure 2.24. In practice, it is difficult to implement a true sinusoidal profile using production welder controllers. Therefore, an equivalent profile of step function can be used. The figure also illustrates a step function of three constant pulses to approximate the half-sinusoidal current. The magnitudes of the steps can be determined by equating the heat generated under these two current profiles. For simplicity, a constant resistance value is assumed during welding for calculation. The currents, both the half-sinusoidal and constant (step) currents, refer to the RMS values of corresponding AC currents.

The constant-current profile is generated by equating it to the half-sine profile in terms of the amount of heat. Because the resistance is assumed constant, it can be omitted from the calculation. The heat generated in a time period dt is

$$q = I^2 \mathrm{d}t \tag{2.29}$$

The half-sinusoidal current can be expressed as

$$I = I_0 \sin \frac{\pi}{2}\left(1 + \frac{t}{\tau}\right) \tag{2.30}$$

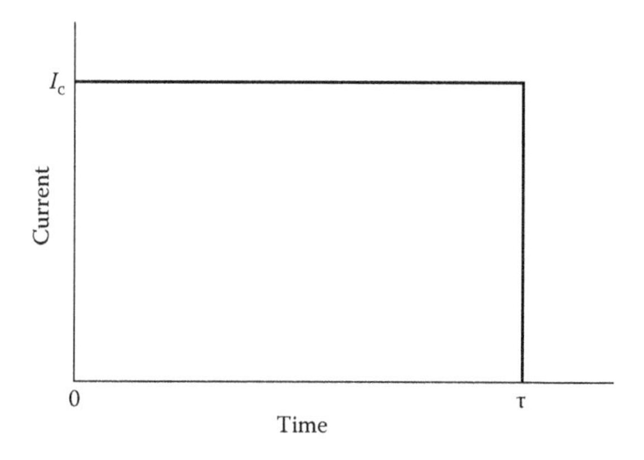

FIGURE 2.23
Constant-current profile.

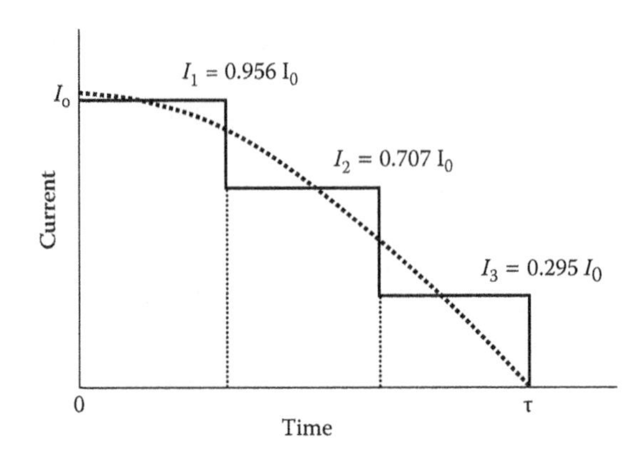

FIGURE 2.24
Half-sine and equivalent pulsed-step current profiles.

If a three-step current profile is used to approximate the half-sine profile, then the (constant) current value at each step can be determined by the heat consideration. The heat equivalence of these two current profiles between $t = 0$ and $t = \tau$ can be written as

$$\sum_{i=1}^{3} \frac{\tau}{3} I_i^2 = \int_0^{\tau} I^2 \, dt \tag{2.31}$$

Then the (constant) current level at each step is determined by equating the heat it generates to that by the sinusoidal curve in the same time interval:

$$\frac{\tau}{3} I_i^2 = \int_{\tau_i} I^2 \, dt \quad i = 1, 2, 3, \tag{2.32}$$

which yields

$$\begin{aligned} I_1 &= 0.956\ I_0 \\ I_2 &= 0.707\ I_0 \\ I_3 &= 0.295\ I_0 \end{aligned} \tag{2.33}$$

If a constant current is used, its RMS value I_c can be linked to the maximum value I_0 of a half-sinusoidal current profile by considering the equivalence of heat generated using these two current profiles:

$$I_c^2 \tau = \int_0^{\tau} I^2 \, dt \tag{2.34}$$

From Equation 2.30, the above equation results in $I_0 = \sqrt{2}\ I_c = 1.414\ I_c$. The comparison is shown in Figure 2.25.

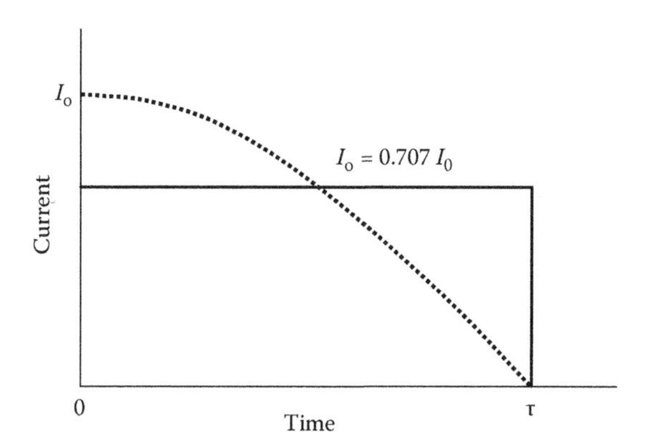

FIGURE 2.25
Equivalence of half-sine profile to a constant-current profile.

2.5.1.3 Sinusoidal Current Profile

If a sinusoidal current profile is applied, there will be an upslope heating at the beginning, in addition to the downslope heating at the end. This allows welding to proceed with an effect similar to pre-heating, welding, and then post-heating to reduce expulsion.

Similar to the case of half-sine current profile, a step function consisting of three constant pulses of current is used instead of a continuous sinusoidal function. Calculations of the magnitudes of the step function are to achieve the same amount of heat input in a time interval from the two functions.

The current can be expressed as

$$I = I_0 \sin \pi \frac{t}{\tau} \tag{2.35}$$

The heat equivalence between $t = 0$ and $t = \tau$ can be written as

$$\sum_{i=1}^{3} \frac{\tau}{3} I_i^2 = \int_0^{\tau} I^2 \,\mathrm{d}t \tag{2.36}$$

Then the (constant) current level at each step is determined by equating the heat it generates to that by the sinusoidal curve in the same time interval:

$$\frac{\tau}{3} I_i^2 = \int_{\tau_i} I^2 \,\mathrm{d}t \quad i = 1, 2, 3 \tag{2.37}$$

and it yields

$$I_1 = 0.707\, I_0$$
$$I_2 = 0.913\, I_0 \qquad (2.38)$$
$$I_3 = 0.707\, I_0$$

The sinusoidal current profile and the corresponding step function are shown in Figure 2.26. The maximum current I_0 can be related to I_c, the magnitude of a constant-current profile, by considering the equivalence of heat generated under the profiles:

$$I_c^2 \tau = \int_0^{\tau} I^2 \, dt \qquad (2.39)$$

From Equation 2.38, the above equation results in $I_0 = 1.279\, I_c$. The comparison is shown in Figure 2.27.

2.5.1.4 Experiments

The effects of current profiles were compared on the operation windows produced using these profiles. A 1-mm galvanized steel was used in the experiments. The variables were electric current profile and electrode force, as they have proven to have a significant effect on the expulsion limits, as demonstrated in Chapter 7.

A current profile with a constant I_c and τ = 12 cycles was used as the baseline. A three-pulse approximated half-sine current profile, with I_1 = 13.52 kA, I_2 = 10.0 kA, and I_3 = 4.17 kA, was derived from Equation 2.33 if I_c = 10 kA. An equal period of four cycles was used for the pulses. Similarly, the current values for a sinusoidal profile were determined as $I_1 = I_3$ = 9.04 kA and I_2 = 11.68 kA if I_c = 10 kA. The electrode force had two levels: 2.8 and 3.2 kN. Truncated, flat face electrodes were used in welding.

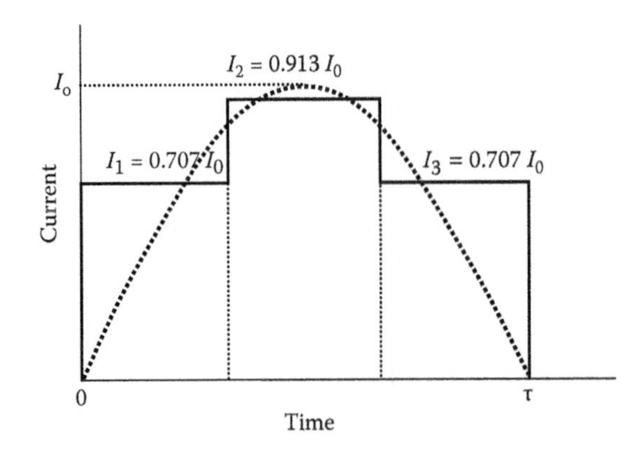

FIGURE 2.26
Sine profile and equivalent pulsed-step current profile.

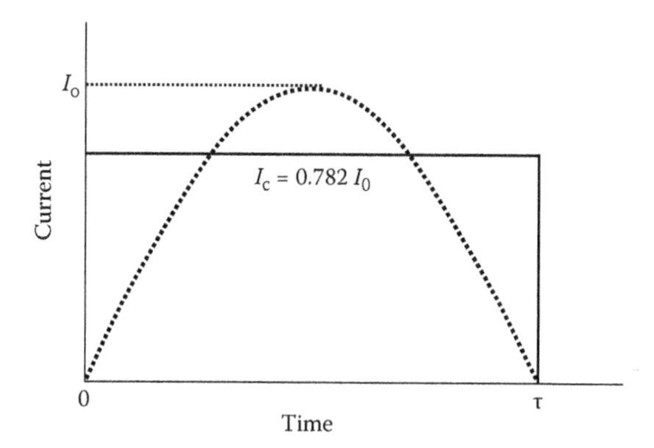

FIGURE 2.27
Equivalence of sine profile to a constant-current profile.

The specimens were peel tested and checked for size and expulsion. The responses were the current for minimum-size welds and the expulsion current, or the operation window, which is the difference between these current limits. The value $4\sqrt{t}$ was used as the minimum weld size. At least five specimens were made and tested when determining the minimum and expulsion currents, and two replicates (windows) were made for each profile. The results are tabulated in Table 2.6.

The results are also illustrated in Figure 2.28. The modified current profiles have clear effects on the minimum current, the expulsion current, and the width of operation window. The minimum and maximum currents are different when different current profiles are used, and they are also affected by electrode force. The expulsion current increases with electrode force for each current profile. The minimum current increases with electrode force for constant and sinusoidal current profiles, whereas a higher electrode force produces a lower minimum current for the half-sine profile. No reasonable explanation has been obtained. In general, the width of operation windows consistently increases with electrode force, and half-sine and sinusoidal profiles are superior to the constant-current

TABLE 2.6
Testing Results for Effects of Current Profiles and Electrode Force

Testing Order	Current Profile	$I_{c,min}$ (A)	$I_{c,max}$ (A)	Window of Operation (A)	Electrode Force (kN)
1	HS	8,600	11,200	2,600	2.8
2	CN	8,500	10,100	1,600	2.8
3	SS	9,100	11,700	2,600	2.8
4	CN	10,000	11,800	1,800	3.2
5	HS	10,600	12,800	2,200	3.2
6	SS	10,300	13,600	3,300	3.2
7	CN	9,000	10,700	1,700	2.8
8	SS	9,400	12,400	3,000	2.8
9	HS	9,000	11,300	2,300	2.8
10	HS	10,200	12,500	2,300	3.2
11	CN	9,600	11,400	1,800	3.2
12	SS	9,700	13,100	3,400	3.2

Note: CN, constant-current profile; HS, half-sine profile; SS, sinusoidal profile.

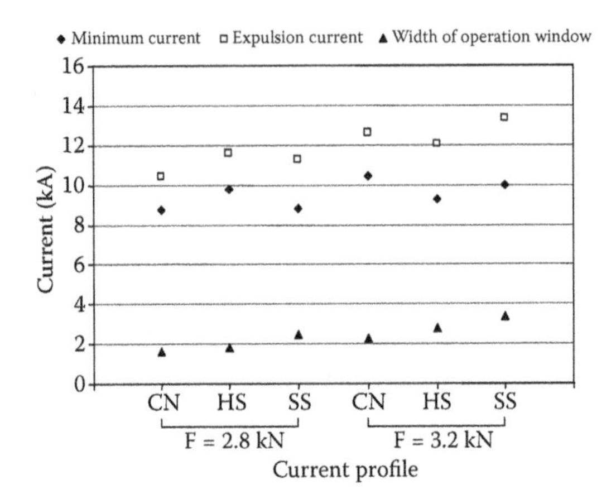

FIGURE 2.28
Effects of current profiles and electrode force.

profile. A 30% increase in the window of operation when using the half-sine profile and a 70% increase when using the sinusoidal profile have been observed, compared with welding using a constant-current profile.

2.5.2 Single-Phase DC

When welding using a single-phase AC welder, there is an "off-time" of current associated with the alternating pulsation, as seen in Figure 2.21. During these periods of time, the current is very small or zero, and the weld being made actually loses heat through conduction to the electrodes and the workpieces. An effective way to overcome this is to use DC, and the simplest DC is single-phase DC. A single-phase DC is generated from a single-phase AC power supply using a rectifier in the welding circuit, and one actual measurement of such a current is shown in Figure 2.29. In actual welding, a single-phase DC cannot be obtained by simply taking the absolute value of a single-phase AC. Because of the impedance of the circuit, the DC current usually has a smoothed and skewed waveform, fluctuating above the zero value. The squared current values, as they are directly associated with the heat generation based on the principles of Joule heating, show clearly

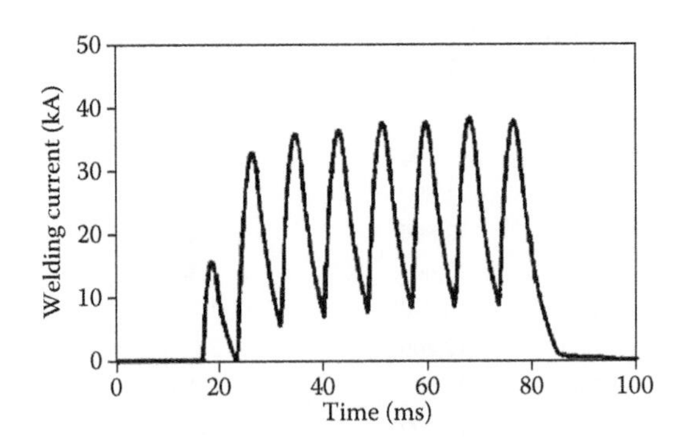

FIGURE 2.29
Single-phase DC welding current profile. (From Spinella, D.J., Aluminum resistance spot welding: Capital and operating costs vs. performance, SMWC VII, Paper A05, 1996. With permission.)

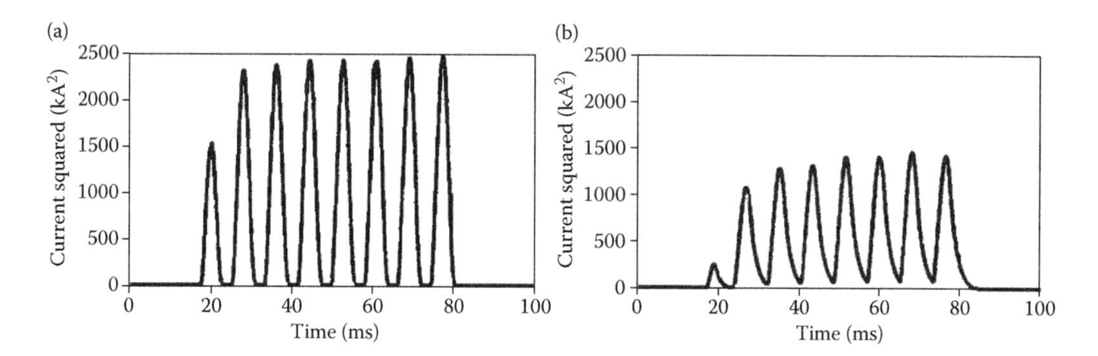

FIGURE 2.30
Comparison of squared current values of a single-phase AC (a) and a single-phase DC (b). (From Spinella, D.J., Aluminum resistance spot welding: Capital and operating costs vs. performance, SMWC VII, Paper A05, 1996. With permission.)

the difference between single-phase AC and DC, as seen in Figure 2.30. The single-phase DC can provide an uninterrupted heating compared to the single-phase AC.

The DC profile shown in Figure 2.29 has a mean current component as well as a ripple current component. The mean current component is of a shape similar to that shown in Figure 2.20, with an upslope and a downslope at the beginning and end of current application. High inductance in the circuit helps smooth the single-phase DC current.[40] As the circuit becomes more inductive, the magnitude of ripple current decreases, whereas the mean current content increases.[51]

One general observation in DC welding is the uneven wear of electrodes of different polarities. The positive electrode tends to wear significantly faster than the negative one, and the electrode life is shorter than that of single-phase AC welding. The energy consumption of single-phase DC supply may be higher than that of a single-phase AC supply when the inductance of the welding circuit is low.[51]

2.5.3 Three-Phase DC

Compared with single-phase AC, single-phase DC avoids the "off-time," and therefore, provides a more uniform heating. However, the large ripple current in a single-phase DC, as seen in Figure 2.29, still implies fluctuation in heat input. An effective and prevalent practice is to use three-phase DC, to significantly reduce the ripple current and supply a more constant heating. The basic process of generating a three-phase DC is very similar to that of single-phase DC. A power convertor transforms the AC of the power supply to a lower voltage, and then rectifies it to produce a DC current. The primary current, however, is a three-phase AC, and it results in a much smoother secondary DC current as shown in Figure 2.19 than a single-phase DC. An actual three-phase DC is affected by the impedance of the welding loop, which often reduces the ripple current and smoothens the output. As the impedance of the secondary loop is affected by the resistance and inductance, which are a strong function of temperature, the resultant DC deviates from the idealized one shown in Figure 2.19. Such an example is shown in Figure 2.31, and it is clearly affected by the welding process as seen in the uneven ripple current portion of the profile. The three-phase rectifier system generally has lower power demand compared to other systems.

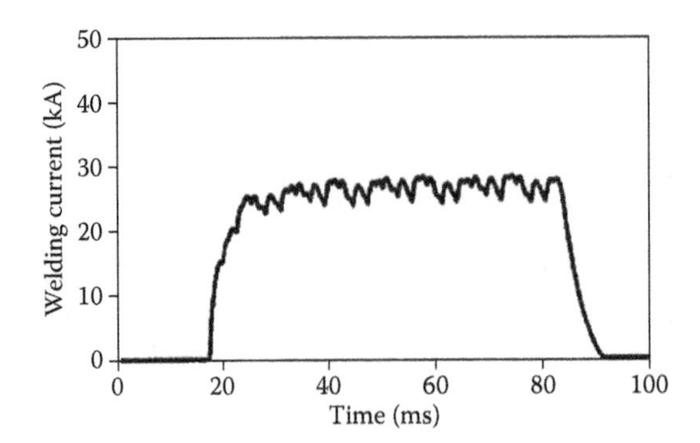

FIGURE 2.31
Three-phase DC weld current profile. (From Spinella, D.J., Aluminum resistance spot welding: Capital and operating costs vs. performance, SMWC VII, Paper A05, 1996. With permission.)

2.5.4 Medium-Frequency DC

The biggest drawback of single-phase AC to AC conversion-based welding equipment is the limited heating rate it provides that is often insufficient when welding zinc-coated steels and Al or Mg alloys. Both single-phase DC and three-phase DC welders lead to improvement over single-phase AC in eliminating "off-time" and providing continuous heating. However, as seen in previous sections, the ripple effect remains and it affects the stability of heating to a weldment.

A significant improvement is possible when welding using MFDC generated by a medium-frequency (MF) transformer. A basic MFDC weld control is illustrated in Figure 2.32. It usually starts with a three-phase AC power supply. Take a 60-Hz, three-phase AC line power as shown in Figure 2.19, as an example. It is first converted to a DC using a six-diode full-wave rectifier. Four high-power electronic switching devices are turned on and off at high frequency, usually between 400 and 1200 Hz, in opposing pairs, and this results in a chopped DC, or square wave AC. The transformer then converts the square wave to a low-voltage, high-current power supply to the secondary circuit. The secondary winding of the transformer then rectifies the AC current and converts it back to DC. Because of the

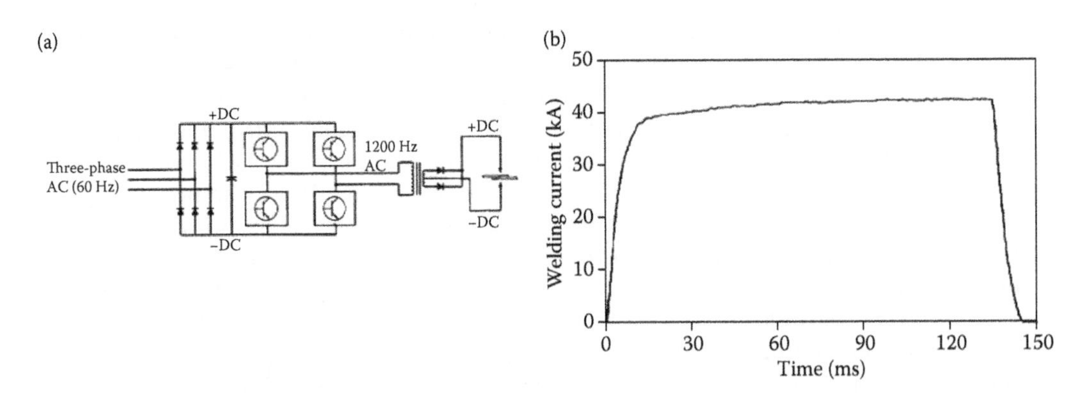

FIGURE 2.32
Control of an MFDC welder (a) and a typical measured MFDC current profile (b). (From Michaud, E. J., Renaud, S.T., A comparison of AC and mid-frequency DC resistance spot weld quality, SMWC VII, Paper A1, 1996. With permission.)

high frequency in the square wave AC, the rectification produces a fairly smooth electric current in the secondary loop, as seen in Figure 2.32. The gradual slopes at the beginning and end of the MFDC current in Figure 2.32b are the results of electric inductance of the welding gun.

The MFDC systems are more expensive than AC systems, but the benefits they bring in certainly justify the higher costs. In general, MFDC systems have the following advantages compared with the AC and other DC systems[41,52]:

- MFDC welding is less demanding on the plant power system. The MF inverter draws reduced and balanced line current from all three phases without phase shift, and operates at a higher power factor and lower inductive losses.
- Because of the higher operating frequency, the reactance of the welding loop only affects the upslope and downslope of the welding current, whereas the peak welding current is limited only by the welding loop's resistance. This results in a smaller transformer suitable for portable and robotic welding guns, which is especially important when welding with large throat, high-impedance guns.
- MFDC systems have significantly lower residual ripples of the welding current than single-phase or three-phase DC systems.
- They provide stable energy input into the weld nugget with the peak welding current almost equal to its RMS value.
- They have smaller controllable units of welding time than other systems, generally in milliseconds, which are a fraction of a regular AC welding cycle. This enables a much faster control of a welding process.

The secondary voltage and current profiles during welding aluminum sheets are shown in Figure 2.33.[53] The shape of voltage waveform is very similar to that of the current except at the beginning of power application. The high voltage and low current at the beginning are the result of high contact resistance possibly due to aluminum oxides existing on the surface. The breakdown of the contact resistive layer is clearly visible in the monitored dynamic resistance profile in the figure. Therefore, although the current profile remains fairly unchanged (often on-purpose), the voltage profile is influenced by the workpiece electrical characteristics. In addition to the changes in electrical resistance of the welding loop, other process characteristics such as expulsion also affect the process signals collected during an MFDC welding except welding current. Because of the simplicity of the process signals of MFDC welding, the evolution of a weld nugget is clearly visible, and interruptions can be captured and remedy can be made if necessary.

The stable heat (current) input an MFDC welder provides enables creating better welds than welding using other transformers. Dupuy and Fardoux[42] directly compared welding using MFDC and AC power sources on hot-dip galvanized, electrogalvanized, and galvannealed low carbon steels. In their study, the hot-dip galvanized steel welding was not affected by the current waveform, whereas the welding current range was significantly enlarged by switching from AC to DC in welding galvannealed and electrogalvanized steels. In the nugget formation tests, however, the DC current is clearly superior to AC current as shown in Figure 2.34 for all the materials studied.

The MFDC welding generally produces larger welds for the same nominal welding current levels. In addition, the weld formation starts earlier and the electrode life is longer than using other current waveforms. The electrodes used in MFDC welding last considerably longer than in AC welding, and the use of current stepping amplifies the difference. The

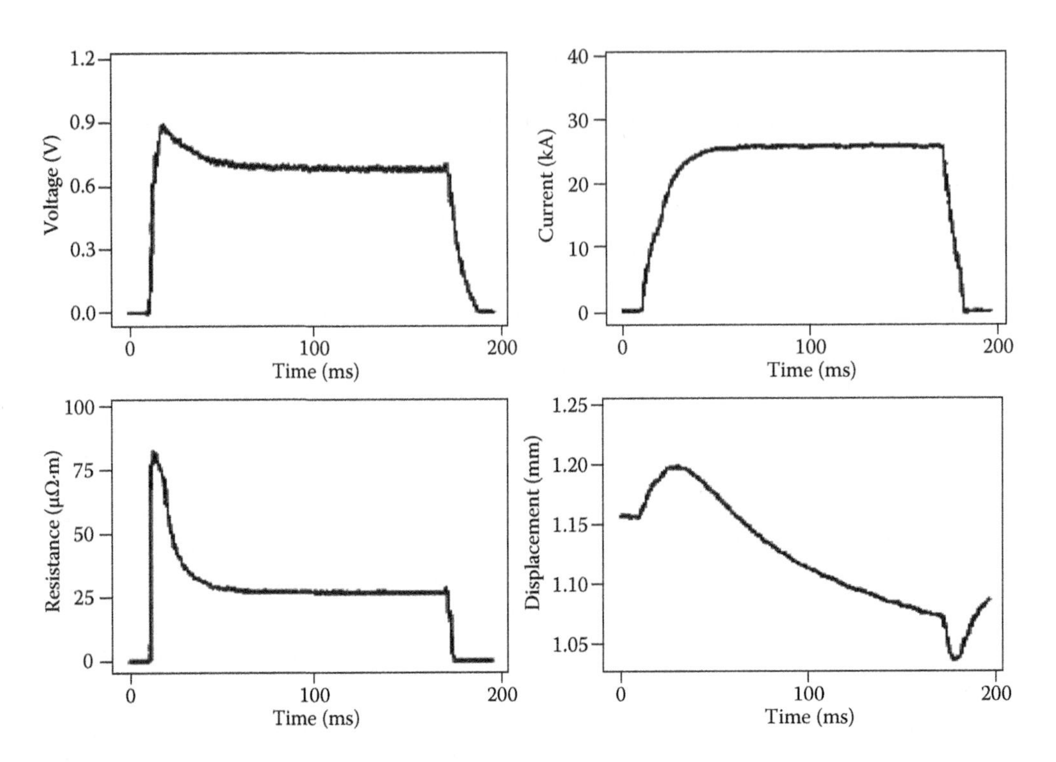

FIGURE 2.33
Signals monitored during an MFDC welding Al sheets. (From Osman, K.A. et al., A comprehensive approach to the monitoring of aluminum spot welding, SMWC VII, Paper B2, 1996. With permission.)

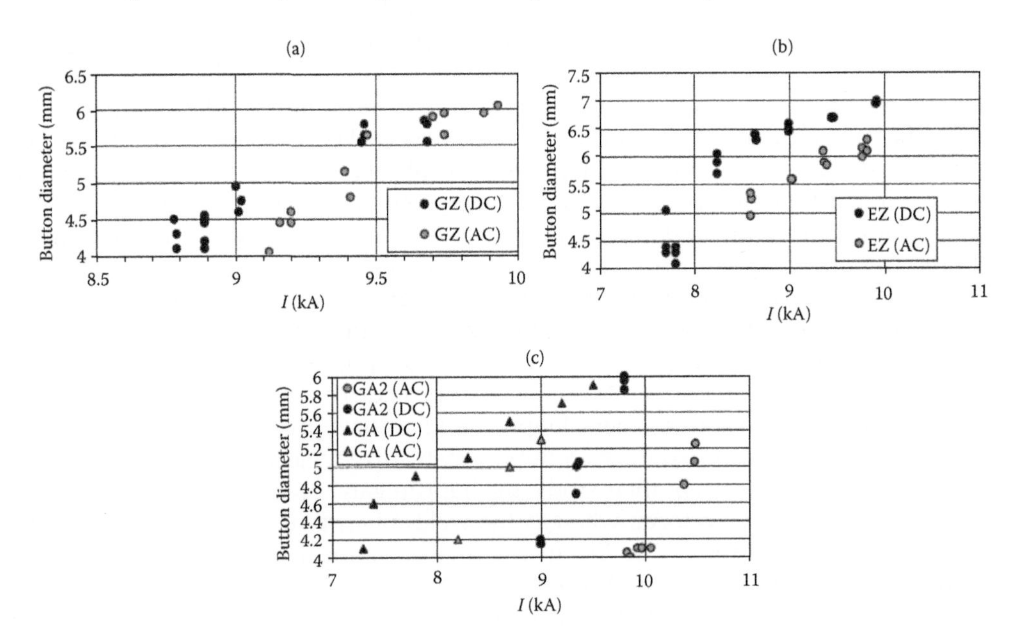

FIGURE 2.34
Nugget formation of hot-dip galvanized (a), electrogalvanized (b), and galvannealed (c) steels. (From Dupuy, T., Fardoux, D., Spot welding zinc-coated steels with medium-frequency direct current, SMWC IX, Paper 1-2, 2000. With permission.)

difference can be attributed to both the different heating rates produced by the DC and AC transformers, and the different current distribution in the weld area, mainly caused by the skin effect in AC as shown in Figure 2.22. Therefore, the stable and uniform supply of heat by an MFDC power source appears to be the reason for its better performance compared to an AC welder.

References

1. Zinov'yev, V. E., *Metals at High Temperatures: Standard Handbook of Properties*, translated and edited by V. P. Itkin, Hemisphere Publishing Corp., New York, 1990.
2. *Aluminum: Properties and Physical Metallurgy*, edited by J. E. Hatch, American Society for Metals, Metals Park, OH, 1984.
3. Baker, H., *Physical Properties of Magnesium and Magnesium Alloys*, Dow Chemical Company, Midland, MI, 1967.
4. *ASM Specialty Handbook, Magnesium and Magnesium Alloys*, edited by M. M. Avedesian and H. Baker, ASM International, Materials Park, OH, 1999.
5. *Handbook of Thermophysical Properties of Solid Materials, Vol. 2, Alloys*, edited by A. Goldsmith, T. Waterman, and H. Hirschhorn, Macmillan, New York, 1961.
6. Tsai, L., Jammal, O. A., Papritan, J. C., and Dickinson, D. W., Modeling of resistance spot weld nugget growth, *Welding Journal*, 47s–54s, 1992.
7. Eager, T. E. and Kim, E., Controlling parameters of resistance spot welding, SMWC IV, Paper 17, 1990.
8. Dilthey, U. and Hicken, S., Metallographic investigations into wear processes on electrodes during the resistance spot welding of aluminum, *Welding and Cutting*, 50 (1), 34, 1998.
9. *Resistance Welding: Measurement of the Transition Resistance in Aluminium Materials*, DVS 2929, Deustscher Verband fur Schweisstechnik e.V, Dusseldorf, Germany, 1985 (in German).
10. Li, Z., Hao, C., Zhang, J., and Zhang, H., Effects of sheet surface conditions on electrode life in aluminum welding, *Welding Journal*, 86 (4), 34s–39s, 2007.
11. Newton, C. J., Browne, D. J., Thornton, M. C., Boomer, D. R., and Keay, B. F., The fundamentals of resistance spot welding aluminum, in *Proceedings of AWS Sheet Metal Welding Conference VI*, Detroit, MI, Paper No. E2, 1994.
12. International Institute of Welding, Procedure for spot welding of uncoated and coated low carbon and high strength steels, draft, Document No. III-1005-93, Section 6.
13. Howe, P., Spot weld spacing effect on weld button size, SMWC VI, Paper C03, 1994.
14. *ASM Handbook*, Vols. 1 and 2, ASM International, Materials Park, OH, 1990.
15. *Specific Heat of Solids*, edited by A. Cezairliyan and A. Anderson, Hemisphere Publ. Corp., New York, NY, 1988.
16. Kimchi, M., Gould, J. E., and Nippert, R. A., The evaluation of resistance spot welding electrode materials for welding galvanized steels, SMWC III, Paper C8, 1988.
17. Gugel, M. D., Wist, J. A., and White, C. L., Comparisons of electrode wear in DSC electrodes having different hardnesses, SMWC V, Paper A03, 1992.
18. Wist, J. A. and White, C.L., Metallurgical aspects of electrode wear during resistance spot welding of zinc-coated steels, SMWC IV, Paper B6, 1990.
19. Kimchi, M., Gugel, M. D., White, C. L., and Pickett, K., Weldability and electrode wear during the RSW of various HDG steels, SMWC VI, Paper D02, 1994.
20. Gugel, M. D., White, C. L., Kimchi, M., and Pickett, K., The effect of aluminum content in HDG coatings on the wear of RSW electrodes, SMWC VI, Paper D03, 1994.
21. Wist, J. A., Gugel, M. D., White, C. L., and Lu, F., Electrode–workpiece sticking on electrogalvanized steel, SMWC VII, Paper E2, 1996.

22. Kusano, H., The importance of electrode management in modern resistance welding, SMWC XIV, Paper 3-7, 2010.
23. Howe, P., Resistance spot weldability and electrode wear mechanisms of ZnNi-UC EG sheet steel, SMWC V, Paper A1, 1992.
24. Thornton, M. C., Newton, C. J., Keay, B. F. P., Sheasby, P. G., and Evans, J. T., Some surface factors that affect the spot welding of aluminium, *Transactions of Institute of Metal Finishing*, 75 (4), 165–170, 1997.
25. Ikeda, R., Yasuda, K., and Hashiguchi, K., Resistance spot weldability and electrode wear characteristics of aluminium alloy sheets, *Welding in the World*, 41, 492, 1998.
26. *Resistance Welding Manual*, 4th edition, Resistance Welder Manufacturers' Association (RWMA), 1989.
27. Okuda, T., Spot welding of thick plates: Part 1. The law of thermal similarity, Welding Technique, *Japan Welding Society*, 21 (9), 1973.
28. Fong, M., Tsang, A., and Ananthanarayanan, A., Development of the law of thermal similarity (LOTS) for low indentation cosmetic resistance welds, in *Proceedings of Sheet Metal Welding Conference IX*, Sterling Heights, MI, Paper No 5-6, 2000.
29. Ando, K. and Nakamura, T., On the thermal time constant in resistance spot welding, Report 1, *Japan Welding Society*, 26, 1957.
30. *Welding Handbook*, 2nd edition, American Welding Society, New York, NY, 1942.
31. Agashe, S. and Zhang, H., Selection of schedules based on heat balance in resistance spot welding, *Welding Journal*, 82 (7), 179s–183s, 2003.
32. *Welding Handbook, Vol. 1, Welding Science and Technology*, 9th edition, American Welding Society, Miami, FL, 2001.
33. *Welding Handbook, Vol. 2, Welding Processes*, 9th edition, American Welding Society, Miami, FL, 2004.
34. Matsuyama, K. and Chun, J., A study of splashing mechanism in resistance spot welding, in *Proceedings of Sheet Metal Welding Conference IX*, Sterling Heights, MI, Paper No 5-4, 2000.
35. Spinella, D. J., Aluminum resistance spot welding: Capital and operating costs vs. performance, SMWC VII, Paper A05, 1996.
36. Dilay, W., Rogola, E. A., and Zulinski, E. J., Resistance welding aluminum for automotive production, SAE Technical Paper 770305, 1977.
37. Brown, B. M., A comparison of AC and DC resistance welding of automotive steels, *Welding Journal*, 66 (1), 18–23, 1987.
38. Roth, D., Power conversion for resistance welding machines, SMWC I, Paper 01, 1984.
39. Moss, L. E. and Bates, F. E., Three phase D.C. flash welding, SMWC I, Paper 05, 1984.
40. Roth, D. K. and Hofman, K. A., Alternating current versus direct current in resistance welding, SMWC IV, Paper D19, 1990.
41. Michaud, E. J. and Renaud, S. T., A comparison of AC and mid-frequency DC resistance spot weld quality, SMWC VII, Paper A1, 1996.
42. Dupuy, T. and Fardoux, D., Spot welding zinc-coated steels with medium-frequency direct current, SMWC IX, Paper 1-2, 2000.
43. http://en.wikipedia.org/wiki/Skin_effect. Accessed in Nov. 2010.
44. *Military Handbook: Grounding, Bonding, and Shielding for Electronic Equipments and Facilities*, MIL-HDBK-419A, Vol. 2, 1987.
45. Tawade, G., Bhole, S. D., Lee, A., and Boudreau, G. D., Robust schedules for spot welding zinc-coated advanced high strength automotive steels, in *Proceedings of AWS Sheet Metal Welding Conference XI*, Sterling Heights, MI, Paper No. 6-3, 2004.
46. Hao, M., Osman, K. A., Boomer, D. R., and Newton, C. J., Developments in characterization of resistance spot welding of aluminum, *Welding Journal*, 75, 1s, 1996.
47. Karagoulis, M. J., Control of materials processing variables in production resistance spot welding, in *Proceedings of AWS Sheet Metal Welding Conference V*, Detroit, MI, Paper B5, 1992.
48. Kimchi, M. J., Spot weld properties when welding with expulsion—A comparative study, *Welding Journal*, 63, 58s, 1992.

49. Schumacher, B. W. and Soltis, M., Getting maximum information from welding lobe tests, in *Proceedings of AWS Sheet Metal Welding Conference III*, Detroit, MI, Paper No 16, 1988.
50. Yadav, K., Study of interactions between electrical, magnetic, and mechanical fields in resistance spot welding, MS thesis, University of Toledo, 2005.
51. Spinella, D. J., Aluminum resistance spot welding: Capital and operating costs vs. performance, SMWC VII, Paper A05, 1996.
52. http://www.isomatic.co.uk/MFtransformers.htm. Accessed in Nov. 2010.
53. Osman, K. A., Hao, M., Newton, C. J., and Boomer, D. R., A comprehensive approach to the monitoring of aluminum spot welding, SMWC VII, Paper B2, 1996.

3

Weld Discontinuities

Improper welding practices in choosing welding schedules, electrodes and welder, etc., may create discontinuities in a weldment in various forms. Some of them have merely a cosmetic effect, whereas others may be detrimental to the structural integrity of welds or welded structures. This chapter discusses the commonly observed discontinuities, their formation mechanisms, and possible means to suppress their occurrence.

3.1 Classification of Discontinuities

The discontinuities in resistance spot weldments are either directly visible to bare eyes or have to be revealed through specialized devices or by sectioning the weld, so they are classified as external and internal discontinuities. However, these two types of discontinuities are often related. For instance, an excessive sheet separation (an external discontinuity) may indicate the possible occurrence of expulsion and existence of large voids (an internal discontinuity) in the nugget. The effects of discontinuities on weld quality can be either cosmetic or structural, or both. Although acceptance criteria of weld discontinuities vary drastically in different industries or companies, there are many common features of discontinuities recognized by the practitioners. The American Welding Society has published a number of standards and recommended practices in this regard.[1,2]

3.1.1 External Discontinuities

Visual inspection of welding quality is the dominant quality check conducted in a production environment. It is practiced by experienced personnel mainly on the appearance of weldments, using simple tools such as calipers and magnifying glasses. In addition to apparent improper welding, a number of other design and manufacturing issues can also be revealed by such inspection.

Most external discontinuities can be found in AWS D8.7: *Recommended Practices for Automotive Weld Quality: Resistance Spot Welding*.[1] Figure 3.1 shows a weld that was made too close to the edge of the sheets. Part of the electrode face, when making this weld, was hanging outside of the sheets. Because the sheets are not contained at the edges, the metal was squeezed out of the sheets. This situation is usually created by insufficient flange width, either by design (to save material), or by stamping. Because the occurrence of such a situation is unintentional and fairly random, usually no efforts are taken in adjusting the welding schedules to compensate for the reduced contact area when it occurs. Therefore, the designated schedule may provide excessive current or current density to the actual, smaller-than-designed contact area, which results in an overheating to the sheets. The figure shows a clear sign of melting of the metal hanging out of the sheet edges. Other possible conditions

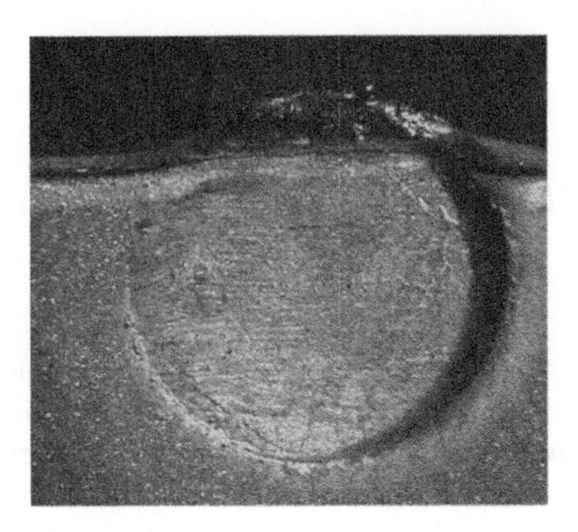

FIGURE 3.1
An edge weld.

that may produce edge welds are misalignment of electrodes and incorrect positioning of welding gun/sheets.

Welds are often made at places different from their designated locations, as shown in Figure 3.2. This is not a problem if the difference is within the tolerance specified by design. However, if large deviations result in edge welds or overlapped welds, as shown in Figure 3.3, measures should be taken to change the practice. As for most problems in assembly, the robotic programming, dimensions of stamped parts, and welding gun should be checked to search for the root causes of the problem. If welds are made too close to each other, either overlapped (Figure 3.3) or with insufficient space between them (Figure 3.4), welding current for the second weld may not be enough to make a quality weld due to the shunting effect. As shown in the figure, the second weld (on the right) appears to have a smaller indentation than the left (first) one in Figure 3.4, similar to comparing the left (second) weld to the right (first) one in Figure 3.3, due to a low heat input, although the same welding schedule was used to make the consecutive welds. The integrity of the welded structures may be compromised as a result.

Indentation is the indent created on the sheet surfaces by electrodes under electrode force during welding. It is a direct indicator of the existence of a weld, and sometimes that of the amount of penetration of a weld. Because it is very difficult to eliminate unless special electrodes and procedures are used, certain indentation is allowed in most practices. AWS

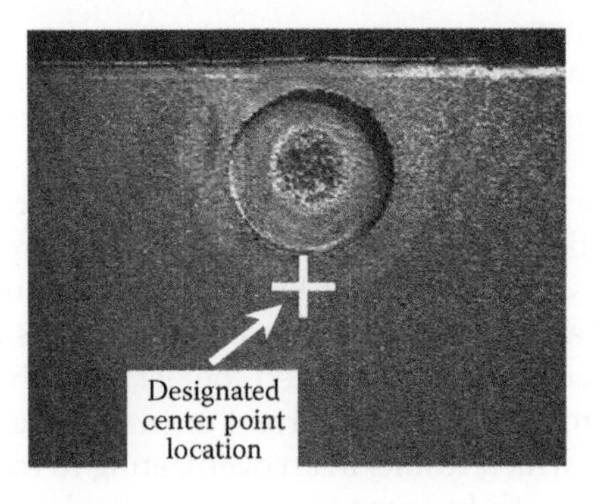

FIGURE 3.2
A mislocated weld.

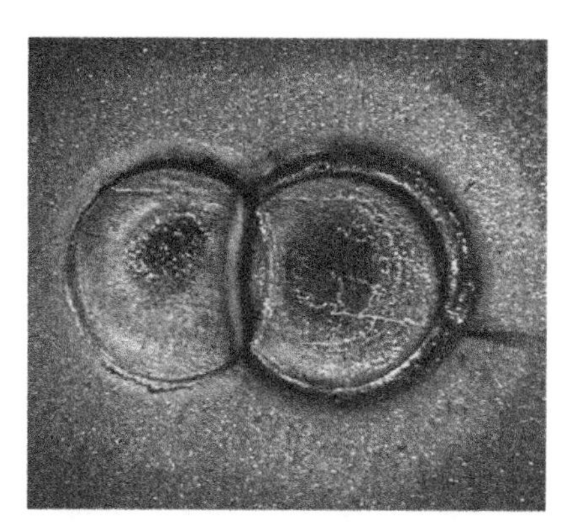

FIGURE 3.3
Overlapped welds.

D8.7 specifies that an indentation up to 30% of the metal stack-up is acceptable. Excessive indentation is not allowed considering its implication in surface finish of the assembled structure, and the load-bearing capacity of the weld. If the surface is exposed to customers of the final products, a deep indentation may create an unfavorable impression. Too much indentation may also create a weak link between a weld and its parent metal sheets because of a reduced thickness in the sheet near the wall of indentation, as seen in Figure 3.5. This is especially true when multiple sheet stack-ups, such as three-sheet stack-ups, are welded, with the outermost sheet as a thin one. The indentation depth is often more than the thickness of the outer sheet, resulting in little joining strength for the sheet attached to the inner ones. In practice, metallographic sectioning is not necessary for measuring indentation, and the weld was sectioned in this figure to make the indentation easier to observe. Some simple mechanical measurement devices, such as dial-gauge meters, will be sufficient for measuring indentation on a weld's surface. Excessive indentation often results from excessive heating, that is, improper welding schedules, and it is usually related to other types of discontinuities. For instance, expulsion and surface melting (possibly resulting in surface cracking and holes) are often associated with large indentation. Excessive indentation may also induce excessive separation, as shown in Figure 3.6. Using correct welding parameters

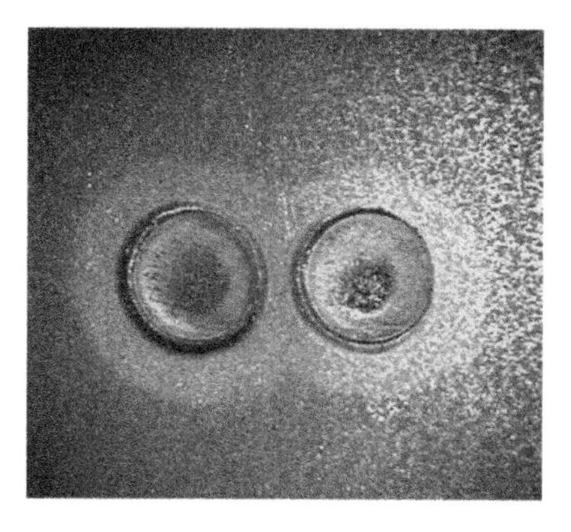

FIGURE 3.4
Potential insufficiently spaced welds.

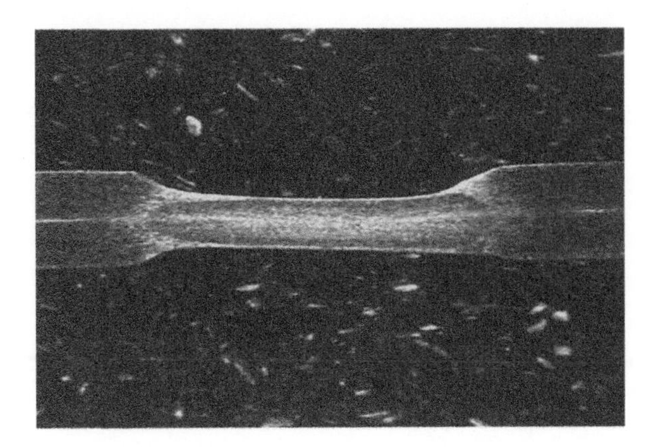

FIGURE 3.5
Excessive indentation.

and welder setup, indentation can be controlled to achieve a weld of sufficient penetration and strength.

Another discontinuity is expulsion, a common phenomenon in resistance spot welding (RSW). Although it is a process rather than a product, it is directly responsible for creating many deficiencies in welding or welds. Its influence on strength of a weld and mechanisms of expulsion are discussed in detail in Chapter 7. In general, expulsion can happen either from the contact interface between electrodes and sheets, as shown in Figure 3.7, or from the sheet faying interface, as shown in Figures 3.8 and 3.9. Expulsion is often associated with other discontinuities, such as excessive internal porosity (Figure 3.17) due to loss of liquid metal. Figure 3.10 also shows a burned hole near the indentation edge. Most welds with expulsion can be identified by visual inspection of the weldment, through the characteristic traces of ejected liquid metal either at the electrode–sheet interface or at the sheet–sheet interface, which requires a peeling action in order to reveal the expulsion trace. Steel welds tend to have a sharp, whisker-like thin layer of ejected and rapidly solidified metal (Figure 3.8). Soft and fan-shaped thin aluminum foils are characteristic of aluminum expulsion (Figure 3.9). For some welds with expulsion, there is no trace of expulsion, such as the ejected and solidified metal around a weld. It is difficult to identify expulsion from such welds without tedious metallographic examination. However, using commonly used sensing devices, such as displacement measurement between electrodes, the occurrence and even magnitude of an expulsion can be accurately recorded, as discussed in Chapter 7.

The hole on the weld's surface in Figure 3.10 is often called a burn-through hole. Associated with this discontinuity are fine whiskers near the hole in the figure. The formation of such holes requires an excessive and often localized heating, and surface melting

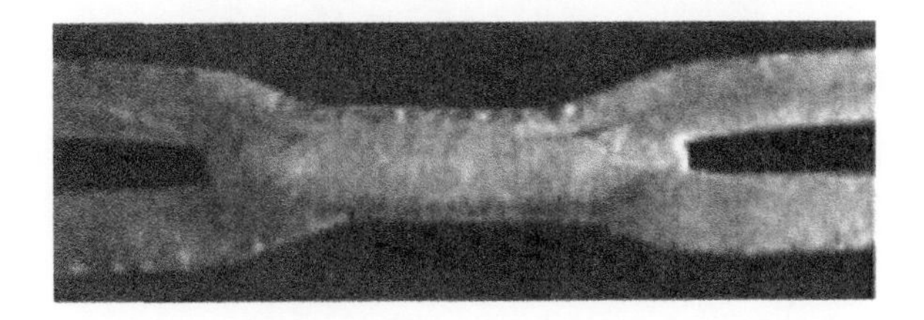

FIGURE 3.6
Excessive separation.

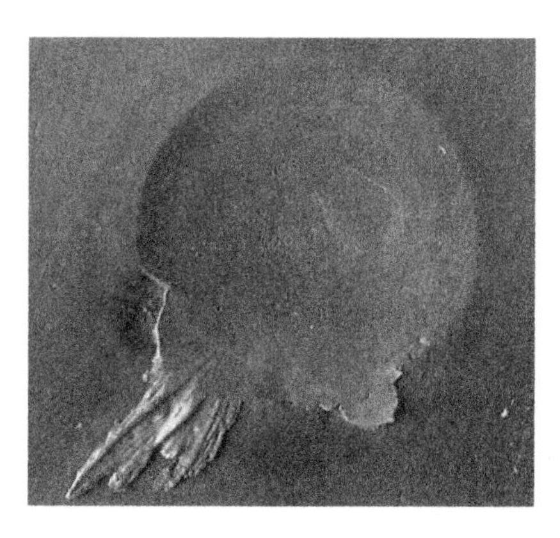

FIGURE 3.7
Expulsion trace on surface (AA5754).

is necessary. Such welds can be expected to have low strength, and remediation is often needed. The electrodes used in making such welds tend to deteriorate rapidly.

Surface cracking is observed in both steel and aluminum welding. Many of such cracks initiate from the melting of surfaces and propagate under stressing due to shrinkage. The crack in a low-carbon steel, shown in Figure 3.11, has a clear sign of surface melting. The crack is large with branches, and it extends into the heat-affected zone (HAZ) from the center. In Figure 3.12, a peeled aluminum weld button shows a crack at the center of the indentation mark. This type of cracking is analyzed in Section 3.3 on its initiation and growth. The nugget of a weldment usually is stressed less than the HAZ under loading, and some discontinuities in a nugget, such as voids, may not affect the strength of a weldment significantly. However, surface cracks, depending on their location and size, may severely weaken the weld and result in premature failure. As shown in Figure 3.29b (Section 3.3), a peeling action may tear off the nugget along the cracks. Cracks often do not appear on the surface as large as they are, or even invisible after welding. However, they may be revealed by a small tensile or peeling action. Some materials tend to be more prone to surface cracking than others, and therefore, special attention is warranted in inspecting surface discontinuities in such materials.

Another type of surface cracking is due to liquid metal embrittlement (LME), as observed in Zn-coated steel welding. LME occurs when the molten metal (zinc in the case of hot-dip galvanized steels) attacks the susceptible steels. It may happen in low-C steels, stainless steels, or advanced high-strength steels. It is usually associated with improper process

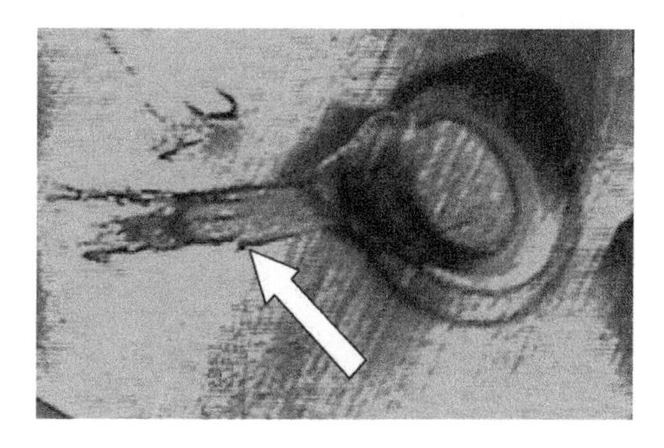

FIGURE 3.8
Expulsion trace at faying interface (low-carbon steel).

FIGURE 3.9
Expulsion at faying interface (AA5754).

setup or welding schedules. The key factors contributing to LME cracking in coated steel welding are electrode misalignment, excessive heat input, excessive electrode wear, and insufficient electrode cooling. The coupling of zinc (from the coating) and copper (from the electrode) may promote LME cracking near the surface of a weld.[3] Figure 3.13 shows a schematic of such a crack. In LME cracking, the induced cracks initiate in the solid phase mainly at grain boundaries that are attacked, such as in the HAZ, which is in direct contact with the electrodes and is at sufficiently high temperature for Zn to melt during welding. Metallographic examination has shown high concentrations of Zn and Cu on the LME fracture surfaces under the microscope. The thermal stresses during solidification are essential to break the embrittled structure. The appearance and location of LME cracks are similar to those of solidification cracks, and they often appear at the bottom of an indentation wall that experiences a large thermal stress during cooling because of the constraint imposed by the electrodes to lateral shrinkage. One test showed that LME cracks do not have a significant effect on the static performance of a spot weld. Their influence on fatigue and impact performances has not been evaluated yet.

Another type of visible discontinuity is excessive sheet distortion after welding. This could occur with improper sheet fit-up or sheet/electrode alignment (Figure 3.14), either

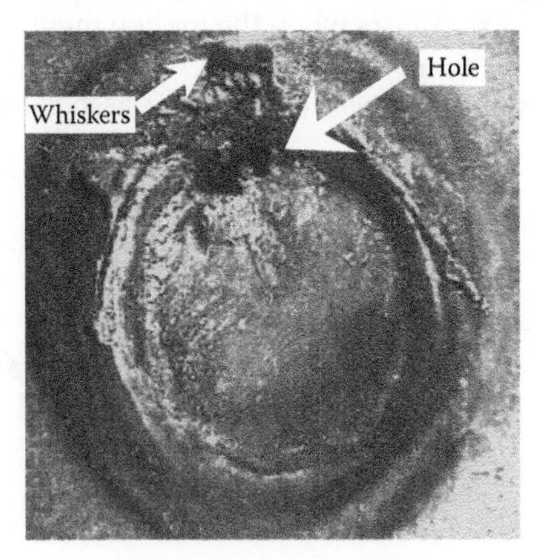

FIGURE 3.10
Holes and whiskers on a weld surface.

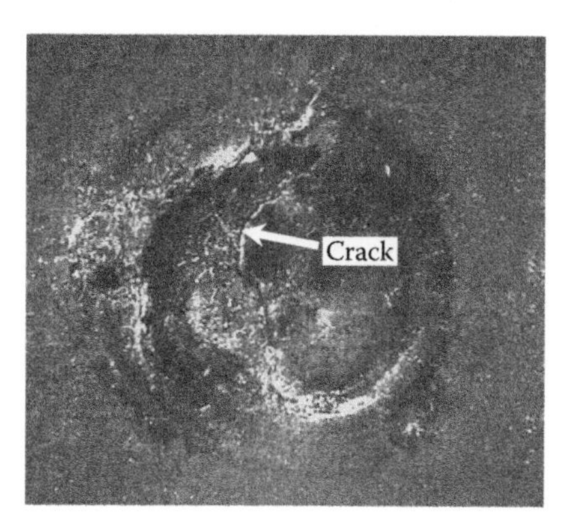

FIGURE 3.11
Surface cracks in a steel weld.

axially or angularly. Although difficult to quantify, such distortion is easy to detect. Remediation can be made by adjusting positioning of the weld gun, using correct electrodes or electrode shanks, replacing worn electrodes, aligning electrodes, or, if necessary, changing designs of the stamped parts. Because of the large weld distortion, the loading mode on the joint may be different from the intended. For instance, an applied tensile–shear loading on a joint shown in Figure 3.14a will result in large bending and tension components, in addition to shear loading at the weld. Large distortion may also affect the dimensional stability of the welded structures.

3.1.2 Internal Discontinuities

Unlike surface cracks, excessive distortion, etc., there are discontinuities that can be revealed only through metallographic examination by cross-sectioning a weld, or by using certain nondestructive devices such as ultrasonic and x-ray imaging. These are internal discontinuities, and they can be divided into two major groups: porosity and cracks. Voids are quite common in weld nuggets, and their formation is discussed in Section 3.2. They result from gaseous bubbles and shrinkage during cooling. As shown in Figure 3.15, there are very few fine voids in an aluminum weld nugget near the fusion line of the HAZ, but numerous large ones near the center. This is related to the distribution and change of temperature in the nugget during welding. Expansion and shrinkage of the weldment

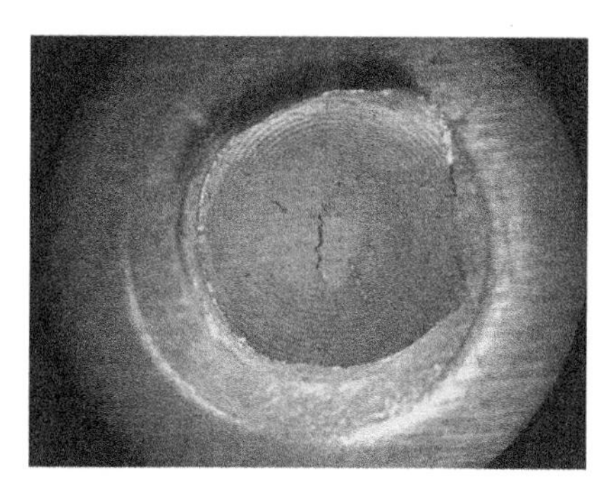

FIGURE 3.12
Surface cracking in an AA6111 weld.

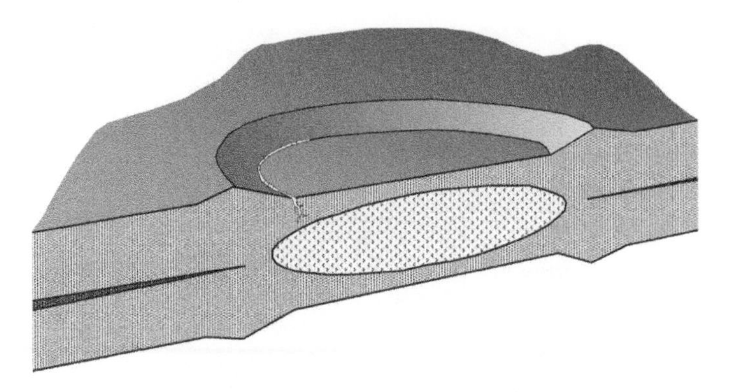

FIGURE 3.13
An LME crack near the bottom of an indentation wall.

and certain associated process such as expulsion are directly responsible for porosity formation. If there is no expulsion during welding, there is no loss of liquid metal from the nugget, and the volume deficit, as well as the amount of voids, will be small. However, very large voids can form in welds that experience expulsion during welding. A sizable void can be seen in Figure 3.16 of a steel weld as the result of expulsion. Similarly, the large voids in Figure 3.17 can be attributed to expulsion from both sides of this magnesium (AZ91D) weld. If the volume or area percent of voids is not too large, and they are located away from the periphery of a weld, the voids may not have a significant impact on the performance of the weld. This is derived from the fact that large stress concentration occurs in the HAZ, rather than in the nugget when a weldment is loaded.

The effect of internal cracks is more complicated. There are several types of cracks in spot welds. The most common one is due to insufficient fusion at the faying interface, as shown in Figure 3.18. In addition to metallographic cross-sectioning, ultrasonic devices can easily

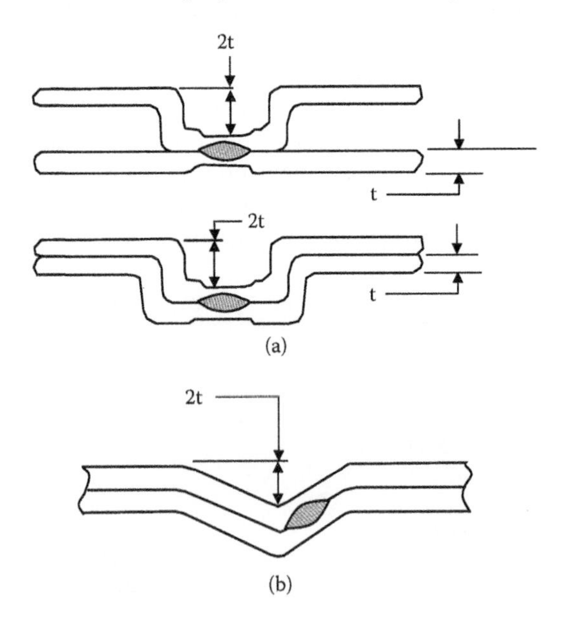

FIGURE 3.14
Weld distortion. (a) Sheets out of alignment. (b) Electrodes out of alignment.

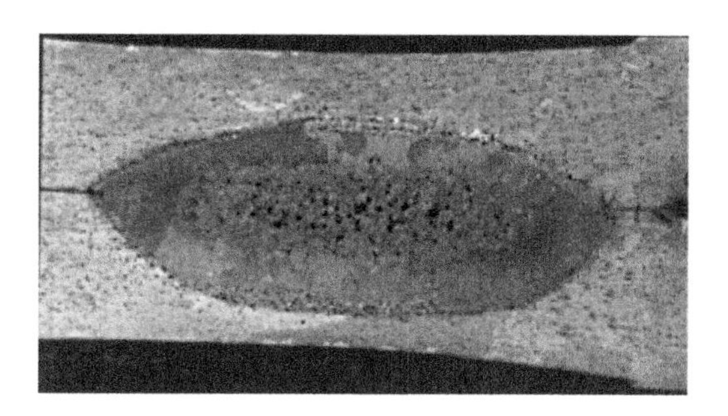

FIGURE 3.15
A weld nugget with voids.

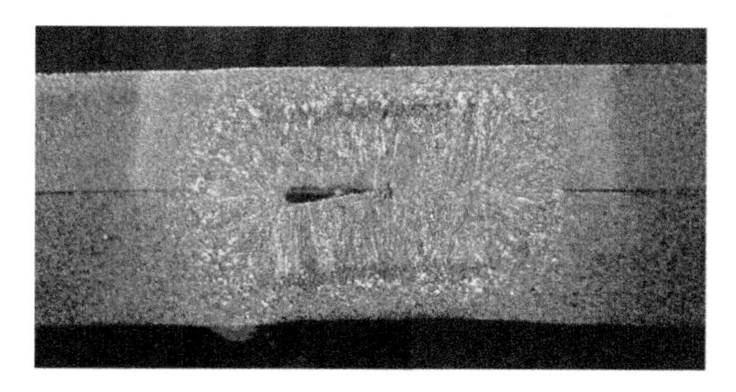

FIGURE 3.16
A mild steel weld with a large void after expulsion.

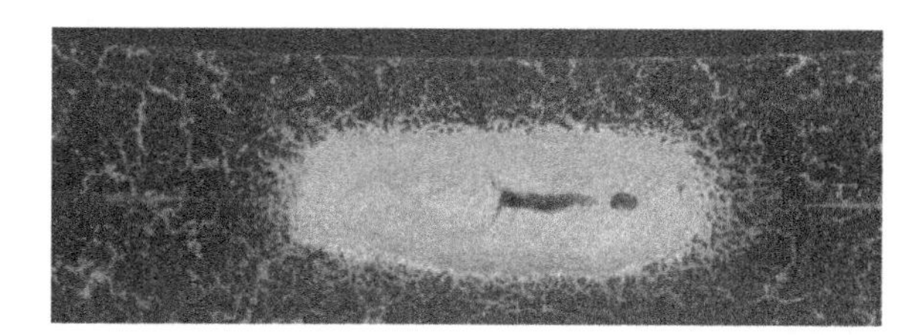

FIGURE 3.17
A magnesium weld with large voids after expulsion.

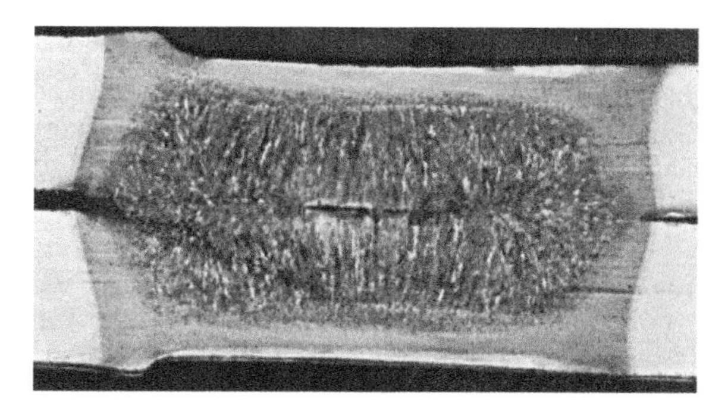

FIGURE 3.18
Cracking in an HSLA steel.

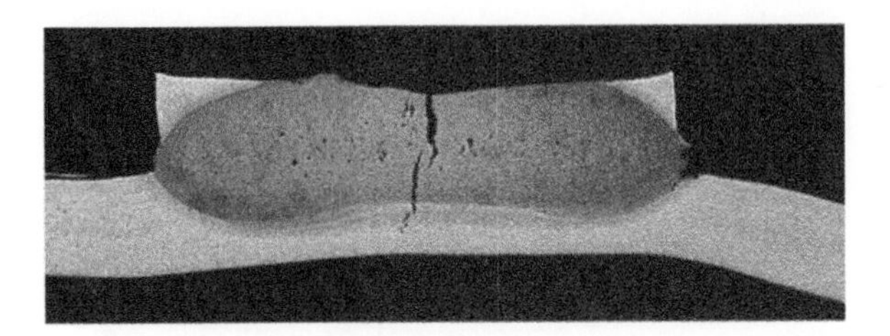

FIGURE 3.19
Cracking in an AA6111 weld.

detect such discontinuities, as they are perpendicular to the incident beams. Such large cracks are associated with small penetration and inferior adhesion at the welded portion. They may behave as cold welds when tested.

Solidification cracking may form under certain conditions. As shown in Figure 3.19, a crack extends from the surface of a weld into its interior, with some voids visible in the nugget. Such cracks may not reduce a weldment's strength if they are confined to the center of the nugget. However, many of them extend to the edges of nuggets and adversely affect the weld quality, as discussed in Section 3.3. The chemistry of the sheet material and electrode geometry are the major factors in such cracking, and corrections may be made based on these considerations.

A less common cracking, as shown in Figure 3.20, occurs in the HAZ, rather than in the nugget. Such cracks originate from the HAZ, due to liquation of grain boundaries and large thermal stresses in the HAZ. Some of them are filled by liquid metal, leaving an outline, rather than an opening/crack. The mechanism and remediation of such cracking are discussed in Section 3.4.

In certain improper welding practices, some adhesion with very little penetration may form at the faying interface (Figure 3.21). Such weldments have very low strength, and they produce interfacial failure when tested with relatively smooth fracture surfaces at the faying interface. They are called cold welds. Ultrasonic devices can easily identify such welds because the joint behaves very similar to single sheets without a welded joint. Insufficient heating is the main reason for cold welds, which may be attributed to insufficient welding current or time, worn electrodes, and shunting. The electrode indentation on surfaces is generally small when a cold weld is made.

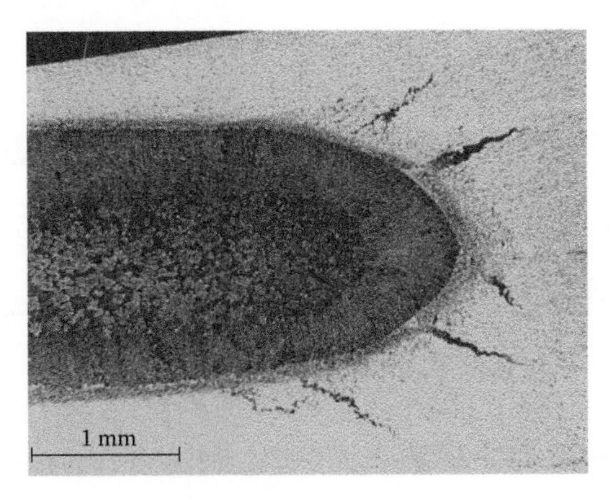

FIGURE 3.20
Cracking in HAZ of an AA5754 weld.

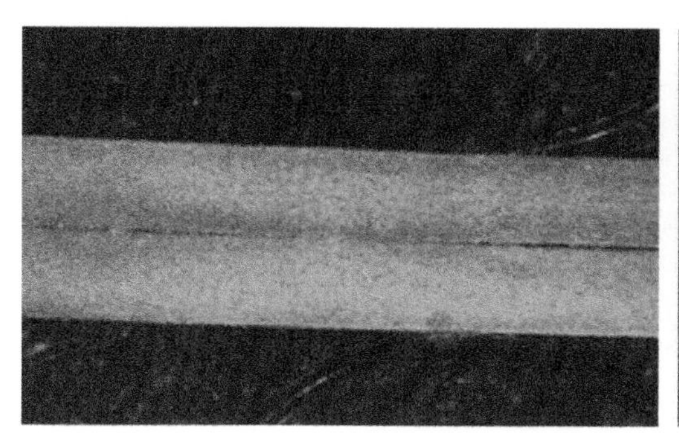
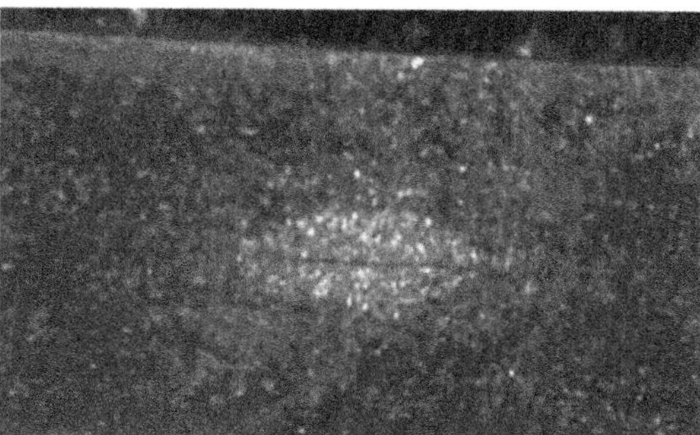

FIGURE 3.21
Cold welds in steel.

3.2 Void Formation in Weld Nuggets

The formation of voids in a spot weld is the result of nucleation and growth processes during solidification of a liquid nugget after the heat source, that is, electric current is shut off. Porosity is more common in Al welds than in steel, mainly due to the differences in their thermal and metallurgical properties. Voids in resistance spot welds have distinctive characteristics in shape, size, and distribution. An example of voids in an Al weld is shown in Figure 3.15. There are usually hardly any visible voids in a nugget near the HAZ, as the periphery of a nugget is the first part to solidify during cooling. Toward the center of a nugget, more voids are observed, and they become bigger in size. In the vicinity of a nugget's center, fewer but larger voids are observed, driven by the tendency of reducing the total energy of the liquid nugget.

There are two types of voids in a solidified weld nugget: one is grown from gaseous bubbles and the other results from solidification shrinkage. These two types of voids have distinctively different appearances in sectioned nuggets, although the two processes often interact with each other in forming the voids. They will be discussed separately for clarity.

3.2.1 Gas Bubbles

The voids from gaseous bubbles usually have smooth surfaces due to free solidification. Under proper conditions, the trace of solidification may be observed directly, as shown in the case of Figure 3.22. In the figure, part of a full interfacial fractured surface of a high-strength, low-alloy (HSLA) steel nugget shows a portion of free solidification, evidenced by the dendritic structure, surrounded by fractured (torn off) surfaces along grain boundaries.

Gas bubbles, existing in the solid states as voids with smooth surfaces, originate from the gaseous pressure in a nugget at liquid state. The gaseous pressure may come from several sources in a liquid nugget. The volatile (mainly light) alloying elements in the liquid may evaporate when overheated and contribute to the pressure in the gaseous phase. Examples of such elements are Zn, which may come from the surface coating, such as in hot-dip galvanized steels, and Mg, which is the main alloying element in the 5XXX series of Al alloys. Table 3.1 lists the alloying elements in AKDQ steel, which has a small amount of light elements. The compositions of aluminum alloys AA5754 and AA6111 are listed in Tables

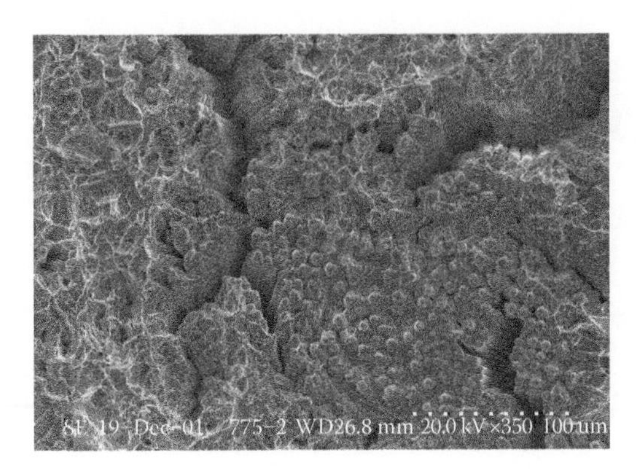

FIGURE 3.22
Surface of an interfacially fractured weld.

3.2 and 3.3. Both contain certain amounts of Mg and Zn. The vapor pressures of some elements are plotted in Figure 3.23.[5] The figure shows that elements such as Zn and Mg in liquid nuggets may impose certain vapor pressures in both Al and steel welds. The pressures are highly dependent on temperature. As welding steels need significantly higher temperature than welding aluminum, it can be expected that light elements such as Zn and Mg existing in a liquid steel nugget as gases contribute to void formation. However, the amount of light elements in steels is not as high as in aluminum alloys, and therefore, bubbles or voids are more frequently seen in Al welds than in steel welds. Another source of gaseous phases in a liquid nugget is the organic elements from surface treatments, greases, etc., on the sheet surfaces before welding. Such elements may be trapped at the faying interface after they are squeezed by the electrodes. The amount of these elements is highly variable, and their contributions to the gaseous pressure in the bubbles are difficult to predict, but can be expected to be small.

The expansion of gaseous bubbles under their internal gas pressure alone is unlikely because any increase in volume (expansion) of a bubble in a liquid nugget dramatically reduces the gas pressure and increases the pressure from the surrounding liquid, as liquid is incompressible. However, small bubbles may merge to form larger ones to reduce the system energy. Assuming the specific surface energy between gaseous bubbles and their surrounding liquid metal is σ, a bubble of radius r has a surface energy $(4\pi r^2)\sigma$. If N identical bubbles merge to form a bigger one, then the new bubble has a surface energy of $(4\pi R^2)$ σ. The size of the new bubble, R, can be obtained by considering the equality in volume, as the total volume of the bubbles can be assumed to remain the same,

$$N\left(\frac{4}{3}\pi r^3\right) = \frac{4}{3}\pi R^3 \tag{3.1}$$

TABLE 3.1
Chemical Composition (wt.%) of an AKDQ Steel

C	Mn	P	S	Si	Cu	Ni	Cr	Mo	Sn	Al	Ti
0.035	0.210	0.006	0.011	0.007	0.020	0.009	0.033	0.006	0.004	0.037	0.001

Source: Provided by National Steel Corp., Livonia, MI. With permission.

TABLE 3.2

Chemical Composition (wt.%) of Commercial AA5754

Mg	Mn	Cu	Fe	Si	Ti	Cr	Zn
2.6–3.6	Max. 0.5	Max. 0.1	Max. 0.4	Max. 0.4	Max. 0.15	Max. 0.3	Max. 0.2

Source: *Automotive Sheet Specification*, Alcan Rolled Products Comp., Farmington Hills, MI, 1994. With permission.

Therefore, $R = N^{1/3}r$. The difference between the total energy of a cluster of small bubbles and that of a large bubble of equal volume can be expressed by the ratio

$$(4\pi r^2\sigma N)/(4\pi R^2\sigma) = (r^2 N)/R^2 = N^{1/3} \quad (3.2)$$

This shows how much the system energy is reduced as the result of merging the small bubbles to a big one. The reduction is not significant when the number of small bubbles is small, and it goes up dramatically when the number is large, as shown in Figure 3.24. Therefore, small bubbles tend to merge into bigger ones to reduce the surface energy of the bubbles, and therefore the system energy. The merge of bubbles requires migration of small bubbles from various locations in a liquid nugget, and an analogy with diffusion can be drawn to understand the process. Similar to a diffusion process, the migration of bubbles is a strong function of temperature and migration path. The bubbles near the nugget center tend to form clusters and larger bubbles at a faster rate than those near the HAZ because temperature peaks at the center. Microvoids entrapped between the dendrite arms during solidification mostly grow from the edge to the center of a nugget and have lower mobility than those near the center. The density of gas bubbles is also lower close to the HAZ than near the center, as the temperature is lower. Therefore, there tend to be scarce, fine gaseous voids near the HAZ and the contact interfaces with electrodes. Large bubbles are usually observed near the nugget center, as this area has the highest temperature during welding and is usually the last part to solidify. Gaseous phases may also migrate to the center as the nugget periphery solidifies and advances to the center.

The size of a void is directly determined by the amount of gases in the bubble and other environmental variables. Consider a spherical gas bubble in a liquid nugget. Its size can be estimated by considering the energy of the bubble, including the surface energy and the volume energy. Assuming the total mass of the gaseous phases in the bubble is m (moles) and the reference radius of the bubble is r_0, then the free energy of the bubble can be written through thermodynamics considerations as

$$\Delta G = 4\pi\left(r^2 - r_0^2\right)\sigma - 3mRT\ln\frac{r}{r_0} \quad (3.3)$$

TABLE 3.3

Chemical Composition (wt.%) of Commercial AA6111-T4

Mn	Mn	Cu	Fe	Si	Ti	Cr	Zn
0.5–1.0	0.15–0.45	0.5–0.9	<0.4	0.7–1.1	<0.10	<0.10	<0.15

Source: *Automotive Sheet Specification*, Alcan Rolled Products Comp., Farmington Hills, MI, 1994. With permission.

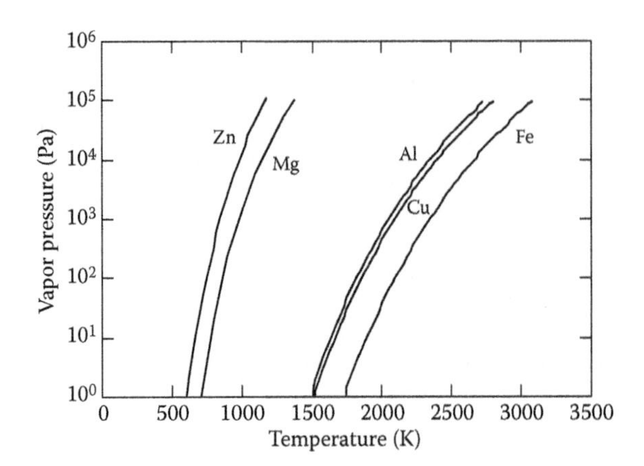

FIGURE 3.23
Vapor pressures of several metals plotting using data from Ref. 6.

This is derived by assuming a closed, constant-temperature system. The stable size of a bubble is achieved when the energy is at the minimum. From $d(\Delta G)/dr = 0$, one can obtain the radius of the stable bubble as

$$r = \sqrt{\frac{3mRT}{8\pi\sigma}} \tag{3.4}$$

It can be seen that the size of a bubble increases with temperature and the amount of gases in it, and decreases with specific surface energy. The amount of gases is determined by the vaporization temperature and solubility of the elements in the liquid metal, and the liquid temperature. Therefore, large amounts of volatile elements in the composition or from other sources and large overheating tend to create large voids.

Voids do not always grow into larger ones. If the thermal stress is high enough, and when there is no sufficient amount of liquid metal to fill in the volume deficit produced by solidification or cooling, voids may grow into cracks to relieve such stresses, as shown in Figure 3.22.

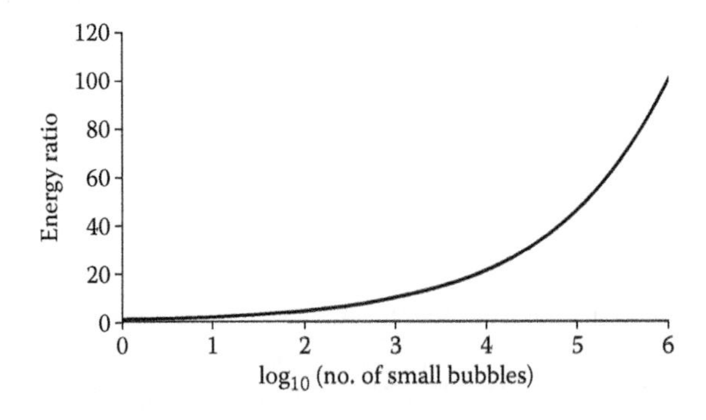

FIGURE 3.24
Surface energy change when small bubbles are merged into a bigger one with equal volume.

3.2.2 Effect of Volume Shrinkage

Under certain conditions, small/microvoids may grow into voids visible under a low-magnification microscope, or even to bare eyes. Besides the void growth driven by the tendency of reducing system energy, the shrinkage of a nugget, through cooling of the liquid phase, solidification, or cooling of the solid phase, contributes significantly to the growth of voids. Figure 3.25 shows the volume changes in pure Fe and an aluminum alloy at various temperature ranges.[5] Cooling at both liquid and solid states generates a reduction in volume, with aluminum shrinking more in the solid state. Solidification induces the largest volume reduction in both materials. For Fe, it is about 3%, and it is about 7% volume reduction in AA5754. The liquid metal and that heated but still in the solid state (such as the HAZ) are easily displaced/deformed by the electrode force during heating. Upon cooling, the volume change is constrained by the electrodes and the cooler parent metal surrounding the nugget area. This restrained shrinkage generates a volume deficit, and the gas-filled bubbles may grow larger in size to occupy the volume. The shrinkage may also generate tensile stresses that may tear off the just solidified structure to form cracks. Figure 3.22 shows solidification cracks along the primary solidification grain boundaries, normal to the fracture surfaces, as the result of solidification shrinkage. Some shrinkage pores grow into microcracks, instead of voids.

Examining an interfacial fracture surface of a DP600 steel weld revealed both shrinkage voids and cracks.[7] Because of the insufficient heat input, the weld failed along the faying interface under fatigue loading. Both macro-voids and cracks are visible in Figure 3.26 as the result of solidification shrinkage. The edge of the large void near the center of the weld in Figure 3.26a is shown in Figure 3.26b. It can be seen that the surface of the void is covered by dendrites formed during cooling, and a crack is located at the border between the free solidification zone and the ductile fractured region. This is another evidence of shrinkage defect due to volume deficit during solidification.

Volume deficit is directly responsible for void formation. Expulsion, which ejects some liquid metal from a nugget, tends to create nuggets with a large volume of voids, as can be seen in Figure 3.16 of a section of a mild steel weld. In addition, large distortion of a weldment tends to produce large voids as the volume deficit created by the deformation

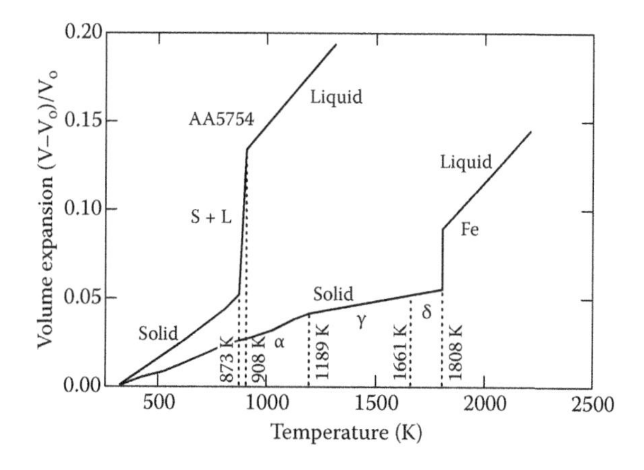

FIGURE 3.25
Calculated thermal expansion for aluminum alloy 5754 and pure iron, in temperature ranges between room temperature and beyond melting point. Part of the data is from Ref. 6.

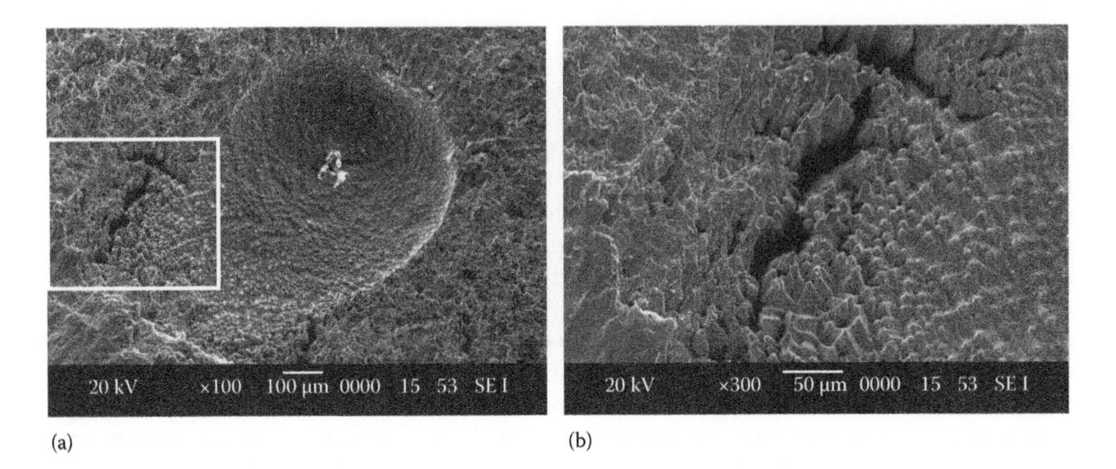

FIGURE 3.26
Views of an interfacial fracture surface of a DP600 weld.[7] (a) Magnified views of center portion of weld nugget, and (b) detail of area near the edge of void as marked in (a).

is large. Because volume change is more significant in aluminum than in steels, aluminum welding tends to create more voids/porosity than steel welding. The shrinkage voids often extend along grain boundaries and form cracks. As the formation of such voids and cracks is often associated with the solidification of the last bit of liquid, they are commonly located near the original faying interface of the sheets, where there is usually a deficit of volume. The sizable shrinkage void in the weld in Figure 3.27a is the result of volume deficit. Cracking along grain boundaries is visible in association with the void (Figure 3.27b).

3.3 Cracking in Welding AA6111 Alloys

An improper welding process may create shrinkage cracks in AA6111, as seen in Figure 3.19. The weld has a clear sign of excessive heating, judged by the nugget width, penetration, and indentation. During welding, the thermal expansion induced by such heating is suppressed by a very large electrode force. The liquid nugget and its surrounding softened solid metal are squeezed to the periphery of the weld by the electrodes at elevated temperature. When cooling starts, the hardening of the solid and solidification of liquid metal

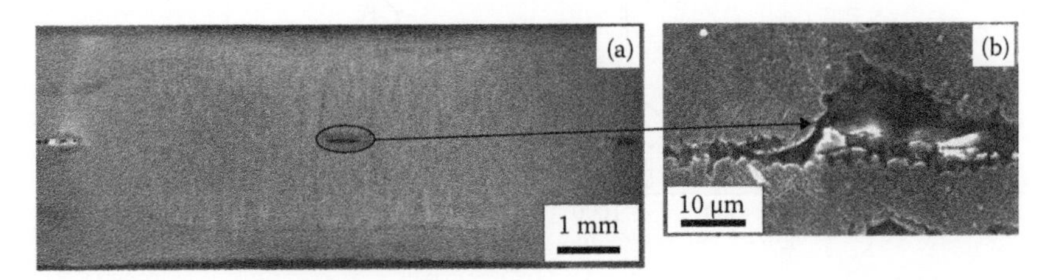

FIGURE 3.27
Cross section of a nugget with a void (a), and details of void (b).[8] (From Wu, K.C., *Welding J.*, 56, 238-s, 1977. With permission.)

start from both the weld edges and the surfaces that are in contact with the electrodes. The volume deficit created due to shrinkage cannot be made up by pulling back the metal previously squeezed out, as it is constrained by the hardened (due to cooling) and newly solidified metal and the electrodes, as shown in Figure 3.28. As a result, openings in the nugget, especially around the center, as it is the last part to solidify, are created. On the surfaces of such openings, the free solidified surface morphology, as shown later in this section, is the evidence of tearing the liquid phase at the middle of the nugget.

The cracks may be visible on the weld surface immediately after welding, or they may appear after certain stressing. In this case, the cracks exist in the vicinity of the center of the nugget, but they do not extend to the surfaces before being stressed. Stressing then makes the cracks propagate to the surfaces. Although a weldment is stressed the most around the HAZ, not the center of a weld, a small amount of stressing will open up the cracks as they are very close to the surface. This is shown in Figure 3.29a, in which cracks, invisible before the weld was peeled, appeared on the surface. With more severe cracking, which appears right after the weld is made, stressing may tear off the weld along the cracks, as seen in Figure 3.29b. About one quarter of the original weld button was totally separated between two branches of cracks. Figure 3.30 shows the fracture of a cracked weld by a similar mechanism as that shown in Figure 3.29b, and it proves that the shrinkage cracking is indeed responsible.

The picture shows the fracture morphology of a weld torn from a cracked weldment. Consider the portion of the fracture surface shown in Figure 3.30a. Figure 3.30c is taken from the edge where the crack surfaces meet in Figure 3.30b. On both sides (crack surfaces) of the edge, freely solidified surfaces are evident, as equiaxed grains are observed (Figure 3.30c-1). The crack tip in Figure 3.30e clearly shows signs of (intergranular) fracture due to crack propagation under stressing, without the trace of free solidification. The top view of the crack tip on the surface (Figure 3.30d), however, reveals a typical columnar structure underneath the weld surface. Therefore, the cracks existed before the specimen was

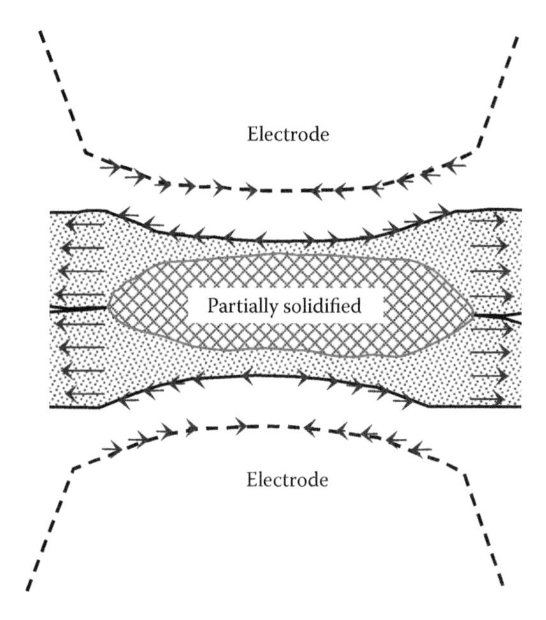

FIGURE 3.28
Lateral stress distribution during cooling.

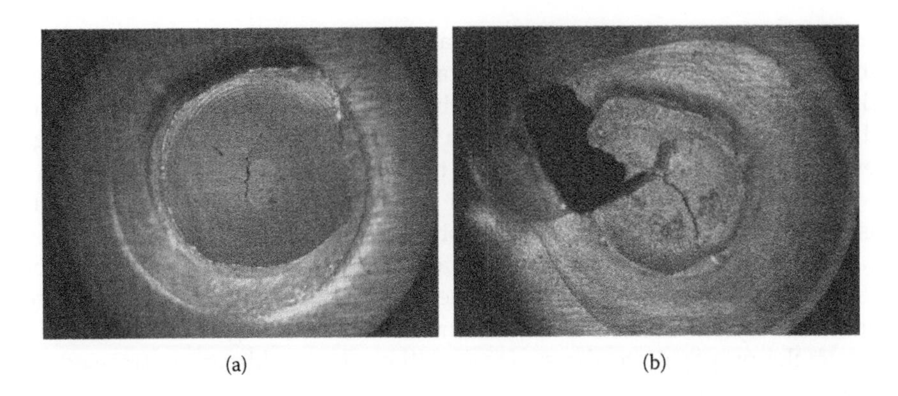

FIGURE 3.29
(a) Cracks appeared after peeling. (b) A torn-off cracked weld after peeling.

peeled, due to solidification shrinkage. The mechanical loading simply widened the cracks and made them propagate in certain directions, such as the sheet thickness direction. The difference in morphology of the fractured surfaces near the weld surface (Figure 3.30d) and near the center (Figure 3.30c) can be understood by drawing an analogy between solidification processes in RSW and casting. Different cooling was experienced by these regions after the electric current was shut off. The weld surface is in close contact with the

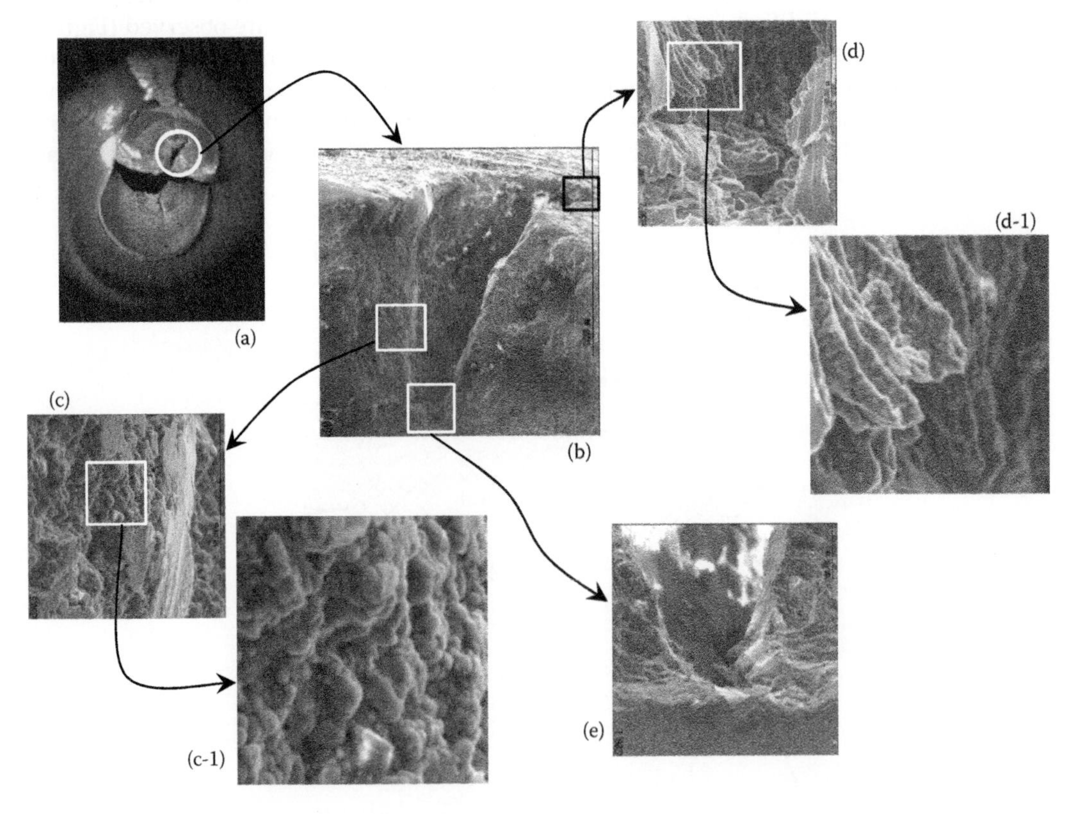

FIGURE 3.30
Fractured surface morphology.

electrode, which is water cooled and works like a heat sink. Therefore, columnar/dendritic structures result, as often occurs near the surface of an ingot. The central portion of the weldment, on the other hand, remains at an elevated temperature for a longer period and equiaxed grains are formed, similar to those observed in a casting. The weld shown in Figure 3.30 was clearly overheated, as evidenced by the expulsion trace seen at the faying interface (Figure 3.30a).

3.4 Cracking in Welding AA5754 Alloys

One of the major problems in fusion welding of some aluminum alloys is the susceptibility to cracking at temperatures close to solidus lines. Such cracking is generally associated with the existence of a wide solidus–liquidus gap, the presence of low-melting-point eutectics (e.g., Al–Cu, Al–Mg, and Al–Mg–Si) or impurities, high solidification shrinkage, large coefficient of thermal expansion, and a rapid drop of mechanical properties at elevated temperatures.[9] Hot cracking during welding at elevated, near-solidus temperatures includes failure of welds (solidification cracking) and cracking in the HAZ (liquation cracking). Cracking has been studied relatively extensively in arc welding aluminum alloys of various working ranges, and high susceptibility to hot cracking during solidification of the liquid pool has been reported. For instance, Lippold et al.[10] investigated hot cracking in two lots of 5083 aluminum alloy (4.28 and 4.78 wt.% Mg, respectively) gas tungsten arc welding weldments. They observed crack initiation and propagation in both the fusion zone and the HAZ, and found that cracking susceptibility depends on the Mg content and weld orientation relative to the rolling direction of the material. Jones et al.[11] reported a fairly low hot cracking susceptibility in continuous-wave CO_2 laser welding and the pulsed Nd:YAG laser welding 5000 Al–Mg alloy series, including AA5754. They observed that the tendency toward cracking increases with Mg content, reaching a peak value at 2 wt.% Mg; high weld strength and low crack susceptibility were found when Mg content was above 4 wt.%. The maximum cracking tendency in Al–Mg alloys was reported earlier at approximately 3 wt.% Mg or 1 to 2 wt.% Mg in two separate works. The observations are consistent with the hot tearing phenomenon in casting of Al alloys, in accordance with the established fact that the peak of hot cracking susceptibility of binary alloys is at about one half of the maximum solubility of the second component in the solid state.

The well-developed theory of hot cracking in fusion welding and casting by the classic works of Pellini,[12] Borland,[13] and Prokhorov[14] helps in understanding cracking in RSW. Cracking in resistance spot welding aluminum alloy sheets was reported first by Watanabe and Tachikawa,[15] followed by Michie and Renaud[16] and Thornton et al.[17] in the 1990s. Watanabe and Tachikawa's investigation reported solidification failure in the nugget or liquation cracking in the HAZ for one of the 5000 series alloys containing more than 5 wt.% Mg. They observed cracking under a wide range of welding parameters and suggested that pre-heating or increasing welding time may decrease thermal stresses, and therefore decrease cracking tendency. The possibility of cracking was also indirectly implicated in spot welding AA5754 by Thornton et al.,[17] Senkara and Zhang,[18] and Zhang et al.[19] as reported in 2000 and 2002. Their work[18,19] systematically studied the cracking phenomenon in the HAZ by considering the thermal, mechanical, and metallurgical aspects of cracking, and proposed measures of containing cracking during resistance spot welding aluminum alloys.

3.4.1 Liquation Cracking in Aluminum Alloys

Cracking in the HAZ of a weldment during welding is related to liquation at the grain boundaries of either the secondary phase or low-melting-point impurities, at subsolidus and supersolidus temperatures of the primary phase. Existing theories of formation and solidification of grain boundary liquid films include equilibrium melting of the vicinity of grain boundaries, constitutional liquation of secondary phases, and the effects of segregation.

A study by Zhang et al.[19] focused on liquation cracking in resistance spot welding 1.6- and 2.0-mm AA5754-O sheets. The base material has a typical rolled structure, as shown in Figure 3.31, with slightly elongated grains of Mg in an Al solid solution, and precipitates of Al_3Mg_2, (Fe, Mn)Al_6, and silicides.

Using correctly selected welding parameters, welds were made with satisfactory appearance. Peel testing confirmed good quality and repeatability of the spot welding process, and a regularly shaped, good-size button is shown in Figure 3.32a.

Despite the normal appearance of a typical weld button after peeling, an amplified side view of the button wall revealed cracks (Figure 3.32a). This was confirmed by optical and scanning electron microscope examination of the sectioned specimens (Figure 3.32b). In these welds, although a certain amount of porosity can be seen on all the cross sections, there are no cracks in these nuggets. Optical microscopy inspection of several welds, however, revealed a few cracks on the sides of the nuggets in many specimens. Cracks were all located in the HAZ. In many cases, the opening of cracks was filled by the parent material, and the cracks were detectable only after etching. In multiple-welded Al strips, cracks were found only on the same sides of the nuggets, with respect to the welding sequence (Figure 3.33a). They are clearly visible from longitudinal cross sections (Figure 3.33b), whereas there are no visible cracks, or only very narrow traces of cracks visible in transverse sections (Figure 3.33c). Cracking always occurred on the opening side of a nugget, not the side adjacent to a previously made weld. If viewed from the longitudinal section, the cracked side (without a previously made weld) has a wider coupon separation gap than the other side, and some squeezed-out material can be clearly seen inside the opened gap between coupons in the vicinity of a nugget (Figure 3.33b).

The appearance of cracks, their locations, and their orientations appear to follow certain rules, as revealed by an examination of a large number of such specimens. The angles

FIGURE 3.31
Microstructure of an AA5754 sheet.

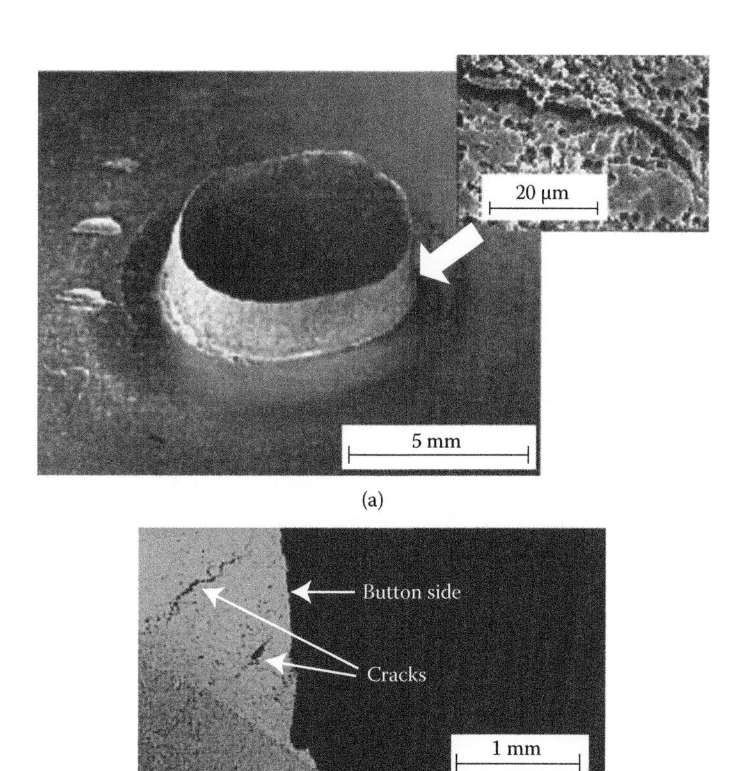

FIGURE 3.32
Appearance of a typical weld button after peel testing: (a) amplified side view of button wall and (b) cross section of the same button.

between the main axes of cracks and the tangents to the fusion line are similar and are equal to about 70° for the specimens examined, as shown by the statistics of the orientations of nearly 80 cracks in Figure 3.34.

A photo of higher magnification shows the intergranular fracture characteristics (Figure 3.35) of the cracks. They initiate near the fusion line in the HAZ and propagate away from the nugget into the base metal. A typical cracking trace is not straight, as shown in Figure 3.35. It tends to follow the grain boundaries, while keeping the overall outward direction. Some of the cracks tilt slightly toward the faying interface as they propagate. Most of them are wide at their bases and become narrower as they enter the base metal. Wide cracks have tree-like structures, that is, large trunks (wide opening of cracks at the bottom) formed from fine roots (grain boundaries). Many of them are fully or partially filled (Figure 3.36). The crack surface has a dendritic morphology (Figure 3.37).

Cracks initiate in a zone where the alloy remains in the solidus–liquidus temperature range during welding at some distance from the fusion line (Figure 3.35). A web of grain boundaries decorated with precipitation is visible around the zone. Grain boundary failure can be clearly seen near the base of the wide-opening cracks. An examination of the

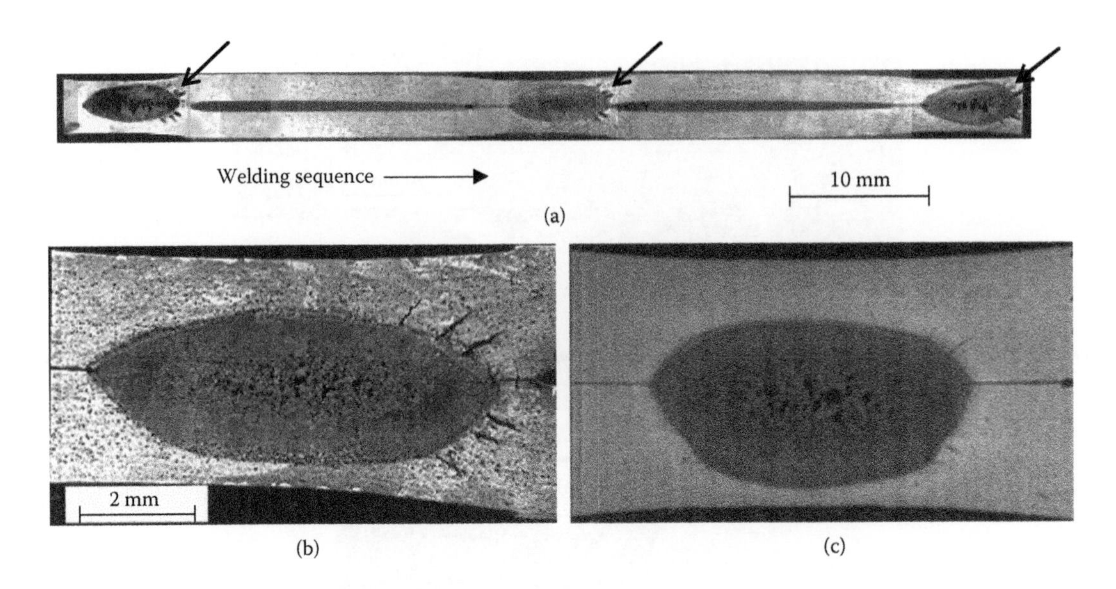

FIGURE 3.33
(a) Longitudinal cross section of a multispot-welded coupon that shows cracks on right side of nuggets. (b) One of the longitudinal cross section's nuggets by higher magnification. (c) For comparison, transverse cross section of a neighboring nugget.

structure in the HAZ reveals large amount of Al_3Mg_2 eutectics at grain boundaries, with a lower melting temperature than the solidus of the alloy. During heating, such precipitates may form a liquid film in the HAZ and adversely affect the strength of the HAZ.

3.4.2 Mechanisms of Cracking

Like other processes of RSW, the formation and propagation of cracks involve interactions of metallurgical, thermal, and mechanical factors. The influence of these factors is analyzed in the following sections.

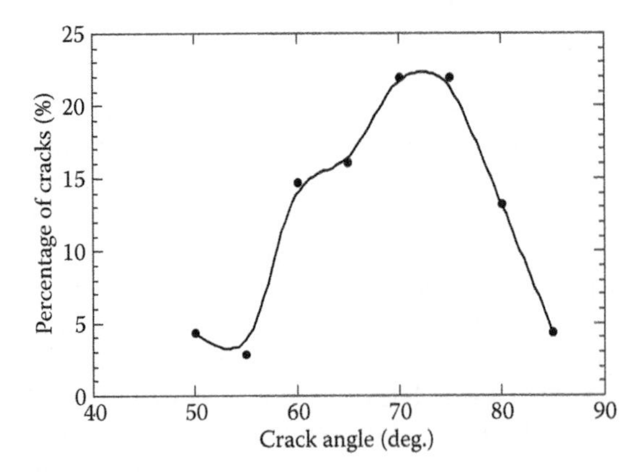

FIGURE 3.34
Statistics of crack angles with respect to fusion line.

FIGURE 3.35
A close look at cracks in HAZ.

3.4.2.1 Metallurgical Effect

The cracking phenomenon in the HAZ during spot welding of AA5754 is consistent with the observations of the susceptibility to cracking of other aluminum alloys containing several percent of magnesium in casting and arc welding.[20] Intergranular characteristics of cracks in the HAZ and dendritic morphology of failure surfaces, as shown in Figures 3.35 through 3.37, are typical features of hot cracking and are evidence of cracking at elevated temperatures. The dendrite structure inside open cracks proves that liquid had to be present at the moment of crack formation, and therefore, it is liquation cracking according to the classification by Hemsworth et al.[21] The microporosity visible in some inclusions of the

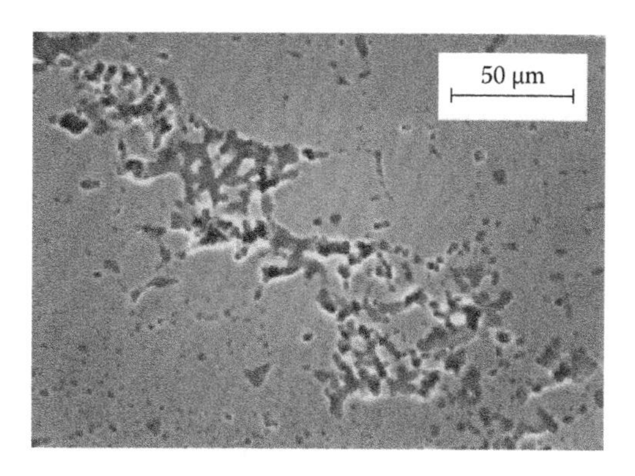

FIGURE 3.36
An almost fully filled crack.

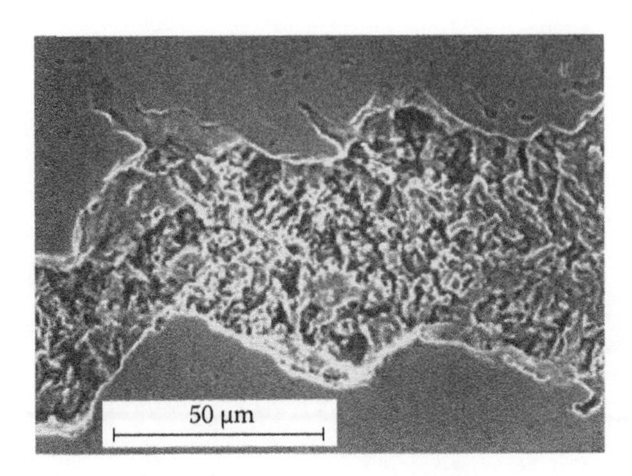

FIGURE 3.37
Intergranular characters of a crack and dendritic morphology of failure surface.

secondary phase serves as additional evidence of the existence of liquid in this part of the HAZ during spot welding.

Generally, there are two possible ways for melting to occur at grain boundaries in the HAZ: at supersolidus temperatures and at subsolidus temperatures. In the heating stage of welding, equilibrium melting of the material near grain boundaries occurs in the part of the HAZ heated to the temperature range between solidus and liquidus (partially melted zone). In addition to partial melting at temperatures above the solidus, liquation of the secondary phase may occur at subsolidus temperatures. During rapid heating, which is a characteristic of RSW, there is no sufficient time to dissolve the Al_3Mg_2 phase in the α-solution matrix, and inclusions of this phase still exist after the alloy is heated over the solvus line. Al_3Mg_2 inclusions melt in the region (next to the partial melting zone) that experiences maximum temperature above the eutectic point, but below the solidus temperature. The existence of liquid below the solidus of AA5754 can also be attributed to other low-temperature melting additions/impurities present in a commercial alloy. The zones around the nugget are schematically shown in Figure 3.38. Structures/zones in the HAZ can be linked to the equilibrium phase diagram via assumed temperature history during RSW. The dynamic effects in the heating/cooling processes, such as overheating and undercooling, may also contribute to hot cracking. For instance, the effective solidus temperature during cooling may be lower than the equilibrium solidus temperature because of the high cooling rate in RSW. This effectively enlarges the temperature range in which the material is weak and susceptible to cracking.

As a result of combined supersolidus and subsolidus melting/liquation, large numbers of grains in these parts of the HAZ are surrounded by liquid during welding. Nearly continuous films of liquid can be formed at the grain boundaries. Therefore, the overall material structure close to the nugget is favorable for crack initiation and growth during the last stage of heating in resistance spot welding Al–Mg alloys.

After the current is switched off, the material cools quickly because of heat transfer through the water-cooled electrodes. The cooling or cooling rate, in addition to several other factors, such as compositional segregation, affects the life of the transient liquid films at grain boundaries. For AA5754, the difference between equilibrium temperatures of liquidus (915 K), solidus (876 K), and eutectic (723 K) temperatures is significant, according to the Al–Mg phase diagram. Lowering of the solidus and eutectic solidification temperatures can be expected due to kinetic effects during cooling. The coexistence of solid and

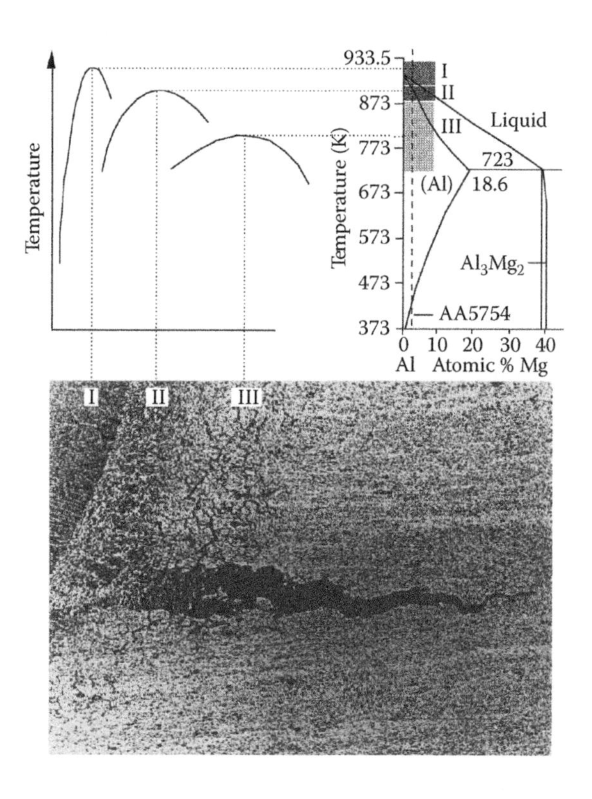

FIGURE 3.38
Schematic diagram of links between equilibrium phase diagram of Al–Mg and solidified structures after welding through possible temperature history of various zones. Zone I, fusion zone; zone II, partial melting zone; zone III, liquation zone.

liquid phases at the grain boundaries is then extended in a relatively wide temperature range during cooling.

High heating and cooling rates, as well as a high temperature gradient in the weldment due to the nature of Joule heating, are the thermal characteristics of spot welding. Therefore, liquid films at grain boundaries may exist for an extended time at elevated temperatures, as shown in the work of Randhakrishnan and Thompson.[22,23] This results from the concentration gradient in the liquid due to rapid solidification, which effectively lowers the solidification temperatures of liquid parts with higher (than equilibrium) Mg concentration.

In general, the metallurgical characteristics of spot welding of AA5754 create favorable conditions for tearing the structures in the HAZ if sufficiently high stresses are developed during welding.

3.4.2.2 *Thermomechanical Effect*

Besides the metallurgical effect, thermomechanical factors also play an important role in the initiation and subsequent propagation and growth of cracks. This section is devoted to describing the mechanisms of crack formation by qualitative thermal and mechanical analyses, using simplifying assumptions, because an accurate calculation/analysis is neither possible nor necessary, due to the complicated nature of the processes and their interactions involved in cracking.[19]

As seen in Figure 3.33a, cracks appear on the leading sides of nuggets that coincide with the welding sequence in a multispot-welded strip. No significant traces of cracking have been found on the trailing sides of the nuggets. Experiments have shown that certain welding conditions may create cracking on both sides of a single weld, as shown in a longitudinal section in Figure 3.39b. The distinct cracking characteristics between the single- and multispot welding strongly suggest that besides the weakened structure due to the metallurgical process, the thermal–mechanical factors play an important role in the initiation and subsequent propagation of cracks.

The initiation and propagation of cracks can be estimated based on the possible thermal history and loading and constraining conditions imposed on a weldment. A thermal–mechanical analysis can start with multispot-welded nuggets because of the characteristic appearance of cracking only on the same sides of the nuggets. From Figure 3.39a, it can be seen that there is a solid material flow on the cracking side, which forms a notch and generates a large sheet separation. On the contrary, no solid deformation and separation can be observed on the left side. This uneven deformation or unsymmetrical geometry is the consequence of uneven thermal–mechanical loading on the two sides during welding.

The loading on the top sheet of a weld can be simplified as a pressure due to the electrode force, a pressure from the liquid nugget, interaction with the other sheet, and constraining from the other sheet and the previous weld. Figure 3.40 shows the outlines of two cracked welds: one is from Figure 3.39a, cut from a multiwelded strip, and the other from a single weld (Figure 3.39b). The figures also show the material flow, as well as the large sheet separation on the free end side.

In Figure 3.40a, the open side (right side) and the constrained side (left side) experienced different stress histories during heating and cooling, and this is the dominant reason for different cracking behaviors on the two sides. The stresses developed on the sides, represented by blocks A and B as taken from the figure near the fusion zone in the HAZ, can explain why the weld shown in Figure 3.39a has cracks on the right side, not on the left. It also helps to understand that cracks may appear on both sides of a single weld without constraining from either side during welding. The possible stress states during heating and cooling are shown in Figure 3.41.

3.4.2.3 Thermal Stress during Heating

The solid phase in Figure 3.40a between the electrode and the liquid nugget tends to expand during heating. Because of the difference in constraining between the left and right sides, the stress developed during heating is different on the two sides. Because of

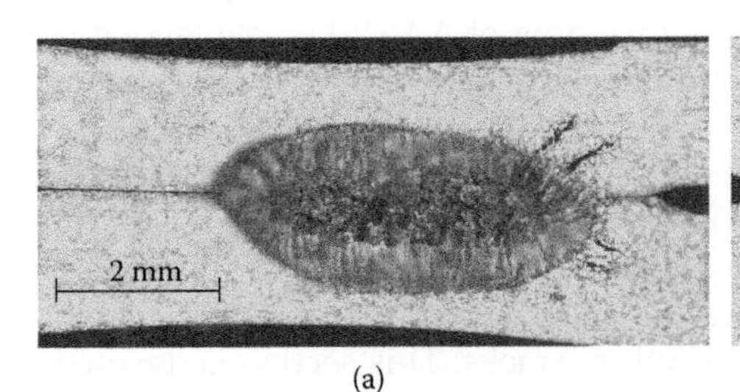

(a)

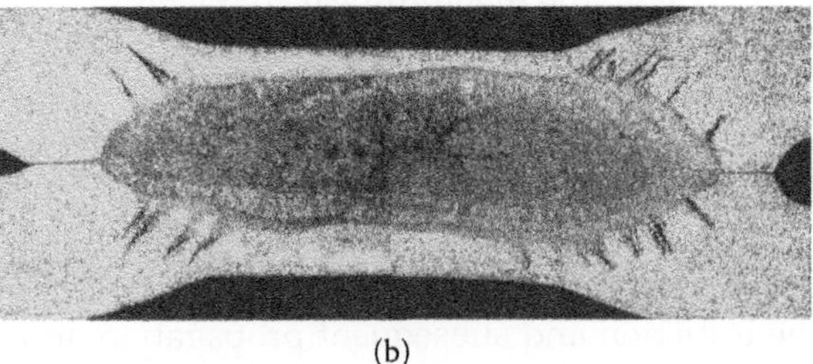

(b)

FIGURE 3.39
(a) Longitudinal cross section of a multiwelded specimen. (b) Longitudinal cross section of a single-welded specimen.

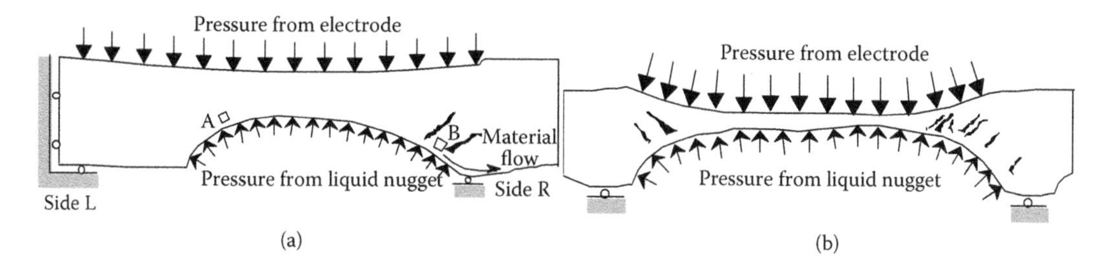

FIGURE 3.40
(a) Loading and constraining conditions in half of weldment of a multiwelded specimen. (b) Loading and constraining conditions of a single-welded specimen. Blocks A and B in (a) are chosen for stress analysis.

temperature gradient, as shown in Figure 3.42, in the HAZ, material near the fusion line tends to expand more than that away from it. In block A, large compressive stress is developed in the direction parallel to the isotherms during heating or expansion, due to constraints from the solid phase of the sheets on side L. In block B on side R, however, there is very little compressive stress buildup near the nugget, because the solid phase can expand more freely and the sheets are free to separate, which effectively releases the stress in B. The stresses are schematically shown in Figure 3.41 (heating).

3.4.2.4 Thermal Stress during Cooling

When the electrical current is shut off, heating is terminated and rapid cooling starts. In fact, cooling is provided by the electrodes and workpiece during the entire period of welding, but its rate is much higher after the electric current is shut off. Because of the high cooling rate, which is similar to one in a quenching process, tensile stresses of similar magnitude develop in blocks A and B within a short period (Figure 3.41, cooling). For B, the solid material displaced during heating can hardly be sucked back due to the lower temperature (or higher yield strength) and the constraint from the solid phase in the HAZ, especially in the region previously squeezed out.

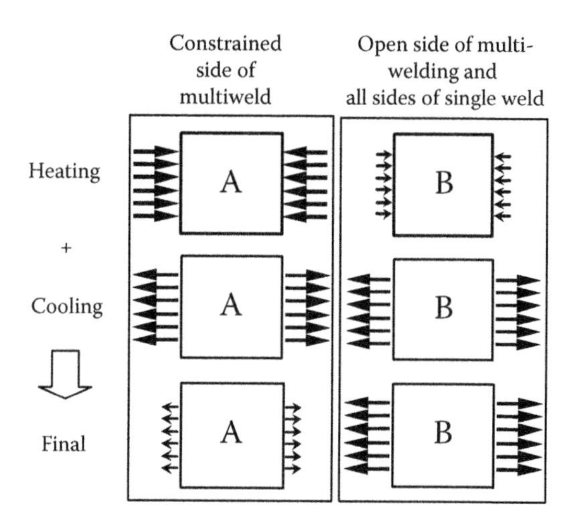

FIGURE 3.41
Possible stress development during heating, cooling, and combined final stress state in material blocks A and B taken from Figure 3.40a.

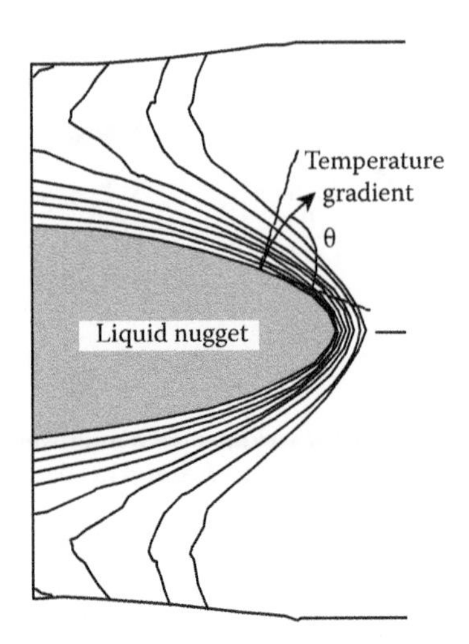

FIGURE 3.42
Temperature distribution by a finite element analysis at the moment when heating is stopped. Temperature is highest in liquid nugget beyond melting point.

The final stress states at A and B can be obtained by superimposing the stresses developed in heating and cooling, respectively. As seen from Figure 3.41, the resultant stress in B is tensile, and that in A is very small, whether tensile or compressive. The net tensile stress in B during cooling is directly responsible for crack initiation in the region. In the case of a single weld, the stress state on both sides of the nugget is similar to that in block B of a multispot weld, which has a final stress in tension.

In resistance welding, under simultaneous heating and cooling, high-temperature gradients are developed and produce thermal stressing in the HAZ. Using a finite element model,[24] a temperature distribution at the end of heating is generated, as shown in Figure 3.42. Because of cooling through the electrodes and workpieces, isotherms in the HAZ appear dense around the nugget, which is similar to the results in the work of many others, such as that by Gupta and De.[25] The figure also shows that the cooling temperature gradient turns toward the radial direction from the fusion line.

The magnitude of thermal stresses can be approximated by a simple calculation. Right after electrical current is shut off, strain parallel to the fusion line in Figure 3.42 can be estimated by considering the elastic and thermal strain components:

$$\varepsilon = \frac{\sigma}{E} + \alpha \Delta T \tag{3.5}$$

where ε is the total strain, σ is the stress; E is Young's modulus, α is the coefficient of thermal expansion, and ΔT is the temperature drop during cooling. During welding, the deformation of any location in the HAZ is affected by its temperature history and the constraints imposed by its surroundings. Considering two extreme cases, the stresses may be

– Lower bound (free to shrink):

$$\sigma = 0 \text{ and } \varepsilon = \alpha \Delta T \tag{3.6}$$

– Upper bound (fully constrained):

$$\sigma = E\alpha(-\Delta T) \text{ and } \varepsilon = 0 \tag{3.7}$$

The actual value of stresses in the HAZ is in the range [0, $E\alpha(-\Delta T)$], depending on the degree of constraint. Assuming $E = 70$ GPa and $\alpha = 33 \times 10^{-6}$ K^{-1} for aluminum, the stress range is plotted in Figure 1.10 as the shaded area. For a temperature drop $\Delta T = -50$ K, the stress range in the HAZ at a temperature just below the solidus temperature is 0–115 MPa in tension. In this range, the thermal stress developed near the solidus during cooling in the solid phase may possibly be high enough to initiate cracks, considering the strength (ultimate tensile strength) shown in Figure 1.10. The rapid drop of ductility near the melting point may also contribute to crack initiation.

The tensile stress developed during cooling is approximately along the tangent of the isotherm near the nugget. Therefore, the orientation of a crack, which is normal to the direction of maximum tension, is approximately normal to the tangent, that is, along the temperature gradient. The orientations of the cracks with respect to the fusion line, in Figure 3.34, are close to the normal direction of the fusion line. The slight deviation of the measured orientation from the normal direction (70° rather than 90°) can be attributed to other effects, such as liquid pressure and the applied vertical load by electrodes. The change in the crack orientations when they propagate away from the fusion line can be explained by the changes in the isotherms, or the temperature gradients, as shown in Figure 3.42.

From the analyses of thermal-mechanical factors discussed above, one can conclude that constraining on the weldment creates compressive stress during heating, which compensates (reduces) the tensile stress generated during cooling. Thus, the side of a nugget to the welded end in multispot welding has less chance of cracking. In contrast, the unconstrained side of the nugget cannot stay compressive during and after heating and will develop tensile stress during cooling, which may result in cracking.

The existence of a tensile stress around a crack root is evidenced by the observed partially filled crack openings, as shown in Figure 3.36. If a crack is present during heating, it may be easily filled because of the high pressure inside the liquid nugget, which is discussed in Chapter 7, and then the structure is healed. As shown in Figure 3.43, under tension, the material near the nugget may open up and form cracks. Between the roots of the cracks and the liquid nugget, there is a zone of low-strength solid, or even a mixture of solid grains and liquid–grain boundary films. Because of the high liquid pressure in the nugget, which can exceed 100 MPa, as shown in Chapter 7, the liquid in the nugget may be pushed at a high speed through the mushy zone into the opening and may follow the extension of cracks. Another source of filling material is the liquid eutectic existing at the grain boundaries near the fusion line. However, eutectics alone are not sufficient to fill the crack openings because of their limited amount (about 6% estimated for AA5754 according to the equilibrium phase diagram). The filling/healing process is not always possible because it depends on the resistance of the mushy zone to liquid metal flow/penetration. If a crack is formed during heating, or close to the nugget, it has a higher chance of being filled. A crack that forms when cooling starts, or that initiates far from the fusion line, has less chance to be filled, or it can be only partially filled. This is supported by a WDX analysis that revealed slightly higher Mg content inside the filled area, but no significant difference between the chemical composition of the material filling the gaps and that of the base metal. This means that the filling is dominated by ejection of liquid metal from the liquid nugget during heating.

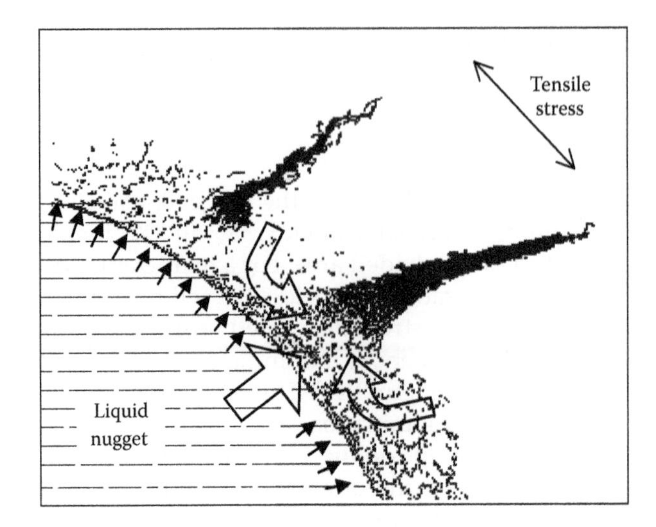

FIGURE 3.43
Schematic illustration of crack filling mechanisms. Arrows indicate possible paths for liquid being ejected into the crack.

3.4.2.5 Influence of Other Factors

Along with stresses developed due to thermal expansion and shrinkage, mechanical loading and constraining caused by other processes also contribute to crack initiation and propagation. The most influential one is probably the effect of liquid pressure generated during melting and volume expansion in the nugget onto its surrounding solid. As shown in Figure 3.40, liquid metal imposes a high pressure onto the solid in the direction normal to the fusion line. The resultant force from the liquid pressure is also partially responsible for workpiece separation when mechanical constraint is insufficient. It was observed that when expulsion happens, the ejection of liquid metal pushes or bends the solid at a high speed. Obviously, this induces a straining at a high rate in the solid surrounding the liquid nugget, and the region under the greatest strain is that near the edge of the nugget, either in the state between solidus and liquidus or where liquid films exist at grain boundaries, with brittle structures of low strength. According to the theory proposed by Prokhorov,[14] hot cracking may occur when the structure is in the brittle temperature range.

From the above observation and analysis, it can be seen that constraining, either from the prewelded end or from the surrounding solid and the electrode, is beneficial for suppressing cracking in spot welding aluminum alloys. Based on the cracking mechanisms discussed, methods are proposed to minimize or eliminate cracking, as described in the following section. The conclusions drawn here can be used in practice to prevent cracking and therefore to make aluminum alloys more suitable for structural use.

3.4.3 Cracking Suppression

A series of experiments were designed and conducted under various constraining conditions, including combinations of various specimen sizes, electrode geometries, and welding sequences. A domed electrode (type A) with a face diameter of 10 mm and a radius of 50 mm for the domed face, which is often used in resistance spot welding aluminum alloys, was used in the experiments. Two additional electrode geometries with drastically different faces were also used. They are type B electrode (with a domed face, called domed electrode hereafter) and type C electrode (with a flat face, called flat electrode hereafter); see Figure 3.44. With very different contacting characteristics, these two types of electrodes represent two extreme cases. The cracking results are examined and represented using the average number and total length of cracks on each weld, obtained over at least three specimens (replicates).

3.4.3.1 Effect of Specimen Width and Electrode Geometry

Three coupon widths (25, 40, and 90 mm) were welded with both flat and domed electrodes and the resulting cracking tendency are given in Figure 3.45. The increase in coupon width can significantly reduce cracking tendency for both domed and flat electrodes. Because specimen width provides a natural constraining on bending or separation of sheets, there is less sheet distortion in wider specimens. In addition, wider specimens provide larger mass to heat, therefore lowering the peak temperature and the thermal stress level. Thus, wider specimens have a lower tendency of cracking.

As for the effect of electrode geometry, specimens welded using domed electrodes show a significantly larger number of cracks and crack lengths than the ones using flat electrodes, for all coupon widths. Metallographic examination indicates that with small-radius domed electrodes, significant material flow occurred and large separation was produced at the edge of the nugget. Compared with flat electrodes, domed electrodes provide a smaller contact area between the electrode/workpiece and the sheets. This can yield two effects: (1) the localized electrode contact produces a higher and nonuniform contact stress at the interface, with less constraining to the specimen distortion, and (2) domed electrodes provide a higher current density and heating rate (or temperature gradient). As a result, larger distortion and cracking tendency are expected. In addition, a domed electrode tends to cause expulsion, which also contributes to cracking.

3.4.3.2 Effect of Welding Sequence

In multispot welding, two welding sequences, sequence A (1, 2, 3) and sequence B (1, 3, 2), were compared, where the numbers are the labels of the weld spots along a longitudinal strip, which also indicate the run order of welding. The difference of sequence B from sequence A is that the third weld was made after two welds were made on the two ends of a strip, rather than in consecutive order. Figure 3.46 shows the cracking tendency (by the total number of cracks) for these two welding sequences with various coupon widths. The total number of cracks is reduced to zero when welds are made with constraining from

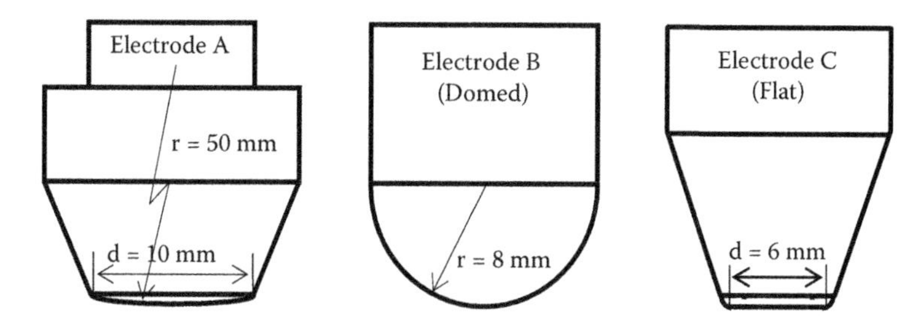

FIGURE 3.44
Three types of electrodes (A, B, and C).

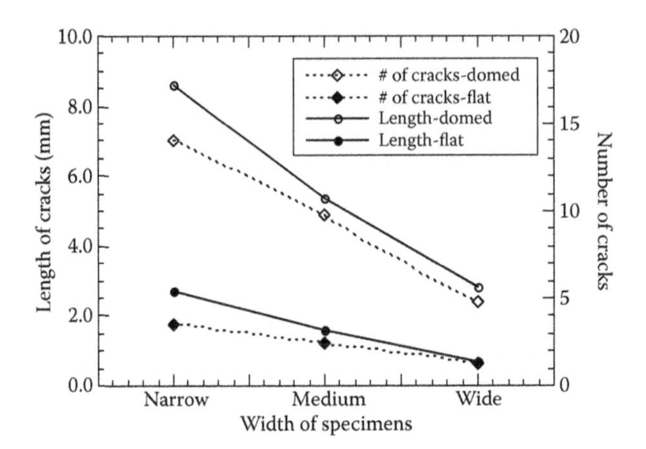

FIGURE 3.45
Influence of specimen size and electrode type on cracking in single-spot welding.

neighboring welds (sequence B), compared with the case of having welds only on one side (sequence A). It was also found that in the both-end constrained case, the cross-sectional area of the nugget was slightly smaller. Apparently, the constraint in sequence B reduces the sheet distortion, and possibly provides an additional electric current shunting path and reduced current density.

3.4.3.3 Effect of Washer Clamping

To further prove the cracking mechanisms and provide a means of cracking suppression, additional constraint was applied, before welding, to the electrodes using two steel washers. Two aluminum coupons were put together with the washers, one on each side. The whole stack-up was held by two C-clamps, as illustrated in Figure 3.47. The electrodes, aligned with the washers, were then put through the openings of the washers onto the sheets, and a weld was made between the sheets inside the clamped area. The comparison

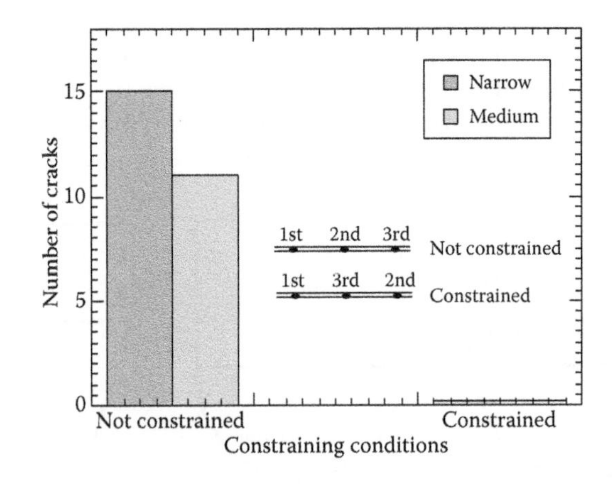

FIGURE 3.46
Effect of constraint imposed by neighboring welds. Welds were made using domed electrodes. Number of cracks was counted on third weld.

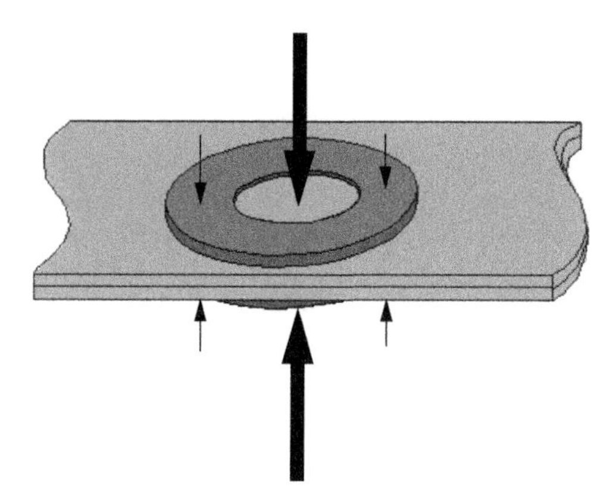

FIGURE 3.47
A schematic diagram for using constraining washers during welding. Thin arrows indicate clamping forces on washers. Electrode forces are shown as thick arrows.

of cracking tendency between specimens welded with and without washers, as well as between medium and narrow coupons, is shown in Figure 3.48. When washers were used, the number of cracks and the total crack length were significantly reduced for both medium-size and narrow coupons, but the reduction was more significant with medium-size coupons. A metallographic examination shows that with washer restraining, the distortion is significantly reduced: only localized sheet separation is observed, within a ring-shaped opening area surrounding the nugget, and it diminishes at the places where sheets are clamped by the washers. Virtually no material flow at the faying interface near the nugget was observed. No significant difference was found when different-size washers were used.

3.4.3.4 Effect of Current Shunting

In order to understand the influence of current shunting due to washer clamping, a comparison was made between welding with and without a thin layer of insulator inserted

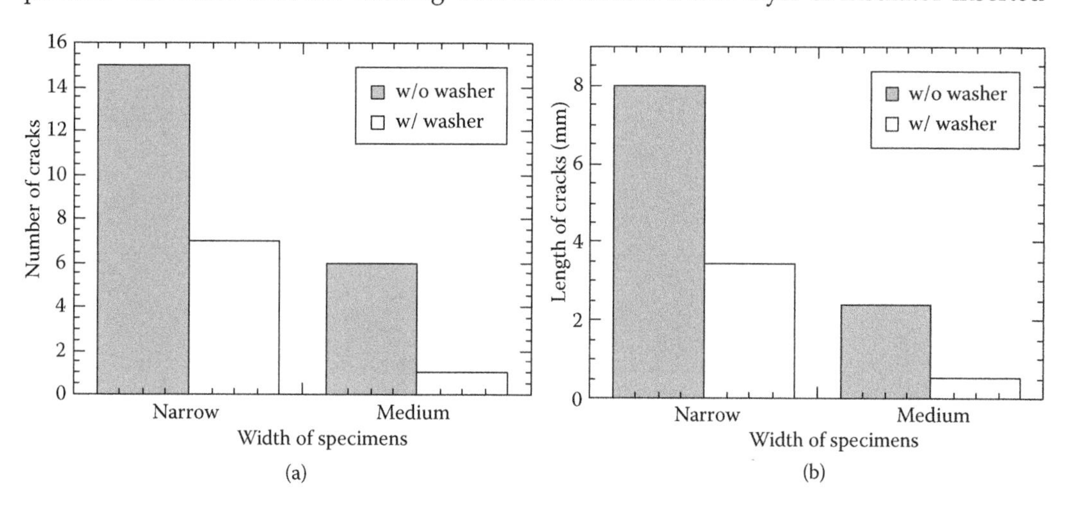

FIGURE 3.48
Effect of constraining with washers on cracking tendency. Domed electrodes were used.

between the two sheets to be welded. As shown in Figure 3.49, the effect of insulator on cracking is slightly different when different electrodes are used. When domed electrodes were used, the use of insulators reduced the cracking tendency for all specimen widths tested, and the same trend can be seen for flat electrodes with wide specimens, but no effect was detected in flat electrodes with narrow specimens. It appears that the shunting effect does exist in most cases, except for the flat electrode narrow specimen case. The local deformation also shows that it is different with or without insulators. When domed electrodes were used, welds with insulators show significantly less distortion at the periphery of the nugget than the ones without insulators. However, metallographic examination shows that there is no significant difference in nugget size and penetration. The insulator inserted between the specimen sheets changes the contact resistance and affects heat generation, yet the exact effect of insulators is not clear.

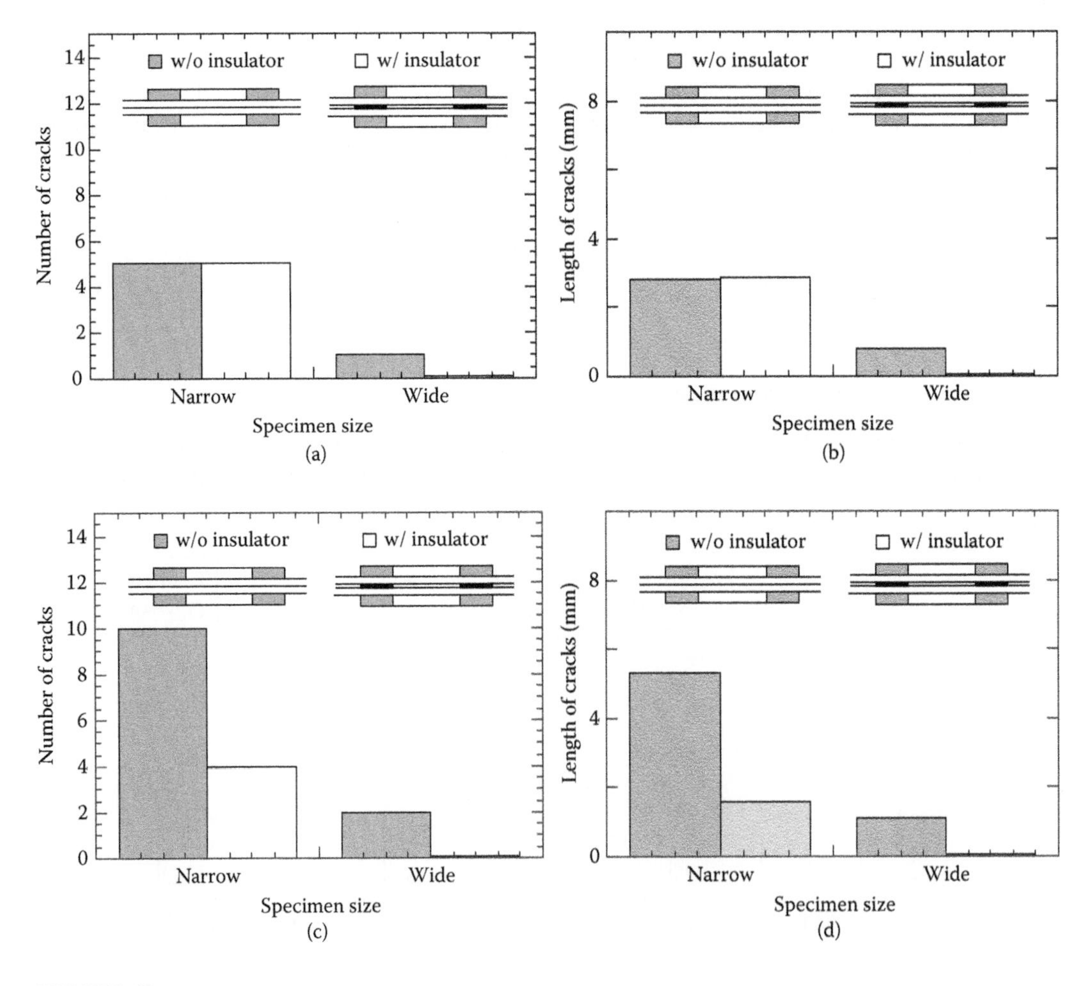

FIGURE 3.49
Influence of insulating between washers and workpieces on cracking. Panels (a) and (b) are for welding using flat electrodes, and (c) and (d) are for welding using domed electrodes.

TABLE 3.4

Dependence of Cracking Tendency on Welding Conditions

Variable	Influence on Cracking
Electrode type	Domed (+), flat (–)
Increase in specimen width	(–)
Constraining by washer	(–)
Constraining by neighboring welds	(–)
Using insulator with washers	(0)

Note: (+) = increasing cracking tendency; (–) = decreasing cracking tendency; (0) = no clear influence on cracking.

The major findings described above are summarized in Table 3.4. The interactions of variables are not listed in the table. In general, this investigation proves that constraining of the sheet deformation around a nugget is an effective means to control cracking in welding aluminum alloys.

References

1. AWS D8.7: *Recommended Practices for Automotive Weld Quality—Resistance Spot Welding*, American Welding Society, Miami, FL, 2004.
2. AWS D8.9: *Recommended Practices for Test Methods for Evaluating the Resistance Spot Welding Behavior of Automotive Sheet Steel Materials*, American Welding Society, draft, 2005.
3. Jiang, C., Thompson, A. K., Shi, M. F., Agashe, S., Zhang, J., and Zhang, H., Liquid metal embrittlement in resistance spot welds of AHSS steels, American Welding Society Annual Convention 2003, Detroit, MI, Paper 9A, April 2003.
4. *Automotive Sheet Specification*, Alcan Rolled Products Comp., Farmington Hills, MI, 1994.
5. Senkara, J., Zhang, H., and Hu, S.J., Expulsion prediction in resistance spot welding, *Welding Journal*, 83, 123-s, 2004.
6. Lide, D. R., *Handbook of Chemistry and Physics*, 74th edition, CRC Press, Boca Raton, FL, 1993–1994.
7. Ma, C., Chen, D. L., Bhole, S. D., Boudreau, G., Lee, A., and Biro, E., Microstructure and fracture characteristics of spot-welded DP600 steel, *Materials Science and Engineering A*, (485), 334–346, 2008.
8. Khan, M. S., Bhole, S. D., Chen, D. L., Biro, E., Boudreau, G., and van Deventer, J., Welding behaviour, microstructure and mechanical properties of dissimilar resistance spot welds between galvannealed HSLA350 and DP600 steels, *Science and Technology of Welding and Joining*, 14 (7), 616–625, 2009.
9. Anik, S. and Dorn, L., Metal physical processes during welding—Weldability of aluminum alloys, *Welding Research Abroad*, XXXVII, 41, 1991.
10. Lippold, J. C., Nippes, E. F., and Savage, W. F., An investigation of hot cracking in 5083-O aluminum alloy weldments, *Welding Journal*, 56, 171-s, 1977.
11. Jones, J. A., Yoon, J. W., Riches, S. T., and Wallach, E. R., Improved mechanical properties for laser welded automotive aluminum alloy sheets, in *Proceedings of AWS Sheet Metal Welding Conference*, VI, Detroit, MI, Paper No. B2, 1994.
12. Pellini, W. S., Strain theory of hot tearing, *The Foundry*, 80, 125, 1952.
13. Borland, J. C., Suggested explanation of hot cracking in mild and low alloy steel welds, *British Welding Journal*, 8, 526, 1961.

14. Prokhorov, N. N., Theorie und Verfahren zum Bestimmen der technologischen Festigkeit von Metallen beim Schweißen, *Schweißtechnik*, 19, 8, 1968.
15. Watanabe, G. and Tachikawa, H., Behavior of cracking formed in aluminum alloy sheets on spot welding, presented at 48th Annual Assembly of IIW, Stockholm, IIW Doc. No. III-1041-95, 1995.
16. Michie, K. J. and Renaud, S. T., Aluminum resistance spot welding: How weld defects affect joint integrity, in *Proceedings of AWS Sheet Metal Welding Conference VII*, Detroit, MI, Paper No. B5, 1996.
17. Thornton, P. H., Krause, A. R., and Davies, R. G., The aluminum spot weld, *Welding Journal*, 75, 101-s, 1996.
18. Senkara, J. and Zhang, H., Cracking in multi-spot welding aluminum alloy AA5754, *Welding Journal*, 79, 194-s, 2000.
19. Zhang, H., Senkara, J., and Wu, X., Suppressing cracking in RSW AA5754 aluminum alloys by mechanical means, *Transactions of ASME—Journal of Manufacturing Science and Technology*, 124, 79, 2002.
20. Rosenberg, R. A., Flemings, M. C., and Taylor, H. F., Nonferrous binary alloys hot tearing, *Transactions of American Foundrymen's Society*, 68, 518, 1960.
21. Hemsworth, B., Boniszewski, T., and Eaton, N. F., Classification and definition of high temperature welding cracks in alloys, *Metallurgy*, 5, 1969.
22. Randhakrishnan, B. and Thompson, R. G., A model for the formation and solidification of grain boundary liquid in the heat-affected zone (HAZ) of welds, *Metallurgical Transactions A*, 23A, 1783, 1992.
23. Thompson, R. G., Inter-granular liquation effects on weldability, in *Weldability of Materials*, edited by R. A. Patterson and K. W. Mahin, ASM International, Metals Park, OH, 1990, 57.
24. Zhang, H., Huang, Y., and Hu, S. J., Nugget growth in spot welding of steel and aluminum, in *Proceedings of AWS Sheet Metal Welding Conference VII*, Detroit, MI, Paper No. B3, 1996.
25. Gupta, O. P. and De, A., An improved numerical modeling for resistance spot welding process and its experimental verification, *Transaction of ASME—Journal of Manufacturing Science and Engineering*, 120, 246, 1998.

4

Mechanical Testing

4.1 Introduction

Mechanical testing is an important aspect of weldability study. Such testing is either for revealing important weld characteristics, such as weld button size, or for obtaining quantitative measures of a weld's strength. As a weld's strength generally refers to its capability of withstanding both static and dynamic loads, mechanical testing of a weldment can be static or dynamic, and it can be either instrumented or not instrumented. Although the dynamic strength of spot welds is recognized as an important quality index because of its implication on the performance of welded structures, static tests have been almost exclusively conducted for weldability. This is mainly because of the complexity, relatively low reliability and repeatability, and high cost associated with dynamic testing. Only fatigue testing has been conducted in limited scale.

As the objective of a mechanical test is to either obtain a quantitative measure or get a qualitative sense of weld quality, the following information is usually collected during a test:

- *Peak load*: The maximum force measured during testing, as illustrated in Figure 4.1 for a tensile–shear test.[1]
- *Failure mode*: This is a qualitative measure of weld quality. As in the case of the chisel test, an operator examines whether the fracture (opening) is brittle or ductile. The morphology of fracture is also examined in tests, such as static tensile–shear tests, to see if there is a weld button pull-out or if the interfacial fracture has a smooth or rough surface.[2]
- *Ductility*: This usually refers to a quantitative measure of weld quality, using quantities such as maximum displacement or energy, as defined in Figure 4.1.[1]
- *Fatigue strength*: This is usually reported in terms of number of cycles to failure under a certain repeated loading pattern, in the form of *L–N* curves.[3,4]
- *Impact energy*: This is the total energy absorbed by a weldment under an impact loading. It has a significant implication on the crashworthiness of a welded structure.
- *Weld nugget width or weld diameter*: As the most commonly monitored quality index, weld size has been, in many cases, the sole measure of weld quality.

Testing welded specimens is different from testing uniform materials mainly because of the geometric characteristics of spot weldments. A weldment is usually considered a unit, and therefore, its strength is often expressed in terms of load instead of stress, and

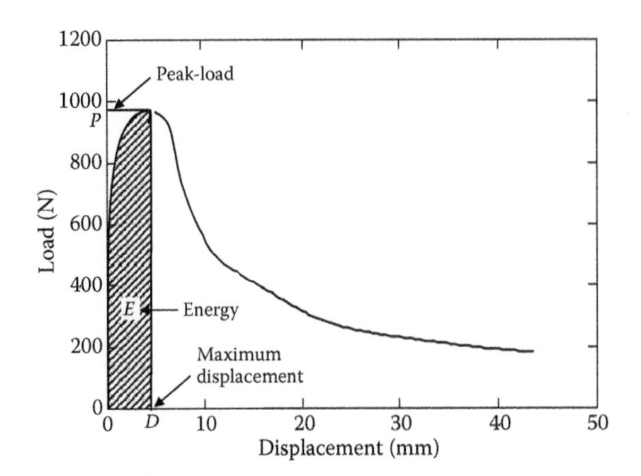

FIGURE 4.1
A typical load vs. displacement curve in quasi-static tensile–shear testing, and definitions of variables that can be monitored.

in displacement instead of strain. Especially for fatigue testing, the conventional expression of fatigue strength using *S–N* curves, or stress range–number of cycles curves, is irrelevant. A more meaningful expression is using *L–N* curves, or load range–number of cycles (at a specified load level before failure) curves, as the total load applied to a spot weldment is more useful in understanding a weld's performance and for design of welded structures. Because a weldment is composed of both the weld and its surroundings, that is, base metal, the measured strength is not solely determined by the weld. Recognizing the contribution of the base metal helps correctly interpret testing results. It is necessary to include the information on base metal (size, strength, etc.) when presenting the testing results of a weld.

An important aspect of mechanical testing is testing procedures and specimen preparation. Although most tests are conducted using standard equipment, many tests, especially impact tests, are generally performed using customized devices. Many difficulties of testing spot welds are related to the inherited configuration of spot-welded specimens. Because a spot weld connects two offset workpieces (or coupons) with a limited-sized joint that is relatively less stiff than the workpieces, it is the place where most bending/rotation occur during testing. The stress concentration caused by the notch-like welded joint determines the deformation and fracture mode when a specimen is loaded. Therefore, testing results are often affected by the restriction imposed by a weld's surroundings, including the dimensions of the specimen, in addition to the weld's strength. The contribution of specimen sizes, however, has not drawn sufficient attention.

The dimensions of testing specimens have not been unified, which may cause confusion in testing and interpreting results. For instance, a survey of standards and specifications reveals significant differences in testing specimen sizes for tensile–shear tests, as in almost every category of testing welded joints. As illustrated in Figure 4.2, both width and length of tensile–shear testing specimens vary dramatically in practice among professional organizations. Plotted in this figure are specimen size recommendations of the American National Standards Institute (ANSI) and the American Welding Society (AWS),[5] the military,[6] and the International Organization for Standardization (ISO).[7] A significant

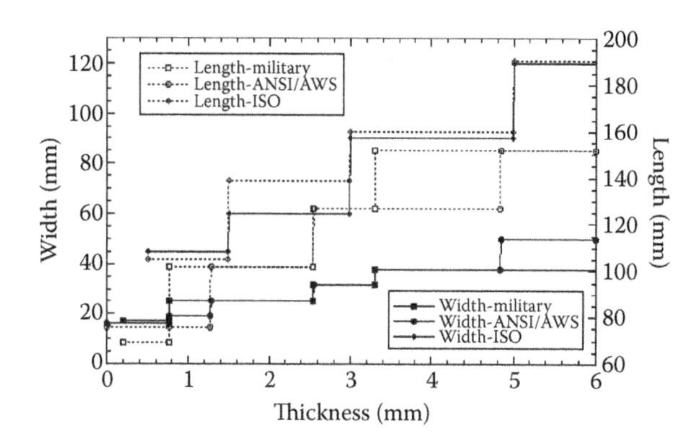

FIGURE 4.2
Selected testing sample specifications.

difference exists in both overlap and width, with no such drastic variance in specimen length. The width required by ISO is 45 mm, which is more than twice that by ANSI/AWS (19 mm) for 0.8-mm-gauge steel sheets. Similar differences can be seen in specimen sizes of aluminum alloys. In general, there is not as much information available for testing and specimen sizes for aluminum alloys as there is for steel, primarily because of a smaller scale application of aluminum in the automotive industry. It is found that with few exceptions, the overlap of the specimens is the same as the width.

In this chapter, specimen preparation and testing procedures, measurement, and data analysis will be discussed. Details will be given for commonly performed tests.

4.2 Shop Floor Practices

A large portion of quality/weldability testing is performed at assembly lines for welding schedule setup and quality monitoring/inspection. Because of the limitation of testing facilities, as well as time restraints, weld quality testing in the production environment is usually limited to measuring the weld button size and monitoring failure modes.

The most commonly conducted tests are chisel and roller tests (Figure 4.3).[8] Although there are no sophisticated equipment or procedures involved, the experience of an operator is extremely important for correct and consistent measurement and interpretation.

4.2.1 Chisel Test

The chisel test is used to measure the ductility of spot welds on welded structures (Figure 4.3a). The objective is to detect brittle (cold) welds, including no-weld. Weld button size is occasionally estimated after the joint is opened up. As the chisel wedge is hammered in between welds, an operator can feel and hear whether a weld is brittle. Because the testing and result interpretation depend heavily on experience/skill, a dedicated person is usually needed to conduct such tests. Repeatability in chisel testing is relatively low. An automated chisel test is used when testing heavy-gauge sheet welding.

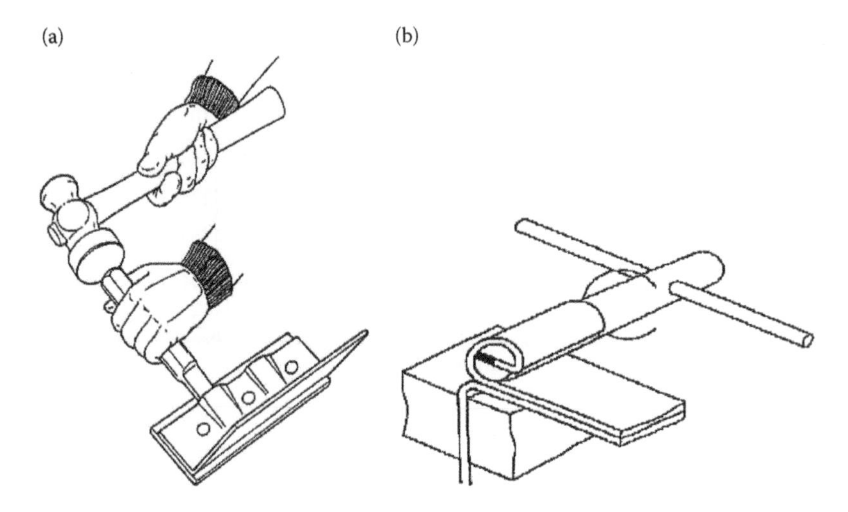

FIGURE 4.3
(a) Chisel test. (b) Peel (roller) test.

4.2.2 Peel (Roller) Test

The peel test is a simple shop test that can be made with a hand tool (Figure 4.3b). It is applicable to sheets of a large range of gauges. In the test, the sheets are first separated on one end of a lap joint, then one sheet is rolled up by the roller while the other is gripped (usually by a vice). As the roller rolls over the weld, one workpiece is torn off of the weld and a weld button is left if the weld is ductile, or the sheets can be separated without much effort if the weld is brittle. In the case of multiple-welded specimens, as in the study of weld spacing or shunting, a specimen is usually cut into pieces with single welds and then tested, although the welds may be peeled in sequence without cutting using customized devices. Caution should be taken when measuring weld buttons of irregular shape, and especially when tails of base metal are left on the button. For details of measurement, refer to various standards such as the *Recommended Practices for Automotive Weld Quality—Resistance Spot Welding*.[8]

Unlike the chisel test, the peel/roller test is conducted on coupons, not on structures. Similar to the chisel test, the roller test is also experience dependent. In general, the testing procedure directly influences the geometry of the peeled button and, therefore, the measurement. The test, however, is not always possible even on coupons, as the strength of the base metal may be too high for the specimen to be rolled, such as in the case of advanced high-strength steel sheets, which have too high a yield stress to be plastically deformed or rolled around a roller.

4.2.3 Bend Test

A bend test is a relatively simple shop floor test to obtain a quick check of spot weld soundness in production, particularly the existence of cracks.[9] Unnotched transverse-weld guided bend tests are frequently used to estimate the ductility of weldments and to qualify welders and welding operators and producers. Bend tests are aimed at detecting weld flaws oriented in the way that can be revealed by the longitudinal sectioning. To obtain a

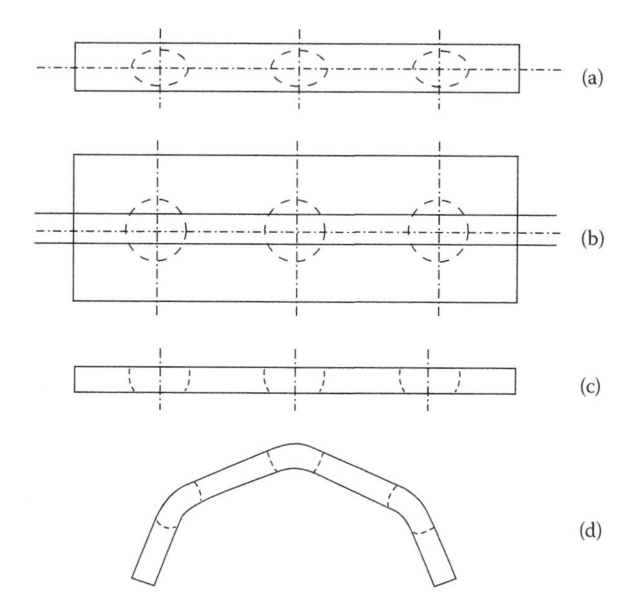

FIGURE 4.4
Bend test.

complete picture of the welds, transverse sectioning and loading may also be needed. In general, a bend test is intended as an aid to process control rather than a requirement. It can be performed with equipment readily available in most shops and requires only visual examination.

The test consists of bending a test specimen from a routine macrosection containing three welds. The procedures of specimen preparation and testing are illustrated in Figure 4.4. The bend test specimen, cut from the center of the weld coupon along its longitudinal direction (Figure 4.4b), is bent along its length to the angles shown in the figure to produce a concentration of bending stresses successively in each of the three welds. Before bending, the edges of the specimen should be rounded and smoothed to remove the burrs. The outside of the bend should be file-smoothed and polished. After bending, the specimen is examined for the presence of cracks or any other surface defects.

The test reveals the presence of defects that are not usually detected in other tests. However, the state of stress in a standard bend specimen varies with the plate thickness because the depth-to-width ratio of the specimen is not held constant. The test is susceptible to nonuniform bending owing to an overmatching strength in the weld. In a longitudinal weld bend specimen, all the zones of a weld joint [weld, heat-affected zone (HAZ), and the base metal] are strained equally and simultaneously.

4.3 Instrumented Tests

Chisel and peel tests are commonly conducted in both the production and laboratory environment because they are simple to perform and can provide instant results. However, a detailed and quantitative description of a weld's strength cannot be obtained by such tests. Although weld button diameter, which is considered quantitative, is routinely measured in the peel test, the measurement is subject to the influence of many random and human factors. The use of weld button size for weld quality is usually based on experimentally developed, nonlinear relations between peak load and button size observed in many tests. It is more of a qualitative measure of weld strength, as the trend, not the magnitude of the dependence, is of more interest. However, accurate information of weld strength is often needed for joint design and welded structure evaluation. For this reason, instrumented tests are conducted to measure a weld's or welded structure's strength in terms of its load-bearing capacity and ductility.

Both load and displacement can be recorded in a static test. Because such tests are usually conducted at a very low speed to minimize the influence of loading rate, they are sometimes called quasi-static tests. Load and displacement are also frequently recorded in dynamic tests. Strain gauge-based and piezoelectric load cells are commonly used sensors for load measurement. The displacement, which is needed to calculate deformation and speed/acceleration in the case of dynamic testing, can be measured using a large variety of sensors, such as linear variable differential transducers and fiber-optic sensors. Besides the range of measurement, response time is also a major concern in selecting sensors for dynamic tests.

Commonly conducted instrumented tests are reviewed in the following sections.

4.3.1 Static Tests

In mechanical testing, there are two basic loading modes: tension and shear. Depending on the engineering requirements, a welded joint can be tested in tension, shear, or a mixed mode of tension and shear. Due to the asymmetric nature of spot-welded joints, a loading often causes significant deformation/rotation of the base metal in the vicinity of a weld button. As a result, the originally attempted loading mode is rarely kept throughout a test. Obtaining a pure loading mode is neither practical nor economical. Therefore, the terminology used for mechanical testing of a spot-welded joint generally describes the main or attempted loading mode. This section focuses on tension tests and the commonly conducted tension–shear or tensile–shear tests.

4.3.1.1 Tension Test

Tension tests are regularly conducted to obtain the basic data on the strength and ductility of materials. In tension tests of uniform materials, a smooth specimen is subject to an increasing uniaxial load, and the load and elongation of the specimen are monitored. The test results are used to make the stress–strain curves, in which the nominal stress and nominal strain are calculated from the original area and original gauge length of the specimen, respectively. Such curves are useful for determining the material's properties, such as Young's modulus, yield stress, and ductility.

In spot-welded specimens, however, the term *stress* is not meaningful for describing a weld's strength for the reasons explained in the introduction. The entirety of a spot weld should be considered when it is measured for quality. Therefore, load and displacement, instead of stress and strain, are usually used to describe a weld's quality.

Besides the peak load and ductility (measured by maximum displacement or energy), the diameter of a weld and the mode of final failure should also be recorded. Two types of tension tests, cross-tension test and U-tension test, provide similar loading onto the welds.

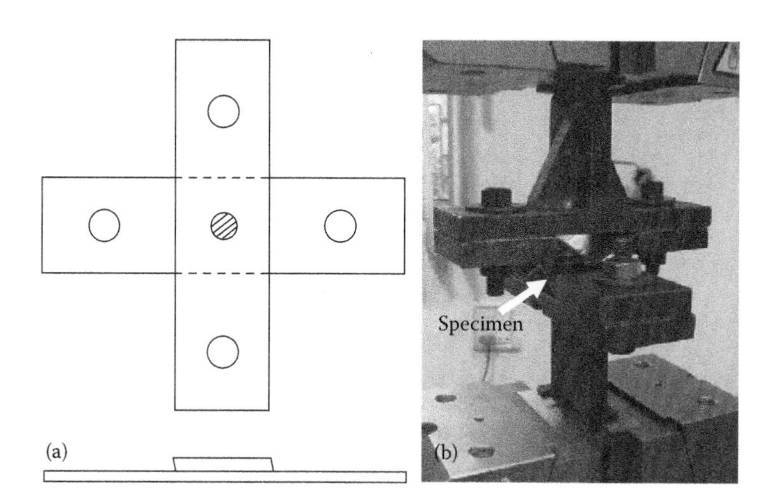

FIGURE 4.5
Cross-tension test specimen (a) and specimen mounting fixture (b).

4.3.1.1.1 Cross-Tension Test

This test is designed to load a weld in a direction normal to the weld interface.[10] The specimen geometry and a specimen mounting fixture (similar to that for drop impact tests as shown in Figure 4.23) for cross-tension tests are shown in Figure 4.5.[11] Special holding jigs are constructed to apply normal tension to the specimen. Various methods of holding the jigs in a testing machine, such as pin connections, wedge grips, or threaded-end testing fixtures, may be used. Precautions should be taken to prevent the specimen from slipping in the holding fixture. The intended pure tension load may not be achievable if the weld is not centered. Sometimes cross-tension specimens with flanges are used, with a set of four holes on the beams bolted to rigid fixtures, to restrain sheet distortion during testing, as shown in Figure 4.6.

4.3.1.1.2 U-Tension Test

A tension test may also be made on a U-shaped specimen, as shown in Figure 4.7.[9] The U-sections are welded and pulled to destruction in a standard testing machine. Support or

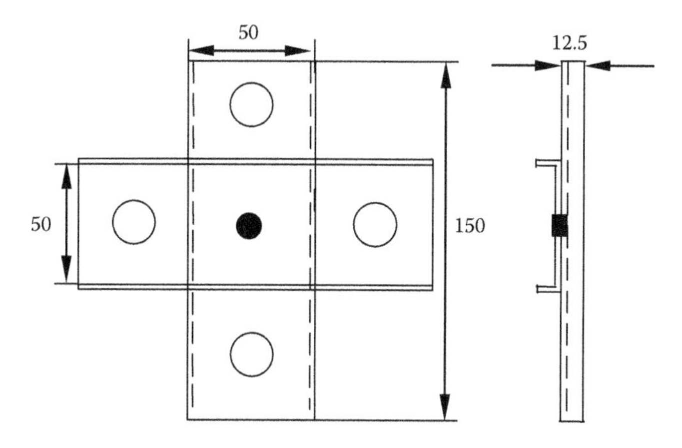

FIGURE 4.6
Flanged cross-tension testing specimens (in millimeters).

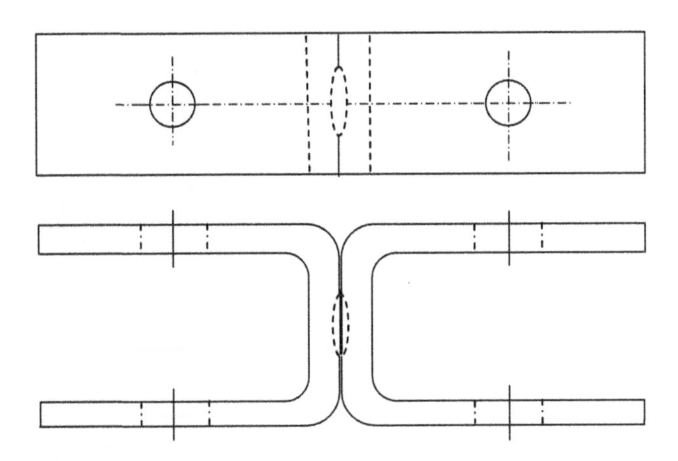

FIGURE 4.7
U-tension test specimen.

spacer blocks must be used for confining the specimen, so that loading takes place at the weld. The U-tension tests are limited to those thicknesses and materials that can be readily bent to the required radius.

The major drawback of tension tests is the unavoidable bending of base metal before the welded joint is fully loaded, and the complication in preparing testing specimens, which involves bending the coupons, drilling holes, and bolting. Like cross-tension tests, U-tension tests are also very sensitive to specimen preparation. If the radii or face size do not match the holders perfectly, which is almost impossible to achieve when high-strength steel sheets are used, a significant amount of deformation of the base metal may precede loading the weld. As the specimens are bent first during loading, corona bonding due to solid-state diffusion between the workpieces in the HAZ outside of the nugget breaks before the nugget is loaded. Although such bonding does not add much load-bearing capacity to the welded joint, it influences the strength measurement, and tension tests tend to be more sensitive to such influences than other tests. In general, tension tests are not commonly conducted in weldability tests because of the difficulties in specimen preparation and fixturing, and large inconsistency in tested results. Accordingly, not much research has been conducted in this area.

4.3.1.2 Tension–Shear Test

Among static tests, tension–shear or tensile–shear testing is most commonly used in determining weld strength because of its simplicity in specimen fabrication and testing.[12,13] In this section, a detailed description of tensile–shear tests, from specimen preparation to testing procedure, and result analysis will be discussed.

This test consists of pulling in tension to destruction, on a standard testing machine, a test specimen obtained by lapping two strips of metal and joining them by a single weld. Occasionally two or more welds are made on a specimen for tensile–shear testing. The testing specimens are shown in Figure 4.8. The ultimate strength of the test specimen and the nature of fracture, whether by shearing the weld material or by tearing the parent material, and whether a ductile or brittle fracture is obtained, should be recorded. In addition, measurement of the diameter of a weld in a tension–shear specimen is desirable after a test. Accurate measurement may be difficult to obtain because the specimen is

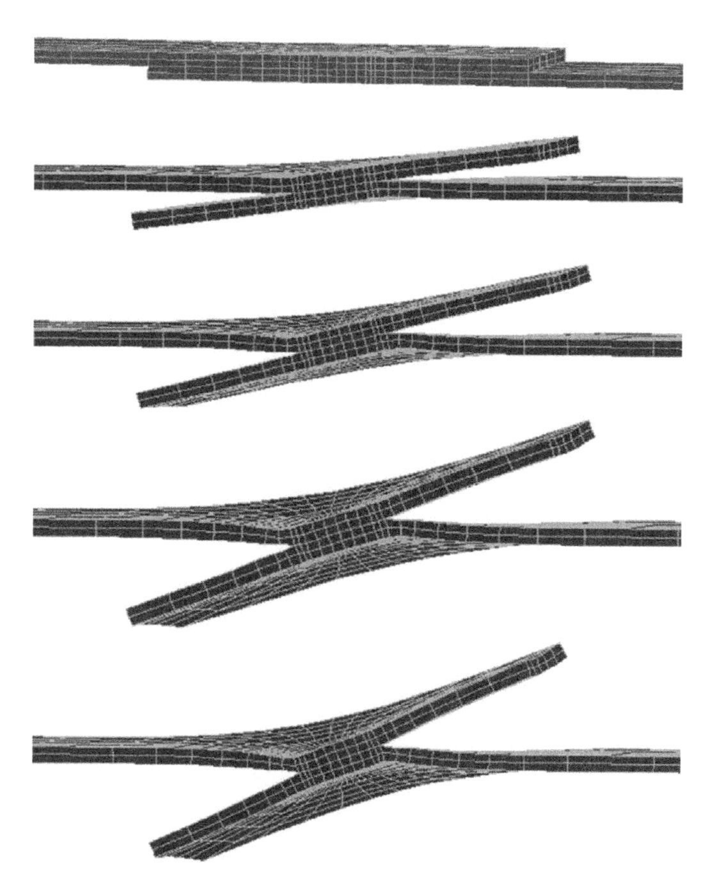

FIGURE 4.8
(Half) tensile–shear test specimen and change of specimen configuration during testing.[14]

distorted in the test and a large piece of base metal is often left around the weld after it is pulled out. When a weld is under a tensile–shear loading, the weldment tends to rotate to align the gripped ends with the welded joint, as shown in Figure 4.8. As the specimen is loaded, sheet separation increases. It is easy to see that the HAZ is the most severely loaded part. The sheet separation/rotation also depends on the specimen width. As shown in Figure 4.9, narrow specimens (with small overlap) tend to rotate more than wide ones (with large overlap) because of different degrees of constraining on deformation of the material around a nugget by different sizes of specimen width.

The rotation of a weld as exhibited in the figures may significantly affect the testing results, in addition to changing the loading mode from pure shear to a mixture of tension and shear. Various attempts have been made to alleviate the rotation effects, by using shims and restraining devices, or simply using wide specimens. Shimming helps reduce the initial rotation at the beginning of loading, but it does not prevent further rotation, as the specimen is asymmetric with respect to the weld. To ensure that tensile–shear tests are consistent and comparable (among researchers/practitioners), certain procedures in testing, specimen preparation, and analysis must be followed. Standard tensile testing machines calibrated according to the American Society for Testing and Materials (ASTM) procedures can be used for such tests. The testing speed should not exceed 15 mm/min to avoid the influence of a dynamic or strain rate sensitivity effect.

FIGURE 4.9
Sheet separation during tensile–shear test. Width of a specimen is the same as its overlap. Only half of specimen is shown.

The most commonly monitored variable in tensile–shear testing is the peak load. However, the displacement at the peak load (maximum displacement) and the corresponding energy should also be monitored, in addition to the peak load. These quantities were defined in Figure 4.1. The maximum displacement indicates the ductility, and the energy is related to the energy-absorbing capacity of a weldment. The displacement and energy should be calculated only to the peak load because the failure of specimens is largely determined at such a moment. After the load reaches its peak value, the displacement and energy are generally not unique because of the uncertainty in subsequently tearing the specimens.

Five types of failure (failure modes), as exhibited in the ISO standards,[2,7] may usually be observed during tensile–shear testing. Typical failure modes, together with corresponding characteristic load vs. displacement curves generated in experiments, are shown in Figure 4.10. After a large number of tests, it was found that mode A is not desirable, because it only tests the base metal,[1] as shown in Figure 4.10a, in which the load vs. displacement curve is very close to a typical uniaxial tensile testing curve of a uniform specimen. Mode B is not desirable either. The specimen fails partially through the periphery of the weld and partially through the base metal. Experimental results show that two similar welds on specimens of different widths result in different strength measurements when the specimens fail in mode B. The reason is that the load-bearing capacity of the specimen is influenced by the width of the base metal on the sides of the weld. In general, failure modes A and B are observed when the testing specimens are too narrow. These two modes are not preferred in weld quality testing because a quantitative measurement of a weld's strength cannot be obtained. Failure modes C, D, and E correspond to weld button pull-out, tearing of base metal, and interfacial failure, respectively. In any of these three, the quality of the welds, rather than the base metal, is tested. Therefore, failure modes serve as a rough indicator of whether a specimen size is adequate. For convenience, failure modes can be grouped as undesirable failure modes (modes A and B) and desirable failure modes (modes C, D, and E). This classification takes into consideration the influence on the measurement of a weld's quality by its strength and the constraint imposed by its surroundings.

Experiments show clearly that the failure modes are a strong function of weld size. Such dependence has not been explicitly mentioned in specifications mainly due to the implicated relationship between weld size (d) and specimen thickness (t), such as $d = a\sqrt{t}$ (a is

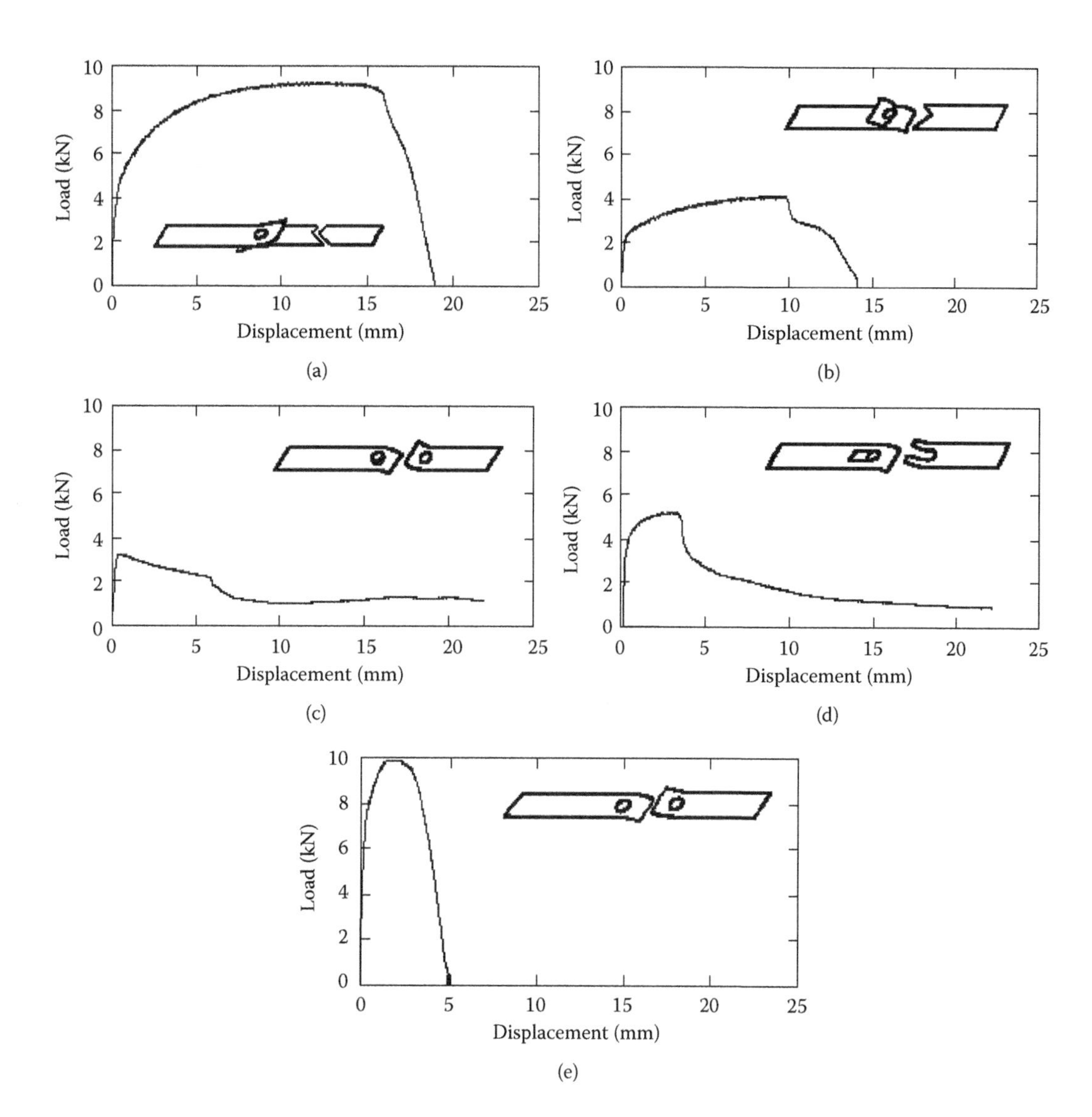

FIGURE 4.10
Typical load vs. displacement curves with their corresponding failure modes in tensile–shear testing.

a constant). Specimens with small welds tend to fail in mode E, that is, interfacial failure, and specimens with large welds tend to produce A and B types of failure. It has been concluded by several researchers that the length of a specimen is far less important than the width of the specimen in affecting testing results.[1,15] For simplicity, a uniform length, such as 150 mm as determined by Zhou et al.,[1] and an overlap equal to the width, can be used.

Through a logistic regression analysis, as detailed in Chapter 10, the probability of getting desirable failure modes can be predicted for a 0.8-mm mild steel by the following equation:

$$\ln \frac{P}{1-P} = a_0 + a_1 d + a_2 w_1 + a_3 w_2 + a_4 d w_1 + a_5 d w_2 \tag{4.1}$$

where P is the probability of getting desirable failure modes, d is the linear effect of weld diameter, w_1 is the linear effect of specimen width, and w_2 is the quadratic effect of specimen width. The last two terms of the equation are for the interactions between the effects. The a_i values are coefficients determined from experimental data; they are $a_0 = 8.673026$, $a_1 = -1.321001$, $a_2 = 9.638809$, $a_3 = -5.555071$, $a_4 = 2.800171$, and $a_5 = -1.642712$ for the sheets studied in Zhou et al.[1]

For fixed weld diameters (4 and 8 mm), the probabilities of getting desirable failure modes are plotted as a function of specimen width in Figure 4.11. Marks 1 and 2 indicate the critical widths for avoiding undesirable failure modes (modes A and B). The difference between the critical values is about 5 mm. One should clearly distinguish between desirable failure modes and critical specimen widths, as the critical values determined in such a way are not necessarily sufficient for testing weld quality. Even if a desirable failure mode is achieved, the specimen may not be wide enough to avoid excessive bending, as the severe distortion may not be confined to the vicinity of a weld. Excessive distortion of narrow specimens may "contaminate" the testing results.

Although experiments may provide important information and a methodology on determining critical specimen sizes, the general minimum widths cannot be obtained through experiments alone. The main reason for this is the random effects in experiments. For example, the peak load does not always stay in a plateau after the initial transition stage, as shown in Figure 4.12, and the variation of measurement is not constant at different widths. Variances in specimen dimension, welding, and testing may also influence the consistency of testing results to some extent. A finite element analysis can be performed to overcome such deficiencies of experiments.

Although failure modes provide important information on the adequacy of specimen size in order to measure weld quality, not that of the base metal, they are more qualitative than quantitative. A more effective, quantitative way of determining critical specimen sizes is to analyze the influence of dimensional variables on strength measurements (peak load, maximum displacement, and energy). Experimental results show that the thickness of specimens and weld diameter influence the peak load the most. The peak load also depends on the overlap of the specimens to some extent.

A work by Zhou et al.[1] systematically studied the effect of specimen size on tensile–shear testing results and derived critical widths for sheets of different properties and thickness.

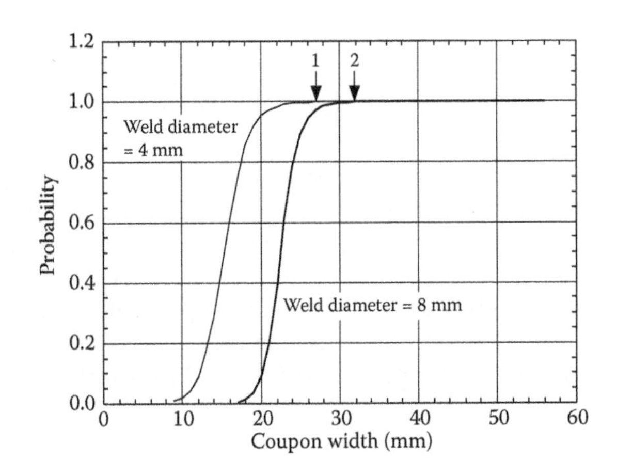

FIGURE 4.11
Probability of getting desirable failure modes.

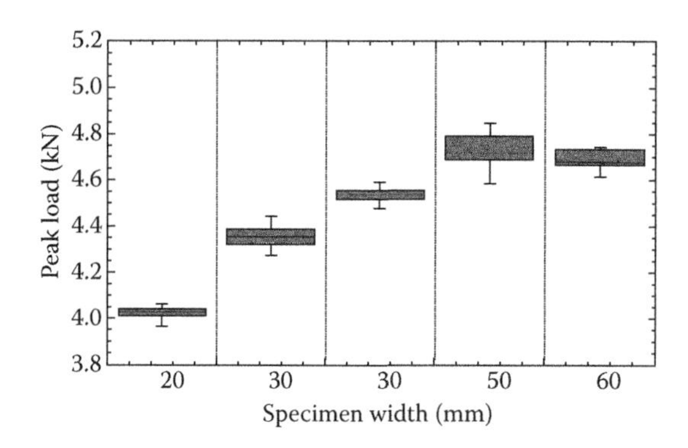

FIGURE 4.12
Peak load vs. specimen width for a 0.8-mm DS.

In their study, specimen width, as the most critical dimension, was determined by plotting the strength measurements vs. the width. The width at which a strength measurement shifts from a rapid change (with large slope) to a smooth change (with small slope) was used as the critical width, as shown in Figure 4.13.

Finite element simulation was used for critical specimen size determination to avoid the uncertainty of experiments and reduce cost. The work by Zhou et al. used different material properties for the nugget, the HAZ, and the base metal in their finite element analysis. A finite element model (FEM) was built to simulate the behavior of spot-welded specimens with the consideration of material and dimensional differences in the nugget, the HAZ, and the base metal. The material properties of various zones in a weldment can be estimated based on relationships between hardness and yield/ultimate tensile stress, as established in Chapter 6. A numerical experiment was conducted by the researchers using a statistical design of experiment for computer simulation (as demonstrated in Chapter 10). The procedures and main conclusions are listed here.

4.3.1.2.1 Finite Element Modeling

The FEM used a fixed length of 150 mm (as determined by experiments) for testing specimens, and the width was taken as a variable. The overlap was the same as the width, and

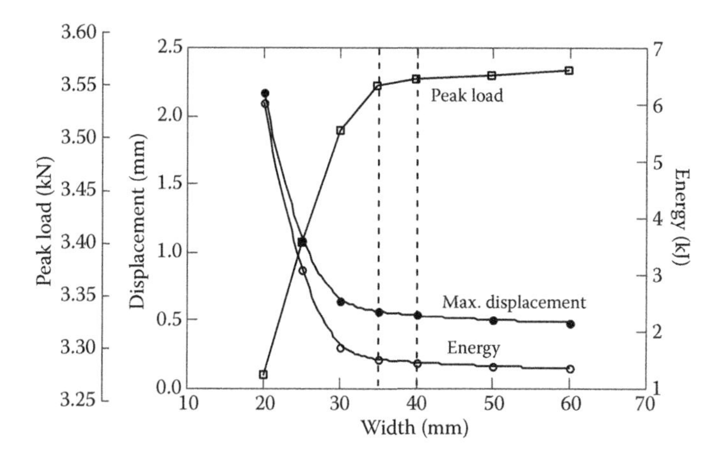

FIGURE 4.13
Dependence of weld strength on specimen width.

two sheets had the same thickness. In the simulation, various zones were distinguished from each other by their mechanical properties. The model was created in a way that the sizes of these zones can be easily changed. The hardness and strength values of various zones can be related to those of the base metal through Equation 6.6 (Chapter 6), which describes the relations between the hardness and mechanical properties.

4.3.1.2.2 Numerical Experiment

To obtain results of a relatively large range of application, both geometric and material properties need to vary in certain ranges. Geometric factors include the thickness (t) and HAZ size (h). The nugget diameter was taken as $6\sqrt{t}$ to overcome possible variations in size and location of the welds in physical experiments. The material properties are Young's modulus (E), yield strength (σ_y), ultimate tensile strength (σ_{uts}), elongation (e), and the hardness ratio between the nugget and the base metal (k).

The critical specimen width ($W_{critical}$) can be expressed as a function of these variables. Through a treatment similar to that in Chapter 6, the number of variables can be reduced and the critical width written as

$$W_{critical} = f(t, h; E, \sigma_y, \sigma_0, e; k) \quad (4.2)$$

The ranges of the variables are shown in Table 4.1. They were chosen to cover a large range of steels of different thickness and material properties. Because of the large number of variables in the function, the concept of design of experiment, which requires a significantly smaller number of runs than conventional experiments, was employed to obtain this relation. The design, simulation, and analysis followed the procedures outlined in Chapter 10 for statistical analysis of computer experiments. Every run consists of obtaining a critical width (through several calculations) for a set of fixed geometric and property variables. A total of 640 runs were conducted in the study.

The statistical model for the critical width contains the main effects of the variables, including t, h, E, σ_y, σ_0, e, and k, and quadratic and interaction terms of the factors (h^2, $e{\cdot}\sigma_0$, $\sigma_y{\cdot}\sigma_0$, $h{\cdot}\sigma_y$, $E{\cdot}\sigma_0$, $t{\cdot}\sigma_y$, $t{\cdot}\sigma_0$, and $k{\cdot}\sigma_0$). The significant effects were chosen through a model selection procedure. As shown in Figure 4.14, the thickness is extremely influential on the specimen width, and its influence is about three times larger than that of yield strength (σ_y). The HAZ size (h) was also found to be very important in determining the critical width. Other effects, such as the interactions, elongation, Young's modulus, and ultimate strength, have much less influence on the critical width. It is interesting to note that the hardness difference between the nugget and the base metal plays an insignificant role.

4.3.1.2.3 Results and Analysis

The statistical analysis yielded a few formulas for critical widths. As the thickness has the largest effect on critical widths, a function that contains only the thickness as a variable was developed in the following form:

$$W_{critical,1} = 13.4044613 + 18.5987839t \quad (4.3)$$

TABLE 4.1

Ranges Selected for Computer Simulation

t (mm)	h (mm)	E (GPa)	σ_y (MPa)	σ_0 (MPa)	e (%)	k
0.5–2.0	0.1–1.5	190–200	205–1725	50–200	2–65	1.0–3.0

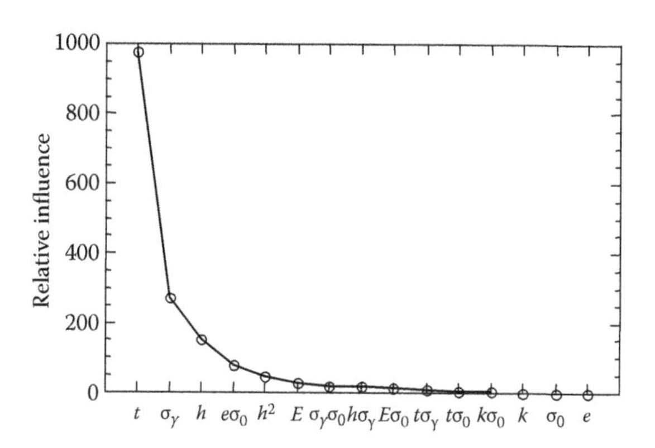

FIGURE 4.14
Effects of various parameters and their interactions on critical specimen width. Parameters shown here are normalized. (From Zhou, M., Relationship between spot weld attributes and weld performance, PhD dissertation, University of Michigan, Ann Arbor, MI, 2000. With permission.)

The statistical analysis revealed that 53% of the total variation of critical width can be explained by this equation. If the second and third largest effects are also included, the formula becomes

$$W_{\text{critical},2} = -6.0291481 + 18.5839362t + 0.0146654\sigma_y + 6.6251147h \quad (4.4)$$

The R^2 value increases to 92.7%. If all of the first six largest effects of Figure 4.14 are included, the critical width is

$$\begin{aligned} W_{\text{critical},3} &= 45.6391799 + 18.5849834t + 0.0146654\sigma_y \\ &+ 21.8791238h + 28.3945601e \\ &+ 0.0811080\,(\sigma_{\text{uts}} - \sigma_y) - 0.0003401E \\ &- 9.5332611h^2 - 0.2280655e(\sigma_{\text{uts}} - \sigma_y) \end{aligned} \quad (4.5)$$

A total of 98.6% of the total variation can be explained using this equation. In the equations, the variables are expressed in natural scale, as shown in Table 4.1, and the critical widths ($W_{\text{critical},1}$, $W_{\text{critical},2}$, $W_{\text{critical},3}$) are in millimeters.

The critical specimen widths calculated by this study lie between those specified by ANSI/AWS and ISO shown in Figure 4.2. For the convenience of practical use, a step function was proposed (shown as dashed lines in Figure 4.15). It is also worthwhile to note that although the dimensions proposed by this study are the minimum widths, they should work fairly well in most cases because they were determined using conservative values, for example, a nugget size of $6\sqrt{t}$ was used, which is rarely exceeded in practice.

In general, the following conclusions can be drawn for tensile–shear testing:

- Using wide specimens in tensile–shear tests is an effective and economical way to restrain specimen rotation and reduce uncertainty in testing results, compared with using other means, such as restraining plates.

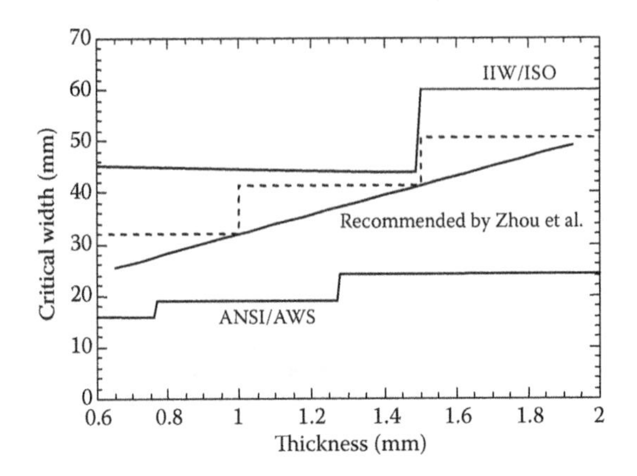

FIGURE 4.15
A comparison of specimen widths calculated by Zhou et al.[1] and specified in standards. The dashed line is the proposed specimen width.

- Failure modes observed in tensile–shear tests provide a direct indication on the adequacy of specimen sizes. Although getting desirable failure modes is not equivalent to the suitability of specimen sizes, getting undesirable failure modes may mean that the testing specimens are not wide enough.
- The specimen width is the most important factor in influencing testing measurements. An overlap that was the same as the width was found to be adequate, and 150 mm for length was determined to be sufficient for thin sheets.

4.3.1.3 Combined Tension and Shear Test

The loading of a spot weld in a structure such as what an automobile may be subjected to during service is rarely of a single type. A combined loading creates a different stress state in a weldment, and ignoring the multi-mode nature of loading may overestimate the strength of a weld, leading to unexpected "premature" failure of welds. Therefore, it is necessary to understand the performance of a weld under biaxial loading which may result from simultaneous tension and shearing. Lee et al.[16] designed a dedicated fixture for loading a spot weldment under combined tension and shear, as shown in Figure 4.16, with the flexibility of changing the loading angle with respect to the weld plane, to vary the proportions of tension and shear loading components.

A specially designed test specimen was made, as shown in Figure 4.17, to be used for biaxial loading tests using the fixture. The deformed specimen under loading is depicted in Figure 4.18. By adjusting the orientation of the pull bars on the lock plates, loading angles of 30°, 50°, 70°, and 90° from the weld plane can be achieved. When the pull bars are attached to the lock plates as shown in Figure 4.16, a load of pure tension without shear component is applied to the weld. Another extreme, tension–shear can be realized using a standard testing specimen and loading procedure as depicted in Section 4.3.1.2.

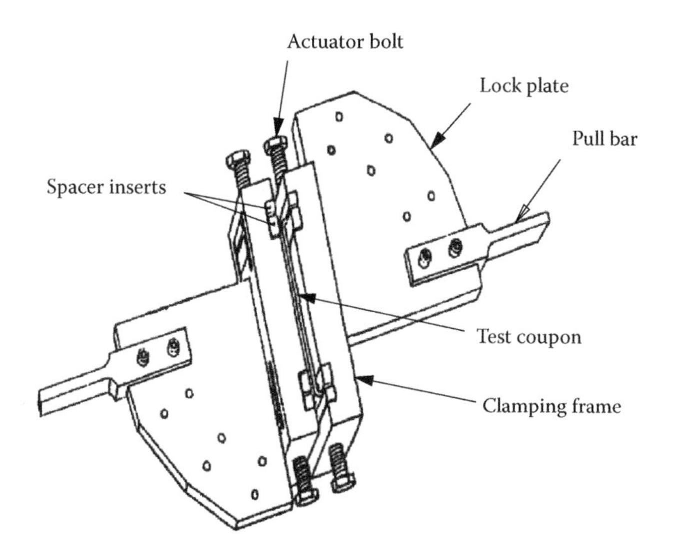

FIGURE 4.16
Fixture for testing spot welds under simultaneous tension and shear loading. (From Lee, Y. et al., Test of resistance spot welds under combined tension and shear, SMWC VII, Paper C2, 1996. With permission.)

Using such a testing fixture, Lee et al.[16] tested spot welds on a mild steel, under different loading angles. It was found that the load carrying capacity of a weldment decreases with the loading angle. As the loading mode changes from tension–shear to cross-tension, an approximately 20% reduction in weld strength was observed. A statistical analysis of the testing results revealed that the weld size and coupon width have significant effect on the strength measurement of a weldment, and the testing specimen length has the least effect.

4.3.2 Dynamic Tests

Although weld quality testing includes both static and dynamic aspects, it is often measured only by static testing, such as tensile–shear tests. Because of the dynamic nature of

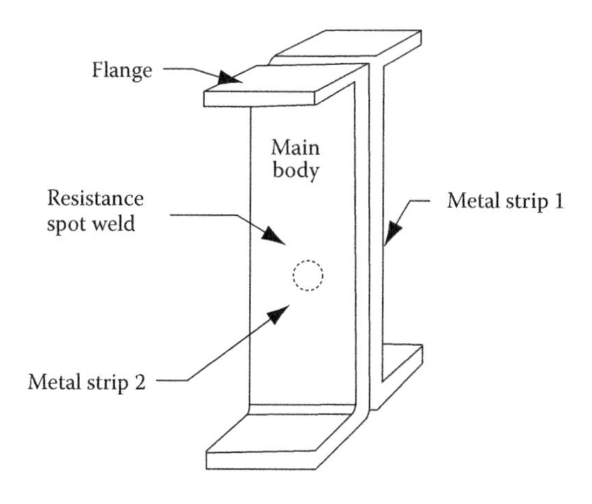

FIGURE 4.17
Testing specimen used for fixture shown in Figure 4.16. (From Lee, Y. et al., Test of resistance spot welds under combined tension and shear, SMWC VII, Paper C2, 1996. With permission.)

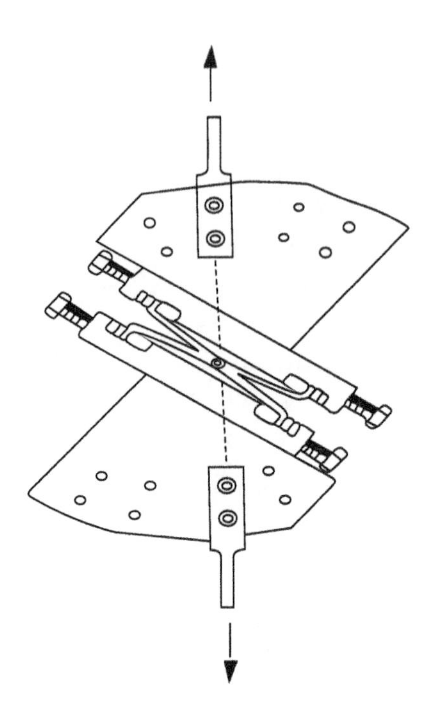

FIGURE 4.18
A weld under simultaneous tension and shear loading using fixture and specimen shown in Figures 4.16 and 4.17. (From Lee, Y. et al., Test of resistance spot welds under combined tension and shear, SMWC VII, Paper C2, 1996. With permission.)

many sheet metal assemblies such as automobiles, dynamic strength of spot welds should be of primary importance in weld quality requirements. It plays a key role in safety, reliability, and integrity of welded structures. Dynamic testing generally refers to fatigue and impact tests. Fatigue testing has been conducted in some cases for dynamic strength measurement of spot welds, whereas impact tests have been largely limited to testing welded structures. Unlike static strength (mostly tensile–shear strength), there are rarely requirements by industries or professional organizations on fatigue strength, and virtually none on impact strength of welds.

4.3.2.1 Fatigue Test

Vibration is observed in all structures, especially in automobiles. In addition to noises and other discomforts induced to the occupants, vibration imposes repetitive loading to the welded joints of a vehicle. An example of force profile collected during a field test of a passenger car is shown in Figure 4.19. Such loading, if imposed on a structure for a considerable period, may deteriorate the strength or integrity of the structure. This process is called fatigue, and it often results in a fracture at which a component completely loses its load-bearing capability. Compared with static test or impact test, failure in a fatigue test has several unique characteristics:

1. Failure occurs only after a large number of loading cycles if the load is low.
2. The failure loads are significantly lower than those normally permitted in structure designs, or static loading limits.

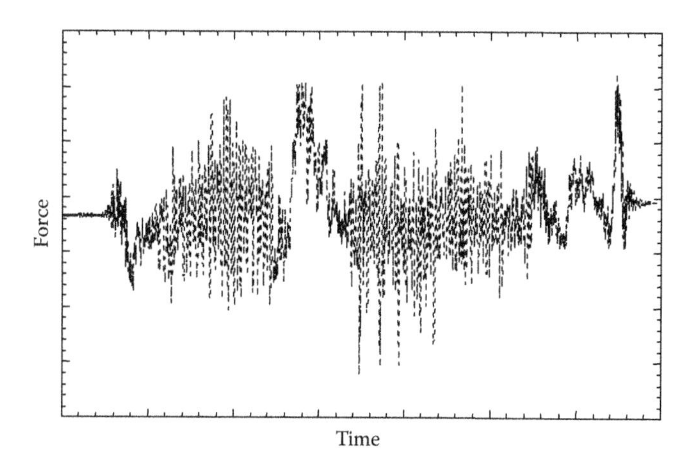

FIGURE 4.19
Force vs. time curve monitored on an automobile component during proving.

3. Fracture surface usually appears brittle; even a static testing of the same weld renders a ductile appearance.

The failure load of fatigue is considerably lower than the static ultimate strength, and it may compromise the safety of welded structures if it is not fully accounted for. The unique geometric characteristic of spot-welded joints, that is, their notch-like shape, makes them stress risers, which are undesirable for fatigue strength.

Because fatigue strength under a specific loading cannot be readily derived from the results of other types of tests, considerable efforts have been devoted to experimentally testing a structure or a welded joint to assess its life of being capable to perform designed tasks under simulated and accelerated service conditions. Although the cyclic loading under actual service conditions varies in magnitude, and often with very rough periodicity, as seen in Figure 4.19, laboratory tests are usually conducted under constant load ranges for simplicity. More severe loading conditions are often adopted to simulate extreme working conditions because it is difficult to predict the stress ranges, or the number of loading cycles in service. Appropriate safety factors are needed to suppress the effects of random factors.[17] In general, it is difficult to make precise calculations of fatigue resistance for structures because of the limited information on actual service conditions and the uncertainties in service of a structure. Laboratory experiments are therefore designed to provide guidance for selecting/designing a joint that is not likely to fail in fatigue under expected conditions of loading.

In a spot weld, the natural notch at the point where the two sheets are joined acts as a sharp crack and is the primary site of stress concentration and fatigue crack initiation. Fatigue testing of a spot-welded joint is discussed in the following sections by considering the influential factors on the fatigue strength of a spot-welded joint.

4.3.2.1.1 Stress Concentration

The notch-like shape makes a spot weldment a natural site of stress concentration around the edge of the nugget. Another important, yet often ignored factor is stress concentration induced by abrupt changes in material properties. Because such material property mismatch mainly occurs in a narrow range from the nugget (or fusion line) to the base metal through the HAZ, a large stress gradient often exists in the submillimeter range when the joint is loaded. Internal discontinuities or flaws may also serve as a source of stress risers. Although they may have little effect upon the static strength of a weldment, their effect on fatigue resistance remains unclear.

4.3.2.1.2 Loading Conditions

The fatigue life (*N*, number of cycles that can be applied at a specified loading range before failure occurs) of a joint is determined by a number of factors of loading. Definitions of the terms can be found in ISO/FDIS 14324.[4]

- *Amplitude and range of applied load*: Although these depend on service conditions with great uncertainties, extreme loading can usually be determined for laboratory experiments.
- *Nature of loading*: The loading mode, such as tension/compression, torsion, or shearing, imposes different stress states to the spot weld joint. Fatigue life tested under one loading mode provides very few clues to the behavior of the same joint under another mode of loading.
- *Rate of loading*: This is more profound for materials with high-strain-rate sensitivity. However, this may also be important for a material with low-strain-rate sensitivity if a low-frequency loading is accompanied with other environmental conditions, such as corrosion.
- *Number of cycles or repetitions*: This directly links to the service life of a welded component, and obtaining this quantity is an important objective of fatigue testing. The total number of repetitions should be realistically estimated based not on the particular component considered, but the service life of the entire structure.

4.3.2.1.3 Residual Stresses

The residual stresses appearing in a spot weldment mainly stem from the thermal–mechanical processes it experiences during heating and cooling. The following information is usually needed to assess the influence of residual stress on a weld's fatigue strength:

- *Nature of stress*: Whether it is compressive or tensile directly affects the fatigue life because it determines, when combined with externally applied loading, the overall stress state.
- *Magnitude of stress*: Because residual stresses are the result of thermal–mechanical interactions, they can be fairly large. The magnitude is often linked to the degree of distortion and indentation.
- *Location*: The location of residual stresses is as important as their magnitude. The primary locations of residual stresses are linked to the geometric irregularities in a spot weldment, such as the area near the nugget. These locations are also those of stress concentration under externally applied loading.

An accurate prediction of residual stresses in a spot weldment may not be possible due to the difficulties in numerical simulation, as discussed in Chapter 9. The restraining of base metal on a weld when welding sheets of a poor fit-up, as often seen in practice, may impose a significant amount of stress on the joint.

4.3.2.1.4 Material Properties

Fatigue strength of both the base metal and the HAZ and nugget should be considered. For obvious reasons, the HAZ and the nugget, which experience metallurgical changes during welding, should be taken into account because of stress concentration and material strength mismatch. In addition, fatigue failure can also be observed in the base metal due to stress concentration in the base metal produced by the welded joint.

Based on experiments of fatigue testing dual-phase and cold-rolled SAE 1006 AK steels, Wilson and Fine[18] concluded that at long fatigue lives, steels of all strength levels behave similarly, whereas at shorter fatigue lives, the higher strength steels exhibit superior fatigue performance. This implies that in the case of spot weldment, high hardness in the HAZ, which is often achievable in steels of high hardenability and in heat-treatable aluminum alloys, may actually provide high fatigue strength. However, this effect may be compromised by the stress concentration induced by material property gradient in the HAZ.

4.3.2.1.5 Environmental Effect

During service, a spot-welded joint may be subjected to changing temperature and other conditions, such as corrosive environment. These factors interact with each other in influencing the fatigue resistance of a spot-welded component. Although these factors are not easy to simulate in controlled experiments and not all factors can be accounted for, efforts should be made to identify and include the dominant ones in the prediction of fatigue life of welded structures.

Fatigue fractures have a unique characteristic appearance, which distinguishes them from other types of fractures. The initiation of cracking usually starts within a very small range from irregularities, either geometric or material in nature, such as internal voids or inclusions where stress concentration may occur. The location of initiation can usually be identified by the characteristic clamshell markings that develop during the various stages of crack growth appearing as concentric circles radiating from the point of initiation of failure. Most of the fracture surface appears brittle, with highly localized plastic deformation appearance in the area where final breakup occurs. One should note that many fatigue cracking characteristics, such as clamshell markings, may not be observed on thin sheets or plates, such as in the case of spot-welded sheet specimens.

Although fatigue failures usually have brittle appearance, they are quite different from low-temperature brittle fracture. Unlike a low-temperature fracture, fatigue resistance increases, rather than decreases, with a decrease in temperature.

There are many theories attempting to explain a fatigue process. They are generally based on the material behavior, either in macro-scale, such as plastic deformation, or in micro-scale, such as the motion of dislocations and atomic debonding. It is difficult to prove the individual theories experimentally, and actual structural problems are too complex to be analyzed using these theories. Therefore, most applications of fatigue data to actual structural problems are based on empirical relationships obtained from laboratory tests. In the last decade or so, computer simulation has been successfully utilized to generate fatigue data economically and efficiently.

The ISO standard on fatigue tests[4] outlined the steps of fatigue testing of spot-welded joints. In general, a laboratory fatigue test involves the following steps.

4.3.2.1.6 Specimen Preparation

A common practice is to follow the ASTM standards for fatigue testing of welded specimens.[19] Proper size and geometry of the specimens should be chosen considering the constraint imposed by the base metal on the weld in a welded component/structure. The dimensional details should be reported when data are presented. Lap joint specimens are commonly used in fatigue tests. Testing other configurations, such as cross-tension test, is occasionally performed. However, such tests require complicated fixtures, which are not preferred, as they may compromise a test's accuracy.

4.3.2.1.7 *Loading Selection*

Cyclic loading, which can be expressed analytically or numerically and implemented automatically, is usually used for fatigue life testing. The common types of loading are:

- Fluctuating tensile loading. Both maximum and minimum loads are tensile.
- Fluctuating compressive loading. Both maximum and minimum loads are compressive.
- Fluctuating loading with maximum load tensile and minimum load compressive.
- Completely reversed loading with maximum load (tensile) and minimum load (compressive) equal in magnitude and opposite in sign.

The loading can be force, moment, or torque.

4.3.2.1.8 *Data Presentation*

For a uniform material, fatigue resistance is usually reported in terms of number of cycles to failure for the applied stress, in the form of an *S–N* diagram. Logarithmic scale is often used for clarity. Experimental data are used to fit the expression of

$$F_n = S\,(N/n)^k \tag{4.6}$$

which appears as a straight line with a slope of k in a log-scale graph. Once this line is determined, the maximum loading (F_n), for a known number of cycles (n), or the fatigue life (n) of a specimen under a known loading (F_n), with similar configuration and under the same loading conditions as the specimens used to obtain this curve, can be determined by knowing the slope k and a point (S, N) on the curve.

For spot-welded specimens, results of fatigue tests are generally presented as the load range (instead of stress range) vs. the fatigue life, or the *L–N* curves. Statistical methods are sometimes employed in determining an *L–N* curve due to the low repeatability of tests. Besides the *L–N* curve, the load range, corresponding endurance limit, load ratio, cycling frequency, and criterion for termination of a test should be reported as well. Because each point in an *L–N* diagram represents the number of cycles to failure under a specific loading, to generate such a curve is time-consuming. As the fatigue strength measured is closely associated with loading conditions, the type of loading must be stated for each curve.

Fracture mechanics concepts are commonly employed for analyzing fatigue behavior of spot-welded joints.

4.3.2.1.9 *Analysis and Correlation of Laboratory Tests with Field Behavior*

Because stress concentration is the single most influential factor in fatigue failure, designers and welding practitioners should be aware of the common causes of stress concentration. Efforts should be made to eliminate or minimize stress risers by avoiding unfavorable geometry and reducing the amount of flaws/defects in a weldment. Controlling unfavorable factors of a joint can be done through proper design and manufacture of structures/components.

There are a number of efforts to predict the fatigue strength of a weld or a welded structure. For instance, Rui et al.[20] developed an expression describing the fatigue life of high-strength and low-carbon steels, correlated by a stress intensity factor. In their work, fatigue life is a sophisticated function of joint geometry, loading condition, nugget diameter, and material thickness. The formula provides a means to minimize the stress intensity factor and to find the optimal weld location for maximum fatigue life or reduce the number of welds needed to meet certain requirements. In an article not directly related to the relationship between weld attributes and weld performance, Wang et al.[3] tried to find, among others, a correlation between the failure mode and fatigue strength for AA5754 welded and weld-bonded joints.

4.3.2.1.10 Examples of Fatigue Strength of Welds

Ma et al.[21] studied the fatigue performance of spot welds of various sizes, created using different welding currents. The room temperature fatigue lives of the welds are shown in Figure 4.20, tested at 50 Hz and $R = 0.1$. The fatigue load drops rapidly between ca. 10^3 to 2×10^6 cycles. The level-off observed beyond that range indicates the presence of fatigue limit. The weld size does not have much of an effect on the fatigue life as shown by the figure if the scattering of the data is considered. The welds with expulsion created using a welding current of 8.54 kA appear to have a slightly lower fatigue limit. From the experiments, the authors summarized the fatigue failure of spot welds as four types of failure modes, shown in Figure 4.21. A correspondence between the configurations of the tested specimens in Figure 4.21 and the regions of the *S–N* curves in Figure 4.20 was established. At a very high load level (Region I in Figure 4.20), the specimens failed either interfacially (Figure 4.21a) or with large amount of plastic deformation (Figure 4.21b). A fracture along the periphery of a weld and then into the base metal, as shown in Figure 4.21c, renders a fatigue life in Region II, the steep drop in fatigue strength in Figure 4.20. When the fatigue

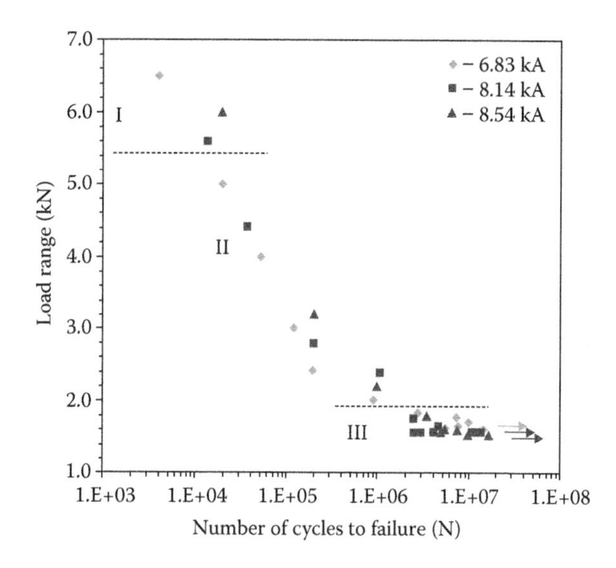

FIGURE 4.20
S–N curves for three groups of DP600 weld samples welded at different current levels. Tested at 50 Hz, $R = 0.1$, and room temperature. (From Ma, C. et al., *Mater. Sci. Eng. A*, 485, 334–346, 2008. With permission.)

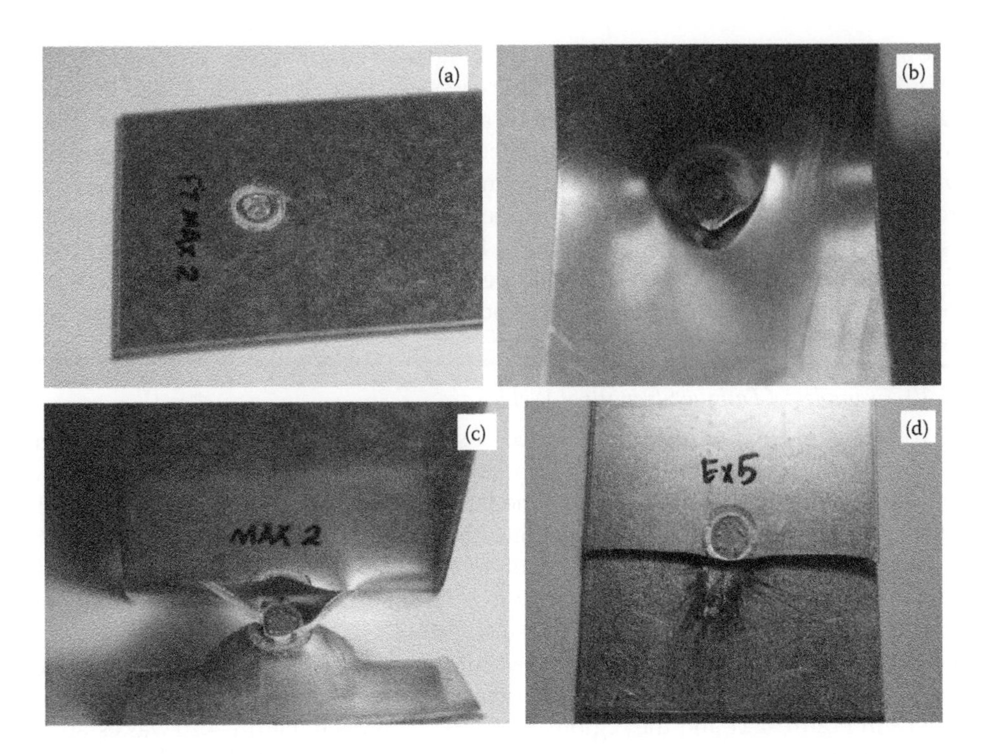

FIGURE 4.21
Typical fatigue failure modes observed in DP600 welds. (a) Interfacial fracture; (b) plastic deformation; (c) fracture along circumference; and (d) fracture along a straight line normal to loading direction. (From Ma, C. et al., *Mater. Sci. Eng. A*, 485, 334–346, 2008. With permission.)

cracking initiates in the HAZ and propagates in the direction normal to the applied loading (Figure 4.21d), the measured fatigue strength tends to fall in Region III in Figure 4.20.

Khan et al.[22] compared their experimental results on dissimilar material combination of HSLA350/DP600 with published data on similar material combinations of HSLA350/HSLA350 and DP600/DP600. From Figure 4.22, it can be seen that when dissimilar steels HSLA350 and DP600 are welded together, the strength of such welds is lower than that of the DP600/DP600 combination, but very close to that of HSLA350/HSLA350 combination. It seems that the steel of lower strength (here, it is HSLA350) dominates the performance in a dissimilar material combination.

4.3.2.2 Impact Test

The impact performance of spot welds is of primary concern in many sheet metal assemblies, as it plays a key role in safety, reliability, and integrity of welded structures. However, the complexity, relatively low reliability and repeatability, and the high cost involved in impact testing inhibit its wide application in weldability testing.

Impact testing can be classified according to the way a welded joint is loaded (loading mode) and the way an impact (shock) loading is applied. The commonly used loading modes are tension, such as in tension impact loading tests, and shear. The impact can be realized by using a modified single-pendulum impact tester, or through a drop weight, as in the case of drop impact tests. These tests are briefly reviewed here. Details of the tests can be found in AWS C1.1-66[9] or AWS D8.9-97.[13]

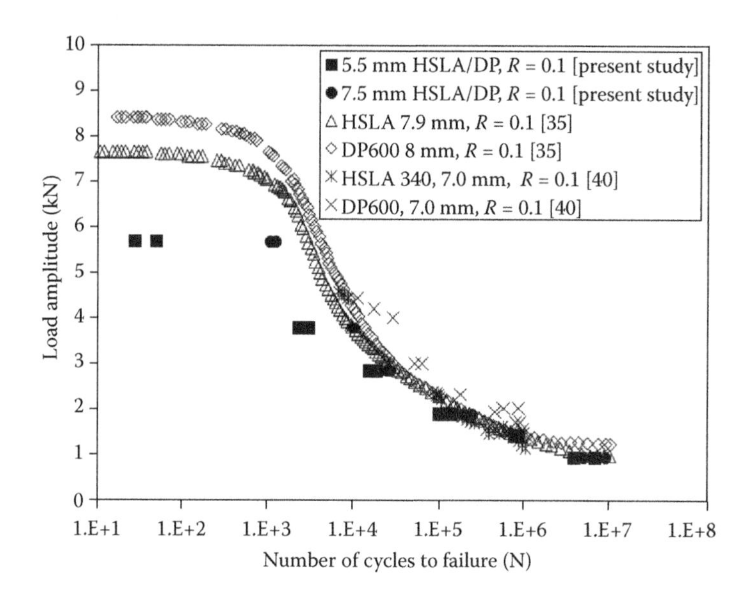

FIGURE 4.22
Comparison of published fatigue strength data. (From Khan, M.S. et al., *Sci. Technol. Welding Joining*, 14(7), 616–625, 2009. With permission.)

4.3.2.2.1 Shear Impact Test

A shear impact test for spot welds is usually conducted on a modified pendulum-type impact testing machine. In this test, the specimen is held by serrated wedge grips in a special pendulum bob and crosshead attachments. When triggered, both the crosshead and the bob, which are connected by the weld specimen, fall until the crosshead is caught by the adjustable anvils at the bottom of the pendulum swing. The pendulum bob is free to continue to swing provided sufficient energy is available to fracture the specimen. The residual swing of the pendulum serves as an indicator of the impact load necessary to break the weld. However, such a measure is not accurate, as it is heavily affected by many factors. For instance, the kinetic energy of the flying anvil is not accounted for, and some of the energy is absorbed by the plastic deformation of the sheets. Other types of modification of the single-pendulum impact tester for welded specimens have been made in individual laboratories.

4.3.2.2.2 Drop Impact Test

In this test, shock loads are applied on specimens usually made of heavy-gauge materials. The configuration of specimens is illustrated in Figure 4.23. The results obtained with the drop impact tests are subject to two types of error even when a "perfect" setup is used. One kind of error comes from the fact that it is not possible to fully restrain the upper and lower plates against bending. The second source of error is the slippage of the specimen in the clamps. Both absorb energy in an uncontrollable manner, and thus a true measure of weld toughness is not obtained. This can be overcome by either providing serrated jaws for clamping to reduce slippage, or placing another plate directly over the lower plate through spot welding to increase the stiffness of the plate for reduced bending. An effective alternative is to make flanges on the plates to reduce bending.

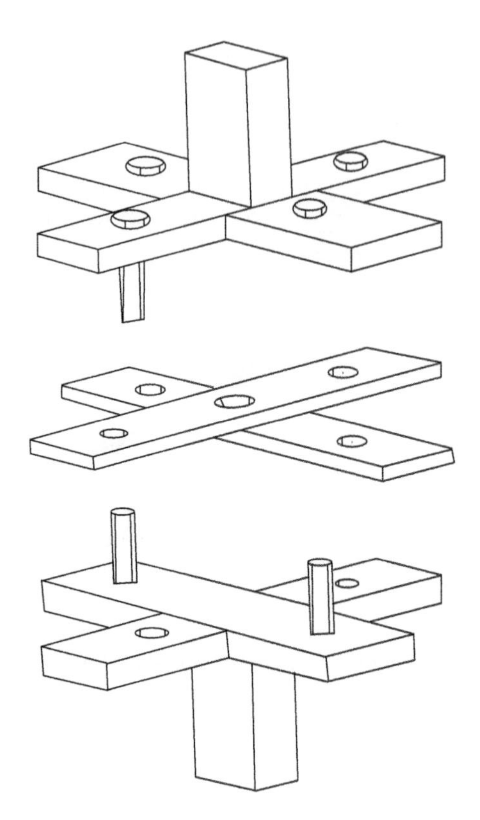

FIGURE 4.23
Testing specimen for drop impact tests.

4.3.2.2.3 Shear Impact Loading Test

This test uses specimens made by joining two U-shaped sections back-to-back by a single spot weld, as shown in Figure 4.24. The specimen is dynamically loaded in a pendulum-type impact testing machine. The test fixture is designed so that the forces applied in fracturing the specimen are essentially in shear. There is a possibility of changing loading mode during testing because of the rotation of the specimen around the bolts. The operation is similar to that of shear impact tests.

4.3.2.2.4 Tension Impact Loading Test

This test also utilizes the U-section test specimens. The test fixture is designed so that the forces applied in fracturing the specimen are in tension (Figure 4.25). Bending of the specimen around the weld is inevitable when the weld is loaded. In all other respects, this test is similar to the shear impact loading test.

Except for the shear impact test, all of the other three tests are generally expensive to perform, and accurate measurements are difficult to make. Complicated fixtures and testing facilities required for these tests prohibit their frequent use in routine quality testing of spot welds. Unlike the other three, the shear impact test has the advantage of being simple to operate, easy to use, and of a relatively low cost.

However, there are several disadvantages in the design of the shear impact testing device that utilizes a modified specimen fixture that attaches the welded specimen to the moving

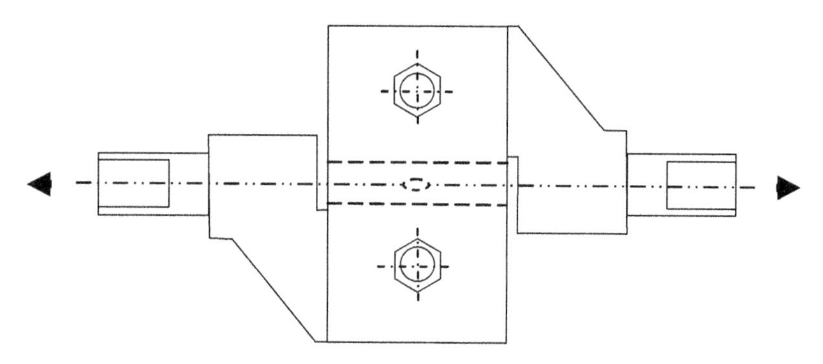

FIGURE 4.24
One type of testing specimens for shear impact loading testing.

pendulum, rather than to the foundation. First, it is difficult to grip specimens using such a fixturing mechanism. Also, because the specimens are mounted to the pendulum, it is difficult to instrument for accurate impact profile measurement. Besides, it is difficult to change input energy because the impact head weight is usually fixed. These disadvantages make such a device impractical, and thus not commonly used for weld quality testing. The widely scattered ranges obtained in some early works of impact testing may be attributed to the aforementioned reasons. There are a few other designs modifying the pendulum-based testing device to accommodate the needs of testing welded specimens, but they generally possess some of the disadvantages.

Because of the need for an easy way to measure the impact strength of spot welds, a device has been developed recently for impact testing of welded joints.[23] This newly designed impact tester minimizes most of the shortcomings of the standard shear impact tester, and preliminary experiments have shown promising results for making impact strength measurement an accurate and routine test for welded joints.

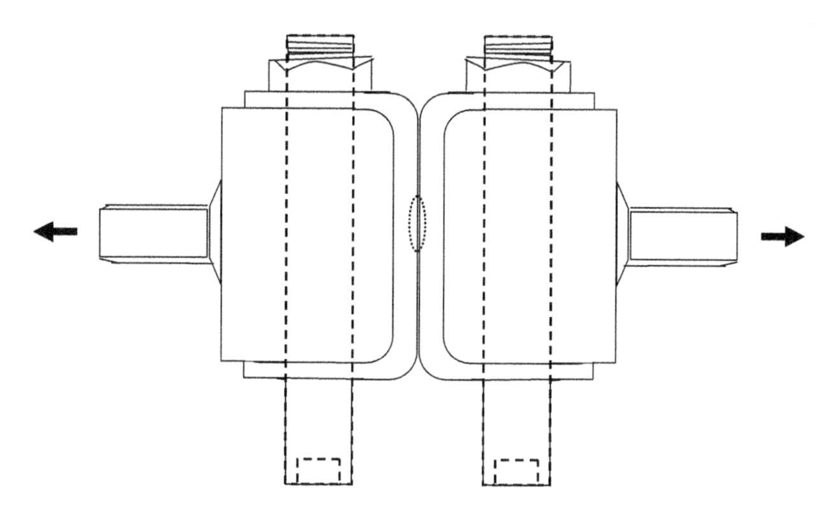

FIGURE 4.25
Tension impact test specimen.

4.3.2.3 A New Impact Tester

The new impact tester is shown in Figure 4.26. It uses pendulums for input and remaining energy measurement, similar to the shear impact tester. However, there are two pendulums on this device instead of one.

One pendulum (the one on the right side in the figure) is the active pendulum because it provides the impact energy; the other is the passive pendulum. They are marked pendulum A and pendulum B, respectively, in Figure 4.27a and b. Before testing, a Z-shaped welded specimen (Figure 4.27c) is mounted on pendulum B at one end and attached to the fixed machine base at the other end. The weld, a spot weld or other type of joint, is placed at the center joining the two pieces of bent coupons. The clamped bends at the ends of a specimen ensure a secure gripping of the specimen during impact.

Figure 4.27 also illustrates the impact procedure. Before testing, the active pendulum is raised to a specific position, for example, the horizontal position ($\theta_0 = 90°$). At this stage, the system has a defined potential energy. Additional weights can be added on the pendulum to adjust input energy. When this pendulum is released, its potential energy is converted into kinetic energy, which reaches its maximum value at the bottom right before impact. After the impact, the struck passive pendulum tends to move in the impact direction and pull the specimen apart. If there is enough input energy, the specimen will be torn into two pieces at the joint, and both pendulums will continue to swing forward because of the remaining energy in the system. The maximum angles (θ_A and θ_B) of movement of the pendulums after impact can be recorded by dial meters.

The energy consumed by a welded specimen can be expressed as

$$M_A gL_A(\cos\theta_A - \cos\theta_0) - M_B gL_B(1 - \cos\theta_B) - E_{rror} \quad (4.7)$$

where M_A and M_B are masses of the pendulums, L_A and L_B are mass centers of the pendulums from their respective pivotal points, g is gravitational acceleration, θ_0 is the initial angle of the active pendulum before testing, θ_A and θ_B are the maximum swing angles after impact, and E_{rror} is the energy consumed by the system, such as that consumed by friction

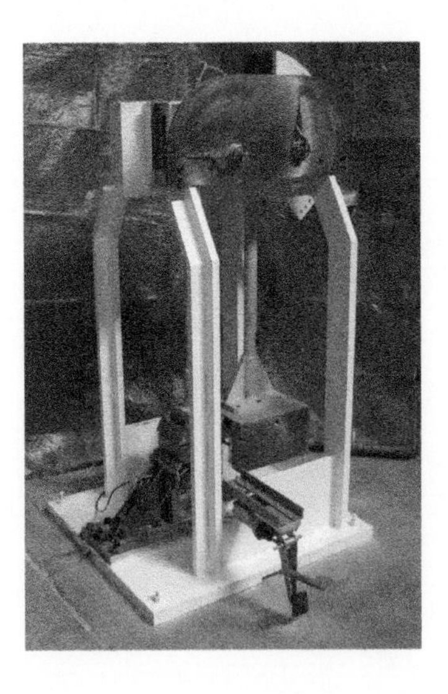

FIGURE 4.26
(See color insert.) A new impact tester.

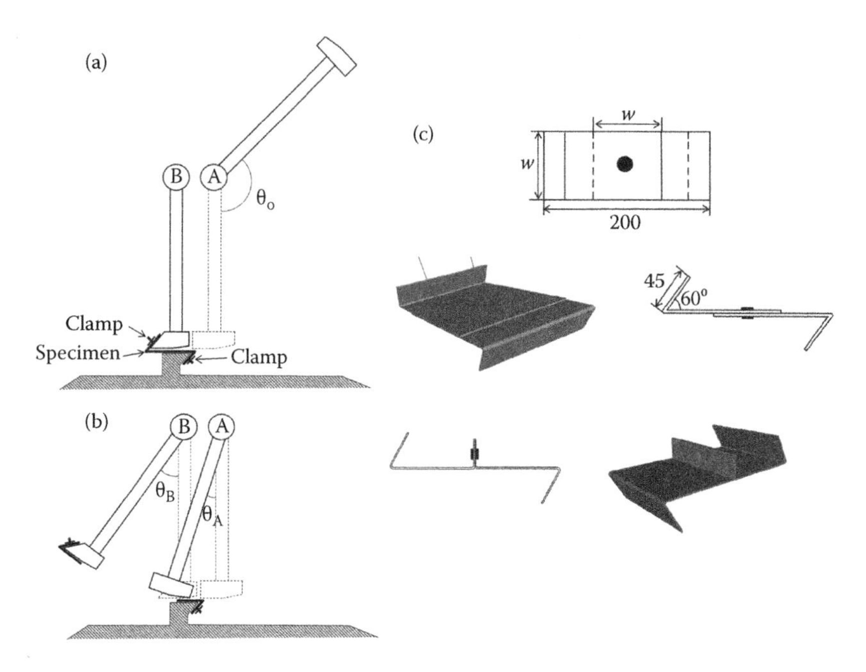

FIGURE 4.27
Schematic diagram of new impact tester testing procedure and specimen configuration.

or aerodynamic drag. After E_{rror} is determined, the maximum angles of the pendulums are the only quantities needed to calculate the energy consumed by the joint.

Figure 4.27c shows the specimens that can be used with this impact tester. One is used for tensile–shear testing the weld/joint, and another is for peeling the joint.

One advantage of this impact tester is the easiness of installing sensors to monitor the impact process, as a detailed knowledge of force and displacement profiles is essential to evaluating the strength and integrity of welded structures. Once the dynamic properties of spot welds are known, more realistic behaviors of welded structures under impact loading, such as crashworthiness, can be predicted. Based on such information, the optimal design can then be achieved to fully utilize the strength of welds without compromising safety.

Impact energy, impact load, and displacement fully describe a weld's impact strength. All three quantities are needed for structural modeling and crash simulation. Figure 4.28a to c shows the signals corresponding to three different failure modes of spot welds under impact.

Measured and calculated results are listed in Table 4.2 for a drawing steel (DS). Specimens 1 and 2 had interfacial failure and weld button pull-out, respectively. Their force and displacement signals look very similar, as shown in Figure 4.28a and b, although the displacement signal for the one with button pull-out is a little noisier than that with interfacial failure. However, for the one that failed by tearing the base metal (specimen 3 in Figure 4.28c), both load and displacement curves are quite different from the other two. The two force plateaus between $t = 0.7$ and 4.5 ms and between $t = 7.6$ and 8.2 ms in the figure correspond to tearing the base metal. In Table 4.2, energies were derived based on Equation 4.7 (marked as direct reading) or directly calculated (marked as calculated) from the signals. Very good consistency has been observed.

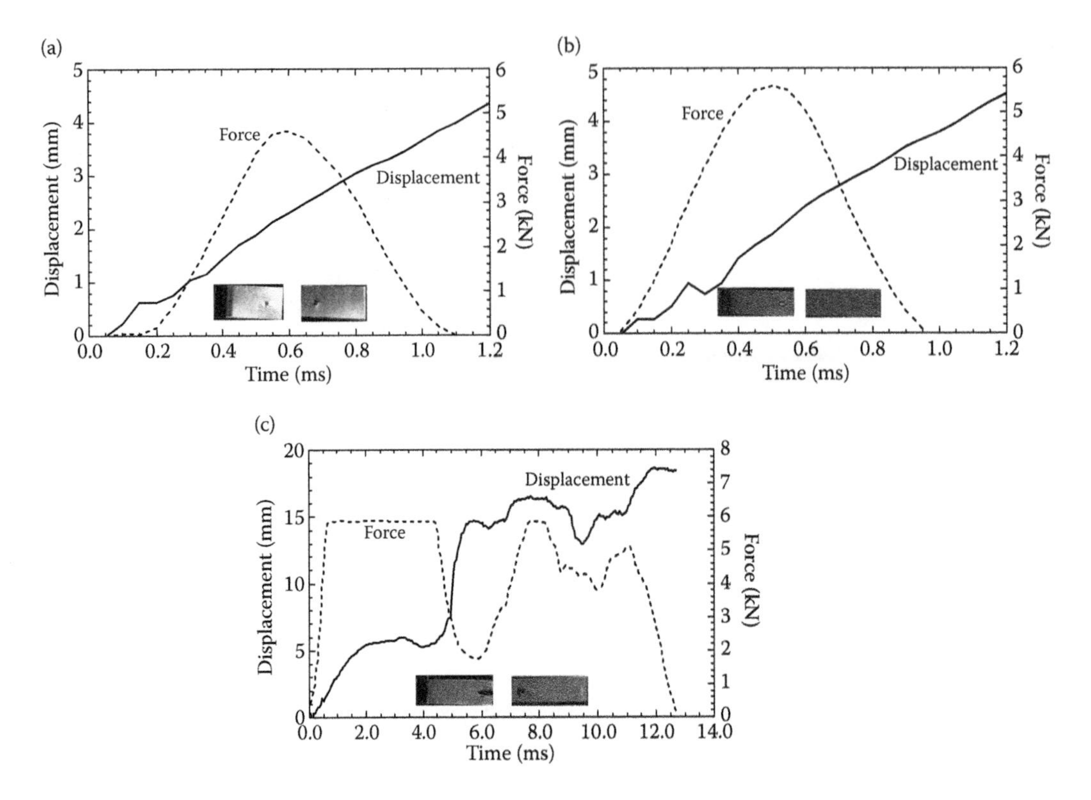

FIGURE 4.28
Displacement and force profiles of impact for various failure modes: (a) failure by interfacial fracture, (b) failure by pulling out button, and (c) failure by tearing the base metal.

In order to compare the impact and static strength of various welds, a study was conducted by testing welds of various sizes, shapes, and orientations, using pairs of electrodes machined to different sizes and shapes, as shown in Figure 4.29. There were three types of electrodes: round (diameter = 4, 6, and 8 mm, respectively), rectangular (3 × 6 mm), and ring (outer/inner diameter = 6/3 mm). The material used was a coated DS of size 150 × 50 × 1.0 mm for static tensile–shear tests, and 170 × 50 × 1.0 mm for impact tests. Five replications were made in each case. Welding time was eight cycles (133 ms), electrode force = 770 lb (3.4 kN), and current = 8.1–15.9 kA, which was adjusted according to electrode face size. Welding current was chosen to make welds that reflect the shape of the electrode faces, just below their expulsion limits. Specimens were prepared exactly the same way for impact and static tests. Static tests were conducted on a MTS testing machine (MTS-810) with a

TABLE 4.2
Measurement of Impact Tests with Various Failure Modes

Specimen No.	Passive Pendulum (°)	Active Pendulum (°)	Energy (J) (Direct Reading)	Energy (J) (Calculated)	Failure Mode
1	75.5	48.5	7.1	7.7	Interfacial shear
2	76.0	47.5	9.9	12.2	Pull-out
3	48.5	36.0	107.6	111.2	Tearing

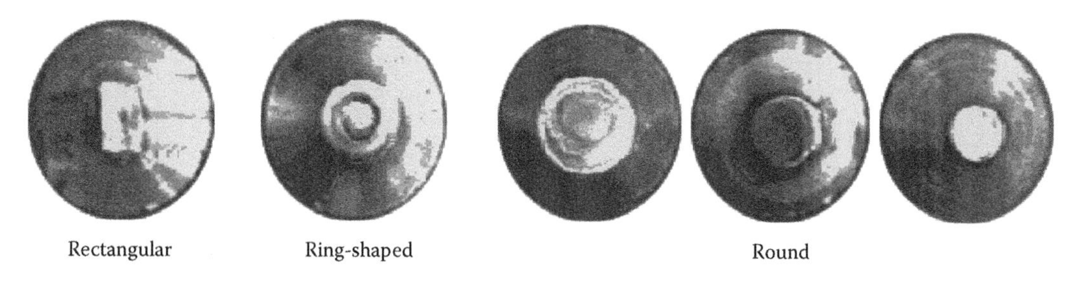

FIGURE 4.29
Electrodes with various faces used in welding.

testing speed of 10 mm/min, and impact tests were performed using the impact tester. Test results are shown in Table 4.3 and in Figures 4.30 and 4.31.

Metallographic examination shows that the weld nuggets generally mirror the electrode face shapes used in welding. The ring-shaped weld nuggets have an unfused region inside. For hot rings, the current was chosen just below the expulsion current, which is about 13.7 kA. In the case of cold rings, the welds were made at 12.3 kA. The welds made by rectangular-faced electrodes were actually oval shaped.

The impact energy was calculated according to Equation 4.7 through the readings of the dial meters. The strength measures of static tensile–shear tests, as defined in Figure 4.1, were calculated through recorded load vs. displacement curves. The impact energy is generally higher than quasi-static tensile–shear energy.

Failure modes in impact and static tests are also noteworthy. There is an almost one-to-one correspondence between impact and static tensile–shear tests, except for the specimens with hot rings (as shown in Table 4.3). Therefore, the same failure modes can be expected in impact testing as in quasi-static tensile–shear testing for the material and procedure used.

It can be seen from Figure 4.30 that impact energy increases with nugget size. For the case of ring-shaped welds, intentionally created internal discontinuities do not affect much of the impact energy for hot-ring welds, whereas they are important for cold rings. No distinguishable difference was observed between the cases of rectangular or oval-shaped welds placed in parallel and perpendicularly with respect to the loading direction. However, one should not conclude from such an observation that loading direction has no influence for these irregular-shaped welds. As a matter of fact, the aspect ratios of these welds are closer to

TABLE 4.3
Test Results (Average Values) for Impact and Static Tensile–Shear Testing

	Impact Tests		Tensile–Shear Tests			
Electrode	Energy (J)	Failure Mode	Peak Load (kN)	Maximum Displacement (mm)	Energy (J)	Failure Mode
$D = 4$ mm	9.39	Shear	5.12	1.06	4.25	Shear
$D = 6$ mm	142.06	Pull-out	6.93	3.64	22.11	Pull-out
$D = 8$ mm	153.85	Pull-out	7.31	4.61	29.86	Pull-out
Hot ring	162.82	Pull-out	7.03	4.07	25.57	Shear
Cold ring	6.14	Shear	5.15	0.48	1.58	Shear
Rectangular, parallel	8.14	Shear	4.48	0.74	2.57	Shear
Rectangular, normal	8.20	Shear	5.19	1.01	4.40	Shear

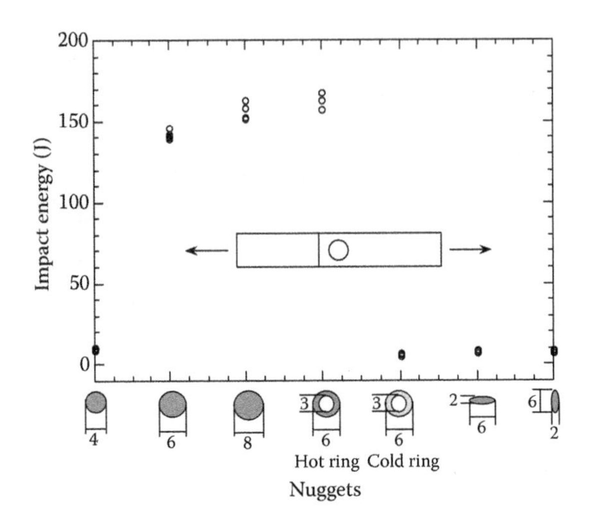

FIGURE 4.30
Impact energy vs. weld size, shape, and orientation. Pulling direction is shown in figure.

unity than they are supposed to. Good repeatability has been observed in impact energy, as shown by the figure, in which all five measures are plotted for each case with small spans.

Impact energy and static energy are plotted together in Figure 4.31. Energies for static and impact tests have similar trends, although their magnitudes are drastically different. Such a difference can be largely attributed to the effect of strain rate sensitivity of the material.

4.3.2.3.1 *Effect of Strain Rate*

During an impact test performed using the impact tester, the maximum strain rate imposed on the weld is approximately proportional to the speed of block B right after it is struck by block A. This pulling on the specimen may cause the connecting part (the joint) between the sheets to rapidly harden, and induce subsequent deformation to occur in the

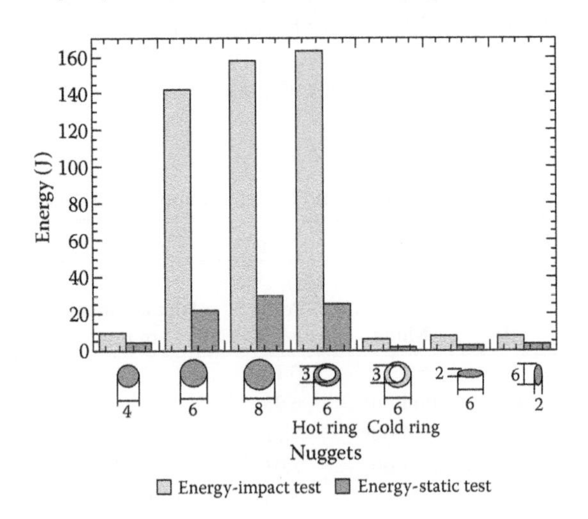

FIGURE 4.31
Impact and static energies vs. weld size, shape, and orientation.

surrounding softer metal. As a result, a significant amount of energy may be absorbed by the surrounding material. In contrast, deformation is limited to the immediate surroundings of a weld during static tests because no significant strain rate hardening occurs. This may explain the phenomenon observed in Figure 4.31, which shows significantly higher energy measurement in impact tests than in static tests of the same specimens. Therefore, possible effects of strain rate on energy measurement should be considered in an impact test and when comparing testing results.

4.3.3 Torsion Test

Although they are not as common as other tests, torsion tests are occasionally conducted on spot-welded specimens. There are two types of torsion tests, as described in the following sections.

4.3.3.1 Twisting

A standard tension specimen such as that shown in Figure 4.7, may be used for twist test to determine the weld diameter, as well as the angle of twist of failure. One end of the specimen is clamped edgewise in a vice, in a horizontal position. A sleeve, having a protractor mounted at its end, is slipped over the protruding end of the specimen so that the protractor is centered on the weld. The sleeve is rotated horizontally (i.e., in a flat arc about the vertical centerline of the weld) until failure occurs. The angle of twist at failure is measured and recorded.

4.3.3.2 Torsional Shear Test

The torsional shear test may be used where a measure of strength and ductility is required. Torsional shear is applied on the weld of a square test specimen by placing the material between two recessed plates, the upper of which is held rigid by means of a hinge, whereas the lower one is fastened to a rotational disk (Figure 4.32). After the specimen is placed

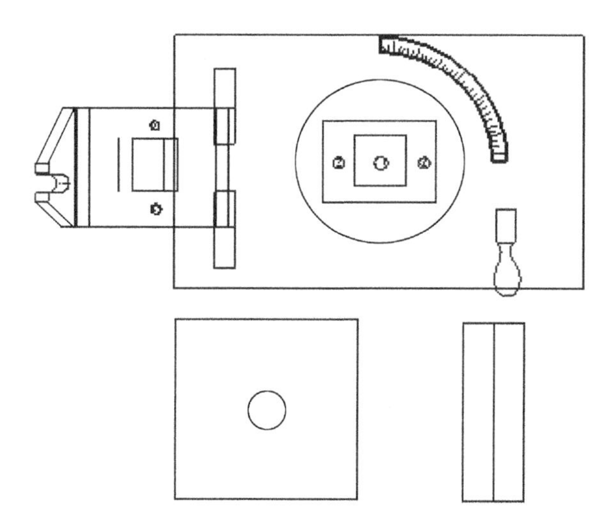

FIGUE 4.32
Torsion test.

in the square recess of the lower plate, the upper is closed over it and locked in position. Torque is exerted by means of a rack and pinion attached to the disk. It is important that the upper and lower sheets of the specimen are engaged separately by the two plates and that the weld be centrally located with respect to the axis of rotation. The following are determined for the weld area from the test:

1. The ultimate torque required to twist the weld to destruction (computed by multiplying the maximum load by the moment arm).
2. The angle of twist at ultimate torque measured by the angle of rotation at maximum load.
3. The weld diameter measured after the test specimen is broken.

The weld strength can be determined by the ultimate torque, the weld diameter, and the ductility by the angle of twist. The values obtained by these tests can be used as a measure of weld quality.

Besides mechanical property-related quality indices, there are other aspects that should be considered when assessing a weld's quality. For instance, the corrosion resistance of a weld, the influence of temperature on weld strength (such as transition), and aging of a weld may have a significant effect on the structural integrity of a spot weld. Testing conducted on a weld under normal conditions may not serve as a complete indicator of its behavior in service.

References

1. Zhou, M., Hu, S. J., and Zhang, H., Critical specimen sizes for tensile–shear testing of steel sheets, *Welding Journal*, 78, 305-s, 1999.
2. *Resistance Welding—Destructive Tests of Welds—Failure Types and Geometric Measurements for Resistance Spot, Seam and Projection Welds*, International Standard, ISO 14329, 2003.
3. Wang, P. C., Chisholm, S. K., Banas, G., and Lawrence, F. V., The role of failure mode, resistance spot weld and adhesive on the fatigue behavior of weld-bonded aluminum, *Welding Journal*, 72, 41-s, 1995.
4. *Resistance Spot Welding—Destructive Tests of Welds—Method for the Fatigue Testing of Spot Welded Joints*, International Standard, ISO/FDIS 14324, Final draft, 2003.
5. *Specification for Resistance Welding of Coated and Uncoated Carbon and Low Alloy Steels*, ANSI/AWS C1.4.
6. *Military Specification—Welding, Resistance: Spot and Seam*, MIL-W-6858D, U.S. Department of Defense, Washington, DC, 1992.
7. *Specimen Dimensions and Procedure for Shear Testing Resistance Spot and Embossed Projection Welds, International Standard*, ISO/DIS 14273, 1994.
8. AWS D8.7: *Recommended Practices for Automotive Weld Quality—Resistance Spot Welding*, American Welding Society, Miami, FL, 2003.
9. AWS C1.1: *Recommended Practices for Resistance Welding*, American Welding Society, Miami, FL, 1966.
10. *Specimen Dimensions and Procedure for Cross Tension Testing Resistance Spot and Embossed Projection Welds*, ISO 14272, 2000.
11. Mukhopadhyay, G., Bhattacharya, S., and Ray, K. K., Strength assessment of spot-welded sheets of interstitial free steels, *Journal of Materials Processing Technology*, 209 (4), 1995–2007, 2009.

12. *Specimen Dimensions and Procedure for Shear Testing Resistance Spot, Seam and Embossed Projection Welds*, International Standard, ISO 14273, 1994.
13. *Recommended Practices for Test Methods for Evaluating the Resistance Spot Welding Behavior or Automotive Sheet Steel Materials*, AWS D8.9-97, SAE D8.9-97, ANSI D8.9-97, Miami, FL, 1997.
14. Zhou, M., Relationship between spot weld attributes and weld performance, PhD dissertation, University of Michigan, Ann Arbor, MI, 2000.
15. Rivett, R. M., Final contract report: Resistance spot welding of steel sheet (for the European Coal and Steel Community), *The Welding Institute*, Report 3570/7/81, March 1982.
16. Lee, Y., Wehner, T., Lu, M., Morrissett, T., Pakalnins, E., and Tsai, C., Test of resistance spot welds under combined tension and shear, SMWC VII, Paper C2, 1996.
17. Maddox, S. J., *Fatigue Strength of Welded Structures*, 2nd edition, Woodhead Publ. Ltd., Abington Hall, Abington, GB, 1998.
18. Wilson, R. B. and Fine, T. E., Fatigue behavior of spot welded high strength steel joints, Society of Automotive Engineers, SAE, Paper No. 810354, 1981.
19. McMahon, J. C., Smith, G. A., and Lawrence, F. V., *Fatigue Crack Initiation and Growth in Tensile–Shear Spot Weldments*, ASTM STP 1058, edited by H. I. McHenry and J. M. Potter, American Society for Testing and Materials, Philadelphia, PA, 47–77, 1990.
20. Rui, Y., Borsos, R. S., Gopalakrishnan, R., Agrawal, H. N., and Rivard, C., Fatigue life prediction method for multi-spot-welded structures, Society of Automotive Engineers, SAE, Paper No. 930571, 1981.
21. Ma, C., Chen, D. L., Bhole, S. D., Boudreau, G., Lee, A., Biro, E., Microstructure and fracture characteristics of spot-welded DP600 steel, *Materials Science and Engineering A*, 485, 334–346, 2008.
22. Khan, M. S., Bhole, S. D., Chen, D.L., Biro, E., Boudreau, G., and van Deventer, J., Welding behaviour, microstructure and mechanical properties of dissimilar resistance spot welds between galvannealed HSLA350 and DP600 steels, *Science and Technology of Welding and Joining*, 14 (7), 616–625, 2009.
23. Zhang, H., Zhou, M., and Hu, S. J., Impact strength measurement and a new impact tester, *Journal of Mechanical Manufacture*, 215B, 403, 2001.

5

Resistance Welding Process Monitoring and Control

5.1 Introduction

The major difficulties in resistance spot welding (RSW) quality evaluation are related to the complexity of the basic processes in welding and their complicated interactions. In addition, variations in materials such as composition and coating, and process conditions such as electrode wear, workpiece fit-up, water cooling rate, and machine compliance, also influence the RSW monitoring. Considerable attention has been devoted to monitoring the process in order to gain information about weld quality, and to control the process to ensure quality welds. Electric current, voltage, force, displacement, and dynamic resistance signals are the most used in a monitoring and control system. For instance, Gedeon et al.[1] presented a work on monitoring these parameters and showed that displacement curves and dynamic resistance provided significant information for evaluating weld quality. However, difficulties are encountered in obtaining these signals because of the strong magnetic interference of the process, especially when alternating current is used in welding. The monitoring and control of a resistance welding process are very closely related to the quality definition of welds, which is discussed in detail in Chapter 6.

The objectives of monitoring and control can be summarized as follows:

1. *Weld size estimation.* The size of a weld (in the form of nugget width or button diameter) is the most commonly used index for quality, as it is closely related to the strength of a joint when there is no expulsion. Because of the multivariate nature of the RSW process, a weld's size is determined by many factors, and much of this dependence is not fully understood. An ideal monitoring system should provide an accurate on-line estimation of a button size based on the signals obtained from the process.
2. *Expulsion detection and its severity evaluation.* As expulsion induces many unfavorable features to a weld, as discussed in Chapter 7, it adversely influences weld quality. Therefore, a monitoring system should be able to detect if expulsion occurs and its severity. Remediation can then be taken if it is deemed necessary.
3. *Process fault diagnosis.* Nonconformable welds (such as undersized welds, stick welds, and welds with expulsion or discontinuities) are often results of faulty process conditions. The study of process faults is especially important, as it is the first step to link laboratory-developed monitoring and control algorithms to real-life production.
4. *Process control.* Effective control algorithms can be developed based on either information obtained from process monitoring or modeling, or both.

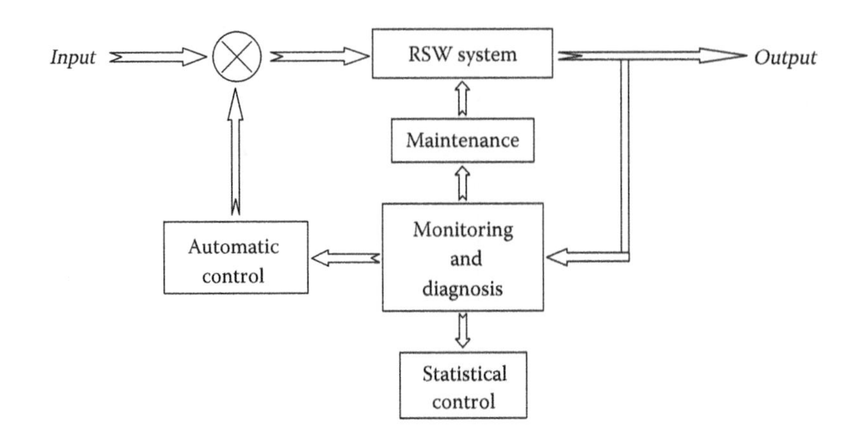

FIGURE 5.1
Schematic of a typical RSW monitoring and control system.

The ultimate goal of monitoring and control is to develop a robust on-line monitoring and diagnosis system for weld quality assurance so as to raise the manufacturer's confidence level and help reduce the cost of welded structures.

A general-purpose RSW monitoring and control system consists of three parts (as shown in Figure 5.1): a welding system, a monitoring unit, and a control unit.[2–4] An on-line monitoring and control system begins with an input to the welder, usually in the form of a welding schedule specifying welding current (or voltage or heat, depending on the weld controller), time, and electrode force. The output of the welder is then fed into the monitoring unit, which comprises data acquisition and signal processing. The processed information is then passed on to the control unit. If an action is warranted, the automatic control unit will modify the input and alter the schedules for subsequent welding processes. The results of the monitoring system can also be used for the purposes of statistical process control of weld quality and process maintenance scheduling. An ideal real-time feedback control system can adjust the welding parameters within a welding cycle if necessary.

In this chapter, process monitoring-related issues, such as signal collection, data analysis, and feature extraction, will be discussed first. Then several control algorithms will be presented.

5.2 Data Acquisition

Collecting process information, such as signals during welding, is the first step in most monitoring and real-time control systems. A data acquisition system can be used to measure tip voltage, welding current, electrode force, and electrode displacement, as shown in a schematic drawing (Figure 5.2). Transducers are usually mounted on a weld gun, at locations as close as possible to the electrode tips to capture signals directly related to the welding process. The tip voltage is measured between the tips of electrodes. A toroid sensor can be used to pick up the induced voltage, and the current is obtained by integrating the toroid voltage. A strain gauge-based force sensor is installed near an electrode. The signal conditioning box provides excitations to some sensors and scales signals to suitable voltage levels for the analog-to-digital converter (A/D) board in the computer. Then

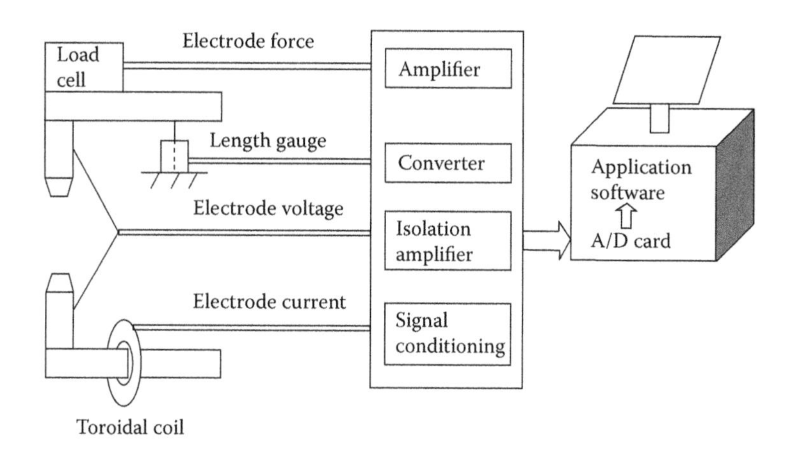

FIGURE 5.2
A data acquisition system for resistance spot welding process monitoring. (From Ma, C. et al., *Sci. Technol. Welding Joining*, 11 (4), 480–487, 2006. With permission.)

computer software such as National Instrument's Labview™ can be used for data acquisition. A common number for the sampling rate is 5000 Hz. Sampling time also needs to be controlled to collect signals only during the desired period; for example, for a typical welding cycle, 2.5 s is usually sufficient.

For monitoring and control, physical models, based on commonly used signals (electric current, tip voltage, dynamic force, and electrode displacement), are often developed to gain understanding of the process. However, the signal collection is not a straightforward procedure, especially when alternating current is used in welding. Because of the high-magnitude alternating welding current, process signals, especially electrode force and tip voltage, are corrupted by strong induced voltage noise. Simple A/D filters are usually not effective in removing this type of noise, and an adaptive signal processing technique may be needed.

Monitoring the electrode tip voltage is one of the earliest and simplest techniques. Although the voltage itself does not directly represent the heat generation or nugget growth, a number of adaptive control units have been developed that shut off the current at some predetermined voltage level (e.g., Nakata et al.[5]). Welding current is another important variable to monitor. Some controllers sense only the welding current and signal a fault when the measured value does not fall within prescribed limits. Similar to tip voltage, the welding current itself does not represent the input heat either. In order to determine the heat input to a weld, both voltage and current must be measured. Consequently, controllers based on the so-called constant-power (heat) control algorithm have been developed and are now commercially available. However, due to the variable energy needs and loss during an RSW process, constant power still cannot guarantee consistent weld quality.

Infrared emission, acoustic emission (AE), and ultrasonic signals have all been attempted for weld quality monitoring. However, the infrared emission method can only measure temperature far away from an actual weld. The variability of emissivity of materials is another reason that this kind of system failed. AE can be used to detect expulsion, but it is not presently useful for monitoring weld nugget growth.[7] The ultrasonic method has also been studied. However, it is intrusive in an assembly process and has not been proven to be reliable as an on-line monitoring means.

Both electrode displacement and electrode force during welding can reflect the nugget growth process. Electrode displacement is generally considered to be a better indication. It is believed that the amount of thermal expansion, melting, and expulsion can be correlated to the slope and magnitude of a displacement curve. Several control strategies have been developed based on monitoring displacement curves,[2,4,8,9] even though many consider it applicable only to pedestal welders and not to portable gun welders. Electrode force during welding can also be correlated to the amount of thermal expansion, melting, and expulsion. However, the correlation may not be consistent due to the variability of welding machine's characteristics. Some researchers have found that the dynamic force reflected the nugget growth process, whereas others have reported that the measurement provided little useful information.[1]

Dynamic resistance is a measure of the electrical resistance change during welding. It can be calculated from the tip voltage and welding current. Dynamic resistance has been shown to have a good correlation to the nugget growth[10] and has recieved increasing attention.

Although electrode displacement is considered the most revealing signal for nugget growth, on-line RSW monitoring and diagnosis systems usually consist of tip voltage, welding current, and electrode force only, to describe nugget growth in a production environment, due to the intrusive nature of electrode displacement sensors.

A significant amount of research has been conducted on instrumentation of RSW processes. However, little has been done to understand the physical origins of the signal waveforms. Some efforts have been devoted to removing the induced noise from corrupted electrode force and tip voltage measurements. Yet it remains an issue for signal processing of the monitoring and control systems.

Most of the work on monitoring and control has been focused on welding under nominal process conditions (e.g., perfect alignment, no edge-weld conditions); very limited work has been done to investigate the effects of different process conditions. In this chapter, the effects of abnormal conditions are also discussed, which serves as a liaison between developed monitoring and control systems and their application in the actual production environment.

5.3 Process Monitoring

Monitoring a welding process provides useful information on the physical processes involved in welding, and it is a necessary step toward successful control of the process. A direct measurement of weld quality, such as weld size or weld strength, can sometimes be used as a means of monitoring welding process. However, such monitoring ignores the details in welding; it considers only the end results of welding instead, and its usefulness is very limited. A process monitoring of real meaning contains detailed observation of the process through the use of various sensors, and it is correlated with weld quality. In this section, common signals collected during RSW are discussed, and their use for welding process monitoring is presented.

5.3.1 Signals Commonly Monitored during Welding

Intuitively, welding voltage and current should be monitored, as they are directly related to Joule heating, or the formation of a weld nugget. In addition, the thermal process during welding is reflected in the expansion or shrinkage of the sheet metal stack-up, or it can be monitored through the changes in electrode force and electrode displacement. Figure 5.3 shows typical signals collected during an RSW. The voltage, current, electrode force, and displacement are each discussed in detail in this section.

The process signals are heavily influenced by the electric and magnetic fields surrounding the welder and, therefore, it is necessary to understand the electrical aspect of a welder. The electrical part of a welding machine involves mainly a transformer, which brings the high line voltage down to a low secondary voltage and provides a high current to the secondary loop.[12–14] The system can be represented by a two-port transformer model, as shown in Figure 5.4. r_{o1} and L_{o1} are the primary resistance and inductance, respectively. Similarly, r_{o2} and L_{o2} are the parameters for the secondary side of the transformer. Because of the high welding current, the single loop formed by the machine throat appears as an inductance element (L) in the circuit. R is the resistance of the weldment, V_1 is the primary voltage, I_2 is the secondary current, and a is the turns ratio. The electrode tip voltage is measured across the resistance R, whose measurement circuit also forms a loop in the magnetic field. The inductance of this loop is L_m.

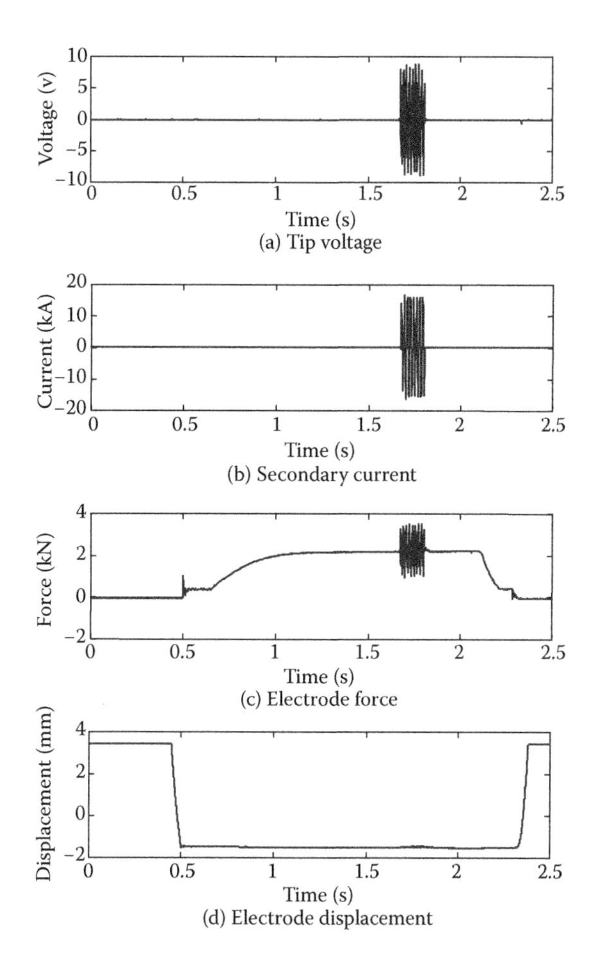

FIGURE 5.3
Typical signals observed during a spot welding. (From Li, W. et al., Signal processing issues in resistance spot welding, Sheet Metal Welding Conference IX, Sterling Heights, MI, Paper 32, 2000. With permission.)

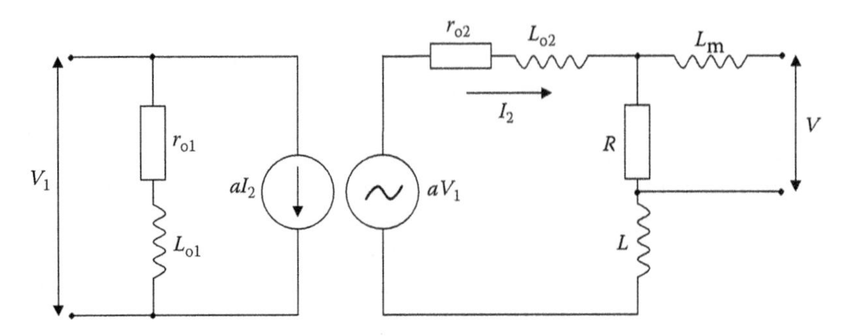

FIGURE 5.4
Schematic of a two-port transformer model.

Based on Figure 5.4, the current and voltage across a weldment are related to other variables in the following form:

$$I_2 = \frac{aV_1}{(R + r_{o2}) + j\omega(L + L_{o2})} \tag{5.1}$$

$$V = I_2 R + L_m \frac{dI_2}{dt} \tag{5.2}$$

The primary voltage of the transformer, V_1, is controlled by a silicon-controlled rectifier (SCR). The SCR starts to fire and lets current pass when the firing command is given. It usually closes when the line voltage reaches zero with a resistive load. If the load is inductive, the SCR will stay open until the current becomes zero. Because of the phase shift between the current and voltage in an inductive circuit, the primary voltage of the transformer is seen to have the waveform shown in Figure 5.5.

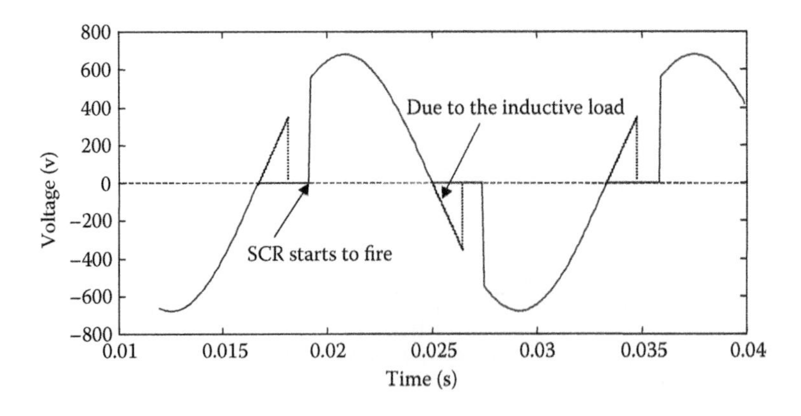

FIGURE 5.5
Simulated primary voltage of a transformer.

5.3.1.1 Electric Voltage

The tip voltage can be measured with two wire leads attached to the electrode tips. As the voltage is kept at a fairly low level at the electrode tips, its value can be directly measured using standard equipment. However, the voltage signal may be corrupted by the noise induced by an alternating current.

Induced voltage becomes a strong noise on electrode tip voltage signals because its measurement has a (unavoidable) wire loop in the magnetic field. It is well known that to minimize the inductive noise, one can use twist pairs to reduce the area of the wire loop. However, the wire loop can never be fully eliminated. For production applications, the two wire leads have to follow the arms and encompass the entire throat of a welding machine. To suppress the induced noise on the tip voltage measurement, a compensating loop can be added. However, adjusting the compensating coefficient is machine dependent and can be time-consuming.

5.3.1.2 Electric Current

The electric current signal is more difficult to deal with than voltage; usually, the current value is very high in a secondary loop, and the measurement is done indirectly. It is usually measured using either a sensor based on the Hall effect or a toroid sensor. Toroid sensors are fairly popular for electric current measurement. As both are based on the induced voltage by a welding current, it is difficult to separate the measurement from the process noise, which is also the result of induction.

A high alternating welding current induces a strong time-varying magnetic field. Any wire loops in the field will pick up induced voltage, whose magnitude is given by Faraday's law:

$$V = \frac{\mathrm{d}I}{\mathrm{d}t} A \cos\theta \tag{5.3}$$

where V is the induced voltage, I is the induced current, $\mathrm{d}I/\mathrm{d}t$ is the time change rate of the current, A is the area of the loop, and θ is the angle of the loop to the magnetic field.

Faraday's law is the basis for the current measurement using a toroid sensor. It is obvious that variations in position or orientation of the toroid can cause variations in the effective area, and hence, in the current measurement. However, the error is usually under 5% of the reading when simply hanging the toroid on the arm of a welding machine. The error can be further reduced by properly fixing the position of the toroid in the magnetic field.

There are two other methods to measure a current using either a Hall effect sensor or a resistive shunt. Hall effect sensors measure the voltage across a semiconductor due to the surrounding magnetic fields. They are small, and thus are more sensitive to temperature change. They are also sensitive to variations in orientation and position. The resistive shunt method directly measures the voltage across a known resistor in the current path. It is a standard means of measuring low amperage or DC currents. However, for RSW applications, electrodes have to be modified to use resistive shunt for measuring electric current.

5.3.1.3 Dynamic Resistance

Resistance of a resistor is usually calculated from the ratio of voltage to current. However, the voltage measured during a resistance welding contains two parts: contributions from resistance and inductance, as depicted in Equation 5.2. In the equation, L_m is the inductance, a (unknown) function of the loop, and it varies with many factors, such as the size and material properties of the workpieces. It can be clearly seen that the ratio V/I_2 produces the resistance value R only when dI_2/dt is zero.

dI_2/dt equals zero when the current reaches its peak points. Therefore, the number of points of a dynamic resistance measurement that can be calculated for an alternating-current (AC) welder is twice the number of cycles. Attempts have been made to obtain a continuous dynamic resistance curve by first removing the induced voltage noise and then taking a direct division of the voltage by current. Considering that there are periodic zero values on welding current signals, dynamic resistance is not attainable in those areas. Close to those areas, the magnitude of the current is small, and thus, the signal-to-noise ratio is very low, compared to the points close to the peaks. Therefore, dynamic resistance can only be reliably calculated at the points around the current peaks.

Without losing much information, the dynamic resistance curve can be obtained with piecewise polynomial curve fitting through the points obtained at the current peaks. Figure 5.6 shows a dynamic resistance curve for a 14-cycle welding made on a 0.8-mm galvanized steel. The dots were calculated on the current peak points, and the curve was fitted by a cubic smoothing spline with the smoothing parameter of 0.5.

The dynamic resistance was measured during welding of a hot-dip galvanized (HDG) DP600 steel,[6] and the effect of applied current is shown in Figure 5.7. The dynamic resistance was measured at peaks of the single-phase AC profile to eliminate the inductive noise, with a fixed electrode force of 3.34 kN. A low dynamic resistance corresponds to a high electric current, possibly due to the large contact area created by the high current or heat input. For all the currents, the dynamic resistance dips first because the molten zinc enlarges the contact area at the interfaces, and then rises when the bulk metal is heated. The following fluctuation reflects the influence of heating, melting/solidification, and deformation of the bulk metal, as well as the change in electric contact at the interfaces. As observed by other researchers, a sudden drop in dynamic resistance occurs when expulsion happens, owing to the decrease in the weld thickness and increase in the contact area at the faying interface when a certain liquid metal is ejected from the liquid nugget.

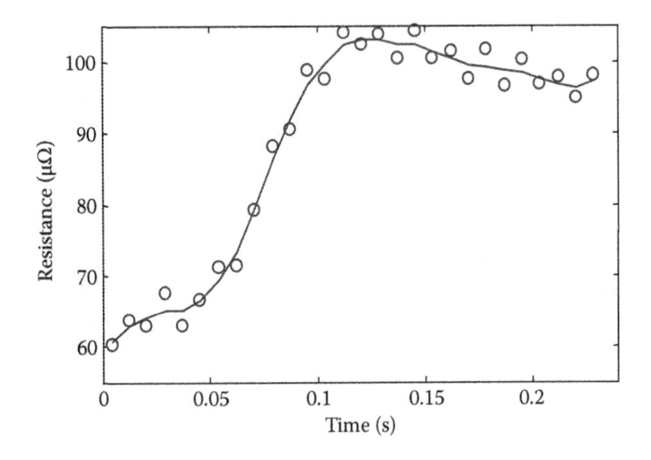

FIGURE 5.6
A typical dynamic resistance curve.

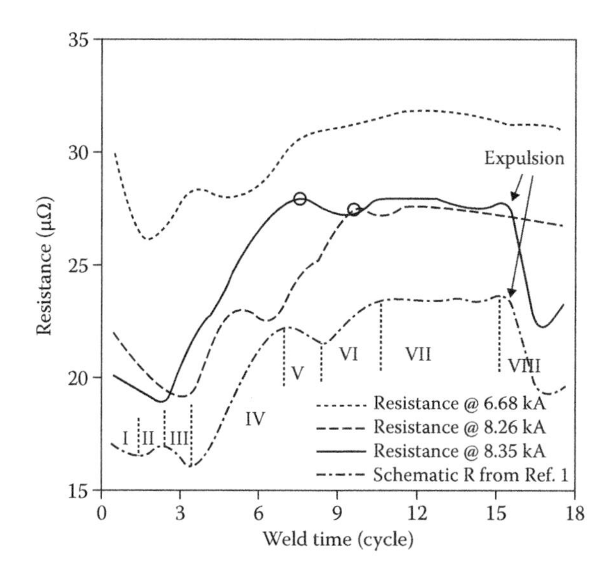

FIGURE 5.7
Dynamic resistance measurements at various current values.[6] (From Ma, C. et al., *Sci. Technol. Welding Joining*, 11 (4), 480–487, 2006. With permission.)

5.3.1.4 Electrode Displacement

The electrode displacement generally refers to the relative motion of electrode tips, which directly reflects the thermal process occurring in a weldment. To avoid the influence of other deformable components of a welder, the displacement sensor should be mounted as close as possible to the electrodes. Commonly used displacement sensors for RSW process monitoring are linear variable differential transducer (LVDT) sensors and fiber-optic sensors.

5.3.1.4.1 LVDT Displacement Sensors

LVDT sensors are commonly used in electrode displacement measurements. They are fairly reliable in catching the signals during both initial touching and welding. The signals collected are usually cleaner than those of current or electrode force. Although the displacement signals collected reveal a significant amount of information about the welding process, the sensor and its fixture are intrusive to the material handling and welding processes. However, the knowledge learned from LVDT sensors, even from a laboratory setting, can be useful for monitoring welding in the production environment.

5.3.1.4.2 Fiber-Optic Displacement Sensors

These sensors are of a reflective type, and they utilize bundled glass fibers to receive and transmit light to and from target surfaces. Various displacement sensitivities can be created via unique combinations of light source, fiber type, fiber bundle shape and size, etc. A fiber-optic displacement sensor can be considered noncontact. However, the distance between the probe and the reflective mirror has to be small enough to be effective. One advantage of fiber-optic displacement sensors, compared to LVDT sensors, is that they are less affected by high magnetic fields during welding. They can be used on most materials, irrespective of the color or conductivity.

As shown in Figure 5.3d, the displacement signals measured using an LVDT sensor directly reflect the physical processes involved in a welding cycle. The initial drop in displacement is the result of electrode close-up. When the electrodes touch the workpiece, depending on the stiffness and damping characteristics of the welder, the electrode displacement may fluctuate during electrode impact. The corresponding electrode force will show fluctuations of (usually) larger magnitude, as shown in Figure 5.3c, as well. In the figure, another cluster of small bumps is recorded when electric current is applied. A magnified view of this period is usually needed when analyzing the data, as the magnitude of electrode motion is fairly small during welding.

The amplified view of the displacement curve during the application of electric current, measured by a fiber-optic sensor, is shown in Figure 5.8. It shows that the thickness of the weldment grows with current due to heating. The drop around the fourth cycle is probably the result of softening of the workpiece stack-up due to initial melting. The number of peaks and valleys in the displacement curve during welding is exactly twice that of the current profile. It is obvious that the half of a current cycle produces both heating (or a peak in the displacement curve) and cooling (corresponding to a valley in the displacement curve). The other half cycle also produces a peak and a valley. After the current is cut off (after the 12th cycle), the fiber-optic displacement sensor shows a region that possibly corresponds to the cooling and contraction of the nugget.

The displacement of electrodes during welding is the result of electrode squeezing and the weldment expansion/shrinkage. Therefore, it is a function of electrode force. A comparison of the effects of various load levels is shown in Figure 5.9. Before current is applied, the workpieces deform under the applied load, as shown by different starting points of the curves. There is a significant increase in indentation from 600 to 1000 lb, resulting from mechanical compression by the electrodes. However, there is no difference between the

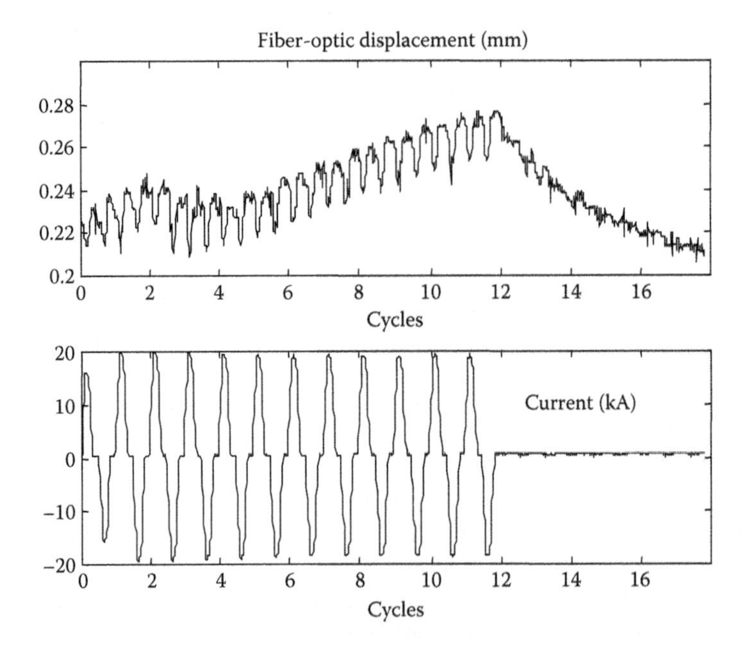

FIGURE 5.8
Fiber-optic displacement compared with current profile. Load level, 1000 lb; current, 11.3 kA; HDG steel thickness, 0.8 mm.

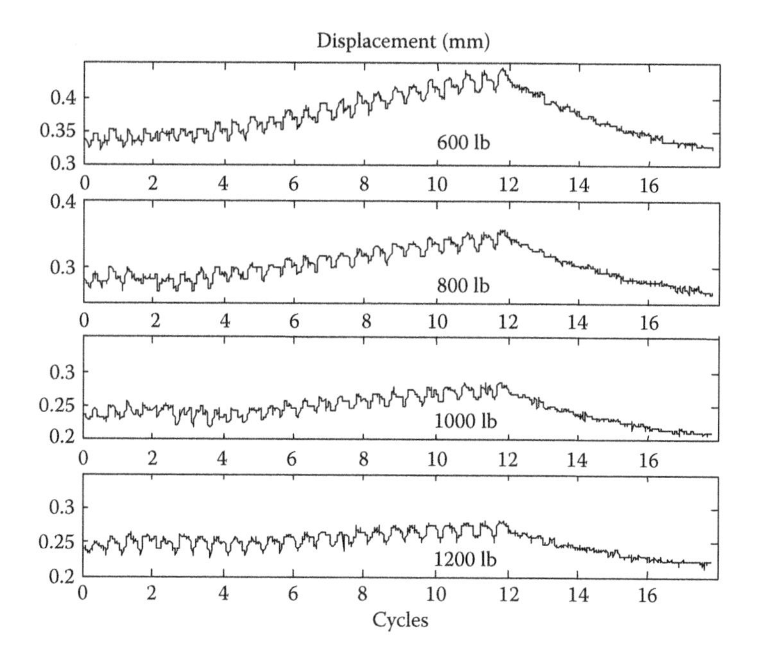

FIGURE 5.9
Fiber-optic displacement under various electrode forces.

observations of 1000 and 1200 lb, indicating the limit of mechanical deformation is reached around an electrode force of 1000 lb. In the welding cycles, a lower applied load results in a larger displacement in magnitude and slope. Rapid nugget growth starts from the third cycle for the cases of 600 and 800 lb, and around five cycles are needed when 1000 and 1200 lb electrode forces are used, because the electric resistance, therefore, electric heating, is inversely proportional to electrode force level up to a certain value. The current is 12 cycles at 58% heat, and the HDG steel thickness is 0.8 mm.

5.3.1.5 Electrode Force

The signals of electrode force are directly related to the interaction between the electrodes and the workpieces, and they reflect the changes in the reaction force by the weldment onto the electrodes. Therefore, electrode force yields useful information on the welding process. There are two types of sensors commonly used for electrode force measurement: strain gauge-based sensors and piezoelectric sensors. During welding using alternating current, the induced voltage produces strong noises on electrode force measurement because the sensor setup has unavoidable wire loops in the magnetic field. Using twist pairs may help reduce the wire loop area and therefore the noise. The induced noise on the strain gauge-based sensors may be amplified hundreds of times as the strain gauge signals are usually in the order of millivolts. Adding a compensating loop may alleviate the problem. However, adjusting the compensating coefficient is machine dependent and can be time-consuming. It is also quite tedious to add compensating loops to a strain gauge measurement device. A typical signal of electrode force is shown in Figure 5.3c, which clearly shows the initial squeezing (or touching), fluctuation during welding, and hold period.

For a strain gauge-based force sensor, the effect of the induced magnetic field also shows orientation sensitivity, besides the noises in the collected force signals. As shown in Figure 5.10 for a strain gauge-based force sensor, the induced voltage in the force measurement is low around orientations 0°, 150°, and 180°. Besides the magnitude of the induced voltage, variability of measurement is also usually a function of orientation.

Compared to a strain gauge-based force sensor, a piezoelectric force sensor is less influenced by the induced magnetic field. Figure 5.11 shows a comparison between a strain gauge-based sensor and a piezoelectric sensor. These two sensors were tested simultaneously with the strain gauge-based sensor mounted above the upper electrode and the piezoelectric sensor mounted under the lower electrode. The piezoelectric load cell shows a force change with each half current cycle and a significantly reduced dI/dt effect, as shown in Figure 5.12, a blow-up of Figure 5.11. However, it did not show the same force buildup during nugget growth as the strain gauge-based sensor.

Before the welding current was applied, the piezoelectric load cell showed measurements similar to those of the strain gauge-based sensor with large fluctuation because of sheet bouncing (Figure 5.11). However, the electrode force measured by the piezoelectric load cell decreases in the welding cycles, as shown in Figure 5.12. The force increases gradually during the hold cycles. This phenomenon should not be considered a true reflection of the actual force, as the force should reach a level similar to that before welding after the current is shut off. This can be explained by the fundamental physical processes involved in a piezoelectric sensor. The quartz crystals of a piezoelectric force sensor generate an electrostatic charge when a force is applied to or removed from them. The electrostatic charge then leaks exponentially to zero through the lowest resistance path, and the resistance and capacitance of the built-in electronics in the sensor normally determine the leakage rate. The decreasing force during the current flow seems similar to the leakage of an electrostatic charge. It is possible that the strong magnetic field during the current cycles might speed up the leakage of the electrostatic charge.

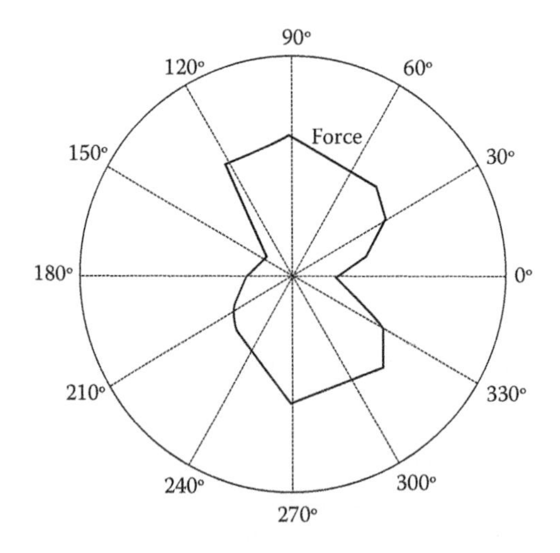

FIGURE 5.10
Polar plot of a strain gauge-based sensor in various orientations. (From NIST—ATP Intelligent Resistance Welding Quarterly Progress Report, No. 103, Ann Arbor, MI, 1996. With permission.)

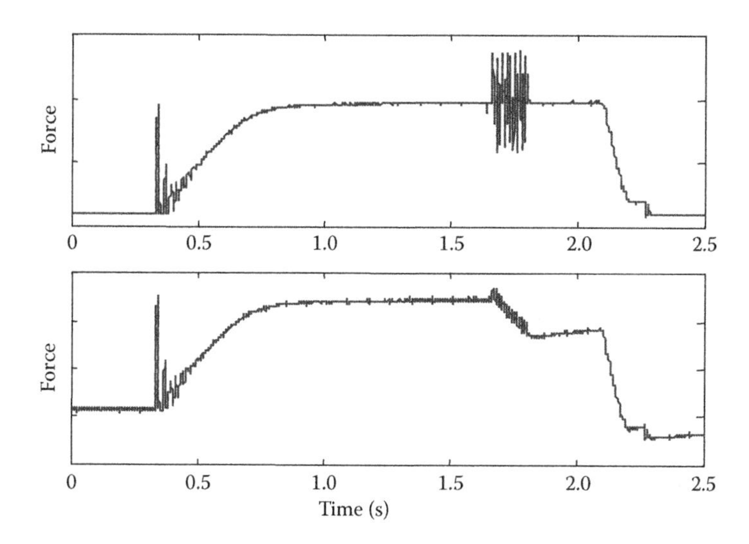

FIGURE 5.11
Force signals by a strain gauge-based sensor (top) and a piezoelectric force sensor (bottom).

5.3.1.6 Acoustic Emission (AE)

AE signals have been used in monitoring a welding process by several researchers (e.g., Ma et al.[6]). A typical AE monitoring system uses a waveform recorder or an oscilloscope to store or view the signals. A detection threshold is set to count the AE energy with gated window from the reference signal. The total AE energy counts have been found to be proportional to the nugget area of a spot weld.

Experiments have shown that when an AE sensor is mounted on the coupons, better signal patterns in the heating cycles are obtained than at other locations. Figure 5.13 shows decimated AE signals when the sensor is mounted on the coupon. High magnitudes at the

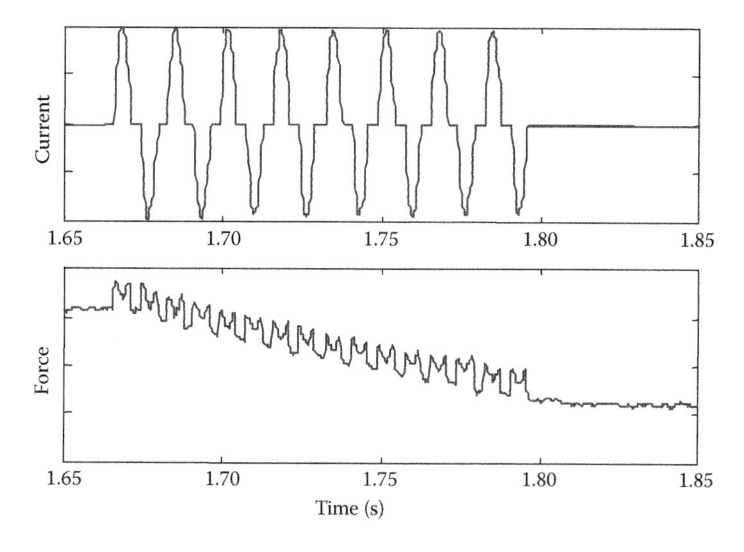

FIGURE 5.12
Detailed piezoelectric output referenced by electric current during welding.

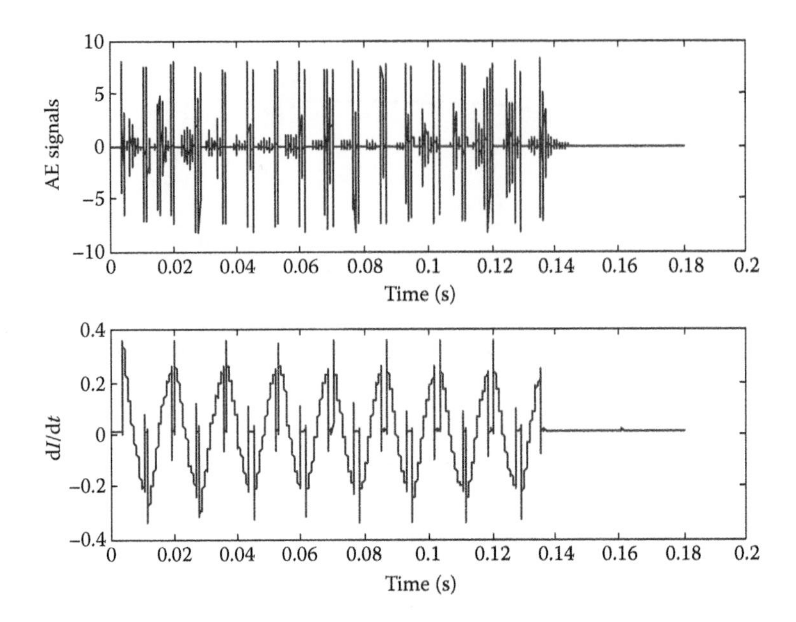

FIGURE 5.13
AE and dI/dt signals measured on coupons.

middle of half cycles are due to higher current input in the middle of the half cycles. Large spikes are shown every time the current is fired or ends. Large signals at the middle of the first few half cycles are related to melting and expulsion of the zinc coatings. Signals from the last half cycles may possibly be related to the nugget formation process.

5.3.1.7 Pneumatic Pressure Fluctuation

Although displacement and electrode force can reveal a significant amount of information about a welding process, they have to be placed very close to the electrodes, that is, they are intrusive. It is preferred if some useful signals can be obtained using sensors far enough from the electrodes and the stack-up. The air pressure change in a pneumatic cylinder during welding may provide certain information on the welding process, as illustrated in Chapter 8. This is expected, as the electrode force and the air pressure are directly related. The working mechanism of a typical air cylinder is shown in Figure 5.14.

Inlet and outlet airflow rates are controlled by the regulating valves. The settings of these two valves affect the moving speed of the upper arm. When the squeeze cycle begins, the air at high pressure from the air supply passes through the regulating valve and flows into the upper part of the cylinder. Because the setting pressure is higher than the air pressure at the lower part of the cylinder, the piston will be forced to move. Meanwhile, the air in the lower cylinder will be squeezed out of the cylinder.

When the upper electrode touches the workpiece and lower electrode, the lower electrode is forced to deform and produce resistive force. Meanwhile, the pressure in the upper part of the cylinder builds up quickly to the set pressure. The pressure at the lower chamber of the cylinder will eventually reach the ambient pressure because the air is released to the environment as the electrode moves down. After the current is turned on, the heated metal sheets start to expand. The two electrodes are then pushed outward. The cooling-induced shrinkage draws electrodes to move closer to each other. Because the electrode

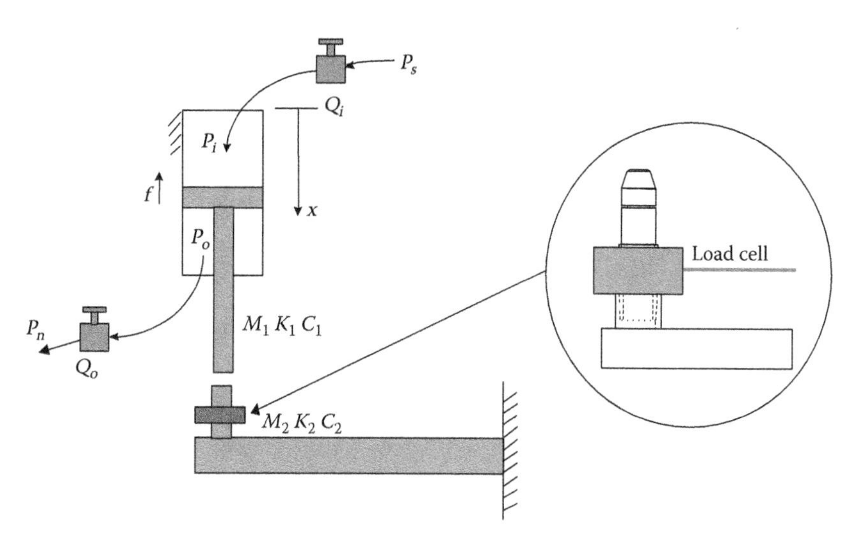

FIGURE 5.14
A schematic drawing of a welding machine air cylinder system.

location and the electrode force are directly related to the position of the piston head in the cylinder chamber, the pressure in the chamber may have similar changes during welding as electrode force and displacement. The results shown in Chapter 8 prove that this is indeed the case.

5.3.2 Adaptive Noise Cancellation

Because of the strong magnetic field induced when welding using alternating current, sensors with wires in the magnetic field may pick up induced voltage, and the raw signals collected on these sensors are the sum of the true signals and the induced noise. For instance, the electrode tip voltage and dynamic force signals are strongly corrupted by the induced voltage noise. The periodic noise, as a result of periodic change in the magnetic field, contains all the odd-order harmonics of the base frequency (60 Hz) of the (secondary) electrical current. The magnitude of the induced noise is proportional to dI/dt (where I is the secondary current). This type of noise is difficult to remove using ordinary analog or digital notch filters, as the true signals also fluctuate with a base frequency of 60 Hz. An adaptive noise cancellation (ANC) scheme may be used for this purpose.

ANC relies on subtracting noise from a received signal in an adaptive manner to improve the signal-to-noise ratio. Ordinarily, it is inadvisable to subtract noise from a received signal, as such an operation could produce even poorer results because it may cause an increase in the average power of the output noise. However, when proper provisions are made (i.e., filtering and subtraction are controlled by an adaptive process), it is possible to achieve a superior system performance compared to direct filtering of the received signal.[15] Figure 5.15 illustrates the ANC scheme.

In Figure 5.15, the primary sensor receives input from both the signal source and the noise source.[3,11] Thus, the output of the primary sensor is the signal corrupted with noise. The reference sensor picks up input only from the noise source. It is obvious that what is picked up by the primary and reference sensors will not be exactly the same. Assuming that the reference sensor picks up the true noise, then it can be assumed that the primary sensor picks up a transformed version of it. The transfer function can be recursively

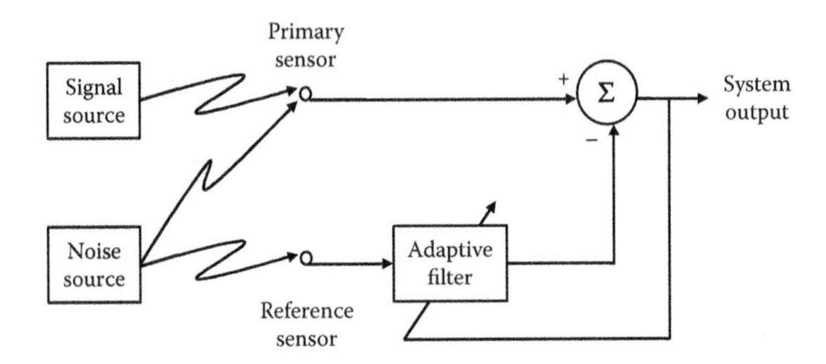

FIGURE 5.15
Illustration of adaptive noise canceling scheme.

estimated by an adaptive filter. Consider electrode force measurement: the primary sensor is a load cell used to measure the force signal, and then the toroid sensor (for electric current measurement) can be used as a reference sensor to pick up the induced voltage. The adaptive filter is realized using a standard recursive least squares algorithm, as shown below.

Initialize the algorithm by setting $\boldsymbol{P}(0) = \delta^{-1}\boldsymbol{I}$, δ = a small positive constant

$$\hat{\boldsymbol{w}}(0) = 0 \tag{5.4}$$

For each instant of time, $n = 1, 2, \ldots$, compute

$$\begin{aligned} \boldsymbol{k}(n) &= \frac{\boldsymbol{P}(n-1)\boldsymbol{u}(n)}{\lambda + \boldsymbol{u}^{\mathrm{H}}(n)\boldsymbol{P}(n-1)\boldsymbol{u}(n)} \\ \xi(n) &= d(n) - \hat{\boldsymbol{w}}^{\mathrm{H}}(n-1)\boldsymbol{u}(n) \\ \hat{\boldsymbol{w}}(n) &= \hat{\boldsymbol{w}}(n-1) + \boldsymbol{k}(n)\xi^{*}(n) \\ \boldsymbol{P}(n) &= \lambda^{-1}(\boldsymbol{P}(n-1) - \boldsymbol{k}(n)\boldsymbol{u}^{\mathrm{H}}(n)\boldsymbol{P}(n-1) \end{aligned} \tag{5.5}$$

The $m \times m$ matrix $\boldsymbol{P}(n)$ is referred to as the inverse correlation matrix. The $m \times 1$ vectors $\boldsymbol{k}(n)$, $\hat{\boldsymbol{w}}(n)$, and $\boldsymbol{u}(n)$ are referred to as the gain, estimated tap-weight, and tap-input vectors at time n, respectively. The scalar values $d(n)$ and $\xi(n)$ are desired output and *a priori* estimation of error at time n. The scalar value λ is a positive constant used to define the forgetting factor. The asterisk denotes complex conjugation, and the superscript H denotes a matrix Hermitian. For the ANC algorithm in RSW, the tap-input vector $\boldsymbol{u}(n)$ is constructed from the toroid voltage signal; the desired output scalar $d(n)$ is set to be the corrupted force signal (or tip voltage signal); and the *a priori* estimated error scalar $\xi(n)$ is the filtered signal, with the induced noise being cancelled.

Dynamic force signals before and after the ANC is shown in Figure 5.16. Their corresponding power spectra are shown in Figure 5.17. It can be seen clearly that the harmonics due to induced noise have been successfully removed, leaving some low-frequency components. The force sensor may also pick up some signals corresponding to metal expansion and contraction during welding. These signals are harmonics of 120 Hz and believed to be due to the effect of heating and cooling and the magnetic force generated by the 60-Hz welding current.

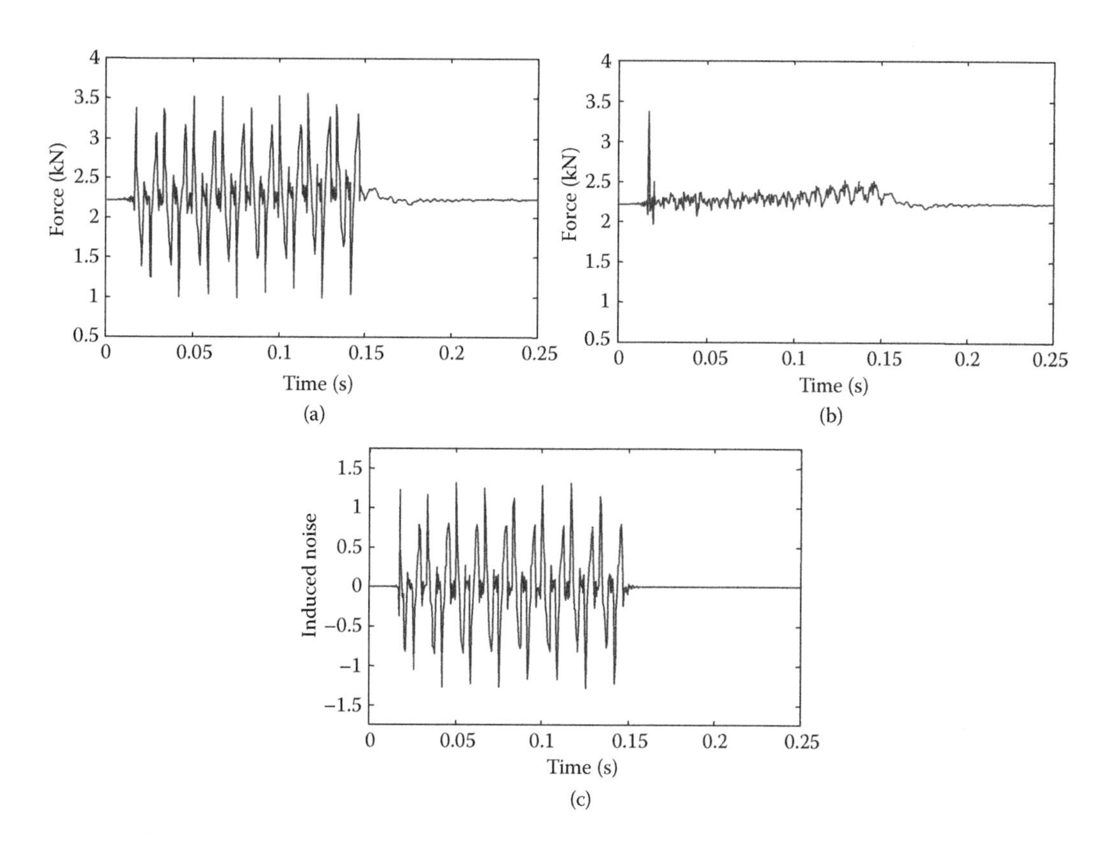

FIGURE 5.16
Adaptive filtering for dynamic force signals. (a) Dynamic force before filtering. (b) Dynamic force after filtering. (c) Noise subtracted.

ANC has certain advantages over the compensation method when applied to electrode tip voltage measurement. When using a compensation wire loop, the loop size has to be adjusted to suit the individual machine. Some researchers used a toroid sensor with a fixed loop size to measure the intensity of the magnetic field. The induced voltage is cancelled in an electronic circuit by adjusting the gain of the measured toroid voltage. Because

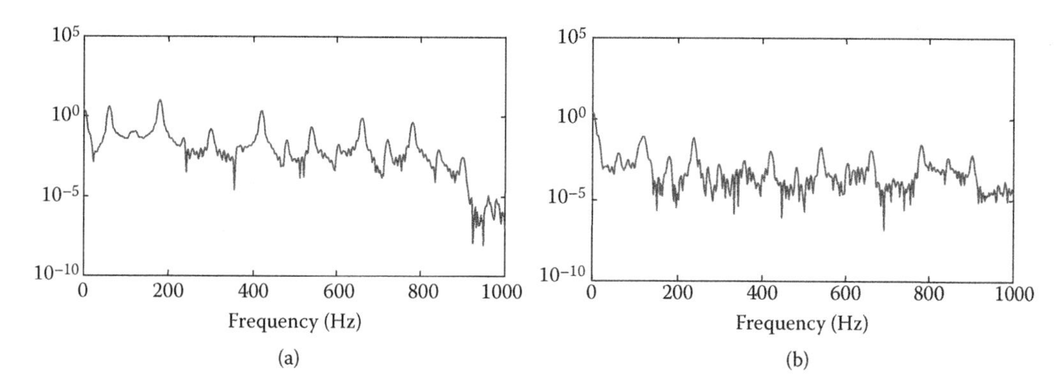

FIGURE 5.17
Comparison of power spectra of force signals. (a) Power spectrum of force signals before ANC. (b) Power spectrum of source signals after ANC.

the toroid sensor is positioned in an arbitrary position in the magnetic field, the underlying assumption is that the noise is linearly induced onto the tip voltage measurement. The gain is fixed once the system is calibrated. Therefore, it cannot deal with situations with time-variant loop inductance. As such, an ANC method has more advantages. It can work with time-variant inductance and eliminate the tedious calibration process. An example of the ANC for a tip voltage is shown in Figure 5.18.

5.3.3 Relationship between Monitored Signals and Welding Processes

Besides a better understanding of the physical process involved in welding, the signals collected during a welding cycle can also be directly linked to the characteristics of welding. The signals collected are usually plotted together, using the same (time) scale, for comparison purposes. A typical plot of signals is shown in Figure 5.19. In this section, various signals are used to monitor a welding process, such as to identify abnormal welding conditions and detect expulsions.

5.3.3.1 Effect of Process Conditions

In general, process signals reflect the effects of not only welding schedules but also process conditions, such as electrode alignment and position of welds on the parts. These conditions can be observed in the measured dynamic resistance, electrode displacement,

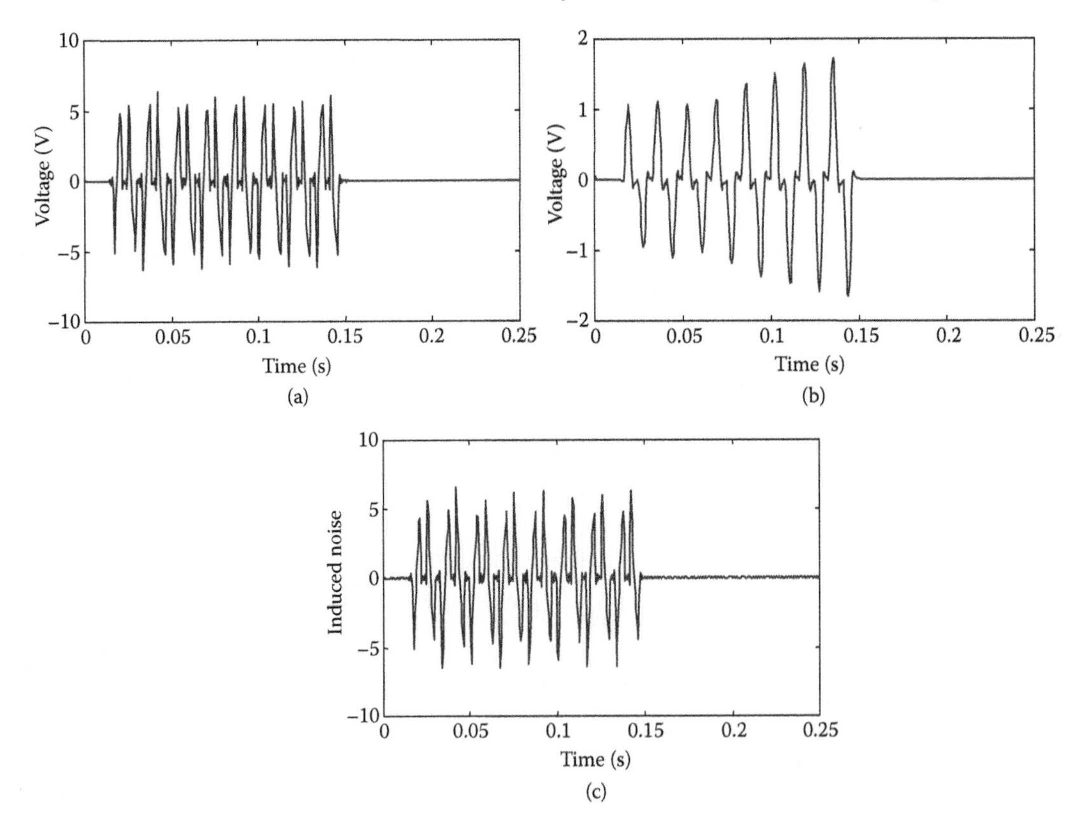

FIGURE 5.18
ANC for tip voltage. (a) Tip voltage before filtering. (b) Tip voltage after filtering. (c) Noise subtracted.

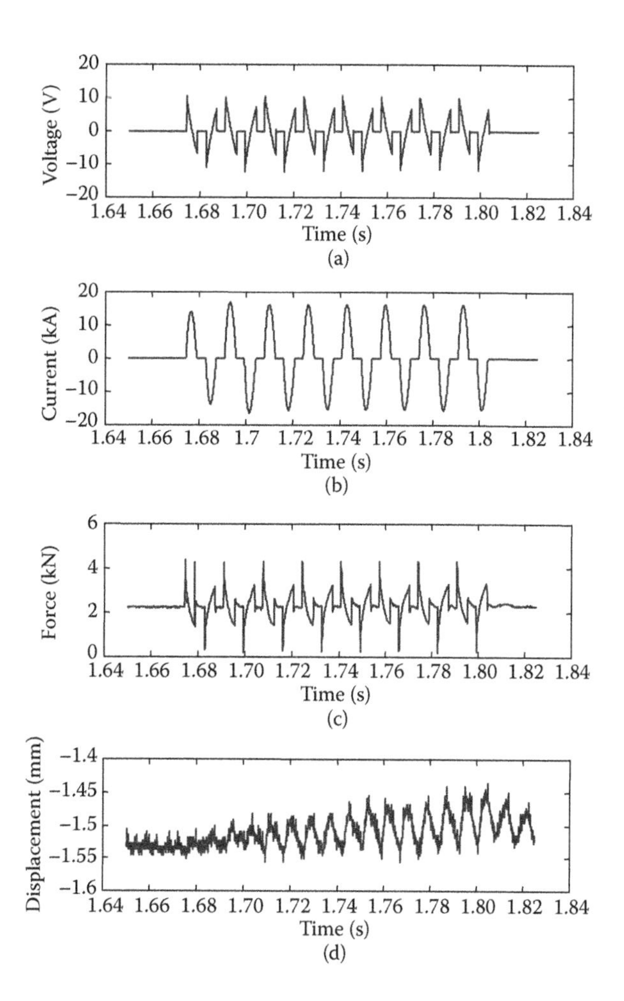

FIGURE 5.19
Typical signals during welding. (a) Tip voltage. (b) Secondary current. (c) Electrode force. (d) Electrode displacement.

and dynamic force. In this section, comparisons are made between normal and abnormal process conditions, using characteristics of process signals.[11]

Process conditions are rarely dealt with in production RSW. For example, an edge-weld condition frequently exists in automobile assembly. Therefore, studying the effects of abnormal conditions is of more practical significance than studying those of idealized welding conditions.

Figure 5.20 shows a comparison between signals of welding under normal process conditions and edge-weld conditions. The dynamic resistance curves remain similar, whereas the electrode displacement and force signals exhibit large differences. This is reflected in the response when the metal is heated. Under normal welding conditions, the surrounding constraining from the solid in the lateral direction is so strong that the metal expansion pushes the two electrodes away along their axial direction. Under edge-weld conditions, there is little solid constraining from the edge because the weld is made on the edge of the parts. Therefore, the melted or softened metal is easily squeezed out of the weld area and the electrodes move inward under the electrode force. Because of its relationship to the

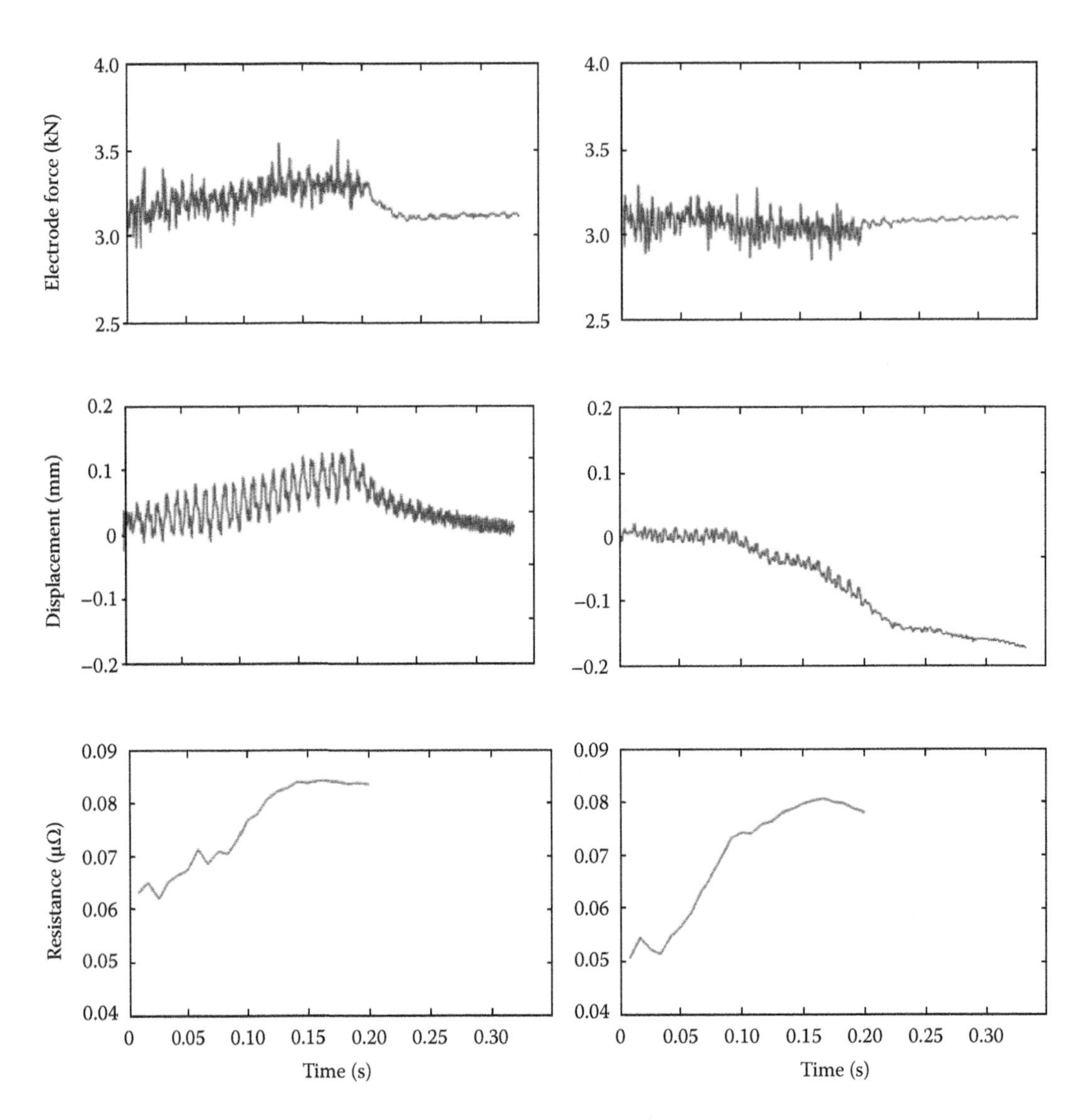

FIGURE 5.20
Signal comparison between welding under normal and that under edge-weld conditions.

electrode displacement, the dynamic force is seen to follow similar trends as the electrode displacement.

5.3.3.2 Fault Identification

The signals collected during electrode touching can be used to identify certain common setup faults. An experiment conducted in the course of the Intelligent Resistance Welding (IRW) program[13,16] has shown that the force signals are directly linked to axial misalignment and faulty fit-up conditions. Axial misalignment exists when the upper and lower electrodes are axially out of alignment, characterized by parameter d in Figure 5.21. A poor fit-up condition in the study is characterized when the two sheet metal parts are initially separated by an insert of diameter D at a distance b from the contact center.

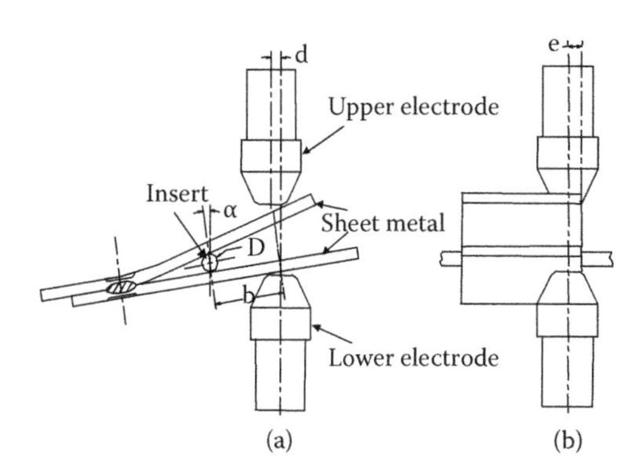

FIGURE 5.21
Structural fault conditions. (a) Front view, and (b) side view.

An experiment showed that it is possible to identify certain common setup faults from signals collected when the electrodes are closed upon the workpieces. Figures 5.22 through 5.24 show the forces measured when no faults, axial faults, and fit-up faults as defined in Figure 5.21, respectively, are present. Each figure shows the forces measured for 10 consecutive welds with the same setup. Clearly, it is easy to distinguish between no-fault and fit-up fault conditions. Examination of Figures 5.22 and 5.23 also shows that there is a clear difference between the no-fault response and the axial fault response. The peak force response during this transient is between 2 and 2.5 kN for the no-fault condition and between 3.5 and 4 kN for the axial fault. Furthermore, the no-fault condition is composed of four exponentially decaying, equally spaced peaks, whereas the axial fault has a greater delay between the first and second peaks, and its third peak is significantly smaller than the second and fourth peaks. The IRW study also shows that the differences in force response for angular and edge faults were not as obvious. However, displacement measurements showed a clear difference between the no-fault and angular fault setups. The occurrence of multiple faults further complicates the detection problem. The force signature during touchdown, while very subtle, has been found to be very repeatable. This observation was possible mainly because the noise in this period is small before the magnetic field is induced by the large welding current. A neural network classification may be able to distinguish between all of the possible setup conditions.

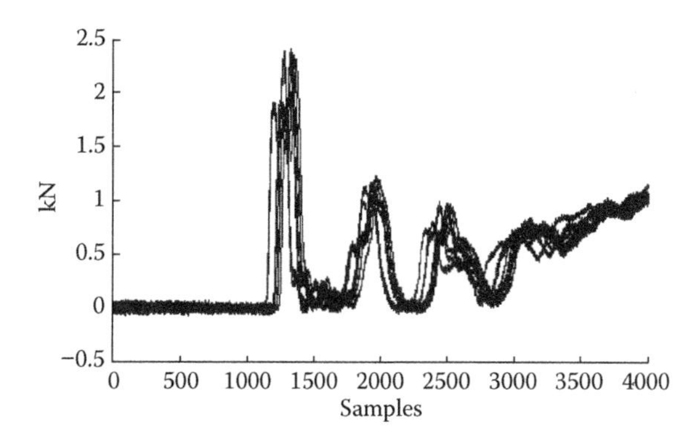

FIGURE 5.22
Force measurements with no faults.

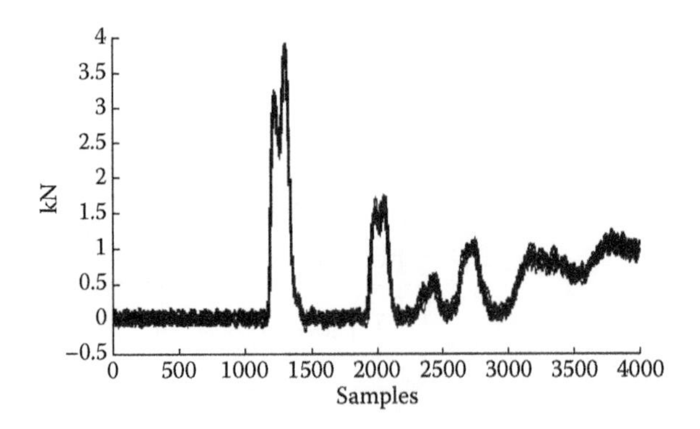

FIGURE 5.23
Force data with axial fault.

Figure 5.25 shows seven traces, each of which represents the mean value of three force signals collected under identical conditions. The figure legend key uses *st* to represent 0.5-mm-gauge steel, *sT* for 1.8 mm steel, *At* for 2 mm aluminum, and *AT* for 3 mm aluminum. Forces of 800 lb (*f*) and 1000 lb (*F*) were used. For the final weld (*pstF*), the mechanical characteristics of the weld gun were changed by stiffening the lower weld arm by adding additional support structure. From the figure, it can be seen that these changes did not alter the responses between approximately the 200th sample and the 2000th sample. The thicker gauges did reduce the initial response. The lower arm stiffening and force affected the later response. Such an approach may be used in the production environment.

5.3.3.3 Expulsion Detection

As discussed in Chapter 7, expulsion is often considered the upper limit of welding current. Because it involves a sudden loss of liquid metal, the occurrence of expulsion is directly reflected in the process signals.

The comparison of the signals for welds with and without expulsion is shown in Figure 5.26, when slightly different amounts of electric current are applied. It can be seen that there is a significant drop in the electrode displacement signal when expulsion occurs.

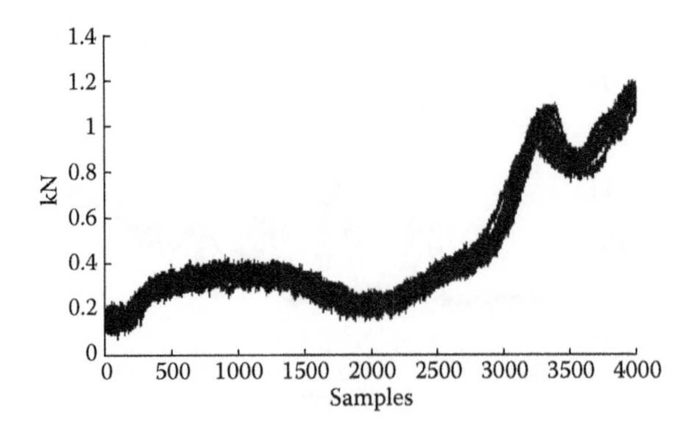

FIGURE 5.24
Force data with fit-up fault.

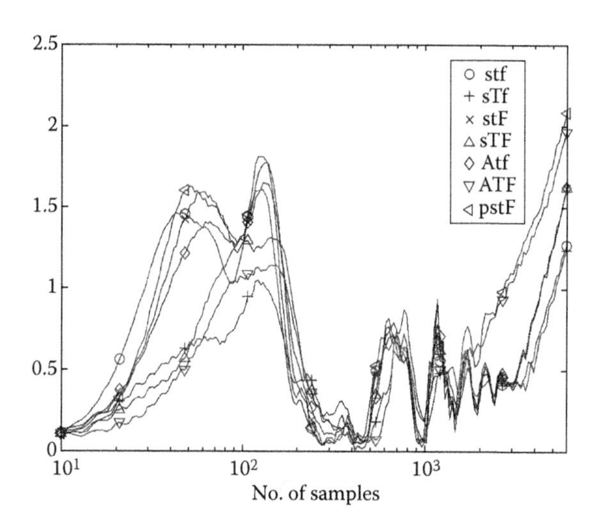

FIGURE 5.25
Force traces using seven perturbed weld parameters.

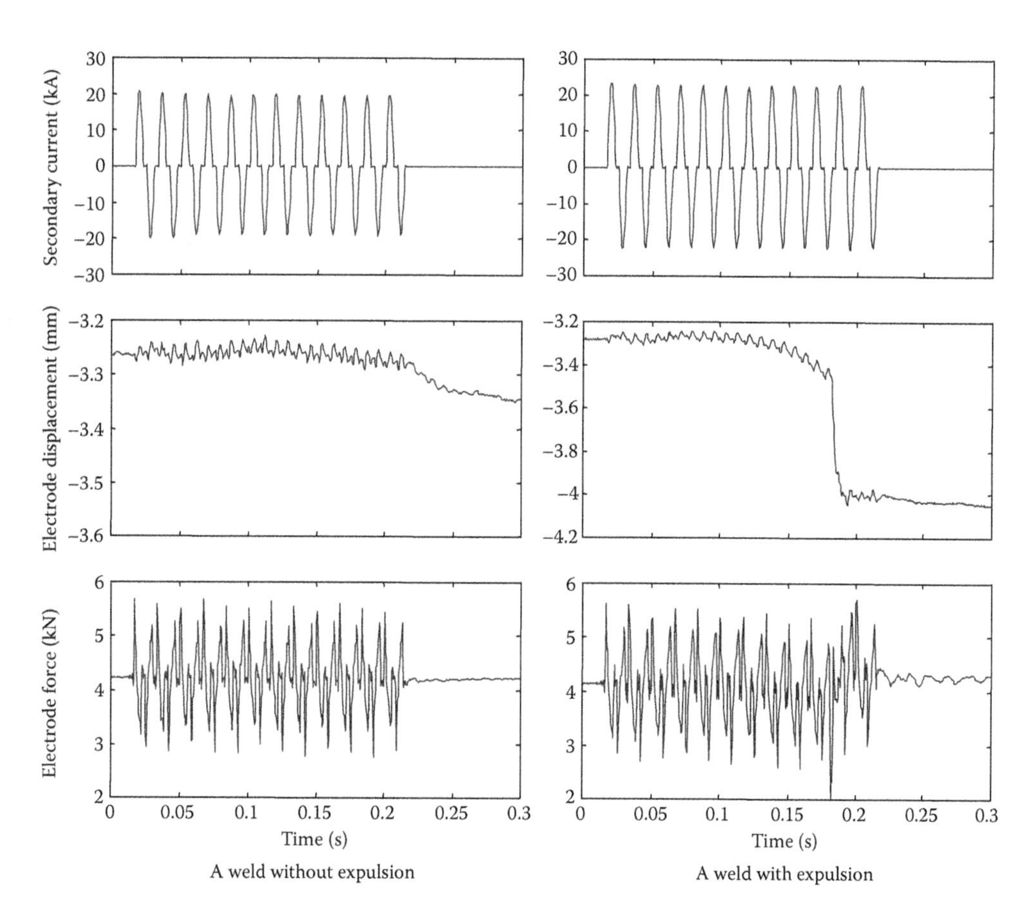

FIGURE 5.26
A comparison of on-line signals with and without expulsion.

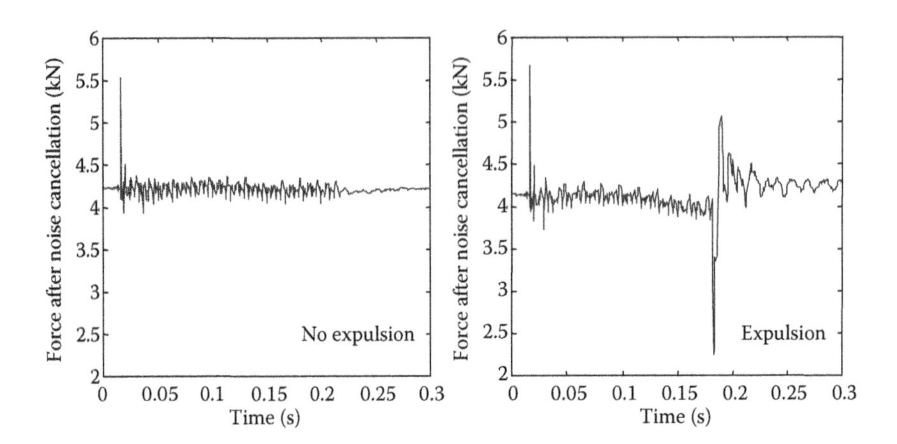

FIGURE 5.27
Force signals using the ANC.

No change can be easily observed on the secondary current (welding current) signal. The electrode force signal showed some difference during expulsion. However, it is difficult to characterize this difference in the raw force signal because of the strong noise. The ANC procedure significantly improves the contrast between the force signals for welding with and without expulsion (Figure 5.27). With the knowledge of expulsion associated with process signals, expulsion can be predicted and ultimately prevented through on-line, real-time feedback control. Figure 5.28 shows the displacement and force signals for a double expulsion.

5.4 Process Control

An important objective of monitoring welding process is to understand the influences of process parameters on weld quality. The commonly used welding process parameters include welding current, time, and electrode force. A few approaches of controlling welding process are presented in this section.

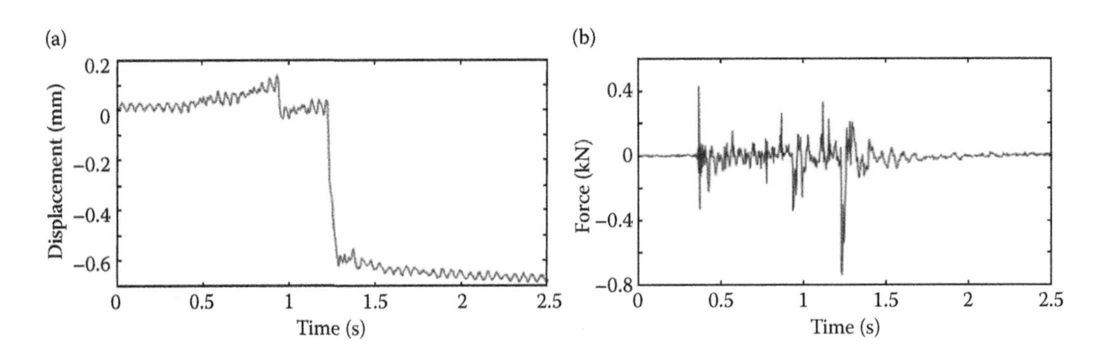

FIGURE 5.28
A weld with double expulsion.

5.4.1 Lobe Diagrams

The most common practice of controlling a welding process is through a so-called lobe diagram, as shown by a schematic in Figure 5.29. Lobe diagrams are plots of welding current and time for fixed electrode forces. They indicate the current–time combinations to achieve acceptable welds. At a fixed electrode force level, a lobe diagram divides the current–time domain, using two lines for the minimum and maximum currents, into three regimes in terms of weld quality: undersized or no weld, acceptable weld, and expulsion. The minimum currents at particular welding times correspond to the acceptable welds meeting the minimum requirements, usually in terms of weld sizes. The minimum acceptable weld sizes depend on the standards used. They are usually in the range of $3.5–4.5 \times \sqrt{t}$, where t is the sheet thickness in millimeters. The maximum currents correspond to the occurrence of expulsion. Occasionally, a current corresponding to a nominal weld size is also plotted in a lobe diagram as a reference for weld current selection. The difference between the maximum and minimum currents is the window of operation or current range. In addition to the minimum weld and expulsion currents, the current corresponding to the excessive indentation or electrode sticking, usually far beyond the expulsion limit, is occasionally used as a boundary in a lobe diagram as well.[17] This is often the practice on a production line where expulsion is commonly used as the indicator of sufficient heat supply to a weld. Welding beyond the expulsion limit has been used by certain welding practitioners to overcome the influence of random factors, and the electrode life usually suffers as one of the adverse results.

For a given material and welder setup, a weld lobe is usually developed in a laboratory environment and then a welding schedule (a setting of welding parameters) is chosen for production based on the weld lobe. There are standard procedures for developing lobe diagrams, such as the recommended practice by the American Welding Society (AWS/SAE D8.9M[18]).

5.4.1.1 *Effect of Process Parameters and Weld Setup Variables*

A lobe diagram basically contains two boundaries for minimum acceptable welds and expulsion, and these boundaries are a strong function of the electrode force, as discussed by a number of researchers, for example, Gould et al.[19] The expulsion limit increases with the applied electrode force, and a large electrode force often means a wide process window.

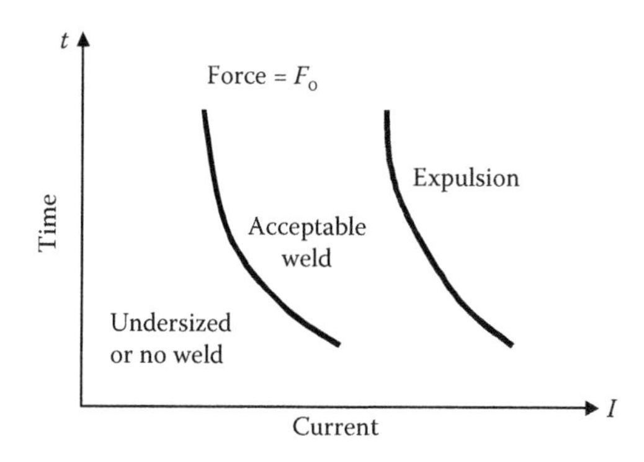

FIGURE 5.29
A weld lobe diagram.

The operation window size is also heavily affected by electric current profiles as presented in Chapter 2. This section discusses the influences of process variables on the shape, size, and location of a lobe. In a laboratory environment, process conditions are usually well maintained to their nominal settings. However, many abnormal process conditions, such as electrode misalignment, electrode wear, part poor fit-up, and changes in electric impedance in the welding loop, are common in production. These process abnormalities alter the relationships between the weld size and the input process variables and thus affect the weld lobes.

Kaiser et al.[20] showed that variations on sheet surface and electrode force could lead to changes in the position and shape of a weld lobe. Nagel and Lee[21] listed several possible abnormalities that need to be addressed in developing an RSW control approach. Karagoulis[22] reported a weld lobe shifting due to electrode misalignment. In a study on the effect of adhesives on weldability, Rivett and Hurley[23] observed changes induced by the use of adhesives on the shape and size of weld lobes. In general, it was found that when adhesives are used, the weld lobes become narrower and the minimum and maximum current limits become steeper, that is, less dependent on welding time.

In a study by Li,[3] the effects of various process conditions were investigated. In the experiment, both normal and abnormal process conditions were considered. Based on the experimental data, weld size and lobes were analyzed as response variables.

Several abnormal process conditions were created and their effects on welding were studied. A single-phase AC welding machine with truncated cone electrodes was used in the experiment to weld a 0.8-mm HDG drawing steel. The weld button diameter was measured through peel test. The current range or weld window was characterized using two response variables: the center of the current range (I_c) and the length of the current range (I_{leng}), as defined in the following:

$$I_c = \frac{1}{2}(I_{min} + I_{max}) \tag{5.6}$$

$$I_{leng} = \frac{1}{2}(I_{max} - I_{min}) \tag{5.7}$$

I_c represents an average current setting determined by the physical experiments. I_{leng} determines the allowable range of the current for making good welds. By definition, I_{leng} is greater than zero. I_c and I_{leng} have been examined and show no correlation. Thus, they can be analyzed separately as two independent responses.

A statistical analysis shows that electrode size and welding time are significant factors, whereas force, fit-up, axial misalignment, and angular misalignment are unlikely to be significant. It is also seen that variables such as electrode size, welding time, and force have quadratic effects on the response. Experimental results showed that the abnormal process conditions significantly reduce the length of the current range, whereas force increases the length of the current range. Electrode size and welding time have little effects on the responses.

Both current and electrode size were seen to have strong influences on the button size. On average, the button size increases under poor fit-up and decreases under angular misalignment conditions. Axial misalignment is not seen to have a strong effect. Electrode force and weld time show minor quadratic effects.

Predicted lobes under various process conditions are shown in Figure 5.30. In general, when poor fit-up and angular misalignments exist, the weld lobe is shifted to the left and becomes narrower. A left shift of the weld lobe implies early nugget formation, which may help increase the weld size. However, a narrower weld lobe indicates that the welding process is less robust under these conditions. It was also found that nuggets grow following different paths under different conditions. The nugget size variation caused by the abnormal process conditions could be different depending on the welding time.

Axial misalignment does not have a strong effect on weld lobe diagrams when the amount of misalignment is small compared to the electrode size. Under small axial misalignment, the process behaves similar to that under normal conditions.

5.4.1.2 Probabilistic Expulsion Boundaries in Lobe Diagrams

The boundaries of both the minimum-sized weld current and the expulsion current in a lobe diagram are usually expressed as lines representing specific electric current values. This implies that when such a current is crossed, an event, weld–undersized (or no) weld or expulsion–no expulsion, will occur. It is well known that this is not true, as a welding process as well as its output is affected by a large number of uncontrollable random factors. The maximum current or the expulsion boundary exhibits more of a probabilistic nature because of the large uncertainty of expulsion. The study by Zhang et al.[24] has shown that a boundary in a lobe diagram should be represented as a range, rather than a deterministic limit (a line), with occurrence probability, as discussed in detail in Chapter 7. As shown in Figure 7.39, an expulsion boundary is considered as a span of expulsion probability from 0% to 100% of chance.

The determination of maximum currents in a lobe diagram is closely related to the detection of expulsion. There are many random factors whose effects are difficult to quantify, such as workpiece fit-up, electrode alignment, and surface condition, in influencing expulsion. Zhang et al.[24] have presented a statistical model taking the random factors into account. Expulsion was treated as a function of both deterministic and random factors, and the boundary for expulsion was expressed as a probabilistic range, not as a line as in a conventional lobe diagram. The model is presented in more detail in Chapter 7. In

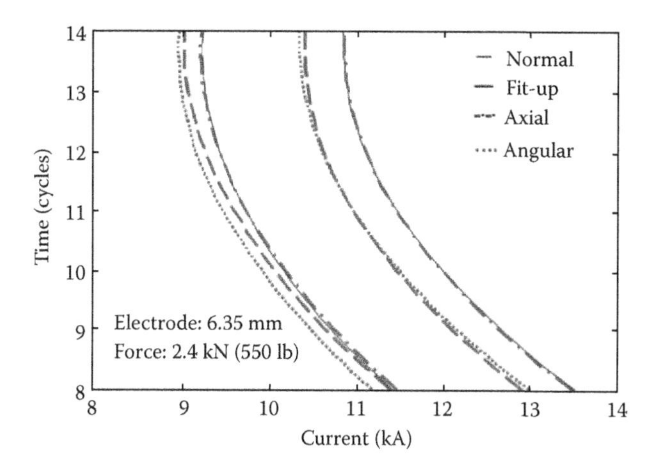

FIGURE 5.30
Weld lobe predictions.

practice, however, it might be difficult to conduct a systematic statistical study that provides a probabilistic boundary for maximum currents. It is more realistic to determine a range of operation, with the minimum and maximum currents as limits, knowing that the limits should not be treated as deterministic values.

5.4.1.3 Effect of Electrode Force

Most of the lobe diagrams are made with fixed electrode forces. In fact, the limits of a lobe diagram, as shown in Figure 5.29 as minimum and maximum welding currents, are a strong function of electrode force. The electrode force affects the contact resistance both at the faying interfaces and the electrode–sheet interfaces, as increasing the force usually reduces contact resistance by creating more actual contact area. Therefore, less heat input results from high electrode force, if the welding current is constant. The minimum welding current needed to create a certain size of weld is raised as well under a high electrode force. The expulsion limit is also pushed up because a high electrode force imposes more confinement onto a weldment and delays the occurrence of expulsion (as discussed in Chapter 7), in addition to reducing the contact resistance and, therefore, heat input. Both effects allow larger value of electric current to be used and welds of larger size to be created. A study on the welding characteristics of an HDG DP600 steel shows that the electrode force affects both the borders of a lobe diagram.[25] As seen in Figure 5.31, both the minimum weld and expulsion limits are raised with higher electrode force, and the welding window size (the difference between expulsion current and that of minimum nugget size, or the operation window) is increased as well.

The influence of electrode force on the lobe diagrams can be better understood by a detailed examination of its effect on expulsion limits. Figure 5.32 shows the change of welding time needed to create expulsion with electrode force. It is derived from the expulsion models developed on a 1.2-mm AKDQ bare steel and a 2.0-mm AA5754 alloy, as

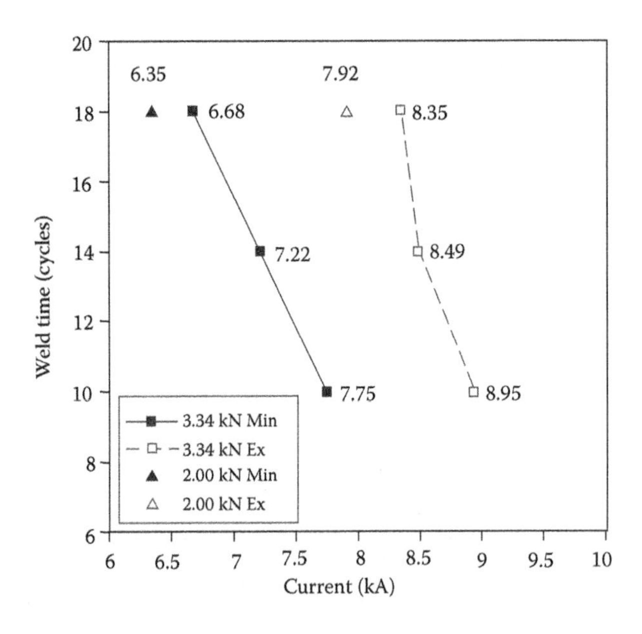

FIGURE 5.31
Lobe diagram of a DP600 steel at different electrode forces.[25]

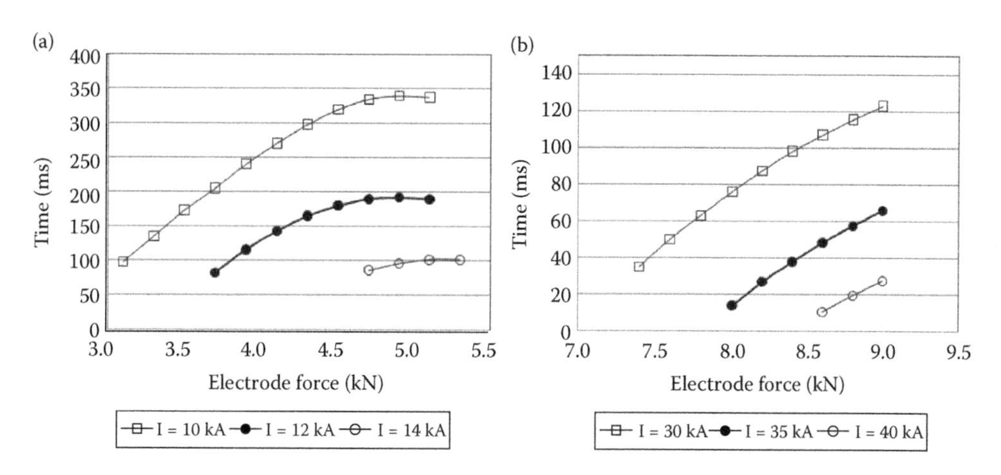

FIGURE 5.32
Effect of electrode force on expulsion time. (a) AKDQ steel, (b) AA5754. Probability of expulsion = 0.5.

described in detail in Chapter 7.[24] An expulsion probability of 0.5, used as the middle value of an expulsion range, is assumed in these figures. For the steel, the welding time to reach expulsion decreases with current, and goes up with electrode force, as electrode force is responsible in most cases for depressing expulsion (Figure 5.32a). However, such an increase in expulsion time stops when the electrode force reaches a certain value, which means that further increasing electrode force will do little to contain expulsion, as explained in the expulsion models in Chapter 7. The effect of electrode force on expulsion limit in welding the AA5754 is similar to that in welding the steel, yet it seems that the welding time for expulsion may still go up if larger electrode force is applied (Figure 5.32b). Such a mathematical extrapolation cannot be justified, however, because in reality, longer welding time may render the weldment totally melted. The electric current at expulsion is also affected by the electrode force. As seen in Figure 5.33, increasing the electrode force allows for larger electric current, and such effect is more profound when welding time is short. In general, welding time has a fairly large effect on expulsion current in the steel, but not much in the aluminum alloy (Figure 5.33b). In the case of AA5754, the expulsion

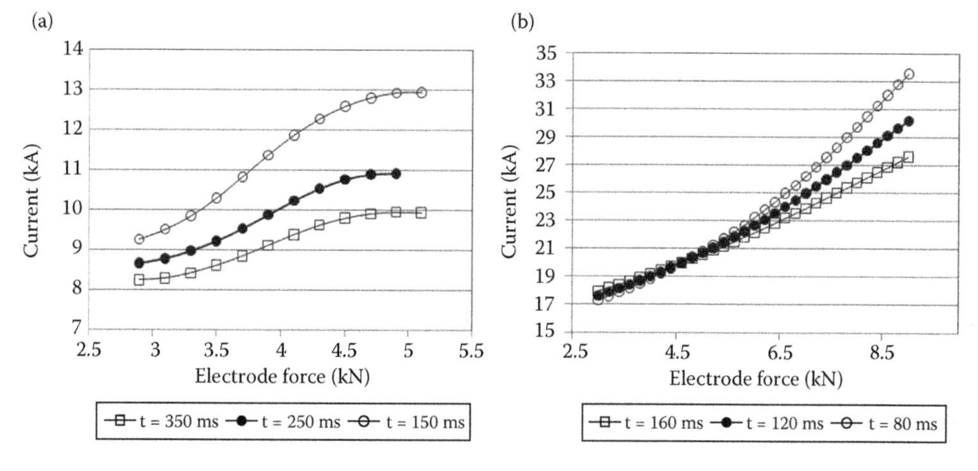

FIGURE 5.33
Effect of electrode force on expulsion current. (a) AKDQ steel, (b) AA5754. Probability of expulsion = 0.5.

current values at different welding times are very close to each other, especially when the electrode force is low. This corresponds to the fact that the curves in the lobe diagrams for aluminum alloys are more linear than those for steels, indicating that there is little dependence of expulsion current on welding time. This could be the result of early expulsion during welding AA5754 when the electrode force is low. When electrode force reaches a threshold value, about 5 kN for this alloy, the occurrence of expulsion is controlled by the force balance models described in Chapter 7. In this case, the size of the liquid nugget is proportional to the amount of heat input, and large electrode forces allow for higher welding current when welding time is short.

5.4.1.4 3-D Lobe Diagrams

Although electric current, welding time, and electrode force are all important welding process parameters, they are rarely represented together in lobe diagrams. This can be partially attributed to the fact that a resistance welding process is generally more sensitive to welding time and electric current than electrode force. The significantly larger test matrix when electrode force is included as a variable, and the difficulties in graphically presenting the responses (limits in a lobe diagram) as a function of electrode force in addition to current and time make 3-D lobe diagrams a rarity.

A 3-D lobe diagram can be created if an analytic expression is available for the limit of minimum-sized weld or expulsion. Using the statistical models developed for predicting expulsion as presented in Chapter 7, Zhang[26] attempted to explain the dependence of expulsion probability on electrode force, electric current, and welding time. Since statistical models describe expulsion probabilities as a dependent of all three welding parameters, it is possible to depict expulsion boundaries graphically in a three-dimensional space. In Figure 5.34, surfaces of expulsion probabilities of 0.05 are plotted using Equations 7.34 through 7.36. The figure shows distinctive characteristics of welding steel and aluminum alloys. For steel welding, the expulsion probability surface covers the corner of (I = minimum, F = maximum, τ = minimum), as shown in Figure 5.34a. Therefore, expulsion is contained when the welding schedule falls in the area surrounded by the surface. According to this model, when current and time are too large, or electrode force is too small, expulsion is likely to happen in the ranges studied.

Unlike the expulsion surface of the steel, AA5754 shows very little dependence on welding time and slight dependence on electrode force (Figure 5.34b). The influence of electric current overshadows those of welding time and electrode force. Because expulsion usually happens at an early stage during this Al welding, total welding time does not matter much. The shape of the expulsion surface also implies a possible critical expulsion current for this aluminum alloy. The expulsion surface for welding AA6111 is different from that for welding AA5754. Electrode force plays a more important role, and expulsion depends strongly on welding time when welding current is low and electrode force is small (Figure 5.34c). In general, welding current is the dominant factor in aluminum expulsion.

The differences in the shape of expulsion surfaces reflect the important differences between welding steel and welding aluminum alloys. These differences stem from the significantly different electrical, thermal, metallurgical, and mechanical properties between these two types of materials. Although these models do not provide exact information on the physical processes, they suggest possible explanations on how physical parameters influence nugget formation and expulsion. As the surfaces represent low expulsion probability (0.05), welding schedules chosen on this surface can produce large nuggets with a low risk of expulsion.

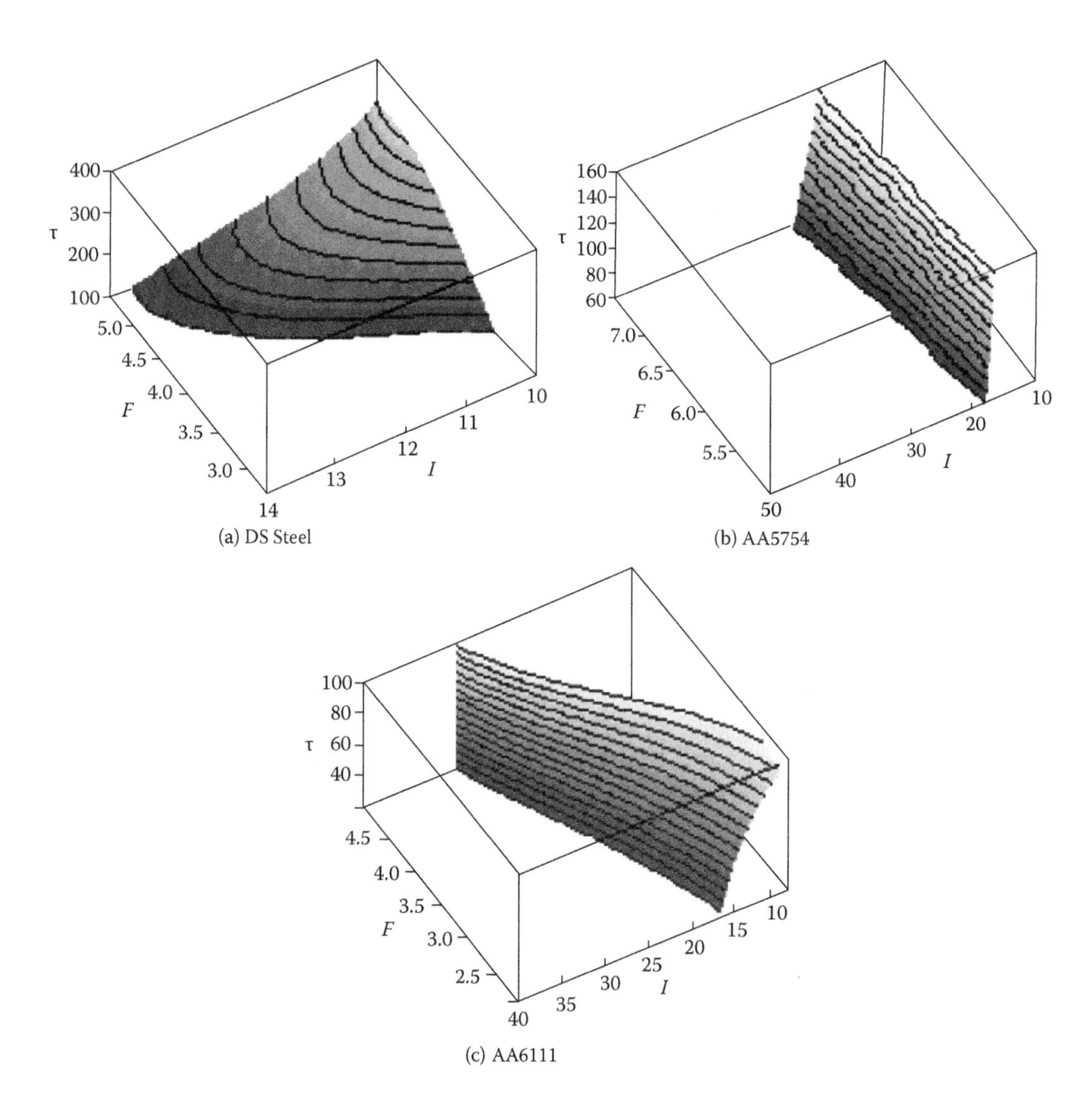

FIGURE 5.34
Expulsion surfaces of probability = 0.05.

5.4.2 Constant-Power Density

The most widely used control strategy in spot welding is constant current.[14] This approach intends to produce a constant heat input. However, the actual average power across a nugget, which is given by ($I^2 * R_d$), varies as welding progresses with time, because in welding steel, the dynamic resistance (R_d) usually increases initially with time (as bulk temperature increases with time) and then decreases, partially because the weldment becomes thinner. Using this algorithm, the effects of many changing factors during spot welding, such as surface and bulk resistivities, and contact area cannot be accounted for. The initial increase in dynamic resistance (and the heat input) after a few cycles may lead to localized heating, and hence early expulsion. The constant-current algorithm may also produce unconformable nuggets under excessively high force levels and faulty conditions such as angular misalignment as they have large influence on resistance.

Therefore, a constant-current control algorithm may not yield optimal output. If the average power, instead of current, across a nugget is considered and properly regulated, the possibility of overheating and hence expulsion may be reduced. This is the motivation of regulating the average power across a nugget as a constant, or the so-called constant-power control algorithms. This approach is based on the assumption that a constant average power density across the nugget will result in a constant heat input.

5.4.2.1 Hypothesis

In the constant-current strategy, a lobe is used to define the region of weldability in the current vs. time space. This two-dimensional plot indicates good current levels for a given time level, which would produce an acceptable nugget given the operating condition. It is a known fact that the lobes corresponding to the different operating conditions may be completely different, without exhibiting any overlap. Therefore, a significant amount of hunting is essential to get a current range for each operating condition.

On the other hand, lobes under different conditions can be defined based on power density in the power density vs. time (in cycles) space. If the overlap of the lobes were significant, a single power density/time level would be sufficient to produce acceptable welds under different operating conditions.

In order to achieve a constant-power control, the contact area has to be accurately estimated. In reality, one of the major consequences of faulty welding process conditions, such as electrode misalignment, is that the contact area deviates from that under normal conditions. Using the same heat input as that under a normal condition could overheat the weldment and induce undesirable features to welding, such as expulsion. Consider the effect of heating, faulty configurations can be modeled as a configuration with a different effective area from that under normal conditions. The effective area affects the dynamic resistance and the heat loss. However, if the average power density (defined as the power divided by the effective area of contact) is regulated, the weld can be made following a heating process similar to that used in making a normal weld. The effectiveness of this strategy depends on a good estimate of the effective area of contact. The algorithm development for constant average power, and power density calculations are briefly explained below.

5.4.2.2 Algorithm

The constant-power density algorithm is a model-based control algorithm. There are two facets to this algorithm: (1) calculating the effective contact area and (2) regulating the input power at the desired level based on the above estimate. The power density is defined as the ratio of the power input to the effective contact area. It can be expressed as

$$\text{Power density} = \frac{I^2 R_\text{d}}{A_\text{eff}} \tag{5.8}$$

where the numerator is the power input to the nugget and the denominator defines the effective contact area.

The total weld time is typically held constant in production to achieve a predictable production cycle. Therefore, the power across the nugget can be treated as a function of time with a fixed time horizon. The power either can follow a time-varying trajectory or be maintained at a constant level. These reference (desired) power trajectories can be calculated off-line using developed models, such as low-order nonlinear thermal models. Once designed, the electrical models can be used to design the control input (firing angle) to the system, as the only signals that can be easily fed back are the electrical signals. The development of control strategy based on maintaining a constant power is outlined in the following sections.

5.4.2.3 Algorithm Implementation

Two issues are involved as far as the real-time implementation is concerned: the implementation of the effective contact area estimator and the control strategy for power regulation.

A thermal model is needed for estimating the effective contact area. This estimator can be implemented in the early stage of welding. For a welding with 16 cycles in an experiment, the first three cycles were used to obtain a contact area estimate. The three cycles were also run at constant-power density, with a certain nominal effective contact value to begin with. This nominal value was ϕ5.4 mm for a 4.4-mm electrode and ϕ6.4 mm for a 6.4-mm electrode. These values were obtained *a priori* from the simulations using experimental data.

The effective contact obtained after three cycles reflects the operating condition of the electrode–workpiece combination. Using this new effective value, the true power required for the operating condition is obtained by multiplying with the predefined power density level. This new power level was updated every cycle to obtain the desired power level for the next 13 cycles in the 16-cycle experiment.

In real-time implementations, a nominal firing angle value can be calculated based on the desired power and a nominal R_d. The measured average power is calculated over a positive half cycle, whereas the control calculations are done based on this measured average power in the negative half cycle. Thus, a firing angle value for the subsequent full cycle is made available at the start of the next positive half cycle. This methodology was implemented in real time under a nominal welding condition with the desired power of 1.8×10^4 W in a laboratory setup. Figure 5.35a shows the measured average power variation for this welding condition, and Figure 5.35b shows the plot for the error in power (which asymptotically goes to zero, or in other words, the actual power is being regulated to the desired power level). It also shows the variation in the firing angle input for every cycle as calculated by the controller.

The weld button size (diameter) measured in this test was 5.5 mm. This control strategy was also tested under poor fit-up welding conditions without changing the desired power level. The obtained nugget diameter was 6.4 mm.

Another implementation issue is the regulation of desired power for the next 13 cycles (for the 16-cycle welding). Because the control input to the electrical subsystem is the firing angle to the bank of antiphase SCRs on the transformer primary, the objective is to design this control input to achieve power regulation across the nugget. In order to achieve this, an electrical model is needed. The input to this model is firing angle and the output is the desired power, with the throat resistance, inductance, and dynamic resistance being the parameters in the model. The dynamic resistance is a time-varying parameter and can be measured during the course of welding.

To achieve regulation, the control input is defined as a combination of feed-forward and feedback terms. The feed-forward term is calculated at the start of welding, assuming a certain nominal dynamic resistance. The feedback term can be derived using a technique called gain scheduling. Because the electrical model is nonlinear, different control gains are needed to achieve good regulation for different operating conditions.

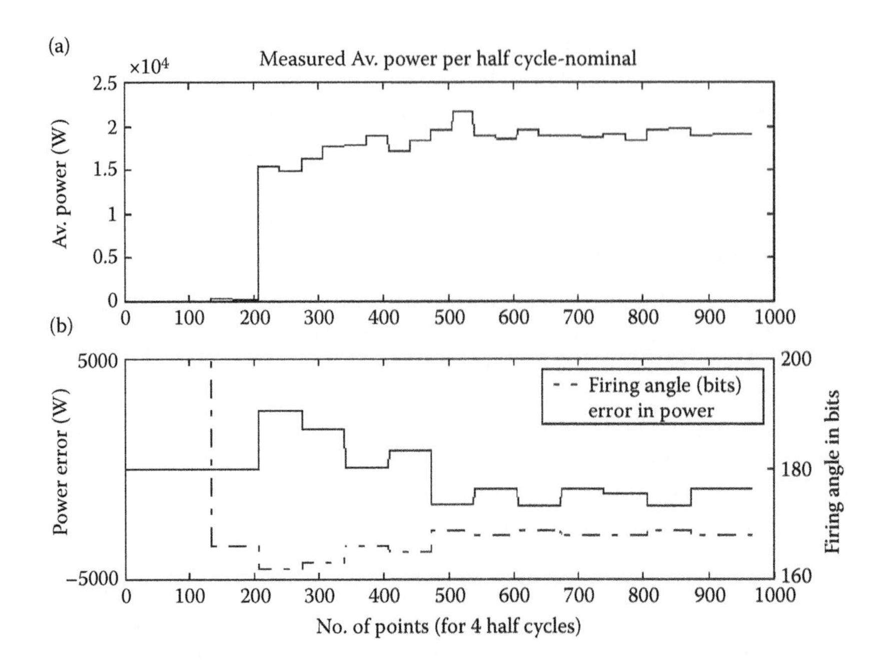

FIGURE 5.35
(a) Measured average power (nominal). (b) Error in power (axis on the right) and firing angle input (nominal).

5.4.2.4 Gain Scheduling

This technique extends the validity of the linearization approach to a range of operating points.[27] The different operating points are characterized by both the desired power level and dynamic resistance (R_d). At each power level (which is calculated by an effective area estimator algorithm) and R_d, which is measured, the input–output model of the electrical subsystem has an associated firing angle input.

A two-dimensional inverse map (Figure 5.36) can be created with the desired power and R_d as the input and firing angle as the output. A linearization at different operating points is achieved by using the difference along the power levels. This map provides the

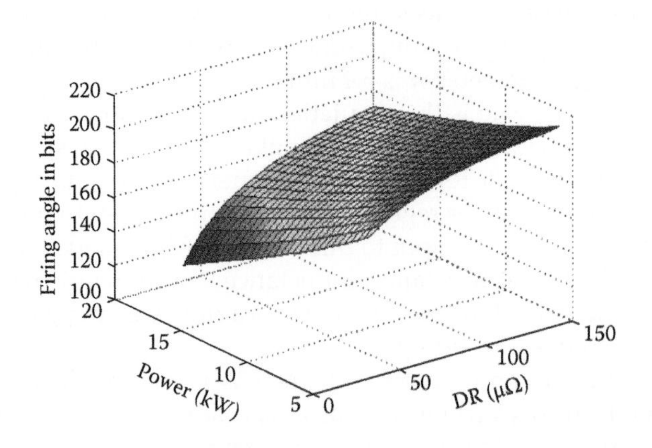

FIGURE 5.36
Two-dimensional map of power/R_d vs. firing angle.

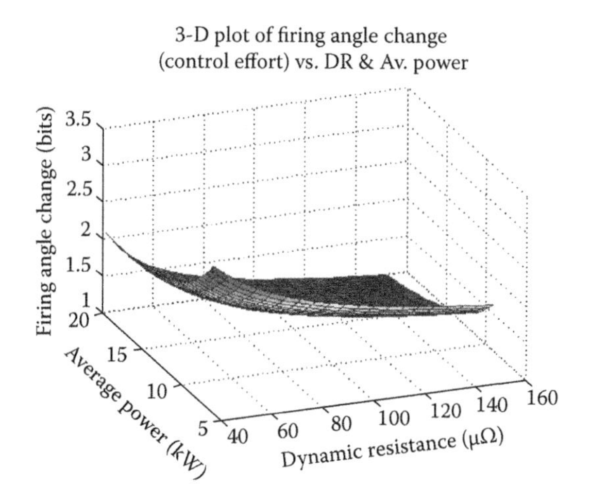

FIGURE 5.37
Firing angle change vs. operating power level and R_d.

firing angle change against a power change of 1000 W at every operating power level and R_d. This map is shown in Figure 5.37. The firing angle calculated can be used to control a welder.

5.4.2.5 *Experimental Results*

Experiments were conducted using the constant-power density algorithm. Constant-power density lobes were created and compared with constant-current lobes done under identical conditions (nominal 4 and 6.4 mm electrodes at 550 lb electrode force). The lobes are plotted in Figures 5.38 and 5.39. From the figures, it can be seen that there is a significant

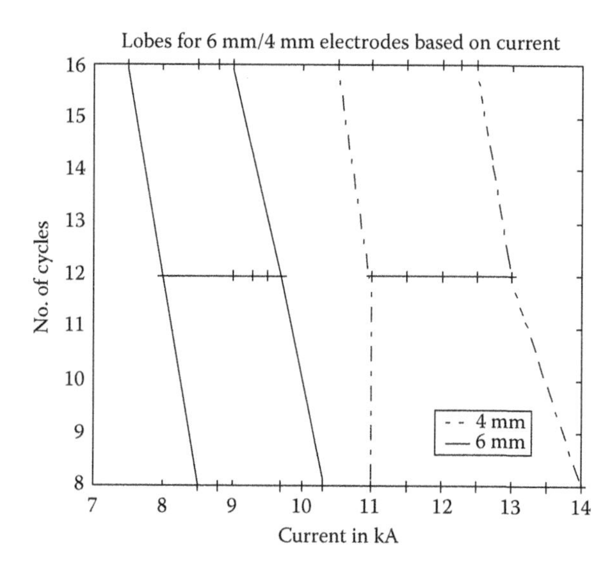

FIGURE 5.38
Constant-current lobes using nominal 4- and 6-mm electrodes (electrode force = 550 lb).

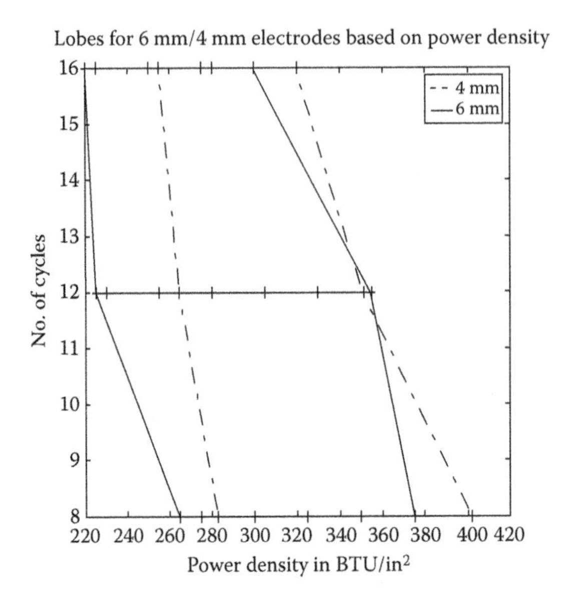

FIGURE 5.39
Constant-power density lobes using 4- and 6-mm electrodes (electrode force = 550 lb).

overlap between the constant-power density lobes and none in the constant-current lobes. The power density level that was chosen to test this algorithm is 280 BTU/in^2 at 16 cycles.

Experiments were also conducted under conditions of combined faults and force levels. The results are shown in Figure 5.40. This figure shows a comparison between the constant-power density and the constant-current algorithms. The fault conditions are denoted in the respective plots. For example, 6FA950W refers to a 6-mm electrode, fit-up-angular (FA) fault combination at 950 lb of force on worn (W) electrodes. The constant-power density strategy performs better than the constant-current strategy by compensating for varying force levels and faults. The constant-power density strategy produced expulsion for one condition, but otherwise produced a weld of acceptable size. The constant-power density algorithm also shows advantages in nugget formation. As shown in Figure 5.41, it promotes early nugget formation, and the nugget formation is more gradual using constant-power density than using constant-current.

5.4.3 Artificial Neural Network Modeling

Artificial neural networks (ANNs) are mathematical representations of how mammalian brains function. Tack[28] summarized the methodology in his research on pavement performance prediction. Within the brain, signal processing and response generation are performed through electric impulses generated and received by neurons, which are special cells that receive impulses through dendrites and transmit impulses through their axons. A neuron will generate an impulse if the sum of the impulses it receives exceeds a certain impulse threshold. For a neuron to receive an impulse, a generating neuron must transmit its charge from its axon across a synapse (space between neurons) to the receiving neuron's dendrite. The length of the synapse controls how much impulse can be transmitted.

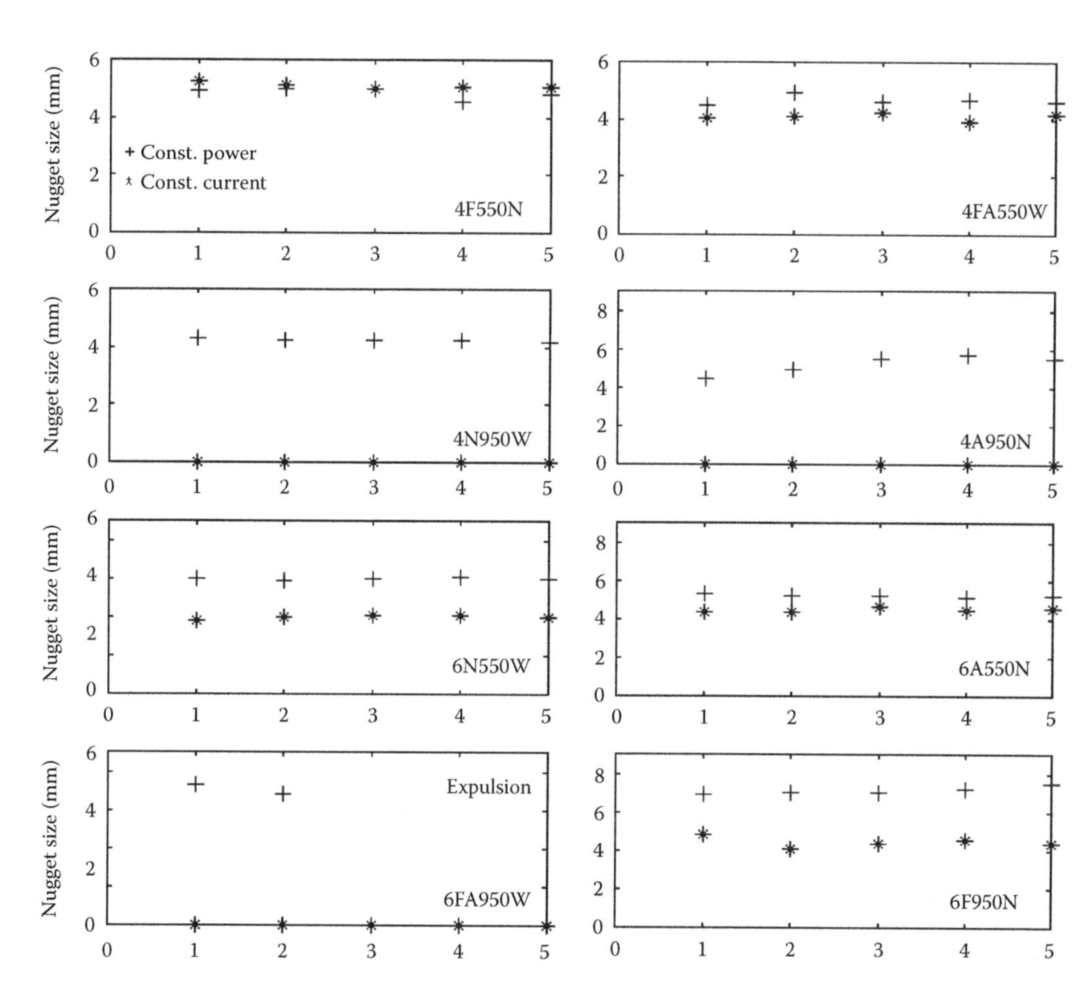

FIGURE 5.40
Comparison between constant-power density and constant-current algorithms for various fault combinations.

Therefore, every neuron will transmit its impulse to other neurons differently. A human brain is composed of approximately 10^{11} neurons, and each neuron is connected to approximately 10^4 other neurons.[29]

After they were first introduced in 1943,[30] neural networks found a broad spectrum of applications, especially after McClelland and Rumelhart[31] introduced new multilayered learning rules and other concepts that removed training barriers that stifled neural network research in the 1960s. Neural networks have quickly gained attention because of their ability to perform the tasks of classification, association, and reasoning. Among many neural network schemes, the multilayered feed-forward ANN using supervised error backpropagation training is the most commonly used network and learning rule for many applications. It has been utilized to diagnose fault conditions of welding processes and predict weld quality.

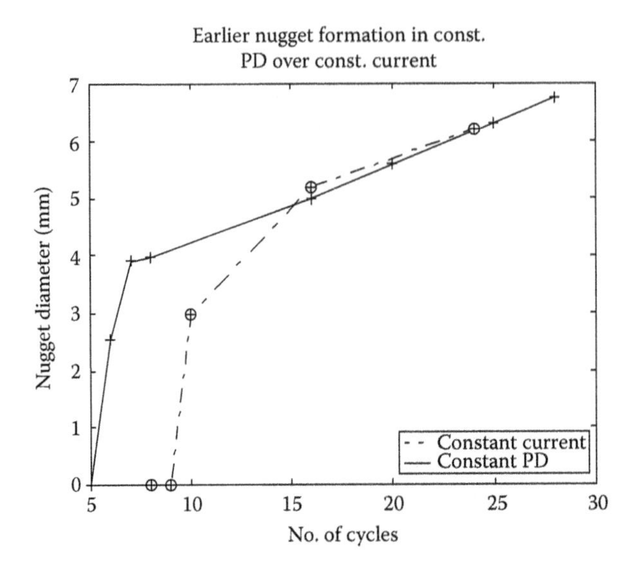

FIGURE 5.41
Comparison of nugget formation between constant-current and constant-power density.

Feed-forward ANNs act as simplified mathematical models of the brain's neural transmitting process. The ANN processes an input vector to generate an output vector. The input vector is used by the ANN to generate the output vector by passing mathematical impulses through a number of unidirectional interconnected perceptrons, which are organized into layers. Each layer of perceptrons between input and output is labeled as a hidden layer. Each hidden layer may consist of a different number of perceptrons. Each perceptron is connected to every other perceptron in both the preceding and succeeding layers. The connections work so that the impulses travel in one direction from input to output. Perceptrons are mathematical processors that function like biological neurons, which process the impulses generated from the previous layer of perceptrons and then generate their own impulse directed toward the next layer of perceptrons. The magnitude of the impulse generated is determined by the use of the magnitude of the total impulse received and an activation function.[32] The magnitude of impulse a perceptron receives depends on the impulse each perceptron generates in the previous layer and the strength of the interperceptron connection, or weights. The weight represents the ability of a neuron in the previous layer to transmit its generated impulse to a neuron in the current layer. A high-valued weight is indicative of an important connection, whereas a low-valued weight would reflect a fairly unimportant connection.

Multilayer feed-forward ANNs consist of a number of hidden layers in between input and output processing layers. Each hidden layer is composed of a number of unidirectional interconnected perceptrons. Komolgorov's theorem states that a neural network can approximate any function if enough hidden layer perceptrons are present and a sigmoid activation function is used.[32] Therefore, a number of tests must be performed to ensure that a sufficient number of perceptrons are present for the function being analyzed to be approximated accurately. Care must also be taken to not use too many perceptrons, because training time is directly proportional to the number of perceptrons present. It is also generally preferable to use only a single hidden layer, because training time increases exponentially with the number of hidden layers.

5.4.3.1 A Case Study of Using ANN for RSW Quality Control

In the work by Li and others,[3,33,34] a neural network with a multilayered feed-forward structure was used for diagnosing fault conditions in RSW. The number of nodes in the input layer for the model was determined by the number of features to be used. Seven features were selected through principal component analysis, and the same number of nodes was used in the input layer. For the output layer, one node was used for the nugget diameter. Between the input and output layers in the network, two hidden layers were used, for which different structures (numbers of nodes) had been tested.

In order to have the same ranges for the input and output, the training data and testing data for the neural network should be normalized. The following normalization formula can be used:

$$x_{\mathrm{ni}} = \frac{2(x_{\mathrm{i}} - x_{\mathrm{min}})}{x_{\mathrm{max}} - x_{\mathrm{min}}} - 1 \tag{5.9}$$

where x_{ni} is the normalized input–output data, x_{i} is input–output data before normalization, and x_{max} and x_{min} are maximum and minimum of the data, respectively. It can be seen that the normalized data are within [–1, 1]. A backpropagation with momentum algorithm can be used for training the model. The quadratic cost function (J) of the error is defined as

$$J = \sum_{t=1}^{M} \left(D_{\mathrm{n}}(t) - \hat{D}_{\mathrm{n}}(t) \right)^2 \tag{5.10}$$

where t is the iteration number, M is the number of training samples, $\hat{D}_{\mathrm{n}}(t)$ is the t-th estimated nugget size, and $D_{\mathrm{n}}(t)$ is the t-th measured nugget size. The weight update at iteration t is given by

$$w_{ji}^{k}(t) = w_{ji}^{k}(t-1) - \acute{\eta} \frac{\partial J}{\partial w_{ji}^{k}(t-1)} + \alpha \Delta w_{ji}^{k}(t) \tag{5.11}$$

where $w_{ji}^{k}(t)$ is weight at iteration t, Δw_{ji}^{k} is the previous weight update, $\acute{\eta}$ is the learning rate, and α is the momentum. In Li's study, $\acute{\eta}$ was chosen to be 0.01 and α to be 0.9.

The neural network model was verified using process input variables and the dynamic resistance, through experiments conducted on a 75-kVA single-phase AC pedestal welding machine. The material tested was 0.8 mm AKDQ (HDG steel). CuZr truncated cone-shaped electrodes were used. The initial face diameter of the electrodes was 6.4 mm.

A total of 170 welding data sets were collected. The welds were made sequentially in two batches. Each batch contained 85 samples. Within each batch, various welding currents, forces, and welding times were used. Between the two batches, a number of welds were intentionally made to wear down the electrodes. The average electrode contact diameters for these two batches were 6.5 and 7 mm, respectively. The ranges of welding settings are shown in Table 5.1.

Based on the data collected from the experiments, three tests were designed, as shown in Table 5.2. These tests used different groups of welding data as training and testing

TABLE 5.1

Ranges of Welding Parameters

Variables	Ranges
Force	3.0–4.0 kN
Current	6.9–13.4 kA
Time	3–36 cycles
Contact diameter	6.4–7.2 mm

samples. In test 1, samples from batch 1 were used to train the model and those from batch 2 were used to test the model. In tests 2 and 3, the training samples were randomly picked from both batches 1 and 2. Test 2 used 85 samples to train and the other 85 to test, whereas test 3 used 120 samples to train and the other 50 to test.

The overall test results are summarized in Table 5.3 and Figure 5.42. The relative error is defined as the fraction of the estimation error over the measured nugget size, that is,

$$\bar{E} = \frac{\sum_{i=1}^{N} E_i}{N} \text{ and } E_i = \frac{\hat{D}_{ni} - D_{ni}}{D_{ni}} \times 100\% \tag{5.12}$$

where $\hat{D}_{ni}$ is the i-th estimated nugget size from the model, D_{ni} is the i-th measured nugget size, and N is the total number of samples for testing.

As an example, the results from test 3 using a hidden layer structure 14 × 5 for the neural networks are plotted in Figure 5.41. The data have been sorted by the estimation results. Both training and testing data sets show a good agreement between the estimated and measured nugget sizes, except when the nugget size is small. The reason for the relatively larger errors when the nugget is small is that the RSW process tends to be unstable in the nugget initiation period, and welding parameters intended for small welds are not capable of overcoming random effects. Another reason is that the number of training data sets around that area was not large enough.

From Table 5.3, it can be seen that although the sum of squared error (SSE) training and σ training for test 1 are much smaller than those of tests 2 and 3, σ testing and the average relative error are much larger. The reason for SSE training and σ training being smaller in test 1 is that the training data sets were collected under relatively consistent welding conditions. The conditions of electrode wear were similar. However, when the trained model is used for the testing samples, where the electrode wear condition is quite different from that of the training samples, the estimations show big errors. In tests 2 and 3, the electrode wear effect is taken into account by randomly picking the training samples in

TABLE 5.2

Design of Training and Testing Groups

Test	Training Samples	Testing Samples
1	Samples from batch 1	Samples from batch 2
2	Randomly pick half of samples from both batches	The other half of the samples
3	Randomly pick 120 samples from both batches	The other 50 samples

TABLE 5.3
Test Results

Test No.	Hidden Layer Structure	SSE Training (20,000 Epochs)	σ Training (mm)	σ Testing (mm)	Average Relative Error (Testing) (%)
1	5 × 5	0.1251	0.1437	1.7755	49.62
	7 × 5	0.2656	0.2092	1.9516	34.15
	14 × 5	0.2638	0.2082	1.9013	38.16
2	5 × 5	0.4419	0.2719	1.0770	14.30
	7 × 5	0.5071	0.2893	0.5156	10.79
	14 × 5	0.7226	0.3462	1.4415	11.20
3	5 × 5	0.3656	0.2074	0.7001	9.98
	7 × 5	0.9863	0.3409	0.4298	9.42
	14 × 5	1.0251	0.3470	0.3616	9.12

both batches. As shown in the table, the testing errors are greatly reduced, indicating an increased robustness of the model. These results have demonstrated that electrode wear must be considered in the process model development for RSW.

Test 3 has shown that the estimation error is less than 10% under current conditions. The accuracy can be further improved, as more training samples become available. More training samples are especially called for during the nugget initiation period.

This ANN multivariate process model was developed for on-line nugget size estimation in RSW. It was tested in the production environment with a certain degree of success. The model's inputs consist of not only features from on-line signals, but also process input variables. Due to the complexity of the RSW process, a multilayered, instead of single-layered, neural network model was used. Principal component analysis was used for a systematic feature selection. Different on-line signals, dynamic resistances, forces, and displacements have been proven to carry similar information. Thus, only dynamic resistance was used because of the easy measurements of current and tip voltage. As a process disturbance variable, the electrode wear effect was explicitly considered in the model

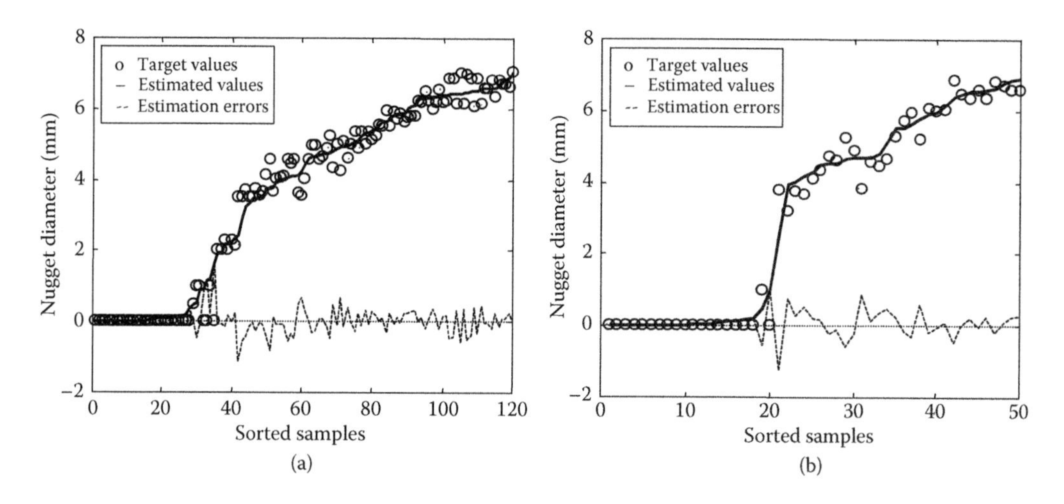

FIGURE 5.42
Estimation results for both training and testing data sets: (a) training data set and (b) testing data set.

training. A variety of welding conditions—current, force, and welding time—had been experimented with. The nugget size estimation model was demonstrated to be successful with an average relative error of the estimation of less than 10% for the conditions tested.

Monitoring and control of an RSW process have attracted much attention in the research and development community. These two tasks are closely linked, and an accurate monitoring is the basis of an effective control. An on-line, real-time recognition of process characteristics using process signals may be used to monitor weld quality, and if necessary, remediation actions can be taken. A repair can be done either right after a particular weld is made or off-line if the welds are properly indexed. A more desirable way of control is probably taking actions before a faulty process actually takes place. This requires knowledge of the characteristics of the process, especially those that may lead to the occurrence of faults. It also requires a rapid process of signals, decision making, and real-time feedback control—all must happen in the order of milliseconds. Such an ideal control of the RSW should become a reality with the advances in understanding the process, and in hardware and software. Although most of the algorithms introduced in this chapter are in the stage of research and laboratory experiment, progress has been made to bring them into the production environment.

5.4.4 Current Stepping

As discussed in Section 2.5, electrode deterioration during welding is inevitable because of the changes in chemistry, resistivity, morphology, and face area of electrodes induced by metallurgical alloying and mechanical squeezing at the electrode–workpiece interfaces. Electrode dressing is an effective means to deal with the electrode wear, yet it involves frequent interruption of the production process in order to perform the dressing operation. Another common practice is the use of current stepper. Current stepping is increasing the weld current input to compensate the electrode wear, usually in the form of mushrooming, which occurs over time.

Current stepping is generally realized through programming using several points in a welding sequence, called steppers, which are determined based on previous experiments or model predictions. In an automatic weld control, the number of welds per step and the amount of current increase per step are programmed. One example is shown in Figure 5.43 where the increase in electric current is expressed in the form of percentage of increased heat input.[35]

The changes in electrode face diameter result in changes in both electric current density and pressure at the faying interface as well as the electrode–sheet interfaces. A larger current density due to the decreasing contact area between the electrode and sheet material

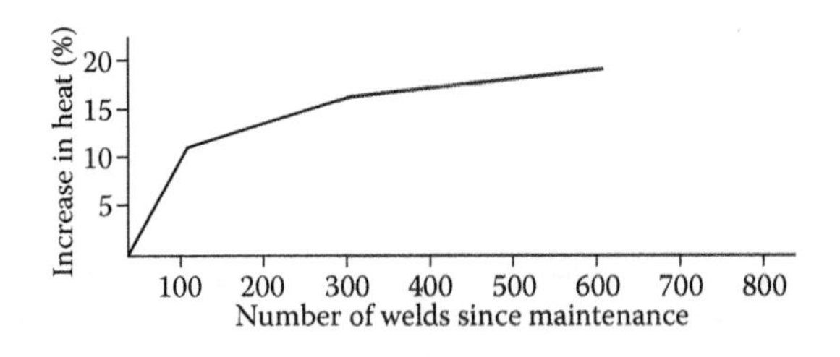

FIGURE 5.43
A schematic of a stepper program. (From Androvich, D.A., New approaches in resistance welding controls, SMWC III, Paper 18, 1988. With permission.)

than the nominal value promotes expulsion; and a smaller current density may produce a subsized weld, cold weld, or even no weld. The inadequate electric current density results from the deviated electrode–sheet contact area from the set value, which may come from electrode misalignment,[25] electrode mushrooming,[36] crater formation on the electrode face, or fracture of electrodes. In addition to changes in electric current density, the change in electrode–sheet contact area also results in a different pressure value that may affect the control of expulsion. A welding process' dependence on electrode force is far less than on electric current. A large electrode force is generally beneficial and therefore, the electrode force value can be preset to encompass a large number of possible electrode–sheet contact areas.

The basic idea of current stepping is to maintain a constant-current density during welding. The steps, or current values, can be determined in two ways. Experiments are usually conducted for a particular type of sheet material, using different sized electrode faces which come either from different electrodes or created by sequential enlargement of the electrode face through machining the electrodes, or natural wear/mushrooming. The proper current densities are determined for various electrode face sizes and the corresponding current values are calculated. A simple mathematical model can be developed as shown in Figure 5.44. The changes in both electric current and electrode force, to maintain a constant-current density and a constant electrode pressure, respectively and simultaneously, are plotted as a function of the change in electrode face diameter, as shown in Figure 5.44. In practice, a step function might be more meaningful than a continuous curve, and one possibility of using such a step function is also plotted in the figure. The correspondence between the electrode face size and current value cannot be directly used in welding control without the information on the instantaneous electrode size. The relation between the progressive change in electrode face size and the number of welds made, as shown in Figure 5.43, can be easily implemented in weld control. The quadratic dependence of changes needed in electric current and electrode force on the electrode face size change in Figure 5.44 implies that a gradually intensified increase is necessary as electrode wear progresses. One also needs to pay attention to the fact that the change in electrode face size is not necessarily monotonic. In addition to mushrooming, which increases electrode face area, material depletion may occur to an electrode surface through electrode

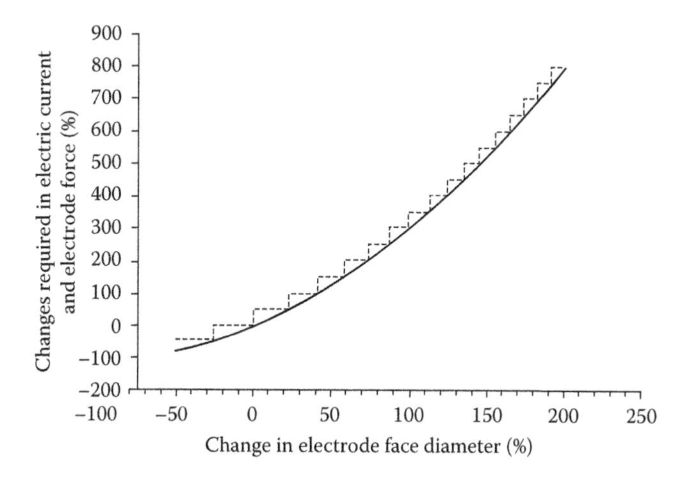

FIGURE 5.44
Changes needed in electric current and electrode force, as a function of the change in electrode face diameter.

sticking to the workpiece, or electrode fracture. Such processes can effectively reduce the electrode face area, and require a reduction in electric current. The fluctuation of electrode face size has been observed by many researchers, and one such observation has been made on galvanized steels of various thicknesses.[37] The weld size generally goes in the opposite direction from the electrode face size. Although fixed-program weld steppers have the advantage of being simple to implement, they do not account for the weld-to-weld variation. Efforts have been made to accurately determine the welding current for a weld based on signals collected on its immediate preceding welds through a feedback control algorithm. For instance, automatic current stepping was achieved by monitoring the dynamic resistance curve or the power factor response to maintain maximum current density without expulsion.[38]

The use of steppers in welding can significantly increase the electrode life, and/or enlarge the intervals of dressing electrodes. Therefore, combining current stepping and electrode dressing provides an efficient way to prolong electrode life.

References

1. Gedeon, S. A., Sorensen, C. D., Ulrich, K. T., and Eagar, T. W., Measurement of dynamic electrical and mechanical properties of resistance spot welds, *Welding Journal*, 66, 378-s, 1987.
2. Tsai, C. L., Dai, W. L., Dickinson, D. W., and Papritan, J. C., Analysis and development of a real-time control methodology in resistance spot welding, *Welding Journal*, 335-s, 1991.
3. Li, W., Monitoring and diagnosis of resistance spot welding process, PhD dissertation, University of Michigan, Ann Arbor, 1999.
4. Haefner, K., Carey, B., Bernstein, B., Overton, K., and D'Andrea, M., Real time adaptive spot welding control, *Transactions of ASME—Journal of Dynamic Systems, Measurement, and Control*, 113, 104, 1991.
5. Nakata, S., Aono, S., Suzuki, M., Kawaguchi, Y., and Inoue, M., Quality assurance characteristics in resistance spot welds by adaptive control system and its field applications, presented at Annual Assembly of IIW, Ljubljana, Yugoslavia, IIW Doc. No. III-720-82, 1982.
6. Ma, C., Bhole, S. D., Chen, D. L., Lee, A., Biro, E., and Boudreau, G., Expulsion monitoring in spot welded advanced high strength automotive steels, *Science and Technology of Welding and Joining*, 11 (4), 480–487, 2006.
7. Cleveland, D. and O'Brien, L. J., Acoustic emission spot welding monitor, Final Technical Report, NSF under Award, No. MEA 82-60345, 1983.
8. Stiebel, A., Apparatus and method for monitoring and controlling resistance welding, U.S. Patent No. 4419558, 1983.
9. Stiebel, A., Ulmer, C., Kodrack, D., and Holmes, B., Monitoring and control of spot weld operations, SAE Technical Paper, No. SAE 860579, 1986.
10. Dickinson, D. W., Franklin, J. E., and Stanya, A., Characterization of spot welding behavior by dynamic electrical parameter monitoring, *Welding Journal*, 59, 170s, 1980.
11. Li, W., Hu, S. J., and Zhang, H., Signal processing issues in resistance spot welding, Sheet Metal Welding Conference IX, Sterling Heights, MI., Paper 32, 2000.
12. NIST—ATP Intelligent Resistance Welding Quarterly Progress Report, No. 103, Ann Arbor, MI, 1996.
13. NIST—ATP Intelligent Resistance Welding Quarterly Progress Report, No. 203, Ann Arbor, MI, 1997.
14. NIST—ATP Intelligent Resistance Welding Quarterly Progress Report, No. 402, Ann Arbor, MI, 1999.

15. Haykin, S., *Adaptive Filter Theory*, 2nd edition, Prentice Hall Information and System Science Series, Prentice Hall, Englewood Cliffs, NJ, 1991.
16. NIST—ATP Intelligent Resistance Welding Quarterly Progress Report, No. 202, Ann Arbor, MI, 1997.
17. Boilard, R. and Farrow, J., Automatic current steppers for improved weld quality, SMWC IV, Paper 8, 1990.
18. AWS/SAE D8.9M, *Recommended Practices for Test Methods for Evaluating the Resistance Spot Welding Behavior of Automotive Sheet Steel Materials*, The American Welding Society, Miami, FL, 2002.
19. Gould, J. E., Kimchi, M., Leffel, C. A., and Dickinson, D. W., Resistance seam weldability of coated steels: Part I. Weldability envelopes, Edison Welding Institute Research Report, No. MR9112, 1991.
20. Kaiser, J. G., Dunn, G. J., and Eagar, T. W., The effect of electrical resistance on nugget formation during spot welding, *Welding Journal*, 61, 167-s, 1982.
21. Nagel, G. L. and Lee, A., A new approach to spot welding feedback control, SAE Technical Paper, No. SAE 880371, 1988.
22. Karagoulis, M. J., Process control in manufacturing: Control of materials processing variables in production resistance spot welding, in Proceedings of AWS Sheet Metal Welding Conference V, Detroit, MI, Paper No. B5, 1992.
23. Rivett, R. M. and Hurley, J. P., Weld bonding of zinc-coated sheet steels, SMWC IV, Paper 4, 1990.
24. Zhang, H., Hu, J. S., Senkara, J., and Cheng, S., Statistical analysis of expulsion limits in resistance spot welding, *Transactions of ASME—Journal of Manufacturing Science and Engineering*, 122, 501, 2000.
25. Ma, C., Chen, D. L., Bhole, S. D., Boudreau, G., Lee, A., and Biro, E., Microstructure and fracture characteristics of spot-welded DP600 steel, *Materials Science and Engineering A*, 485, 334–346, 2008.
26. Zhang, H., Expulsion and its influence on weld quality, *Welding Journal*, 78, 373s, 1999.
27. NIST—ATP Intelligent Resistance Welding Quarterly Progress Report, No. 403, Ann Arbor, MI, 1999.
28. Tack, J., Pavement performance prediction using pattern recognition: Artificial neural networks and statistical analysis, PhD Thesis, The University of Toledo, Toledo, OH, 2002.
29. Ballard, D. H., *An Introduction to Natural Computation*, MIT Press, Cambridge, MA, 1997.
30. McCulloch, W. S. and Pitts, W. H., A logical calculus of the ideas imminent in nervous activity, *Bulletin of Mathematical Biophysics*, 115–133, 1943.
31. McClelland, J. L. and Rumelhart, D. E., *Explorations in Parallel Distributed Processing*, MIT Press, Cambridge, MA, 1988.
32. Zurada, J. M., *Introduction to Artificial Neural Systems*, West Publishing Company, New York, 1992.
33. NIST—ATP Intelligent Resistance Welding Quarterly Progress Report, No. 204, Ann Arbor, MI, 1996.
34. NIST—ATP Intelligent Resistance Welding Quarterly Progress Report, No. 301, Ann Arbor, MI, 1997.
35. Androvich, D. A., New approaches in resistance welding controls, SMWC III, Paper 18, 1988.
36. Stiebel, A., Ulmer, C., Kodrack, D., and Holmes, B. B., Monitoring and control of spot weld operations, SMWC II, Paper 4, 1986.
37. Kuo, M., Kelly, D., Boguslawski, V., Liu, P. S., Lario, T., Orsette, C., and Tann, L., Methodology development of tip dresser application in the production environment, SMWC IX, Paper 3-3, 2000.
38. Boilard, R. and Farrow, J., Automatic current steppers for improved weld quality, SMWC IV, Paper 8, 1990.

6

Weld Quality and Inspection

Weld quality evaluation lies at the center of all aspects of welding. A welding schedule is judged by whether it can produce welds of acceptable quality; a material, before it is used in production, needs to be qualified as weldable: using standard welding equipment and schedules would yield welds of sufficient size and strength or quality. However, there are no universally accepted standards of weld quality, as reflected in an American national standard (*Standard Welding Terms and Definitions*, ANSI/AWS A3.0:2001[1]), which defines an acceptable weld as "a weld that meets the applicable requirements." Therefore, determining a quality weld is largely at the manufacturer's discretion. This chapter will discuss the common practices for evaluating a weld's quality.

6.1 Weld Quality

The quality of a weld is usually expressed by its measurable features, such as the physical attributes and the various strengths, when inspected in either a destructive or nondestructive manner. In this section, the commonly measured quality attributes and requirements are discussed first, then the relations between the required measurements and the weld strengths are presented.

6.1.1 Weld Attributes

A weld's quality can be described in three ways: by its physical or geometric features, its strength or performance, or the process characteristics during welding. Depending on the specific needs, usually more than one quality attribute is needed in order to evaluate a weld.

6.1.1.1 Geometric Attributes

These refer to the geometric features either directly visible after a weldment is made or revealed through destructive tests, such as peeling or cross-sectioning, or nondestructive tests using, for example, ultrasonic or x-ray devices. The commonly referred weld attributes are:

- Nugget/button size
- Penetration
- Indentation
- Cracks (surface and internal)
- Porosity/voids

- Sheet separation
- Surface appearance

Among these weld attributes, weld size, in terms of nugget width or weld button diameter, is the most frequently measured and most meaningful in determining a weld's strength. When two sheets are joined by a weld at the nugget, its size determines the area of adhesion and its load-bearing capability. However, the nugget/weld size alone is often insufficient in describing a weld's quality, as it does not necessarily imply the structural integrity of the weld. Other features of a weld, such as penetration, may complement the nugget size and provide useful information on the degree of adhesion. *Weld* and *nugget* are considered exchangeable by many, especially in oral presentations. Although closely related, however, they are not the same by definition or measurement. In fact, a weld is meant to contain all parts of a weldment, such as the heat-affected zone (HAZ), in addition to the nugget. Another confusion is the use of *button diameter* and *nugget diameter.* As a nugget and its size are usually revealed by metallographic cross-sectioning, a nugget is exposed for measuring its width, not diameter, as shown in Figure 6.1.[2] The figure also shows other features that can be revealed by cross-sectioning a weldment.

There are other less common attributes of a weld, which may require a substantial amount of work and interpretation to use. For instance, the hardness distribution across a weldment, measured diagonally on a cross-sectioned surface, often provides important information on the heating and cooling history of the weldment, and on the inference on the performance. The structures and dimensions of the nugget and HAZ are other such weld attributes through which a direct link can be established between the process parameters and the performance.

There is very limited research about the influence of other attributes on weld quality, such as the HAZ, indentation, penetration, and material properties of the weld nugget. There are certain efforts to distinguish the material properties in the nugget, HAZ, and base metal (as shown in Refs. 3 and 4).

6.1.1.2 Weld Performance

Besides occasional cosmetic considerations, spot welding quality is mostly about the performance. Welding performance characteristics usually refer to static strength and dynamic strength of a weld, and the ones mostly relevant to the sheet metal industries are listed here:

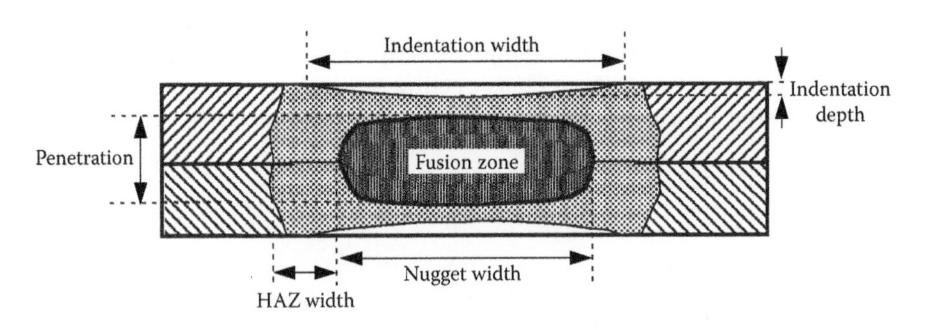

FIGURE 6.1
Weld attributes revealed by metallographic sectioning.

- Tensile–shear strength
- Tensile strength
- Peel strength
- Fatigue strength
- Impact strength
- Corrosion resistance

The most commonly measured is tensile–shear strength, because it is relatively easy to measure, and many structures are designed to bear tensile–shear loading. The details of testing these strengths are presented in Chapter 4.

6.1.1.3 Process Characteristics

There are a few process-related characteristics that may serve as indicators of the welding process, and often implicate the weld quality. The most frequently noticed is expulsion. Expulsion, which is the ejection of liquid metal during welding, is a clear indication of possible weakening of a weld. As a matter of fact, the trace of expulsion is usually clearly visible even after the weld is made. Expulsion in steels leaves a burning mark on the sheet surface, in addition to the metal debris, as discussed in Chapter 7. In aluminum welding, however, no sign of burning can be seen, either during expulsion or after welding. Expulsion, when it occurs, can be detected using sensors. Signals such as acoustic emission and electrode displacement can easily show the occurrence of expulsion. An operator can easily hear or feel an expulsion without much experience or training.

Another process characteristic, which should be constantly monitored in a production environment, is welding consistency. This is more meaningful for large-volume production than the quality of a particular weld, as it directly implies consistency of production quality. It is actually a collective of all quality measures.

6.1.2 Weld Quality Requirements

For practical use, easily measurable requirements of weld quality are quantified, mostly in the form of tables. Many of the standards and recommendations are developed by individual companies, such as Ford Motor Company and General Motors. Professional organizations such as the American Welding Society (AWS) and International Organization for Standardization also contribute to a significant portion of the standards.

Because of the drastic differences in design, understanding and perception of weld quality, and production and testing environment, automobile manufacturers and others tend to have very different requirements on weld quality. As shown in Figure 6.2, the requirements on weld nugget sizes are significantly different. For the same thickness, the largest weld size required can be more than twice that of the smallest one. However, in general, they are enveloped between $3\sqrt{t}$ and $6\sqrt{t}$ (t is the thickness of the sheets).[5] Most of the requirements are located between $4\sqrt{t}$ and $5\sqrt{t}$, and many nominal weld sizes are set in this range when determining operation windows. The requirements on weld sizes are probably the most commonly used criteria regarding weld quality. A possible confusion in applying such requirements is the use of *weld size*. Both weld button diameter and nugget width are used as *weld size* in practice, depending largely on convenience. However, very different values can be expected in some cases for the weld button and nugget. A nugget

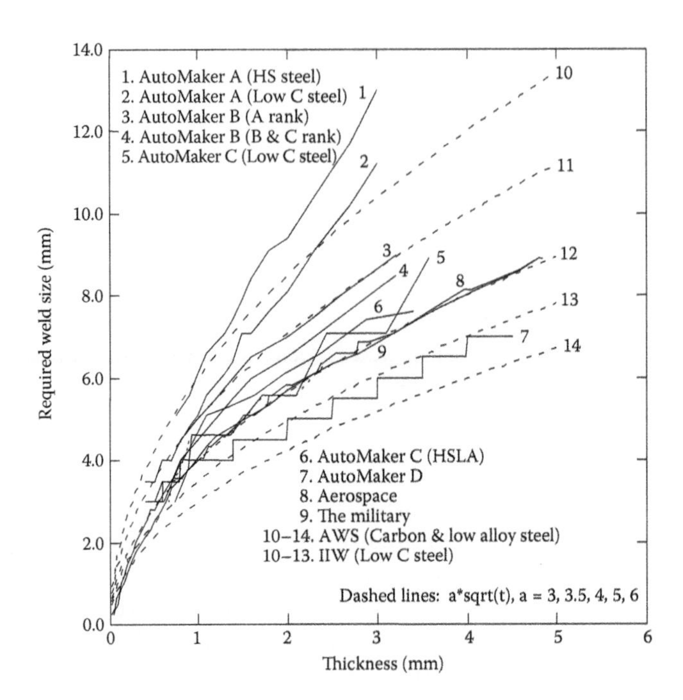

FIGURE 6.2
A comparison of nugget size requirements.

can be smaller than, similar to, or bigger than a weld button for the same weldment. A weld button diameter is more variable than a weld nugget width, as the testing method, the geometry of the tested specimen, the fracture mode, and the measurement can all make a significant difference in measured weld button diameter. Therefore, a measurement using nugget width is more consistent and comparable. However, the sectioning in order to measure nugget width prohibits its use in daily production. Some of the available nondestructive evaluation methods presented later in this chapter can be used to make simple, yet accurate measurements of nugget width. The common measure in production is still the weld button diameter, as it can be produced using a chisel and a caliper, and the criteria for weld size usually refer to weld button diameter.

Another quality-revealing feature of a weld is penetration. It describes the amount of material that melts during welding in the thickness direction. A small penetration may mean insufficient heating and indicate a cold weld. In general, large penetration is preferred. However, penetration is directly linked to the amount of heating of the sheets between the electrodes and, therefore, large penetration means softening of the sheet metal and large indentation. The requirements on penetration are usually very loose; as seen in Figure 6.3, it goes from 20% to 90% or more. One of the manufacturers compared in the figure did not specify the upper limit, yet this does not mean 100% penetration. In general, a large penetration is acceptable if it does not create an excessive indentation. The requirements on penetration are often applied together with those on weld size.

Besides the physical attributes, such as weld size and penetration, requirements on the performance characteristics, such as tensile–shear strength, have been specified in standards. Again, there is a vast difference among manufacturers and professional organizations. The values, as shown in Figure 6.4, are closely related to the strength of the base

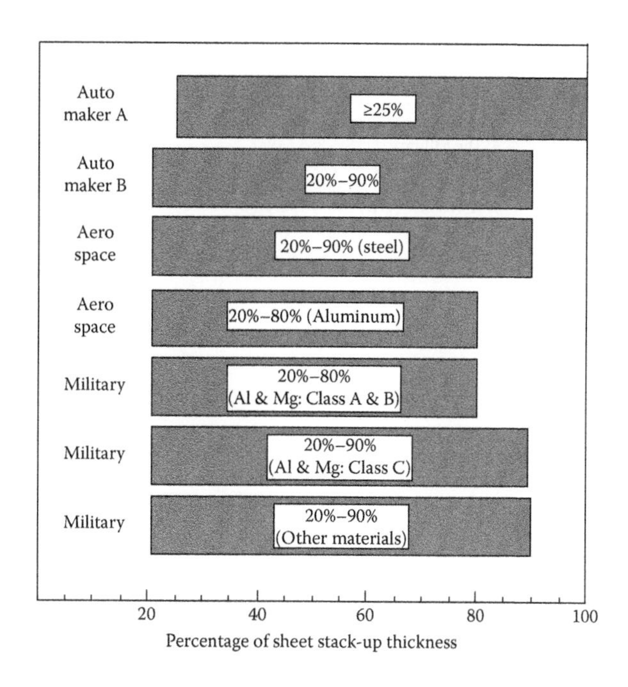

FIGURE 6.3
A comparison of nugget penetration requirements.

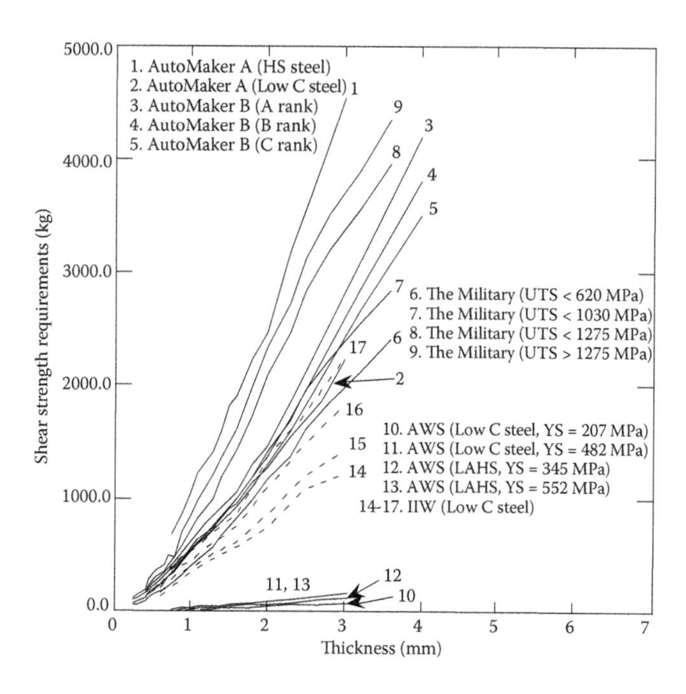

FIGURE 6.4
A comparison of shear strength requirements in steel welding.

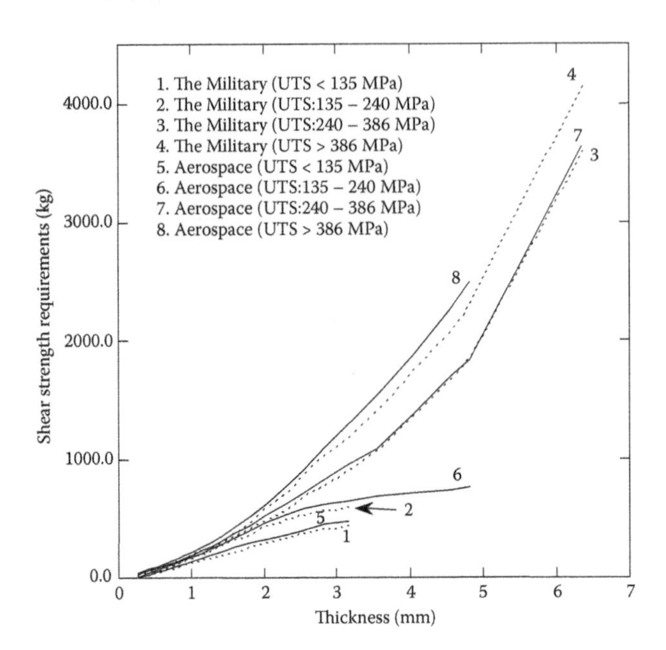

FIGURE 6.5
A comparison of strength requirements for Al and Mg alloys.

metals. Unlike the requirements on weld size, in which materials are not specified, different materials have different requirements on strength. Aluminum and magnesium alloys usually have lower (base metal) strength than steels, and the welds made in these alloys have lower expected strength values for sheets of similar gauges (Figure 6.5).

Sometimes the nugget or button size is used as a sole parameter to describe the weld quality. This is because intuitively, nugget/button size has the largest influence on weld strength. Another reason is that it is convenient to measure. However, more often in practice, mixed weld attributes are used to reflect the concerns of a manufacturer. Spinella[6] defined good welds as those with the large nugget and high tensile strength, and without expulsion or partial interfacial failure. Newton et al.[7] define a good weld as one with a full-size nugget and larger than the minimum strength, and without cracks, flash (expulsion), or porosity. They also tried to define a bad weld as one with an undersized nugget, cracks, excessive porosity, excessive expulsion, and damaged adhesive in the case of weld bonding.

6.1.3 Relations between Weld Attributes and Strength

As most weld quality inspection obtains only measurable geometric quantities, such as weld button size, it is desirable to learn from such measurements the strength level of a spot-welded joint. A very common way to quantify weld quality is to build the relationship between weld attributes and spot weld strength. Because weld size is the most common measurement and tensile–shear testing is mostly conducted in practice, the majority of the work is therefore on the relationship between weld size and tensile–shear strength. Some of these relationships are given in this section.

First, the strength attributes of a quasi-static tensile–shear test should be defined. The peak load in such a test is the most popular measurement. It describes the maximum amount of loading a weld can take, so it provides useful information to designers and other users. However, no information can be obtained from such measurements on the ductility, nor the performance of such a weld under dynamic instead of static loading.[8] A quantity related to the toughness of a weld should be defined in order to fully describe the strength of a weldment under either static or dynamic loading, especially impact loading. Figure 6.6 shows that a typical load-vs.-displacement curve during tensile–shear testing has an increase in load at the beginning of testing; then after reaching a peak value (peak load), the load decreases with additional displacement. A weld can be strong (tough) with large displacement, or a cold weld may have a high peak load but little displacement to final fracture. Therefore, displacement can be used to describe the ductility of a weld. The total displacement when the specimen finally fails is not quite relevant to a weld's strength—it reflects more on the influence of the specimen rather than that of the weld. With identical welds, a specimen with a large (long and wide) overlap may experience a high total displacement value when torn off, than a specimen with a narrow or short overlap. Therefore, the total displacement does not accurately indicate a weld's strength. From Figure 6.6, it can be seen that the displacement when the load reaches the peak is independent of the fracture of the base metal and, therefore, can be used for the weld's strength. It is termed maximum displacement. A related quantity is the energy absorbed by the specimen up to the peak load. It can be calculated by the area under the load-displacement curve and, therefore, like the maximum displacement, it may also serve as an indicator of the toughness. The peak load, together with the maximum displacement or the energy, should fully describe a weld's strength.[8]

When tensile–shear loaded, the HAZ in a welded specimen usually experiences the highest stresses. As a result, the specimen bends around the HAZ and cracks often initiate in this area, as seen in the one-quarter model of a lap-joined specimen in Figure 6.7, obtained from a finite element simulation.[9] In fact, the properties of the HAZ are significantly more influential than those of the nugget in determining the performance of a weldment. Figure 6.8 compares the effects of material property changes in the nugget and the HAZ, on a

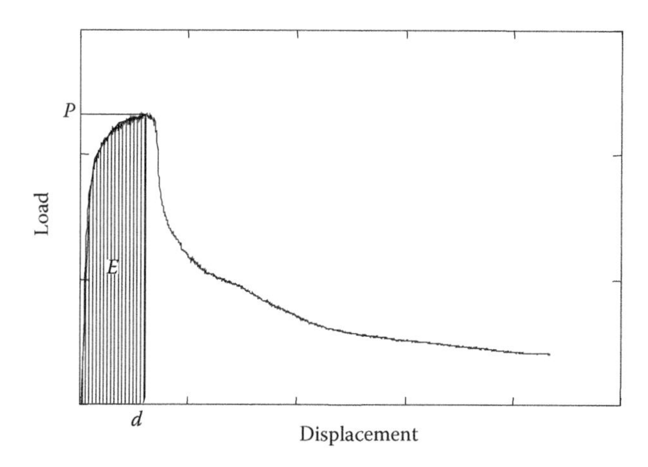

FIGURE 6.6
Definition of peak load, energy, and maximum displacement in a quasi-static tensile–shear test (same as Figure 4.1).

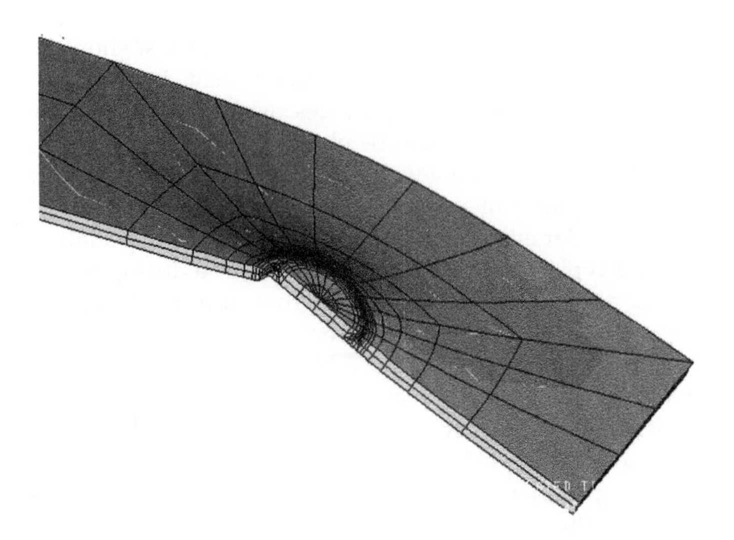

FIGURE 6.7
Failure at the HAZ.

lap joint under tensile–shear loading. With all other dimensions and properties fixed, an increase in the nugget size results in increases in the peak load, maximum displacement, and energy, as defined in Figure 6.6. Therefore, the finite element simulation results are consistent with experimental observations on the effect of nugget size. However, increasing the ductility of the nugget has no effects on the strength, and changes in ultimate tensile strength and yield strength of the nugget have negligible influence. As the only influential factor, the nugget size determines the size of the connection/joint between the sheets, which intuitively affects the strength of the weldment. The interior of a nugget, however, is not stressed when a specimen is loaded in tensile–shear mode. Therefore, changes inside a nugget have no effects on the overall performance of the specimen.

The HAZ, on the contrary, is very sensitive to all changes in mechanical properties. In Figure 6.8, increases in the HAZ size, ductility, and ultimate and yield stresses result in higher peak load, maximum displacement, and energy. Unlike the nugget, which is stressed around its periphery when the specimen is loaded, the HAZ, because it is located at the edge of the nugget, is under constant loading. High ductility and UTS/yield stresses make the HAZ strong and, therefore, the weldment strong. Because the HAZ serves as a transition zone between the base metal and the nugget, which are distinguished by their different material properties, a wide HAZ produces a smooth transition, and less stress concentration due to strength mismatch between the nugget and base metal. In practice, however, the HAZ size is difficult to control, and when the size varies, its properties usually will not stay constant. This study shows the idealized cases to illustrate the individual effects.

There are a number of efforts of relating the tensile–shear strength of a weld to the weld button diameter. In the 1940s, a very simple expression of strength vs. weld diameter was found by Keller and Smith[10] and McMaster and Lindrall[11]:

$$P = 120d^2 \tag{6.1}$$

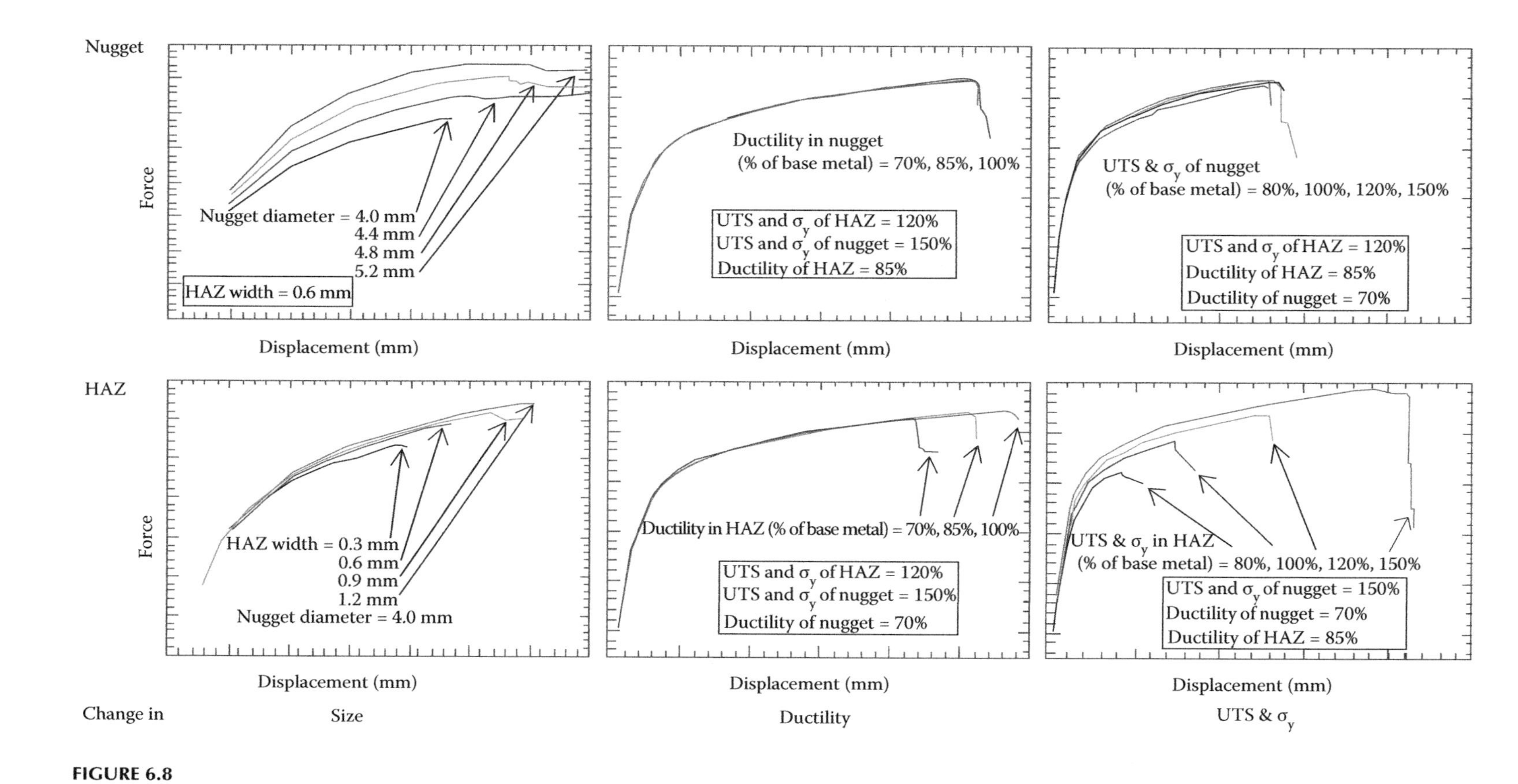

FIGURE 6.8
(See color insert.) Influence of dimension and mechanical properties of nugget and HAZ.

where P is shear load (in Newton) and d is weld diameter (in millimeters). Heuschkel[12] proposed the following linear empirical relationship for tensile–shear strength:

$$S = t{\cdot}S_0{\cdot}d{\cdot}[\alpha - \beta(C + 0.05Mn)] \quad (6.2)$$

where S is tensile–shear strength, S_0 is base metal strength, d is weld diameter, t is sheet thickness, C and Mn are weight percent values of the alloying elements, to reflect the contribution of base metal chemistry, and α and β are functions of thickness t. Following Heuschkel's work, Sawhill and Baker[13] proposed another similar formula for rephosphorized and stress-relieved steels:

$$S = F{\cdot}t{\cdot}S_0{\cdot}d \quad (6.3)$$

where F is a material-dependent coefficient, F = 2.5–3.1. By considering fracture mode, an expression was proposed for aluminum alloys by Thornton et al.[14]:

$$P = (0.12t - a)d \quad (6.4)$$

where P is expressed in kilonewton, t and d are in millimeters, and a is the coefficient of fracture mode. These equations provide valuable empirical relationships between weld diameter and tensile–shear strength. However, they are suitable only for specific weld joint geometry rather than an arbitrary weld joint and often only for particular materials. Ewing et al.[15] tried to develop a relationship between spot weld failure load, base metal strength, testing speed, joint configuration, and welding schedule. Various tests such as tensile–shear, cross-tension, and coach–peel testing with various materials used in automotive bodies were conducted to establish the relationship. However, they finally concluded that "the spot weld failure process is such a complex process that it is very difficult to isolate variables."

In the study by Zhou et al.,[8] through numerical simulation, attempts were made to link the weld strength to the geometry of a specimen, HAZ, and nugget, and the material properties of the weldment. They proposed to express the strength of a weldment in terms of the peak load, and corresponding energy and displacement at the peak load. Intuitively, they can be expressed by

$$P = f_P \text{ (geometry; material properties of base metal, HAZ, and nugget)} \quad (6.5a)$$

$$U = f_U \text{ (geometry; material properties of base metal, HAZ, and nugget)} \quad (6.5b)$$

$$W = f_W \text{ (geometry; material properties of base metal, HAZ, and nugget)} \quad (6.5c)$$

where P is peak load and U and W are corresponding displacement and energy. In general, these relationships are complicated, and it is impossible to derive them analytically. To establish the relations, a new approach was adopted using numerical simulation aided by the concept of design of experiments (DOE or DOX) (Koehler and Owen).[16]

In Zhou et al.,[8] two sets of variables were studied. One is for geometrical variables, including sheet thickness, specimen width, HAZ size, and indentation depth. The other group of variables includes material properties, which are Young's modulus, Poisson's ratio, yield strength, ultimate tensile strength, and elongation. Because the material structures in nugget, HAZ, and base metal are different, different material properties should be used for each part of the weldment. However, the material properties of other parts can be linked to those of the base metal by hardness with the following relations:

$$\sigma_{uts} = \sigma_0 + k_1 * H_\nu \tag{6.6a}$$

$$\sigma_y = k_1 * H_\nu \tag{6.6b}$$

$$e = k_2 / H_\nu \tag{6.6c}$$

$$H_\nu = k * H_{\nu BASE} \tag{6.6d}$$

The number of material parameters was significantly reduced by using these equations. For simplicity, only steel was considered with the Young's modulus and Poisson's ratio fixed at $E = 210$ GPa and $\nu = 0.3$. Hence, in the design, only the base metal properties (yield strength σ_y, ultimate tensile strength σ_{uts}, and elongation e) and the hardness ratio (k) of nugget and base metal were left as material variables. Geometric attributes were sheet thickness t, sheet width W, HAZ size h, and indentation t_i.

Therefore, the expressions of Equation 6.5 can be simplified as

$$P_{max} = f_P\,(t, W, h, t_i; \sigma_y, \sigma_{uts}, e, k) \tag{6.7a}$$

$$U_{max} = f_U\,(t, W, h, t_i; \sigma_y, \sigma_{uts}, e, k) \tag{6.7b}$$

$$W_{max} = f_W\,(t, W, h, t_i; \sigma_y, \sigma_{uts}, e, k) \tag{6.7c}$$

Table 6.1 lists the ranges of each design variable needed in the statistical design.

An optimal Latin hypercube design[16,17] for eight variables was chosen based on the maximum distance criterion. The design points are distributed fairly uniformly in the design space to eliminate the randomness, and ensure that all the points are neither too far from nor too close to each other.

In order to effectively conduct the experiment, a generic finite element model (Figure 6.9) is developed so that it is easy to implement parameter changes of geometrical variables (width, thickness, nugget size, HAZ size, indentation) and material variables (elastic and plastic properties in base metal, nugget, zones in the HAZ). A stress distribution using the model is shown in Figure 6.10, which shows that the largest stresses are located in the HAZ.

TABLE 6.1

Values of Design Variables

t (mm)	h (mm)	W (mm)	t_i	σ_y (MPa)	σ_0 (MPa)	e (%)	k
0.5–2.0	0.1–1.5	30–50	0–20%	205–1725	50–200	2–65	1.0–3.0

Note: σ_{uts} was replaced by σ_0, which is the difference between the ultimate tensile strength σ_{uts} and yield strength σ_y. This is to ensure that the ultimate tensile strength is always greater than the yield strength. Otherwise, σ_{uts} might be smaller than σ_y.

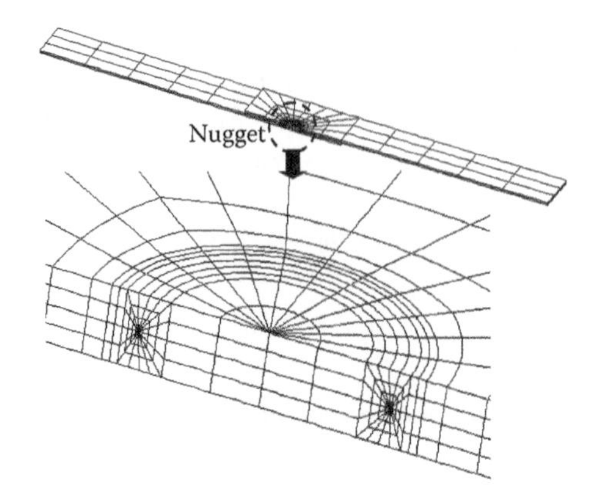

FIGURE 6.9
A generic finite element model for tensile–shear testing.

Using computer modeling, the significance of each effect was determined and ranked, to determine which effects should be included in the regression model.

Figure 6.11 shows the influences of variables on the maximum load P_{max}. It indicates that the yield strength and sheet thickness are more influential on P_{max} than other variables. The size of HAZ also plays an important role on P_{max}. Therefore, P_{max} can be expressed in terms of sheet thickness t, yield strength σ_y, and size of HAZ h only without much error:

$$P_{max} = -6.42 + 3.02t + 0.01576\sigma_y - 10.06h + 15.34t^*h \text{ (kN)} \tag{6.8}$$

It has a high coefficient of determination (94.5%).

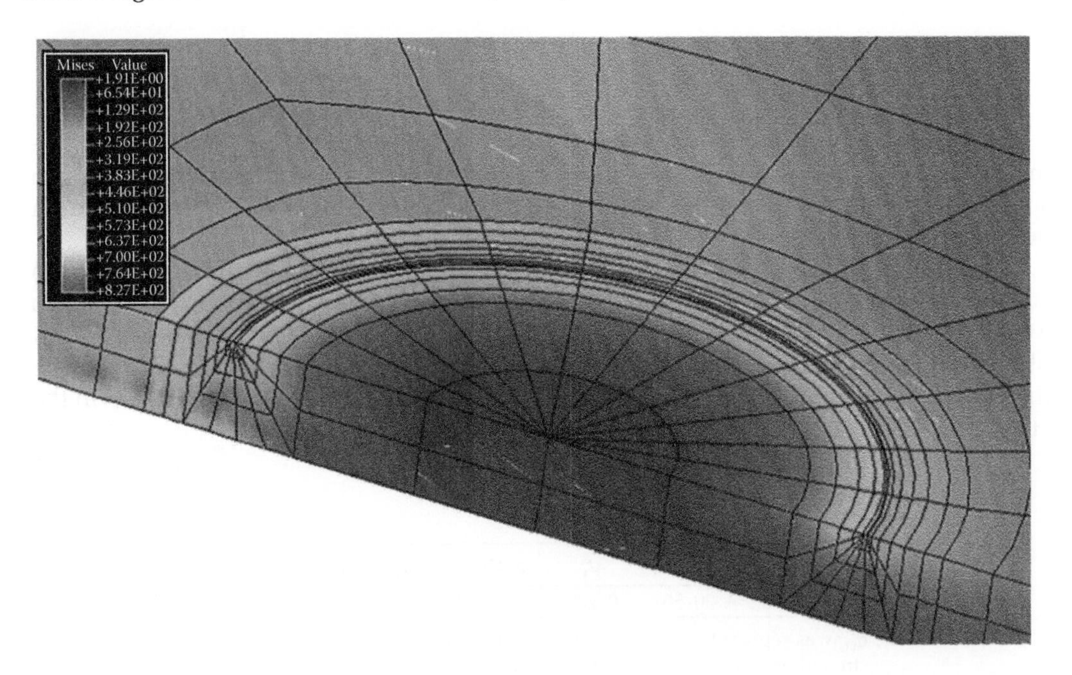

FIGURE 6.10
(See color insert.) von Mises stresses distribution at weldment.

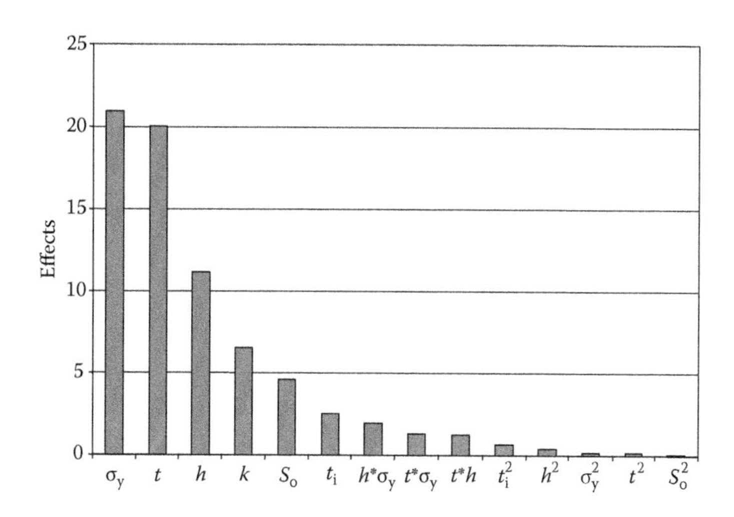

FIGURE 6.11
Variable effects on maximum load P_{max}.

Following similar procedures, the expressions for W_{max} and U_{max} are obtained, based on the significance of each effect on these two outputs as shown in Figures 6.12 and 6.13. They are shown in Equations 6.9 and 6.10 and have coefficients of determination of R = 97.6% and 97.0%, respectively:

$$W_{max} = 126966 - 414160t + 325520h - 106.718\sigma_y + 70.452\sigma_{uts} + 3288k - 6898.8t * h + 22.50t * \sigma_y + 26.916h * \sigma_y + 164950t^2 - 204840h2 \quad (6.9)$$

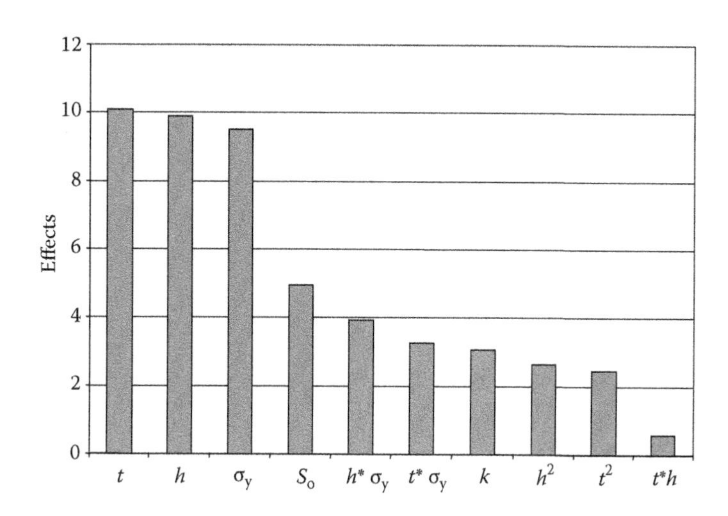

FIGURE 6.12
Variable effects on maximum energy W_{max}.

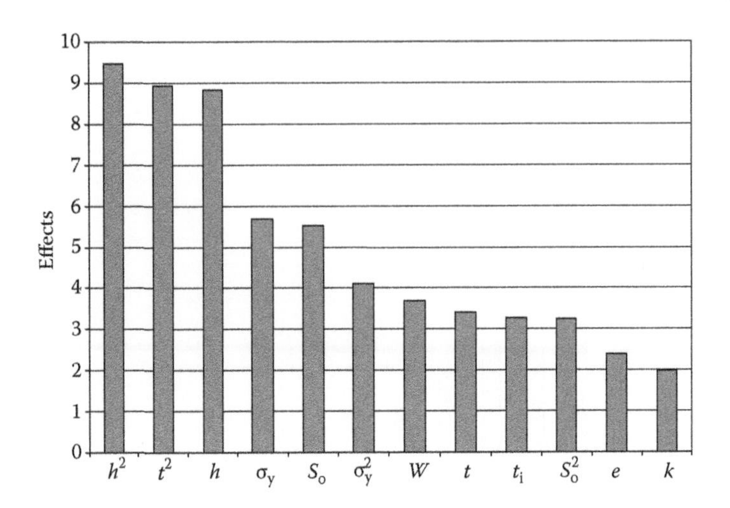

FIGURE 6.13
Variable effects on maximum displacement U_{max}.

$$U_{max} = 3.4129 - 12.485t + 10.255h - 0.012032w - 1.0705t_i - 0.0525\sigma_y + 0.048391\sigma_{uts} + 0.34688e + 0.064421k + 5.0534t^2 - 6.1509h^2 + 0.0000022609\sigma_y^2 - 0.00018391(\sigma_{uts} - \sigma_y)^2 \quad (6.10)$$

Figure 6.12 shows that sheet thickness t, HAZ size h, and yield strength σ_y have large effects on energy. For maximum displacement, however, the most important variables are the quadratic terms of h and t, and the linear term of h and, therefore, the most important variable is the size of HAZ (Figure 6.13). In both cases, some other terms, including quadratic and interactive terms, cannot be neglected.

Based on the analysis, it can be observed that the sheet thickness (and, therefore, the weld size, as it is linked to the thickness), HAZ size, and material yield strength are the most important attributes in determining spot weld quality.

In general, a weld's quality has been described using a set of selective attributes of weld performance and physical characteristics, and welding process. It is unlikely that a single parameter will be sufficient to define a weld's quality, yet efforts are still needed to create a unified measurement for weld quality, based on proven relations between weld attributes and performance characteristics.

6.2 Destructive Evaluation

Weld quality is usually monitored through destructive testing, such as peel and chisel tests. The shape, size, and other features of a weld button are used to judge if the weld meets the requirements. Instrumented destructive testing is common in the laboratory environment, and it provides quantitative strength information, in addition to the measurable features, as can be revealed by uninstrumented tests. Such a test requires certain experience in specimen preparation, equipment setup, testing, and data analysis. Commonly conducted instrumented tests are discussed separately in Chapter 4.

Peel and chisel tests are frequently conducted in both laboratory and production environments. These tests provide rapid and valuable information on weld quality and serve as the primary tool for setting up welding schedules.

6.2.1 Peel Test

The peel test (Figure 4.3) consists of peeling apart, to destruction, a weld specimen to determine the weld button size and weld fracture mode. A spot weld is considered acceptable if the peel test reveals a weld button size greater than or equal to values from specific requirements.

6.2.2 Chisel Test

A chisel test consists of forcing a tapered chisel into the lap on each side of the weld being tested until the weld separates, resulting in a pulled button or interfacial failure. The edges of the chisel must not touch the weld being tested (see Figure 4.3). This type of test should be used when a peel test is not practical. The weld acceptance is based on the same criteria as those for the peel test. A chisel test can be conducted either manually or mechanized, that is, driven by hydraulic or other types of power.

Unlike in a peel test in which a tensile load is exerted on one side of a weld, a chisel test imposes tension from both sides of a weld. In practice, a chisel test tends to produce more interfacial fracture than a peel test, especially when the constraint on the weld from the sides is high. Depending on the material and type of loading, a button may not always result from peel or chisel testing, even if it is an acceptable weld. In the case of an interfacial fracture, if a visual examination cannot decide the size of fusion at the faying interface, a macrosectioning of the fusion zone must be performed to determine a weld's acceptability.

6.2.3 Metallographic Test

A metallographic test is used to determine the geometric features of a weld such as nugget width, penetration, indentation, and HAZ width (as shown in Figure 6.1). It can also be used to detect cracks, porosity, and nonmetallic inclusions. In this test, weld sections are cut from product samples, polished to the weld centerline, chemically etched to reveal the microstructure, and then optically examined. An acceptable weld has a fusion zone equal to or greater than the required values specified, without excessive internal discontinuities.

The appearance or features of fractured welds by peel or chisel tests are currently correlated with strengths measured by instrumented tests, and the correlation serves as the basis for acceptance criteria for materials with yield strengths below 420 MPa. Welds of sufficiently large size are considered acceptable since such welds often produce satisfactory strength performance. Because such a correlation does not exist at the moment for higher-strength steels, defining or checking weld quality by uninstrumented destructive tests, often the only tests in production, is a major challenge in adopting advanced high-strength steels (AHSSs) or other high-strength steels.

Peel or chisel testing individually made coupons can be done following standard procedures. On production parts, the spot welds are tested for the required size, as explained

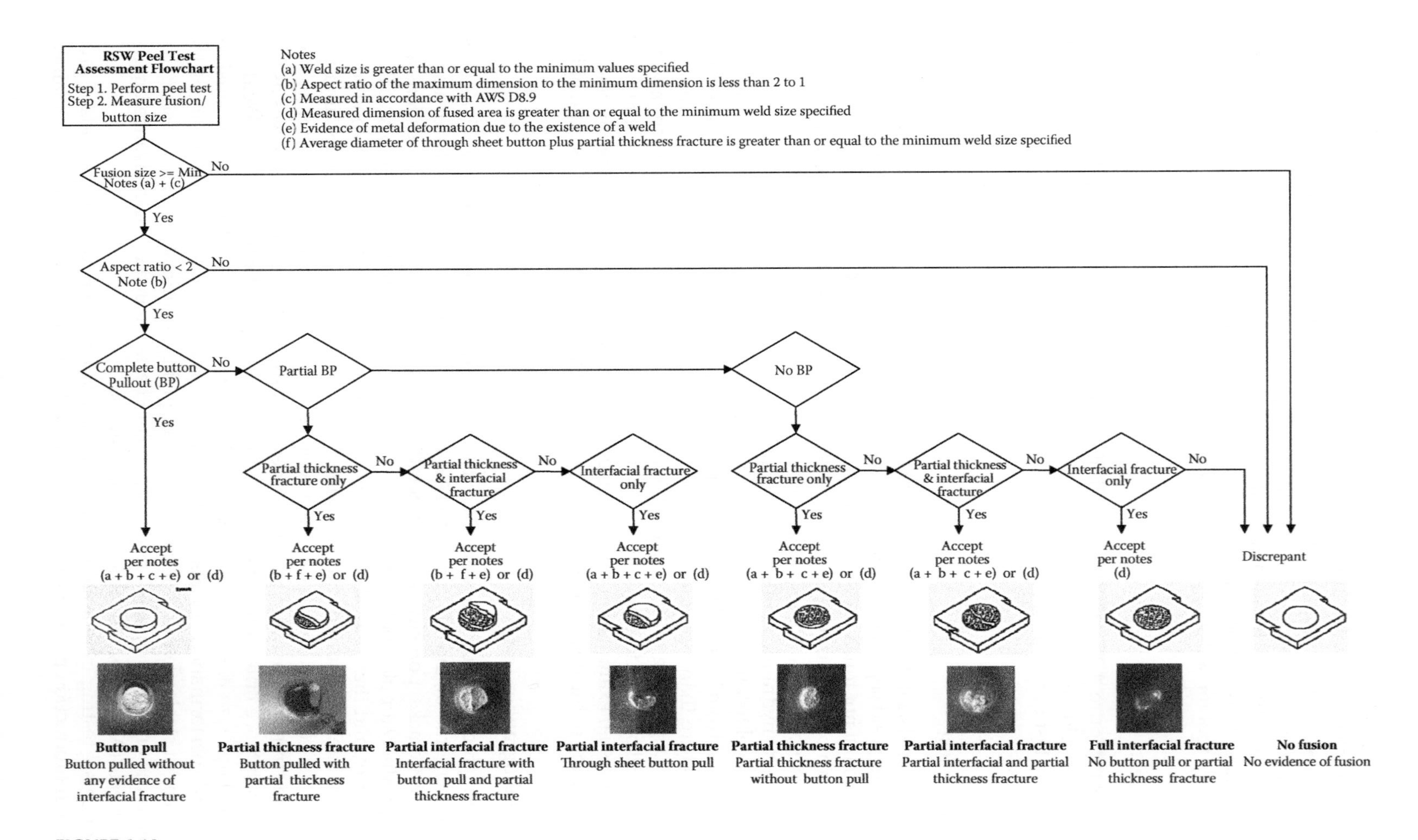

FIGURE 6.14
Weld evaluation procedure.

in Section 6.1, after the completion of all spot welding on the assembly or subassembly. Destructive testing should be made on as small a spot-welded production unit as is practical.

The difficulty lies in determining the weld quality when interfacial fracture, including partial interfacial failure, is observed when a weld is destructively tested. Whether an interfacial fracture or a button pull-out is obtained on a weldment depends mainly on the following factors:

1. Weld strength, its shear and tensile resistance
2. Loading mode, the shear and tensile loading components
3. Stiffness of the weldment near the weld edge
4. Loading rate

Both tensile and shear loading components are present in all types of testing of spot-welded specimens, including peel, tension, tensile–shear, and twisting. Stiffness of the weldment and loading mode together determine the magnitudes of shear and tensile loading components on the weld. Interfacial failure may result if the shear loading component is large or the weld's shear resistance is low, or both. Tensile–shear and twisting impose larger shear loading to a weld, and have a higher tendency of producing interfacial fracture, than other types of loading, such as peel and cross-tension. When a weldment is tensile–shear loaded, the rotation of the jointed part is small if the specimen is stiff, as in the case of high-strength steels or thick aluminum sheets. As a result, the shear component on the weld is large and an interfacial fracture may result. The opposite is also true that the rotation of a less stiff weldment tends to be large, resulting in a weld button pull-out if the weld's tensile resistance is low.

A high loading rate may produce interfacial failure in a weld that might otherwise fail with a pulled button if quasi-statically tested. In general, interfacial or partial interfacial failure may be observed in both good welds and substandard welds, depending on weld quality, loading method, and specimen stiffness. Welds that failed with interfacial or partial interfacial failure mode are considered acceptable if they have a clear sign of fusion in the designated weld area, and their size is larger than or equal to the minimum size requirement. Such welds usually have reasonable strength and should not be considered discrepant. An evaluation procedure developed by the American Welding Society's Technical Committee on Automotive Resistance Welding and the Auto-Steel Partnership accounts for the interfacial fracture often observed in testing AHSSs and aluminum alloys,[18] as shown in Figure 6.14.

6.3 Nondestructive Evaluation

Traditionally, weld quality is monitored through destructive testing. Final assurance of weld quality requires that a percentage of the assembled parts be destroyed to verify the welding process. The disadvantages of such a procedure are obvious. It takes time, during which corrections to the weld controls and process are not possible. Remedial measures can be taken only after a number of welds are made using the set welding schedules. Substandard welds may be produced before necessary adjustment of welding schedules is made, and detecting and repair of the welds are generally costly. Many efforts to nondestructively evaluate welds have been made to save the cost of scrap parts that result as a percentage of the assembled parts destroyed in the verification process.

Nondestructive evaluation of resistance welds can be made in a number of ways. Acoustic emission, eddy current, and x-ray are some of the techniques that have been attempted for resistance spot weld quality inspection. Many of them have limitations in demonstrating effective solutions in the manufacturing workplace. Most current nondestructive techniques are heavily dependent on the experience or skill level of an operator, or may require expensive integration. The ideal nondestructive solution to resistance spot weld quality inspection would be one that requires minimal changes in the workplace-required skill levels and work procedures, and provides accurate information on the quality of a weld.

Using ultrasonic techniques for inspecting resistance weld quality has proven to be a viable means in the production environment.[19–23] Therefore, they are the focus of this section. Ultrasonic techniques were first used in 1978 to detect defective resistance spot welds through a procedure that is commonly known as a ring-down technique. The ring-down name aptly describes the waveform observed as a series of echoes as the ultrasound being bounced between the surfaces of the welded plates. This technique is known as A-scan. The time intervals of the echoes are used to measure the various thicknesses in a weldment. Incomplete welds produce echoes that have roughly half the interval, whereas less overt weld defects exhibit both a series of echoes having the half interval and a series of echoes having the longer intervals. The difficulty lies in relating the amplitude of the half-interval and longer interval echoes to the physical parameters of the weld, specifically to the quality of the weld. To reduce the difficulties in interpretation, the transducer is matched to the diameter of the weld. Ambiguities in the echo amplitude result from the shape and surface finish of the top and bottom of the weld, and from the location and type of weld defects that might be present. Low-amplitude echoes that result from an important and common defect, cold welding, are observed by the use of the natural focus of the transducer (e.g., the one by Krautkramer and Krautkramer[19]), which increases the sensitivity of the ultrasonic transducer as an echo passes through the transducer's natural focal region. When used by experienced and skilled hands, the technique reportedly produces satisfactory results.

Advanced processing techniques such as artificial intelligence techniques have been applied to processing the ultrasonic signals; these processing techniques are sometimes known as advanced learning networks,[24] neural networks,[25] and others. These artificial intelligence techniques have been introduced to assist in interpreting the ultrasound waveforms. Such techniques rely on a set of actual defects called training sets to obtain good and bad waveform data sets needed to set up the instrument. A problem with such a procedure is that if the defects are verified by destructive testing, then the samples are no longer available for use in training or retesting. Similar artificial intelligence techniques were utilized in the nuclear power industry, and most of the equipment built on such principles has been phased out in favor of that using techniques based on more solid physical principles. Most of the ultrasonic nondestructive devices are based on the A-scan technique, as described in the previous paragraphs. B-scan- and C-scan-based techniques for evaluating spot welds have been developed in the past two decades.[22]

6.3.1 Ultrasonic A-Scan

Ultrasonic imaging techniques for testing resistance spot weld quality have attracted significant attention because of their robustness and accuracy proven in other industries.

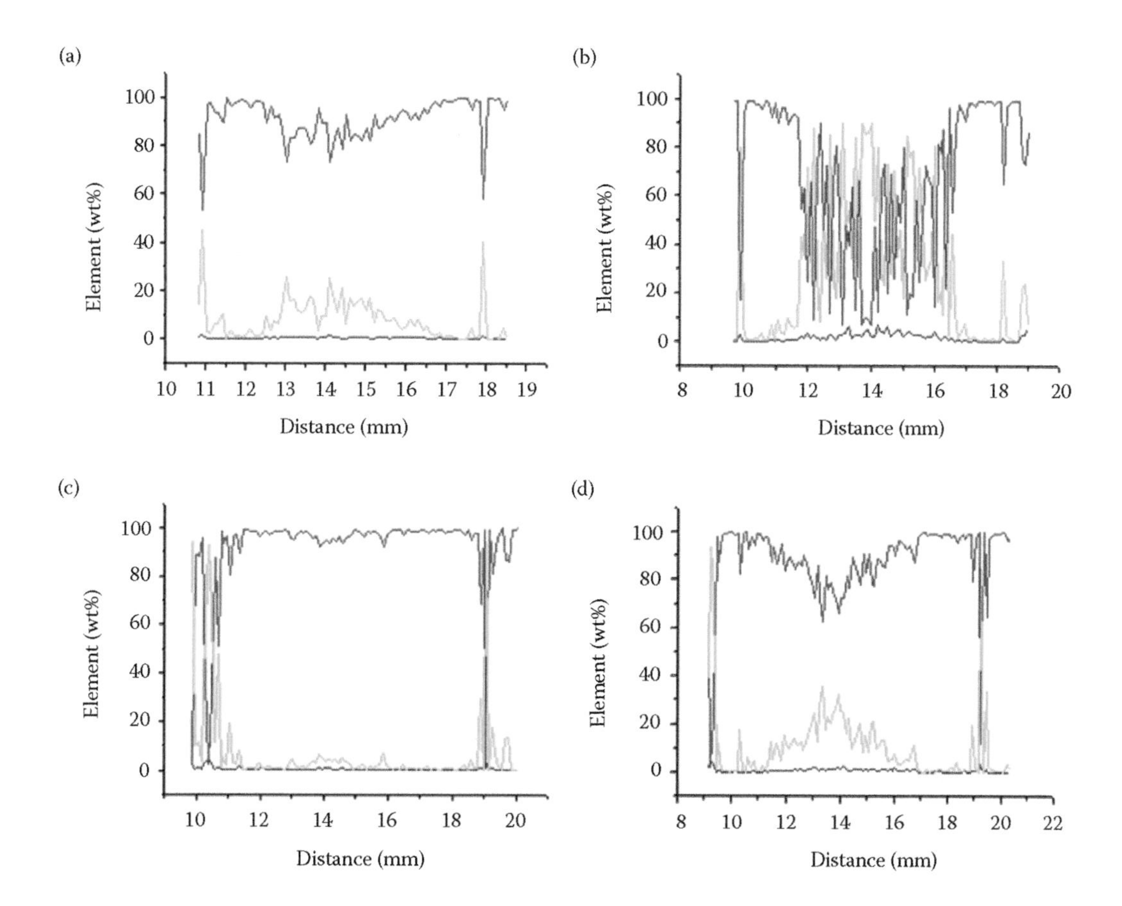

FIGURE 1.28
Composition profiles of electrode surfaces after 60 welds using schedules of (a) $F = 4.5$ kN, $\tau = 60$ ms; (b) $F = 4.5$ kN, $\tau = 180$ ms; (c) $F = 9.0$ kN, $\tau = 60$ ms; and (d) $F = 9.0$ kN, $\tau = 180$ ms. Red line is for Cu, green is for Al, and blue is for Mg.

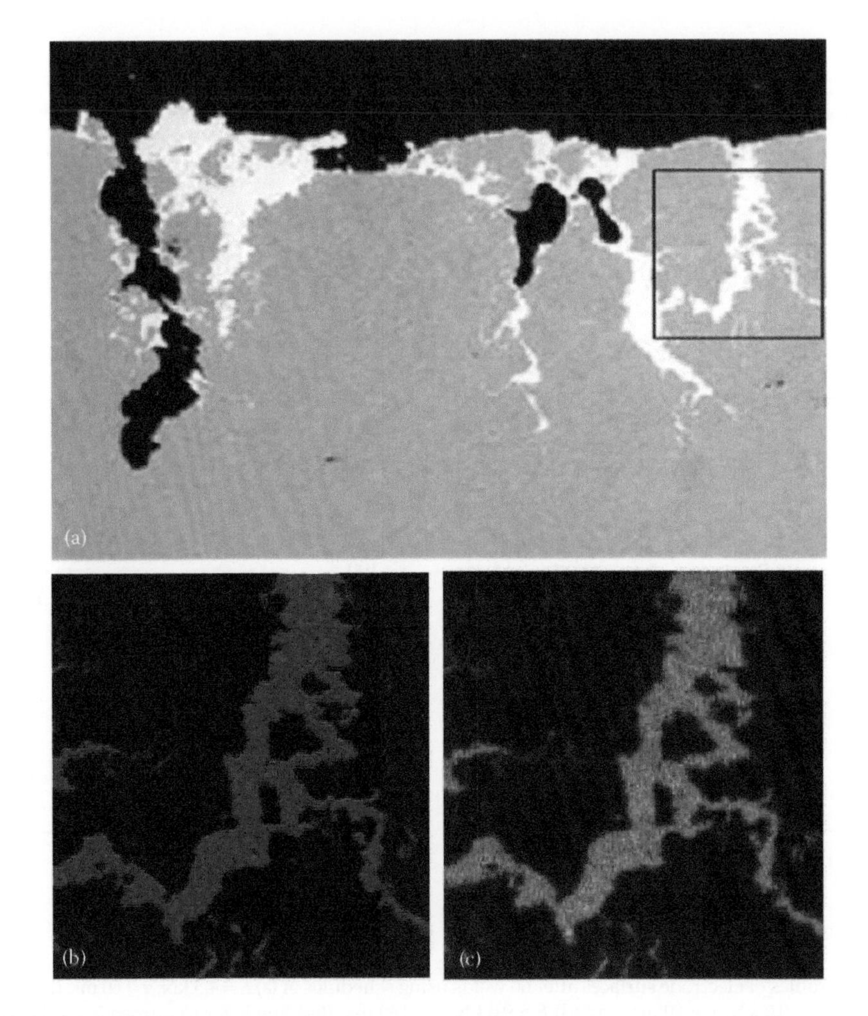

FIGURE 1.34
Liquid metal embrittlement cracking in a Zn-coated HSLA steel spot weldment (a), and x-ray maps of Cu (b) and Zn (c) of area outlined in (a). (From AET_Service_Capability.pdf. Available online at http://www.aet-int.com/capability/AET_Service_Capability.pdf. Accessed in Nov. 2010. With permission.)

FIGURE 2.10
Electrode wear in welding steel.

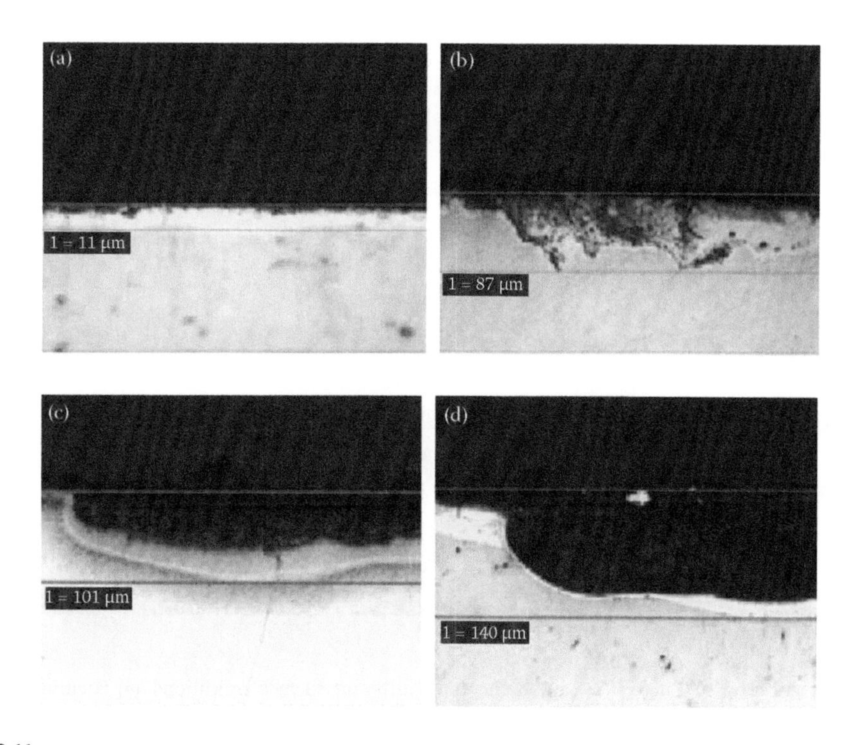

FIGURE 2.11
Crater formation and growth on electrode surface as a function of number of welds.[22] (a) Alloy layer after 100 welds (11 μm); (b) crater depth after 300 welds (87 μm); (c) crater depth after 400 welds (101 μm); (d) crater depth after 500 welds (140 μm).

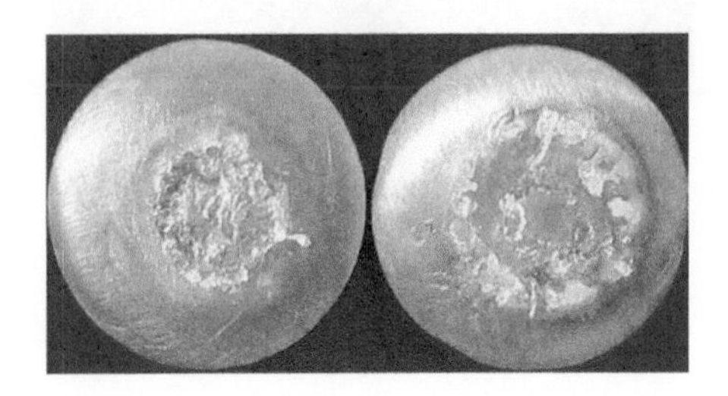

FIGURE 2.12
Electrode wear in aluminum welding.

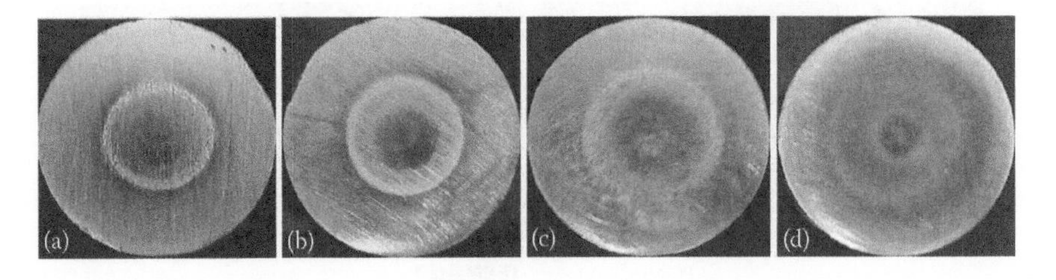

FIGURE 2.13
Electrode surfaces after making 60 welds on sheets of different surface conditions: (a) chemically cleaned, (b) degreased, (c) electric arc cleaned, and (d) untreated.

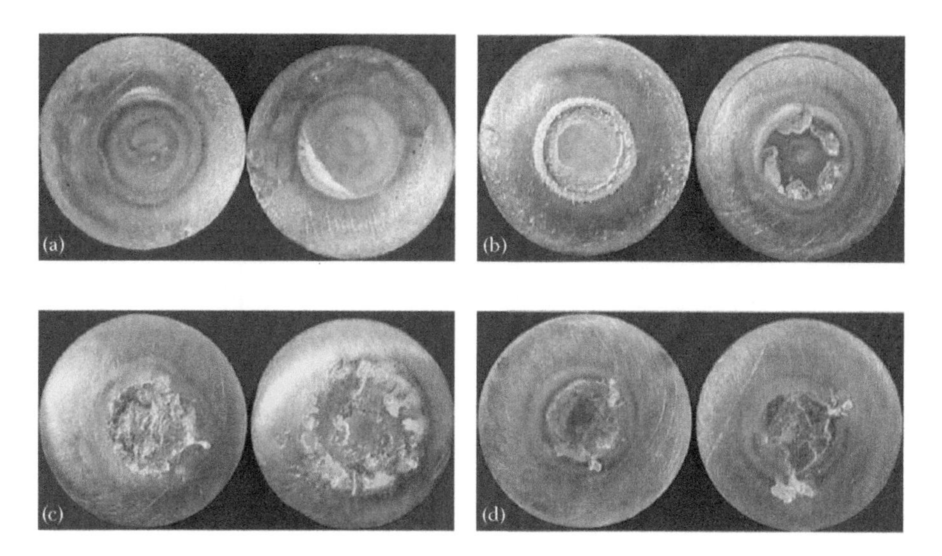

FIGURE 2.15
Electrode surface morphology after life tests using (a) chemical cleaning, (b) degreasing, (c) electric arc cleaning methods, and (d) untreated aluminum sheets. Electrodes on left side are from the lower arm of welder (negative), and those on right side are from the upper arm (positive).

FIGURE 4.26
A new impact tester.

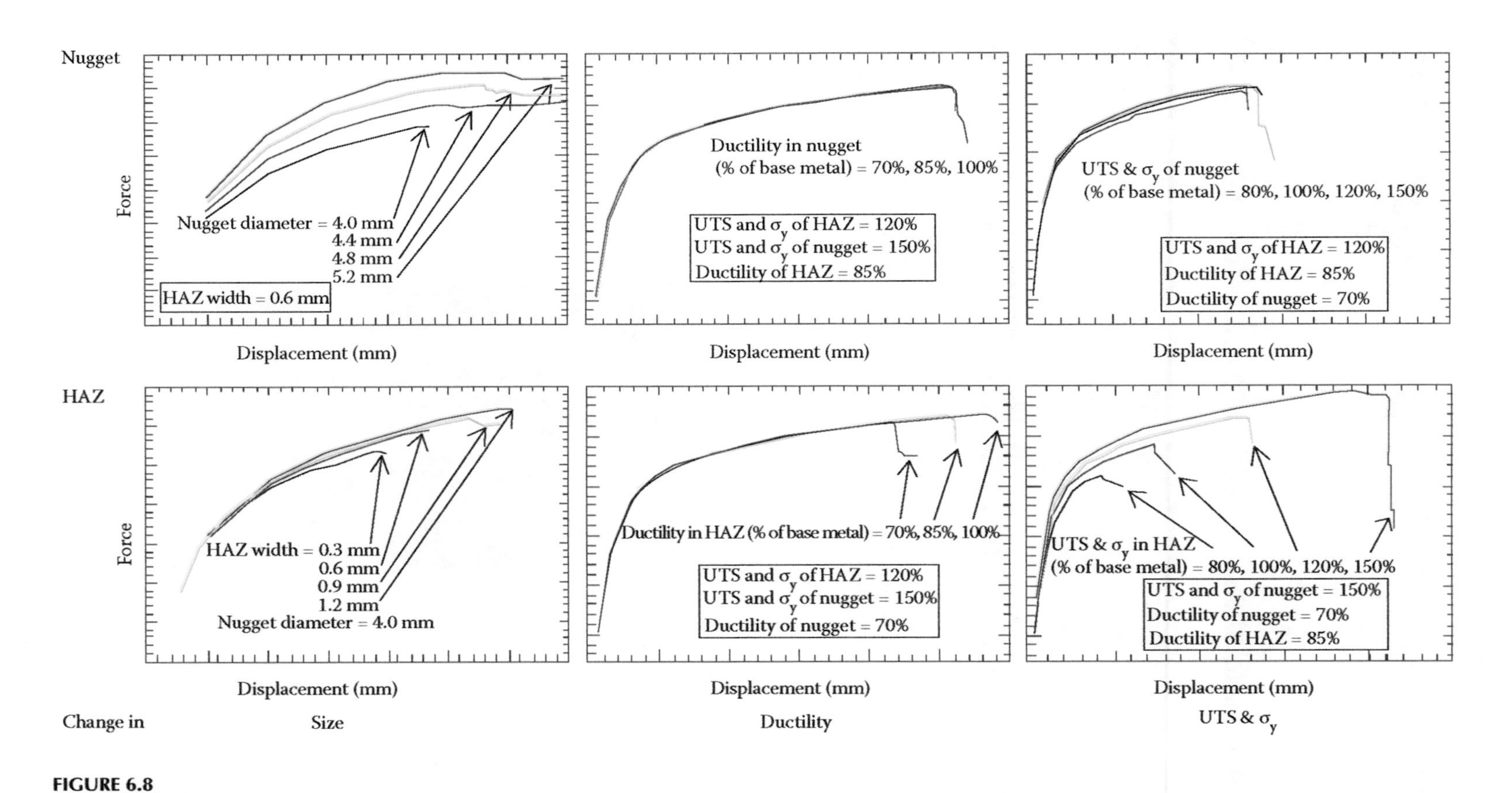

FIGURE 6.8
Influence of dimension and mechanical properties of nugget and HAZ.

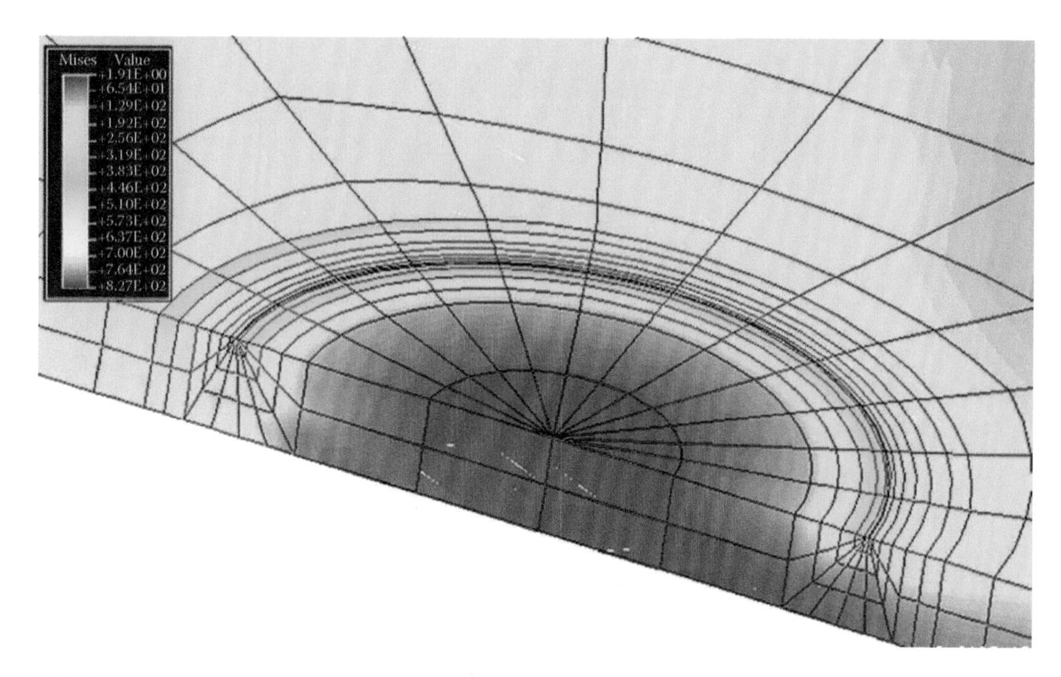

FIGURE 6.10
von Mises stresses distribution at weldment.

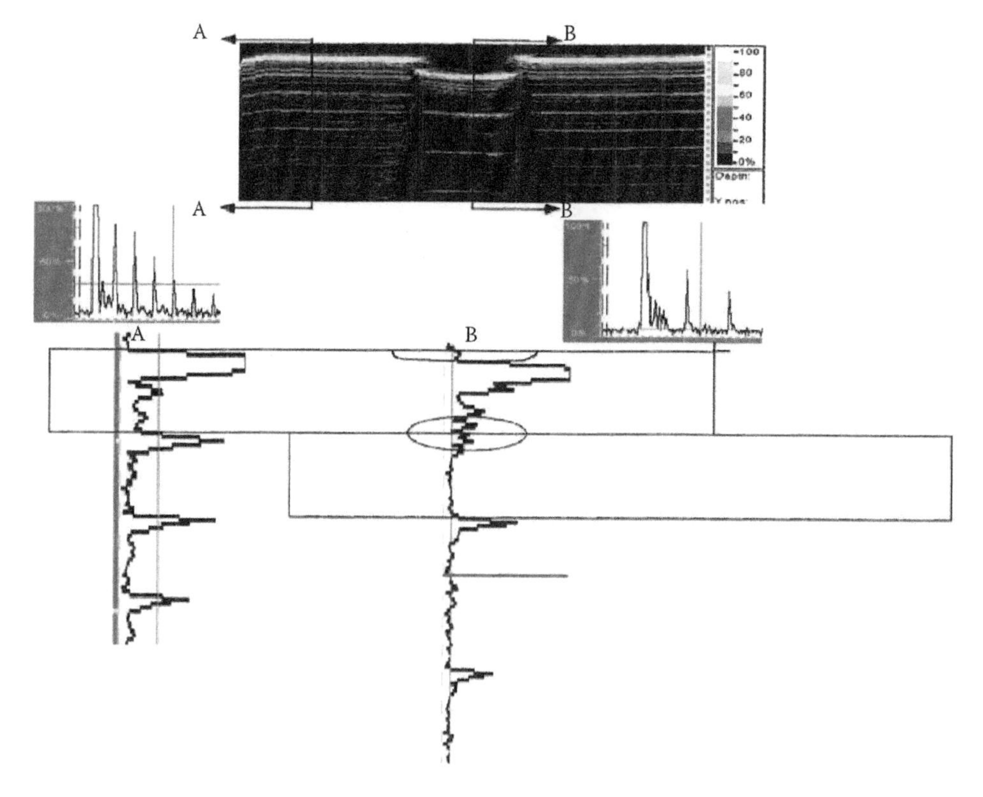

FIGURE 6.24
Formation of a B-scan image of a spot weld.

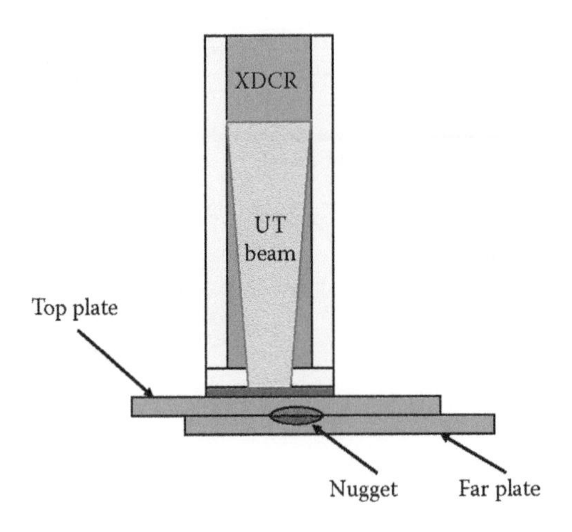

FIGURE 6.26
Schematic diagram of probe setup.

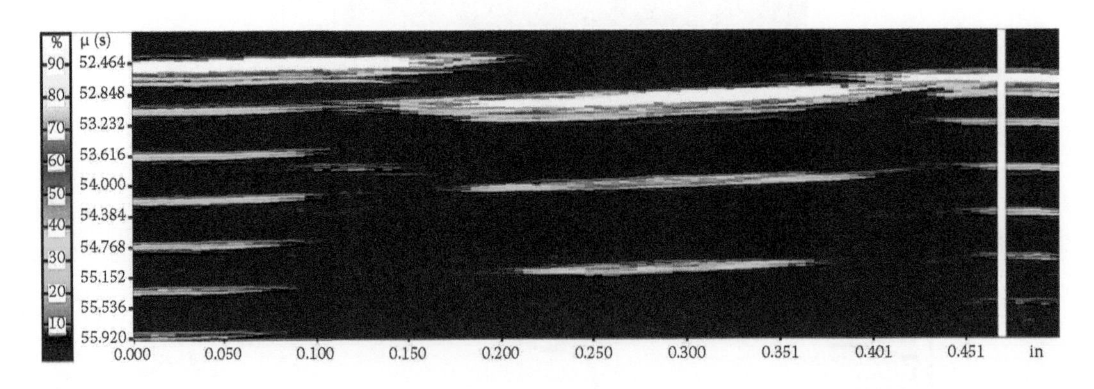

FIGURE 6.27
B-scan image of a good weld.

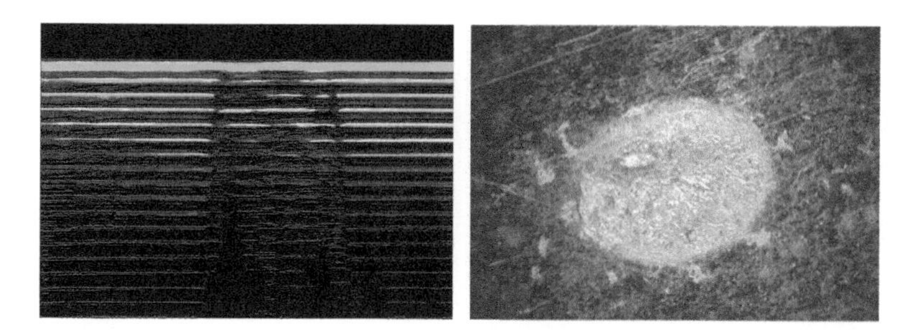

FIGURE 6.34
B-scan and faying surface of a cold weld (current = 5000 A).

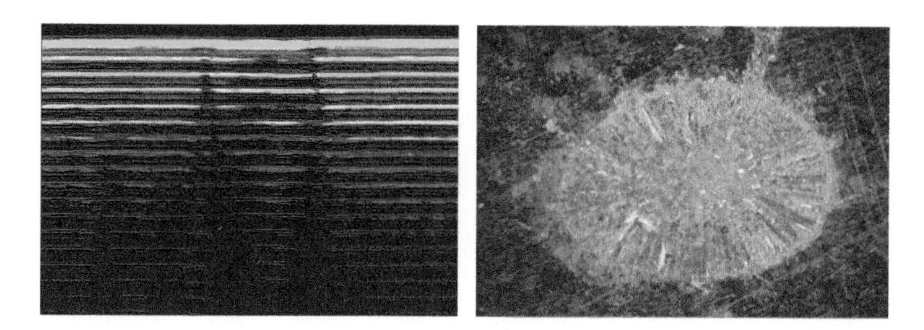

FIGURE 6.35
B-scan and faying surface of a cold weld (current = 5500 A).

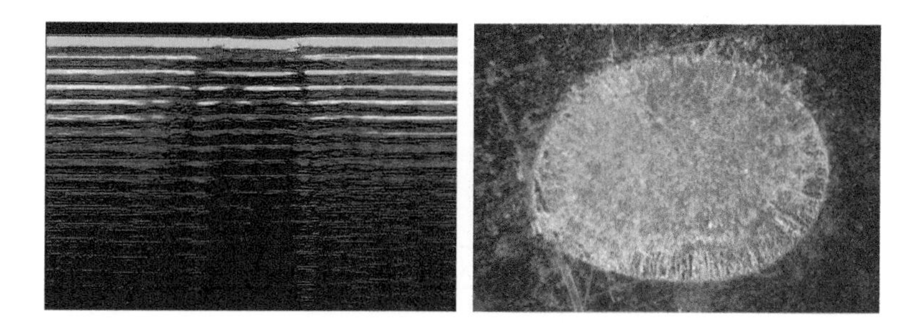

FIGURE 6.36
B-scan and faying surface of a cold weld (current = 6500 A).

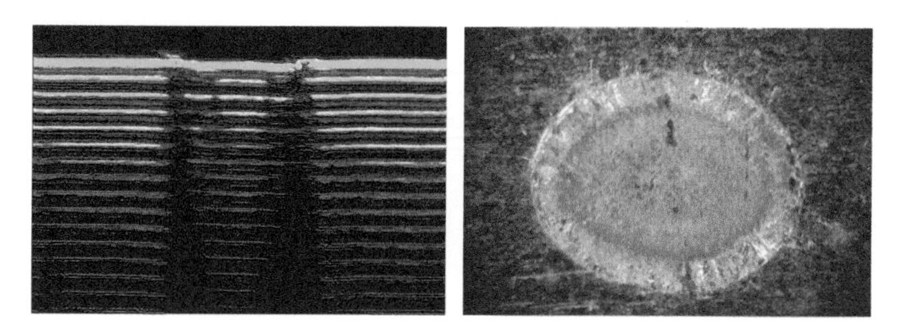

FIGURE 6.37
B-scan and faying surface of a peeled weld (current = 8000 A).

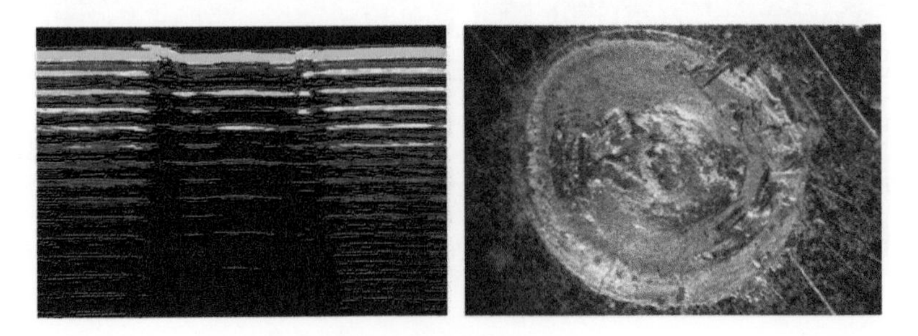

FIGURE 6.38
B-scan and faying surface of a peeled weld (current = 9000 A).

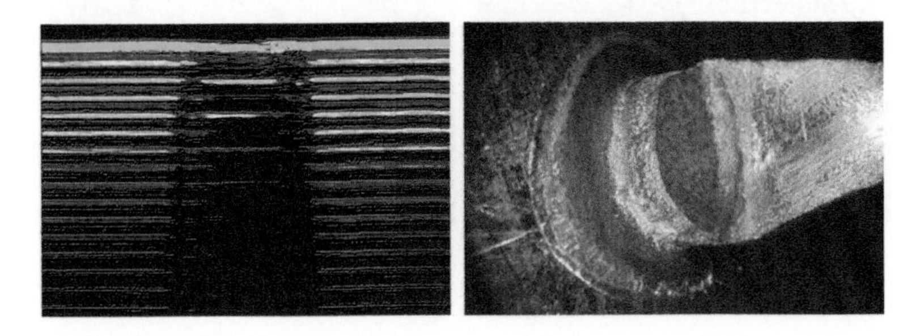

FIGURE 6.39
B-scan and a peeled weld (current = 9500 A).

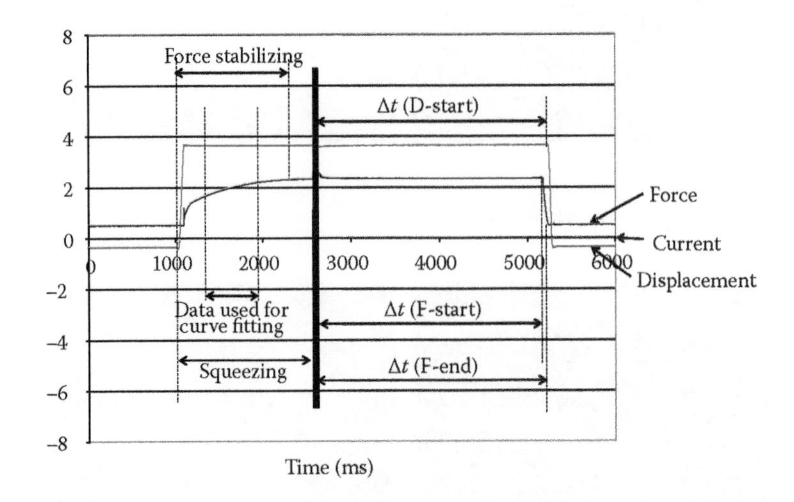

FIGURE 8.12
Signals obtained for a typical welding cycle using DAQ system.

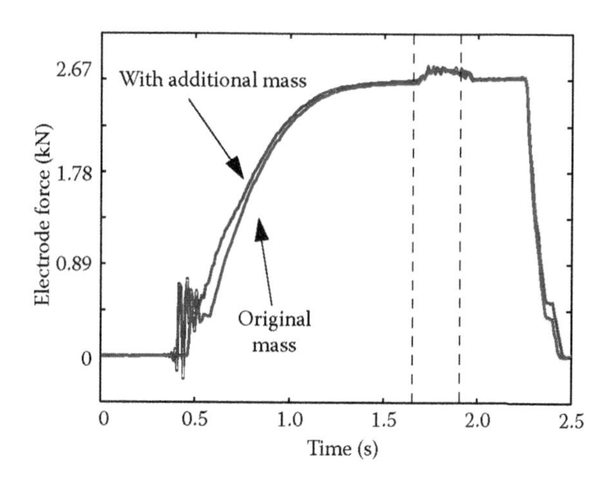

FIGURE 8.25
Effect of moving mass on electrode force.

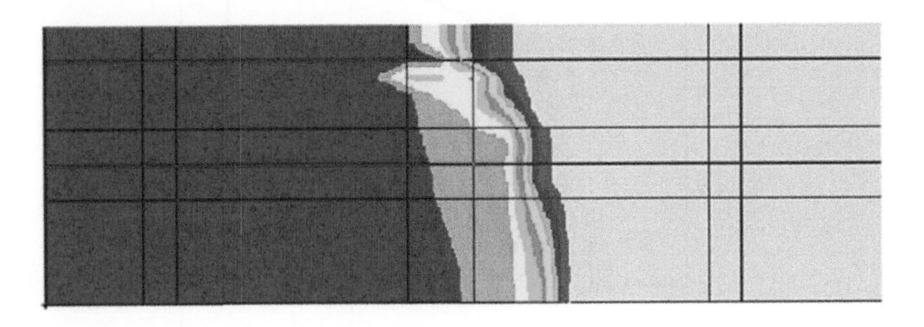

FIGURE 9.12
Hardness gradients predicted in weldment.[25]

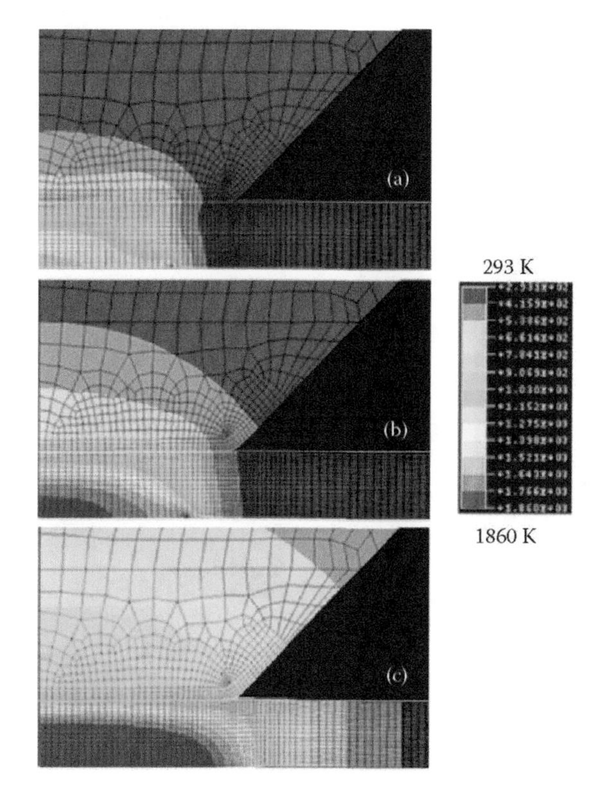

FIGURE 9.14
Simulation results of nugget formation using conditioned electrodes.

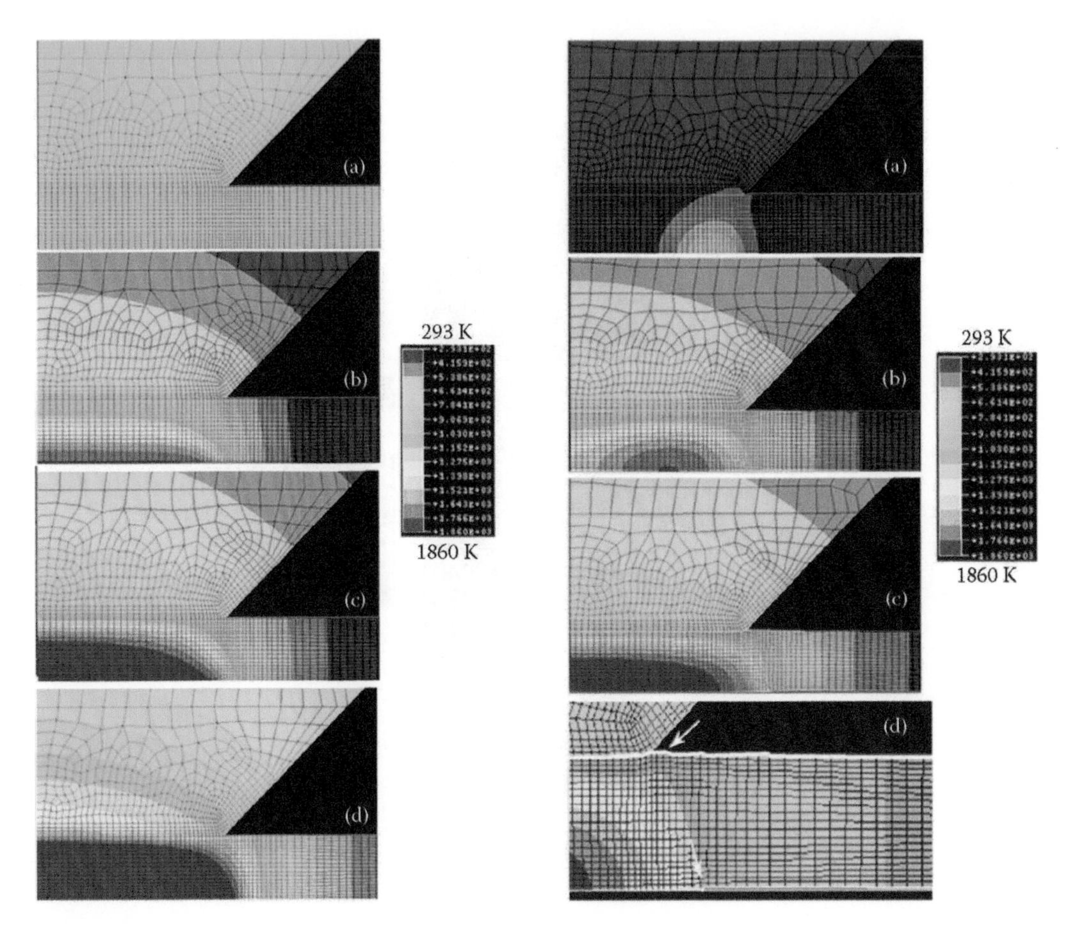

FIGURE 9.15
Nugget growth simulation using uncoupled and coupled electrical–thermal–mechanical algorithms.

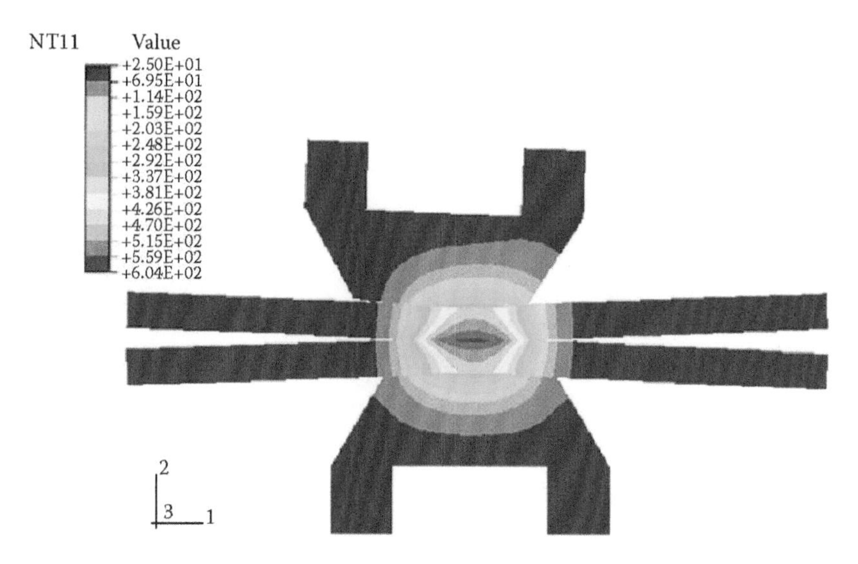

FIGURE 9.18
Temperature distribution after second cycle.

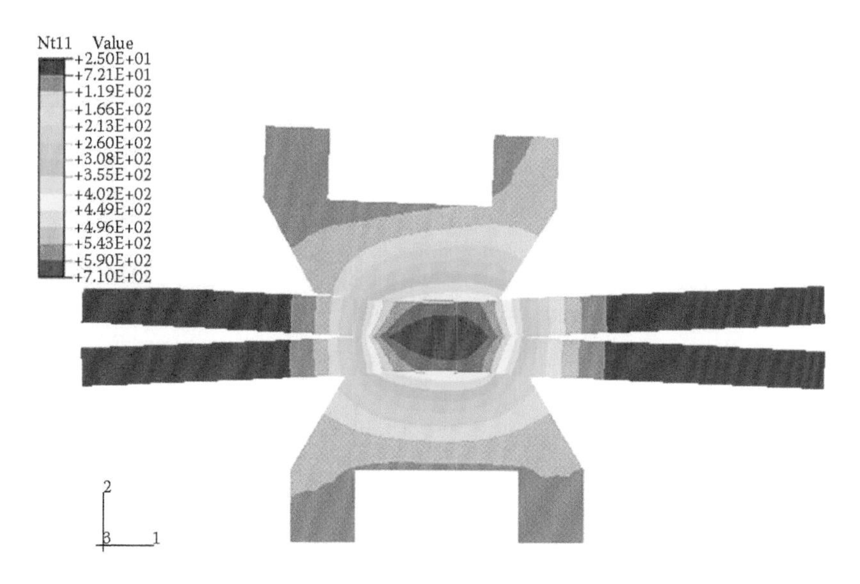

FIGURE 9.19
Temperature distribution after 5.5th cycle.

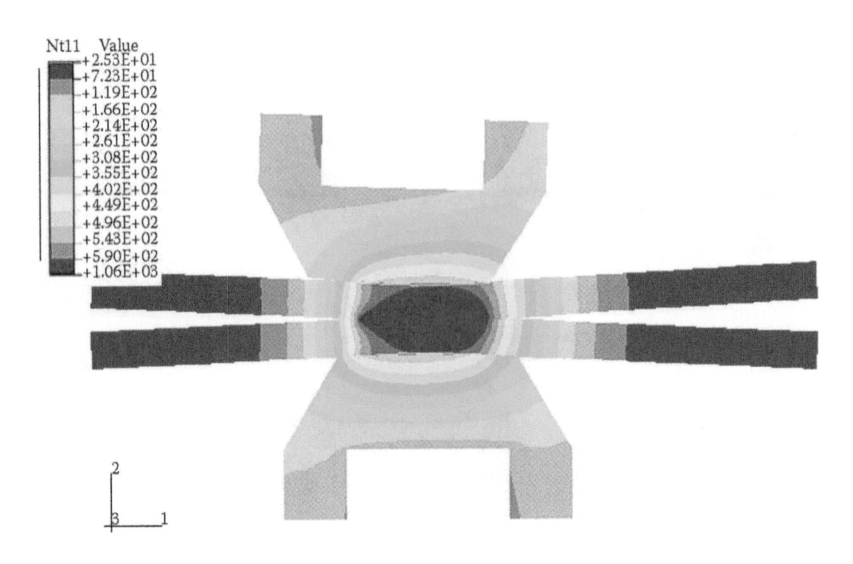

FIGURE 9.20
Temperature distribution after 8 cycles.

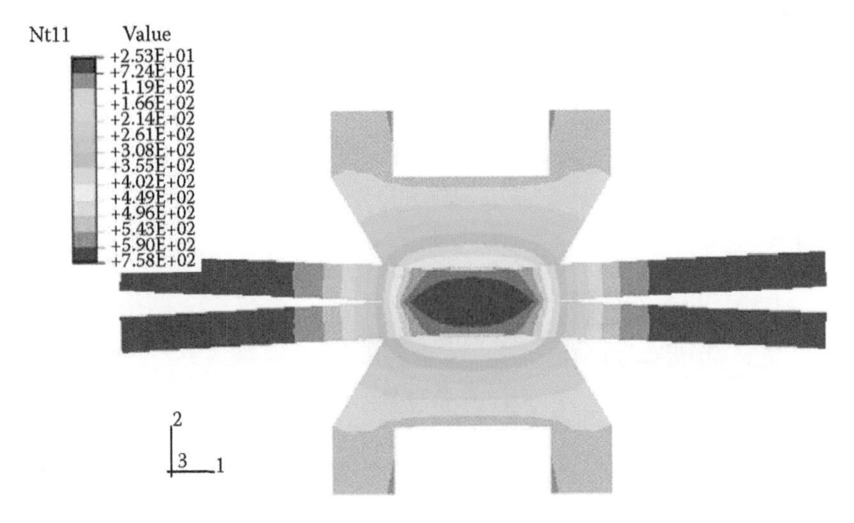

FIGURE 9.21
Temperature distribution after 8 cycles with perfectly aligned electrodes.

In ultrasonic A-scan, the internal structure of a solid is characterized through a series of echoes of the ultrasound reflected by the interfaces. The sound waves, of frequencies in the range from a few to hundreds of megahertz generated by a piezoelectric transducer, travel through the solid. When the sound encounters an interface, such as the back surface and the surface of an internal defect, it is reflected back to the transducer. The location of the reflecting interfaces can be determined by the echo positions on the time scale. In addition, the attenuation characteristics, phase inversion, and other parameters are also used to characterize the structure of the solid.

It is commonly known as a ring-down technique when ultrasonic A-scan is used to test resistance spot welds. The time intervals of the echoes are used to measure the various thicknesses in a weldment. The basic principles of ultrasonic A-scan used for resistance spot weld evaluation are illustrated in Figure 6.15. A resistance spot weld is classified according to its ultrasonic characteristics.[27,28]

The quality of the weld can be derived from the shape of the echo sequence. A good weld with sufficient fusion and large size generally demonstrates reflections from the back surface of the stack-up and a rapid attenuation of the signal (Figure 6.15a). The amplitudes of the subsequent echoes drop quickly because of the microstructure of the weld that has a high sound attenuation characteristic. The echo intervals, equally spaced, reflect the thickness of the weld. With a small weld, the acoustic beam is split into two: the central part of the beam goes through the welded area, and the lateral part is reflected by the intermediate faying interface. Therefore, small intermediate echoes are observed, in addition to the normal echo sequence as shown in Figure 6.15a, resulting from echoes from the faying interface between the sheets (Figure 6.15b). This happens when the weld nugget size is smaller than the diameter of the sound beam, and probes of various diameters are used for testing welds of different nominal diameters. Although ultrasonic A-scan cannot quantitatively measure a weld's size, it provides valuable information on whether a nominal size is reached. For a cold weld, a reflection similar to that of a good weld is observed (Figure 6.15c), with two subtle differences: the intervals between two adjacent echoes are slightly larger than those for a good weld, as the distance between the top and back surfaces in a cold weld is slightly larger than that of a weld with sufficient fusion and, therefore, smaller weld thickness due to indentations on both sides. In addition, a cold weld has a longer echo sequence because of smaller amount of the fused metal and usually finer structures of the re-solidified metal at the vicinity of the faying interface, which has a lower sound attenuation than the coarse structure in a good weld. In the case of no weld, multiple reflections

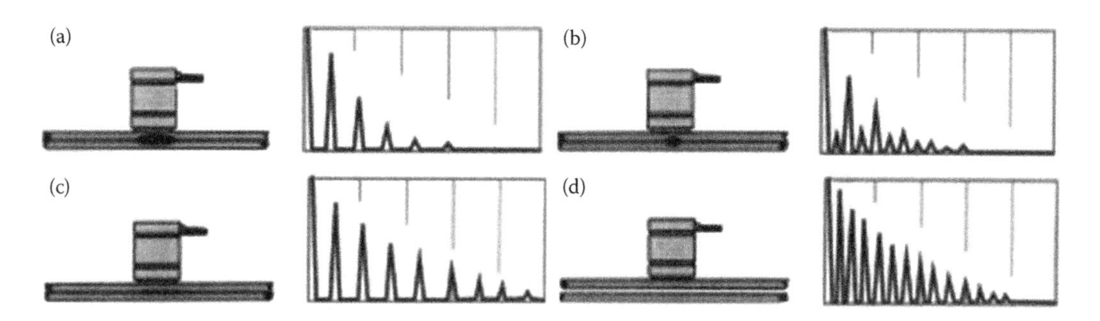

FIGURE 6.15
Principles of ultrasonic A-scanning. (From Roye, W., Ultrasonic testing of spot welds in the automotive industry, Special Issue no. SD 298, Krautkrämer GmbH & Co. With permission.)

of the wave only within the upper plate are observed, with very slow attenuation (Figure 6.15d).

Testing a spot weld using ultrasonic A-scan is, in many cases, not trivial because of the ambiguities in interpreting the echoes in both amplitude and sequence, resulting from the large variety of spot welds of complicated geometric and physical features. The difficulty also lies in relating the A-scan signals to the physical parameters or the quality of a weld. The echoes are affected by a large number of factors, for example, the shape and surface finish of the top and bottom of the weld, the location and type of weld defects that might be present, and microstructure of the weld. In addition, the operator's knowledge and level of expertise in evaluating spot welds play a vital role. It is for this reason that efforts have been made to computerize the scanning process and data analysis. Special algorithms have been developed by various manufacturers to automate the process, and minimize the training requirements and dependence on experience of the inspection personnel.

6.3.1.1 A Case Study on R&R of an Ultrasonic A-Scanner

The accuracy of ultrasonic A-scan testing depends on the apparatus, the components/welds tested, and the operators. A study was conducted on the repeatability and reproducibility (R&R) of an industrial ultrasonic A-scanning system at the University of Toledo's Material Joining Laboratory.[29] In addition, the accuracy of the ultrasonic testing was also verified by correlating the ultrasonic inspection results with those of destructive testing. Such a link between the ultrasonic signals measured on welds and the process parameters used to make these welds may serve as a valuable guidance to welding process monitoring and control. The major findings are presented in the following sections.

6.3.1.1.1 Experiments

A 0.7-mm mild steel (MS) and a dual-phase high-strength steel (DP) of two thicknesses, 1.0 and 1.5 mm, were used in the study. Because same thickness/material welding is not common in the sheet metal industry, the following different-material/thickness combinations were used in order to reflect the actual automobile practice:

1. Combination 1 (C1): MS 0.7 mm/DP 1.5 mm
2. Combination 2 (C2): DP 1.0 mm/DP 1.5 mm

Welds of various features representative of a large range of possible welds in a production environment were made on the steel strips. The following types of welds were made:

1. Good weld: a weld with a diameter $d \geq 4\sqrt{t}$, where t is sheet thickness
2. Undersized weld: a weld with a diameter $d < 4\sqrt{t}$
3. "Very" cold weld: a weld with no or very little fusion at the faying interface
4. "Medium" cold weld: a weld with more fusion than a very cold weld, but insufficient bonding between the sheets
5. Weld with internal discontinuities: a weld with voids and/or cracks
6. Weld with excessive distortion: a weld created using wire inserts between the sheets
7. Weld with excessive expulsion: a weld created using excessive welding current

Extensive preliminary experiments were conducted on the various material combinations to determine the desirable welding parameters. For each of the weld types, several welding schedules were used, and specimens were peeled for visual examination to determine if they fit the definition of the specific category. Certain variation was observed in making every type of weld, and it was larger for some types than for others. Several welds created on different material combinations using the same welding schedule have quite different characteristics as shown in Figure 6.16. Among the welds created for developing desirable welding schedules, only the one in Figure 6.16a can be classified as a good weld. In general, different welding schedules are needed to create good welds on different material combinations. The testing strips were made for each weld type using the corresponding welding schedules developed for each material combination.

For each of these 7 types of weld, 11 welds were made on each strip with a 1.5-in pitch using identical welding parameters. Four replicate strips (with 11 welds on each strip) were made for each type of weld. Two of the four replicates were first ultrasonically tested (noted as UT hereafter) and then peeled (noted as DT hereafter for destructive testing) for verification. The buttons were inspected and measured. The results of the examination of the peeled button and the ultrasonic testing were directly compared, producing an accuracy estimate of the ultrasonic testing. Selective cross-section examination was conducted on the third strip, and the last one was saved as backup.

There were 2 (combinations) × 7 (weld types) × 4 (replicates) = 56 welded strips. In ultrasonic testing, welds were examined by four operators in a random order. The UT testing was conducted in the following order:

1. Setup of the ultrasonic A-scanner, with the help of ultrasonic testing professionals
2. Operator training, performed using the ultrasonic device and its software following the manufacturer's standard procedure
3. Ultrasonic inspection (by four operators)
4. Data collection and reduction

6.3.1.1.2 *Accuracy of UT Measurements*

For each type of weld, after ultrasonic testing was conducted on the strips, they were peeled for direct measurement and inspection. A weld's diameter was measured in two perpendicular directions for an average value. An approximation was made for an incompletely peeled weld.

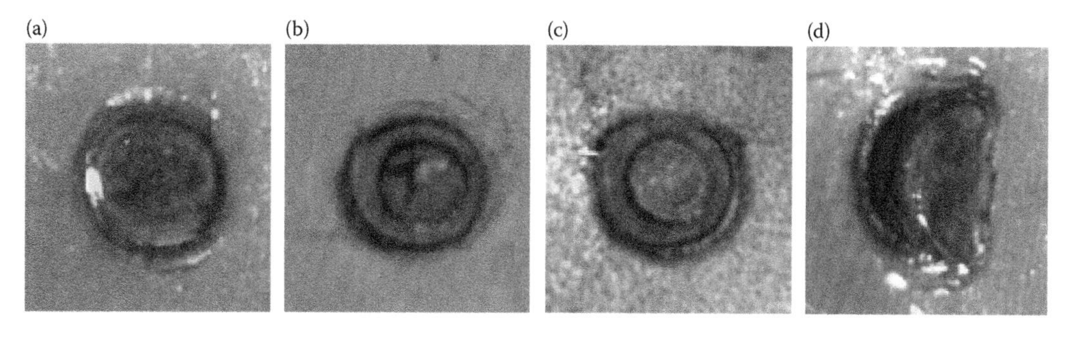

FIGURE 6.16
Welds created for the "good weld" category.

The comparison between UT and DT tests shows clearly that the accuracy of a UT test depends on the type of weld, material combination, and operators. The accuracies of UT testing for two combinations are depicted in Figures 6.17 and 6.18. The operators appeared to be fairly consistent in testing combination C1. The "very cold" welds clearly confused the UT device and the operators, as a large disagreement exists among the operators on this type of weld. In fact, cold welds are difficult to detect by any nondestructive technique, as a thin layer of melting at the faying interface, although providing very limited adhesion, may make the welded joint possess most of the physical characteristics detectable to a nondestructive testing device. The "medium cold" welds, on the other hand, were detected correctly most of the time (more than 90%). This can be largely attributed to the significantly higher consistency of this type of weld, than that of the "very cold" welds that were highly variable. The lowest accuracy was observed on welds with internal voids, as the voids were small in size and few in amount, making them difficult to detect.

Large discrepancies are observed in testing material combination C2. As shown in Figure 6.18, except on "very cold" and "expulsion" welds, operators generally disagreed in making judgment. The accuracy of UT measurement is also strongly affected by the types of weld tested. The UT device was the least accurate when used on "undersized" welds, which is quite different from what was observed in testing material combination C1. For this combination, the testing yielded higher accuracy on welds with voids and lower accuracy on "medium cold" welds than for combination C1. The significant difference between Figures 6.17 and 6.18 can be attributed to the difference in the ultrasonic characteristics between mild steels and high-strength steels. The ultrasonic signals generated from DP steel welds are more difficult to distinguish than those from the MSs. Therefore, the experiences learned in UT testing of one type of material may not be extended to other types of materials.

6.3.1.1.3 Repeatability and Reproducibility

An R&R study is a necessary step to qualify a measurement device. Such tests were conducted on the UT device and results are shown in Figure 6.19. The repeatability of testing (Figure 6.19a), that is, performed by the same operator is affected by the material combination, as the measurement on combination C2 is slightly less repeatable than that on combination C1. Operator 3 was consistent at both combinations, whereas operator 1 showed the largest discrepancy.

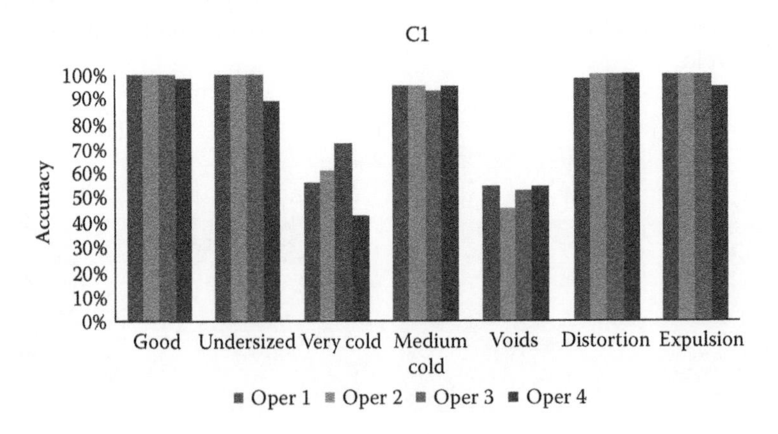

FIGURE 6.17
Accuracy percentage of UT test for various weld types of combination C1.

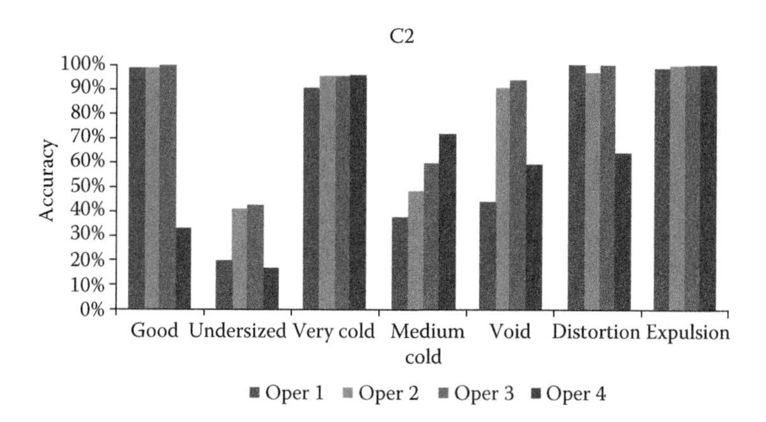

FIGURE 6.18
Accuracy percentage of UT test for various weld types of combination C2.

The overall repeatability and reproducibility are presented in Figure 6.19b for both combinations. The repeatability of the measurements is higher than the reproducibility for all the operators, which means that the same operator was fairly consistent and different operators may make significantly different judgments using the same device. Combination C1 is slightly higher in both repeatability and reproducibility than combination C2, which is consistent with what was observed in testing accuracy shown in Figures 6.17 and 6.18.

6.3.1.1.4 *"Go and No-Go" Accuracy*

Instead of classifying welds into the seven categories as in the previous sections, in practice, it might be more meaningful to classify them into the categories of "Go" and "No-Go." An acceptable weld is put in the "Go" group, and an inferior one in the "No-Go" group. Whether a weld is acceptable or not depends on the companies or industries that may have significantly different opinions/standards in this regard. In the study, the "Go" category contains good welds, and "No-Go" is for undersized and cold (very cold and medium cold) welds as defined in Section 6.3.1.1.1. The other types of welds, that is, those with internal voids, excessive distortion, and expulsion were not included in either group as the effects of these defects on weld quality are generally case dependent. Based on such a classification, the success rates of determining the respective welds were calculated. For the "Go" group, the success rate was defined as the ratio of the number of good welds

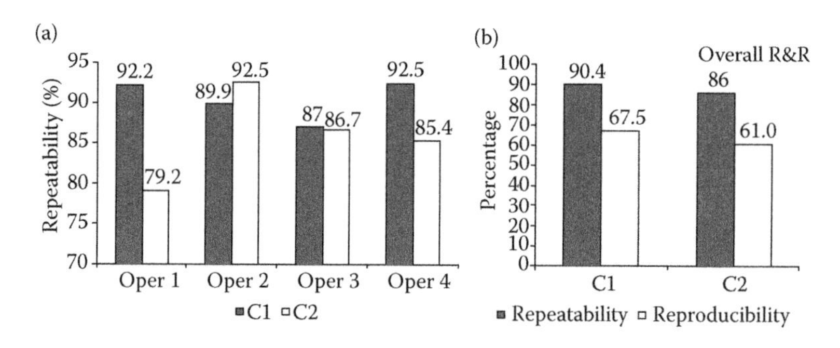

FIGURE 6.19
Repeatability of operators for combinations C1 and C2 (a); overall R&R for combinations C1 and C2 (b).

determined through the DTs, to that of welds measured as "good" by the operators using the device (UT). The success rate for the "No-Go" group was calculated as the ratio of the welds that were actually inferior (revealed through DT) to the total welds determined by UT as undersized or cold.

For the simplicity in classification, the welds were put in the following four groups:

1. UT test yields "good," DT test yields "good"
2. UT test yields "good," DT test yields "bad"
3. UT test yields "bad," DT test yields "good"
4. UT test yields "bad," DT test yields "bad"

Therefore, the "Go" includes groups 1 and 2, and groups 3 and 4 belong to "No-Go." Groups 1 and 4 contain accurate measurements by the UT test; regardless, "good" or "bad" welds and groups 2 and 3 correspond to incorrect UT results. This classification makes it easy to calculate the accuracy of UT test: in "Go," it is the number of welds in group 1 divided by the total number of welds in groups 1 and 2. All the welds determined by the UT as bad welds are in groups 3 and 4. The correct measurements, that is, the actual bad welds, are in group 4. Therefore, the accuracy ratio of "No-Go" can be calculated as the ratio of the number of welds in group 4 to the total in groups 3 and 4.

The accuracy of a UT testing is affected by both operators and material combinations. In the study, the accuracy rates of "Go" and "No-Go" were calculated for operators both as individuals and as a group, and for each of the two material combinations. Table 6.2 lists the UT and DT measurement results.

All the operators were more accurate in testing good welds, in the "Go" group than in the "No-Go" group on combination C1 (Figure 6.20). Figure 6.21 shows the results for combination C2, with a similar trend yet a significantly larger difference between the "Go" and "No-Go." The accuracy ratios of the individual operators in the "Go" and "No-Go" groups, combining the two material combinations, are presented in Figure 6.22. It shows the same trend as in Figures 6.20 and 6.21. In general, a low accuracy rate in the "No-Go" group does not affect the safety and integrity of a structure as a tested "bad" weld is actually a good weld. The real impact is the unnecessary repair work and unwarranted changes to the welding schedules that may be created by the false alarms. Therefore, it is beneficial to

TABLE 6.2
UT and DT Testing Results

	Destructive Test	Ultrasonic Test							
		Operator 1		Operator 2		Operator 3		Operator 4	
Combination	DT (good)	UT (good)	UT (bad)	UT (good)	UT (bad)	UT (good)	UT (bad)	UT (good)	UT (bad)
C1	168	164	4	163	5	163	5	154	14
C2	120	72	48	84	36	81	39	62	58
	DT (bad)	UT (good)	UT (bad)	UT (good)	UT (bad)	UT (good)	UT (bad)	UT (good)	UT (bad)
C1	45	6	39	5	40	7	38	3	42
C2	51	19	32	9	42	25	26	6	45

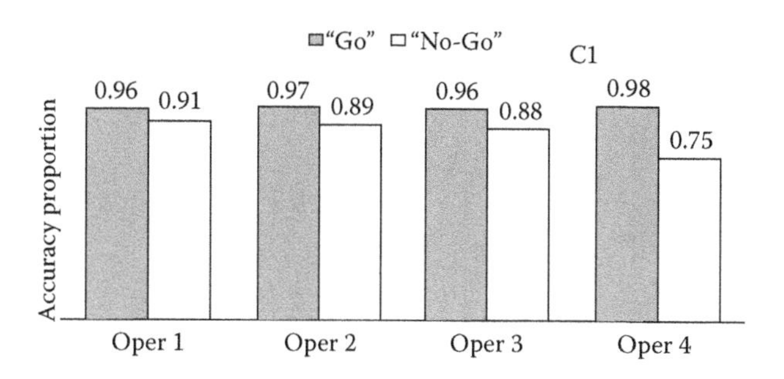

FIGURE 6.20
Accuracy ratios of "Go" and "No-Go" for combination C1.

maintain the accuracy ratios of both "Go" and "No-Go" groups beyond certain levels in order to ensure quality and avoid disruption to production.

6.3.1.1.5 AHP Analysis of Operator Hierarchy

It is clear that the measurements of a UT device are operator dependent. The testing results can be utilized to obtain essential information regarding the skill level of individual operators, and to determine the suitability of an operator for conducting UT measurement. An analytic hierarchy process (AHP), a multicriteria decision making method developed by Saaty and Vargas,[30,31] was used to understand the influence of operators.

AHP is useful in setting priorities and making optimal decisions. Based on paired comparisons, an AHP derives decisions in terms of ratio scales through evaluation of alternatives that satisfy a certain set of criteria. In an AHP, pairs of alternative solutions are compared, and the intensity and level of preference toward one alternative in relation to the other are produced according to the criteria. In AHP, a complex decision making is effectively reduced to a pair comparison between alternatives, which also leads to a rational decision.[32]

In the study, an AHP analysis was carried out, using an online AHP software,[32] on the accuracy of the measurements and repeatability of the operators. The results of the two material combinations were pooled together, and the importance of repeatability (= 0.5) and accuracy (= 0.5) of the operators were balanced. The operators were ranked based on their highest accuracy and repeatability in UT measurement and the results are presented in Figure 6.23. This hierarchy decision through the AHP analysis ranked the operators in the order of operator 3, operator 4, operator 2, and then operator 1, according to the quality

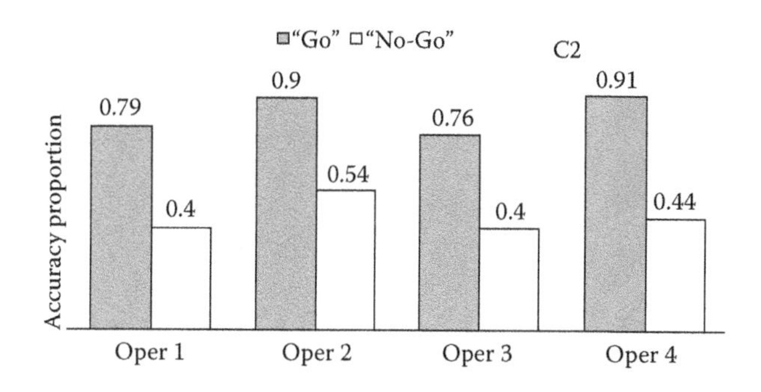

FIGURE 6.21
Accuracy ratios of "Go" and "No-Go" for combination C2.

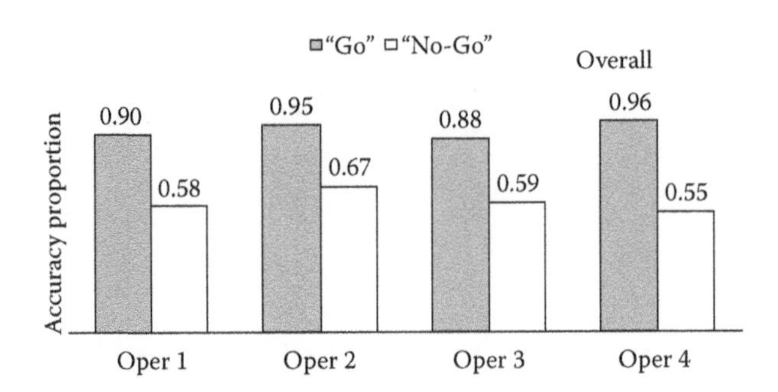

FIGURE 6.22
Overall accuracy ratios of "Go" and "No-Go."

of their work. Such an analysis can be used to determine the most qualified operator for a UT testing position, and if operator training is needed.

In general, the UT tests produced valuable results on the dependence of UT measurement on material/combination, weld type, and operator. Particularly, the study found that the accuracy and consistency of UT measurements in mild steels are higher than in DP steels, and the operator has a large influence on the repeatability of measurement, which is also a function of the material combinations.

6.3.2 Ultrasonic B-Scan

Historically, the concept of B-scan was developed in the time of the first radar systems in the early 1940s and there are many versions in use today. A B-scan image is often thought of as a cross-sectional image of a weld similar to an image obtained from metallurgical cross-sectioning; it represents a series of echo responses as the transducer is moved across the weld. An echo response is associated with the shape of an ultrasonic beam, which may cover a significant portion of a weldment. It is convenient to describe a B-scan as being formed by rotating a series of A-scan signals into a display and presenting the amplitude of signals with a color code, as illustrated in Figure 6.24. The figure shows the A-scan signals at different locations on a spot weldment. At each location, the transducer produces a series of echoes between the top and back of the plate, and they are scaled to fit the plate. Then the echo amplitudes are color coded, and the times of flight are represented as a line of colored pixels in a computer image, as shown in the figure. The image formed by a sequence of such echo lines is called a B-scan. Figure 6.24 clearly shows the cross section of a weldment, including the base metal (top sheet), indentation, and weld nugget area.

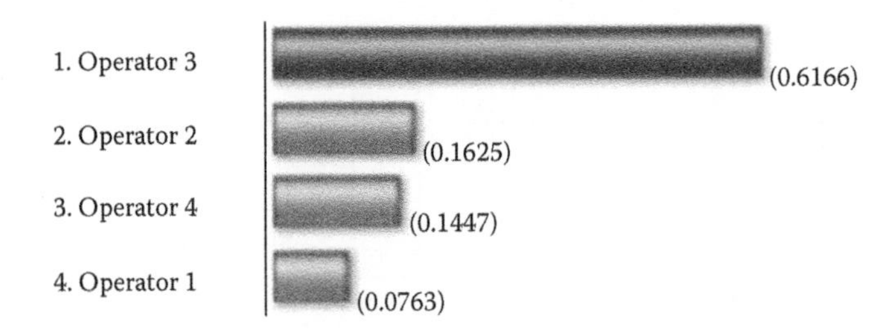

FIGURE 6.23
Ranking of operators derived by AHP analysis based on repeatability and accuracy data.

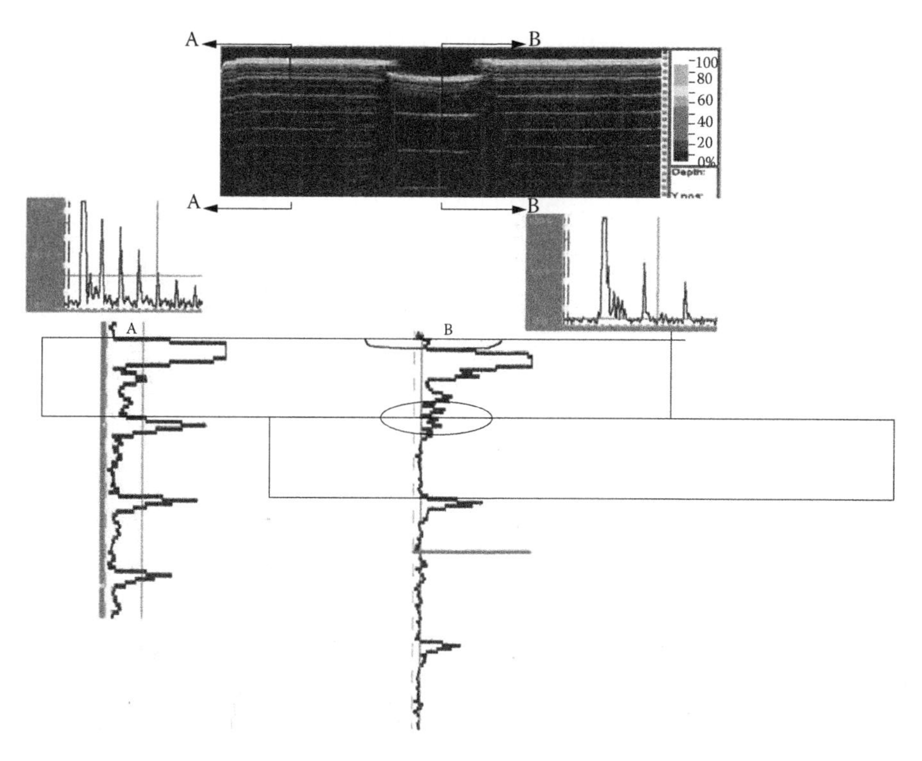

FIGURE 6.24
(See color insert.) Formation of a B-scan image of a spot weld.

The important features of a B-scan are found in the accuracy of the location and size of the ultrasound reflections from the top and bottom surfaces of the plate, the top and bottom surfaces of the weldment, and the location of any anomalous reflectors within the weldment. A schematic of a B-scan-based spot weld inspection system manufactured by Applied Metrics[22] is shown in Figure 6.25. It consists of a high-resolution scan encoder and has the capability of resolving high-resolution thickness data in real time. The scanner was

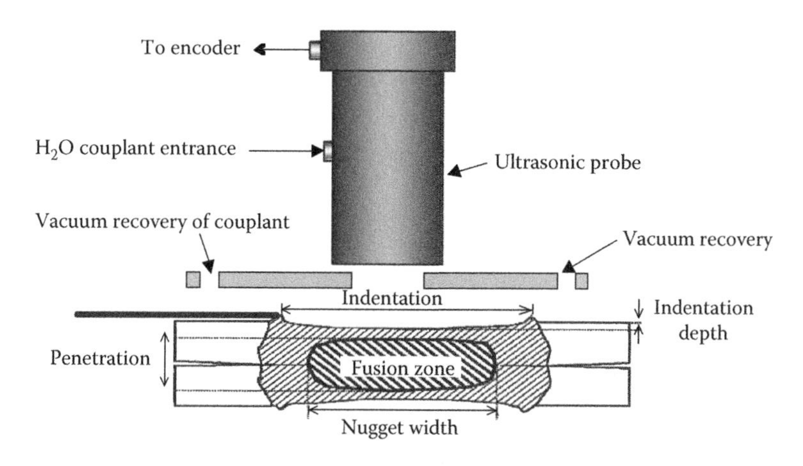

FIGURE 6.25
A B-scan-based spot weld inspection system.

developed for the automobile manufacturing environment and uses the same probe for measuring weld nuggets ranging from 4 to 12 mm in width. The couplant flow is recovered by a unique vacuum system, and the transducer footprint allows examination of the resistance spot weld within 6 mm of an interfering wall.

The probe setup is shown in Figure 6.26, where a standard immersion transducer (XDCR) is mounted in a tubular case. The length of the case fixes the distance off the plate needed to set the ultrasonic beam focus at the correct distance. A bushing that tolerates the weld surface irregularities is used between the tubular case and the welded plate. Additionally, the bushing has features to recover the water used as couplant. In this form, the transducer is easily scanned manually or attached to a robotic scanner.

An image of a good weld generated using this system is shown in Figure 6.27. It is noted that the top plate surface is clearly shown on both the left and right sides of the weld. Below the top plate are shown five echoes that can be used to accurately measure the thickness of the top plate. The thickness can be used to calculate the minimum weld size for an acceptable weld. The reflection on the top plate is used for *in situ* calibration of the ultrasonic system. The front surface of the weld is shown in the center of the figure, where the image shows a deeper electrode impression on the left side. The shape of the electrode face is nearly flat and shows essentially no wear. Immediately under the electrode impression are two echoes that measure the thickness of the weld and the length of the fusion zone.

6.3.3 Examining Various Welds Using a B-Scan System

The B-scan system has been successfully used to measure the geometric attributes of spot welds with various characteristics.[23,33] Figure 6.28 shows the image of a good weld made on DP600 steel sheets. It has the characteristics of complete fusion at the contact interface, a sufficiently large nugget width, an indentation mark of adequate depth, and no sign of expulsion. The image of a good weld has a reflection from the top sheet surface and the back sheet surface only in the nugget zone. No reflections/echoes in the nugget zone are found in the image, which means that there are no unfused areas, inclusions, or voids in the nugget. A good weld on low-carbon steel has similar characteristics (Figure 6.29).

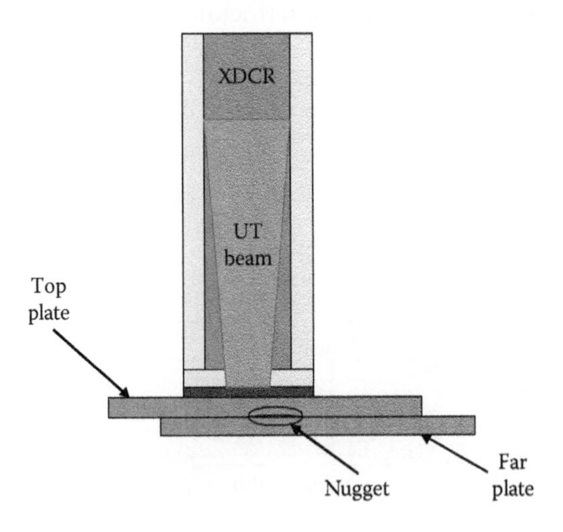

FIGURE 6.26
(See color insert.) Schematic diagram of probe setup.

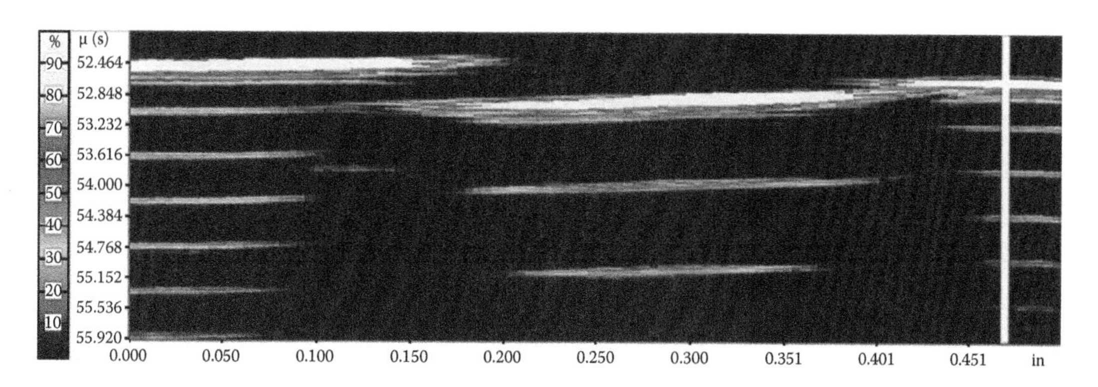

FIGURE 6.27
(See color insert.) B-scan image of a good weld.

After scanning, the specimens were sectioned to reveal the structures of the weldments. A direct comparison with the B-scanning results is shown in Table 6.3.

A cold weld is characterized by partial or incomplete fusion of the sheet metal at the interface. The fused area is too small and too shallow to be considered a weld nugget. Sheets joined by a cold weld can be pulled apart easily, sometimes leaving small bumps on the sheets at the interface. A B-scan image of a cold weld usually has, as observed from Figure 6.30, extremely shallow depth of electrode indentation, a clear reflection from the bottom surface of the top sheet in the area under the electrode during welding, and no reflection from the bottom surface of the second sheet. Unlike other types of welds, cold welds may have very different amounts of fusion and, therefore, different B-scanning signals. They are discussed in more detail later in this section.

The B-scanning technique is capable of detecting the projection of a discontinuity in the direction of the ultrasonic beam. Figure 6.31 shows the image of a weld with a crack (unfused interface area) created by using hollow electrodes. A weld with a void/crack inside may have a pulled button on peeling. The presence of a void/crack can usually be easily detected by B-scanning through the reflections from the top of the void/crack. There are usually no echoes from the portion under the first reflection of the void/crack in a B-scan image.

An undersized weld is one that is formed by complete fusion of the sheets in contact at the interface, but has a very small nugget (smaller than the minimum nugget diameter recommended for the particular sheet thickness). As the B-scan provides a direct dimensional measurement of the nugget width, undersized welds can be easily detected. One

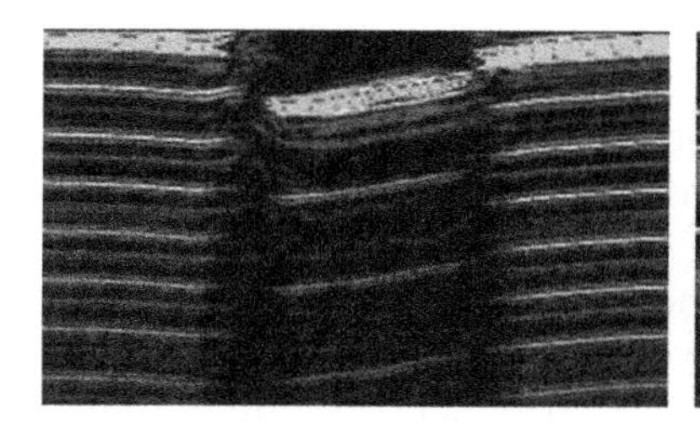

FIGURE 6.28
A good weld (DP600 steel).

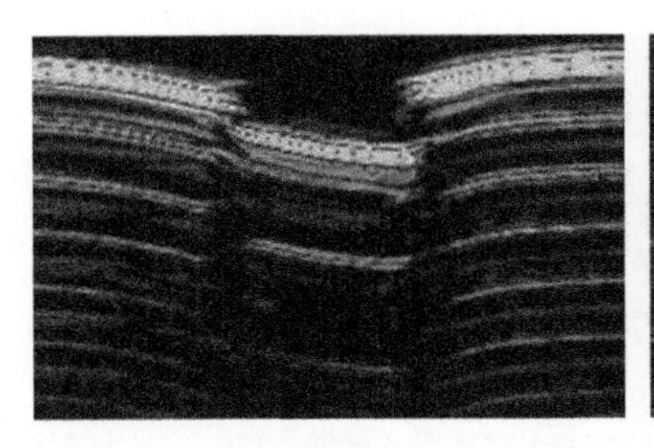
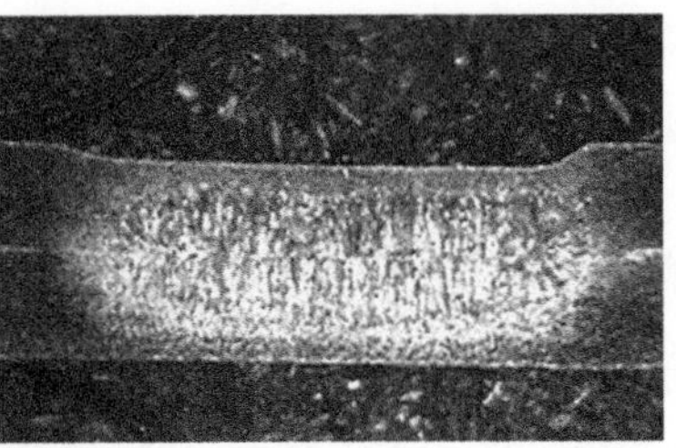

FIGURE 6.29
A good weld (low-carbon steel).

such example is shown in Figure 6.32. An ultrasonic image of an undersized weld has a shallow depth of electrode indentation and a narrow width of indentation.

The B-scanner can also detect other geometric characteristics of a weld. For instance, indentation and separation of the sheets and excessive distortion can be quantitatively described. Figure 6.33 shows a weld with excessive expulsion, resulting in large-sheet separation and electrode indentation, as observed from the image.

The fairly good agreement, shown in Table 6.3, between direct metallographic measurements and B-scan images shows that the B-scan technique can be applied to nondestructive evaluation of resistance spot weld quality.

6.3.4 Identification of Cold Welds

As a special class of substandard welds, cold welds pose a challenge to ultrasonic evaluation of weld quality. Cold welds generally have poor fusion between the sheets, with a thin nugget or no nugget at all. The difficulty in ultrasonic scanning lies in the fact that these welds have different degrees of coldness, depending on the amount of fusion. This section links the features of B-scan images to the degrees of coldness of cold welds, and it should help interpret ultrasonic B-scan images in practice.

A set of welds of different coldness was created using 1-mm galvanized mild steel sheets. The welding time was fixed at 10 cycles (167 ms), hold time was 50 cycles (836 ms), and electrode force was 600 lb (2.67 kN). The only variable was electric current. By controlling the amount of heat input, different amounts of fusion are generated between the sheets, resulting in welds of different coldness.

The intensity of ultrasonic signals is indicated by various colors in a B-scan image, and it provides a visual identification of the adhesion between the sheets. The minimum welding current used was 5000 A. The separated interface and the B-scan image are shown in Figure 6.34. The center part squeezed between electrodes during welding shows very little difference from the base metal, indicating no or very little fusion in this area. The peeled specimen confirms this, showing melting of the zinc coating only.

When welding current is increased by 500 A, very little difference results in the B-scan image, as shown in Figure 6.35. The fractured surface, however, shows that in addition to the zinc melting, some of the base metal has melted near the center of the contact area, although it is very limited.

When the current is raised to 6500 A, the B-scan image starts to show a difference between the base metal and the supposed weld area, although it is still a cold weld. The color lines in the B-scan image in Figure 6.36 are no longer continuous, indicating a fused area at the contact interface. The fractured surface shows a larger melting area, and it is rougher than in Figure 6.35.

TABLE 6.3

Comparison between B-Scan Measurements and Direct Metallographic Measurements

	Good Weld		Cold Weld		Weld with Void		Undersized Weld		Weld with Distortion	
	B-Scan	Measured	B-Scan	Measured	B-Scan	Measured	B-Scan	Measured	B-Scan	Measured
Sheet thickness (mm)	0.7		0.7		0.75		0.75		0.7	
Nugget width (mm)	4.25	3.526	2.80	2.116	4.98	4.794	2.83	2.290	4.78	3.410
Indentation width (mm)	4.51	4.048	3.16	2.621	4.76	4.612	2.56	2.285	5.92	4.586
Indentation depth (mm)	0.19	0.093	0.03	0.026	0.08	0.074	0.11	0.079	0.79	0.634
Void size (mm)	N/A	N/A	N/A	N/A	0.95	0.794	N/A	N/A	N/A	N/A

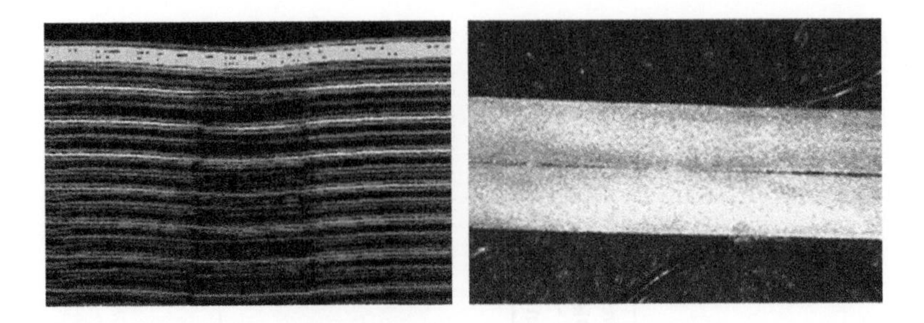

FIGURE 6.30
A cold weld (DP600 steel).

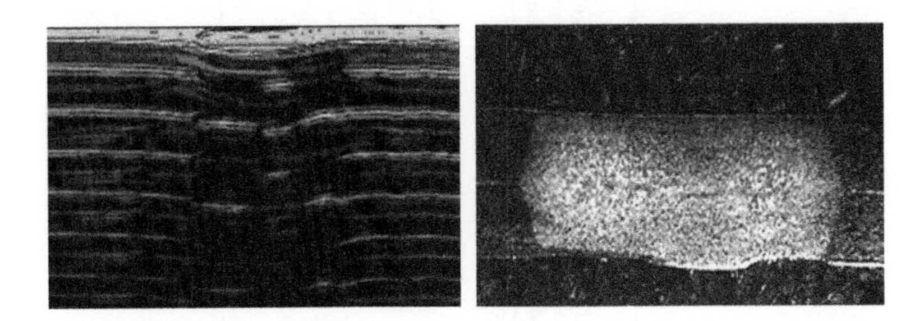

FIGURE 6.31
A void/crack at center of a weld (low-carbon steel).

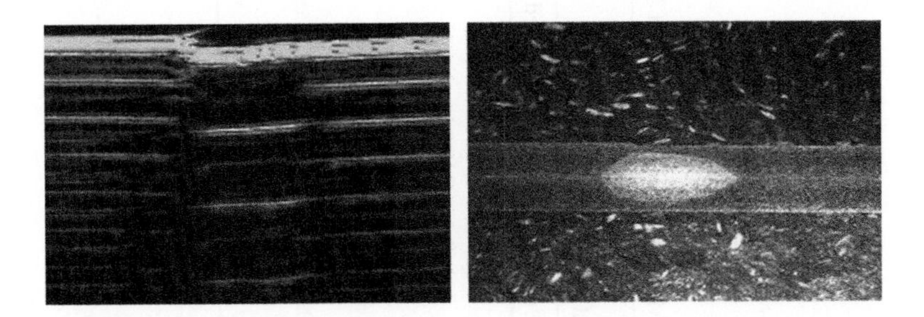

FIGURE 6.32
An undersized weld (low-carbon steel).

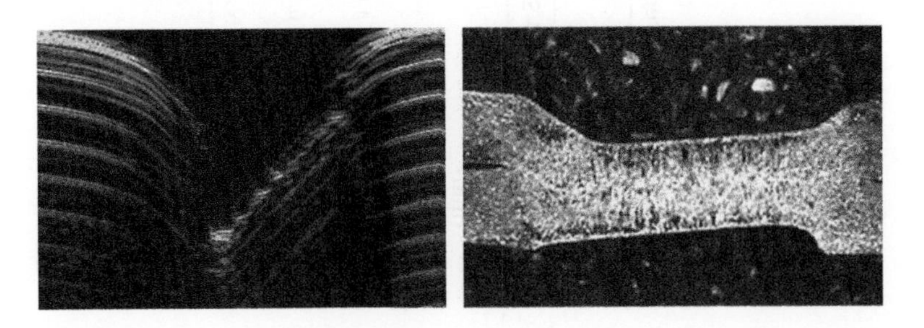

FIGURE 6.33
A weld with sheet separation, deep indentation, and interface expulsion (DP600 steel).

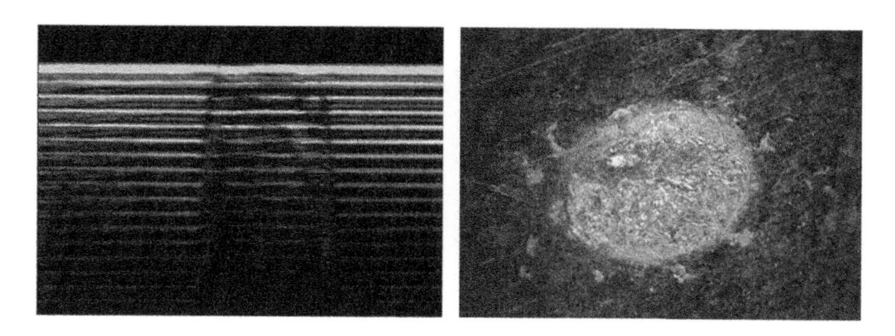

FIGURE 6.34
(See color insert.) B-scan and faying surface of a cold weld (current = 5000 A).

When the current is increased to 8000 A, a weld is actually created. This is clearly shown in Figure 6.37, which shows a (approximately) double-sheet thickness at the area squeezed by electrodes during welding. The indentation is more obvious than previous welds. An unfused area appears at the center of the weld, surrounded by a ring of fused metal. This is confirmed by the fractured interface in the figure, which shows a large smooth area at the center. Because of the insufficient fusion and the presence of a large crack in the weld, an interfacial fracture was produced when peel tested.

A further increase in current to 9000 A produced a small weld nugget, as shown in Figure 6.38. The B-scan image has the characteristics of a regular weld, but it is clear that the size of the fused area, or nugget width, is very limited. A rough or grainy surface was created when peel tested, indicating a certain fusion of the base metal. When 500 A more current was added, a regular weld was created (Figure 6.39). A weld button was pulled out when peel tested, and the coldness no longer existed. From the aforementioned analysis, it can be seen that the B-scan signals/images can be effectively used to indicate the degree of adhesion, or coldness, of a weld.

By comparing the B-scan images of various welds created with different welding currents, the following observations are made on identifying the features of welds from a B-scan image:

1. It is important to differentiate the color lines of reflections from different interfaces. This includes both the color (intensity) and thickness of the lines. Consider the color lines in a typical B-scan image of a spot weld, such as that in Figure 6.39. Because the indentation is small, it is convenient to use the reflection lines in the base metal as a reference to describe the structure in the weld area. The first line is the sheet (indentation) surface. The line is bright and thick, due to the repeated

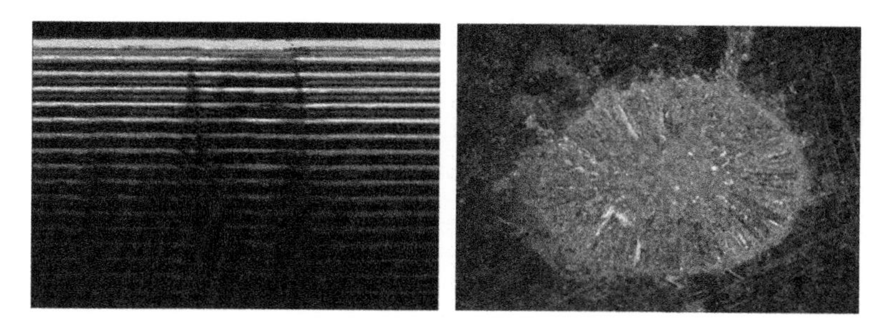

FIGURE 6.35
(See color insert.) B-scan and faying surface of a cold weld (current = 5500 A).

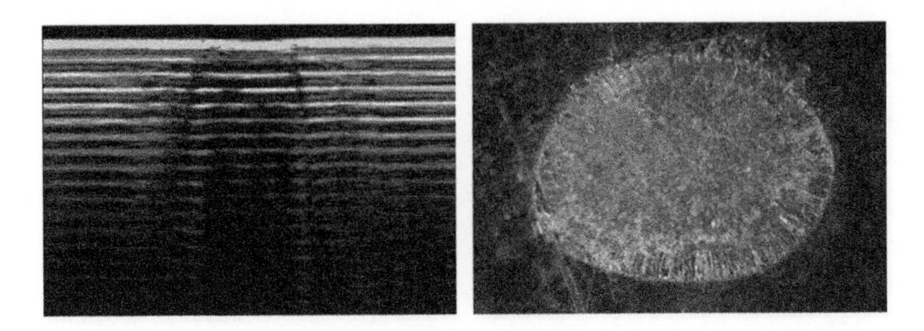

FIGURE 6.36
(See color insert.) B-scan and faying surface of a cold weld (current = 6500 A).

reflections from the surface. It slightly curves at the edge of an indentation mark. Comparing Figures 6.34 and 6.39, a cold weld (Figure 6.34) tends to have a fairly straight first line. A curved line indicates fusion in the weld area. The second line is a reflection of the ultrasound from the lower surface of the top sheet. It continues, with or without changing color/thickness, into the weld area when the weld is cold (such as shown in Figure 6.34), indicating little or no fusion at the faying interface. As the amount of heating increases, the portion of the line in the weld area turns darker and blurrier.

2. Judging only from the second line for fusion is not sufficient. The line can be bright even when there is sufficient fusion at the weld area, due to a structural difference of this portion of the weld, resulting from the original faying interface and solidification process. However, when there is fusion, the fused (original) faying interface does not reflect the ultrasonic waves the same way as an unfused or slightly fused faying interface does, as seen in the fourth line. Comparing Figures 6.36 and 6.39, the second lines in these figures are both clearly visible, but the fourth lines are drastically different. For the cold weld (Figure 6.36), the fourth line is bright, with increased thickness due to reflection from the deformed sheets and some fusion in the weld area. A good weld (Figure 6.39), on the other hand, has a very thin or even no fourth line. Therefore, the fourth line serves as a good indicator, together with the second line, for the degree of fusion at the faying interface.
3. The attenuation rate in the B-scan echo pattern reflects the coarseness of the microstructure and the smoothness of the interfaces. By studying the attenuation of the sound waves from B-scan images, it is possible to gain a rough idea about the structural characteristics of the adhesion.

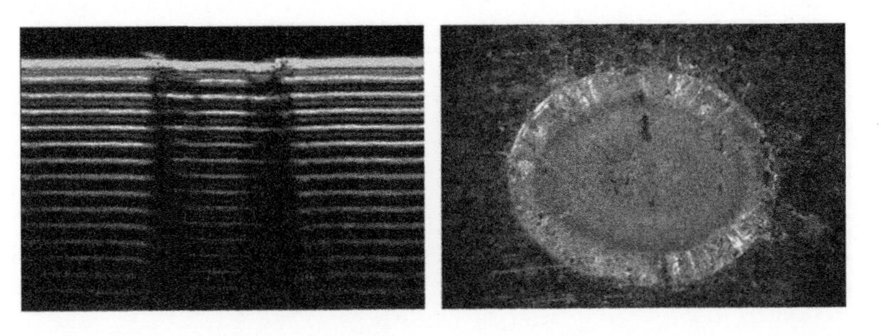

FIGURE 6.37
(See color insert.) B-scan and faying surface of a peeled weld (current = 8000 A).

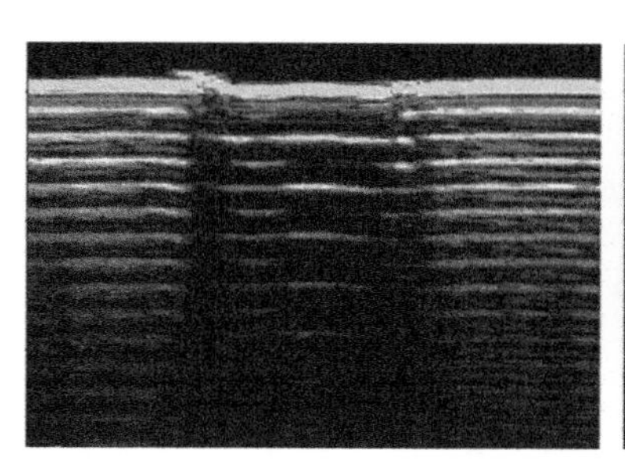 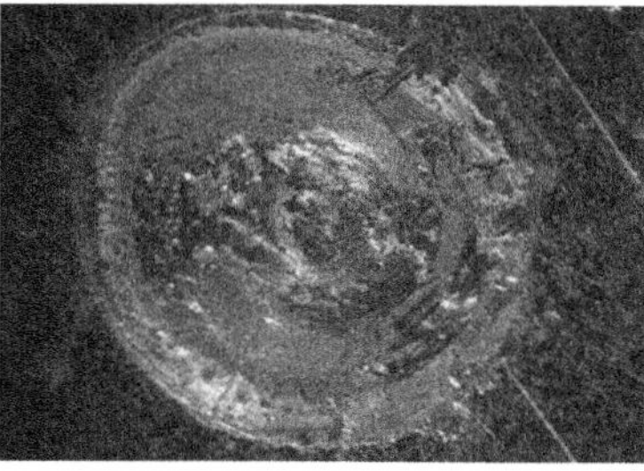

FIGURE 6.38
(See color insert.) B-scan and faying surface of a peeled weld (current = 9000 A).

6.3.5 Relationship between Weld Attributes and Weld Strength

When weld attributes, such as weld button size and indentation, are measured for weld quality, it is implied that these geometric features of a weldment correspond to the performance or strength of a weld. Therefore, a relationship between weld attributes and weld strength is needed in order to make a valid judgment of weld quality based on the geometric dimensions measured destructively or nondestructively. Ultrasonic B-scanning provides an opportunity to establish such a relationship by measuring the weld attributes before destructively testing the weldments. One such relationship was attempted through impact testing an AHSS.[34,35]

The AHSS was welded using four different schedules to create welds of various characteristics. The specimens were scanned, using a B-scanner made by Applied Metrics, for the physical attributes, such as nugget width, indentation depth and width, before being tested under an impact loading. The specimens, testing device, and testing procedure are described in Chapter 4. The measured dimensions of the weldments were then correlated to the impact performance, such as the impact energy and peak load during impact.

For this material, there is no clear dependence between nugget width and impact energy, as shown in Figure 6.40. Unlike what is often observed in static tensile–shear testing, large welds do not necessarily imply high-impact strength. However, the strength is closely related to the failure mode of the tested specimens. For welds of similar sizes, a complete button separation from both sheets produces the highest impact energy, whereas an interfacial fracture is associated with low-impact strength. This strongly suggests that the weld structure, in addition to the weld size, determines a weld's performance. A weak dependence of impact energy on weld thickness is also observed (Figure 6.41). In general, a small (or thin) weld thickness corresponds to a large nugget penetration, which is needed

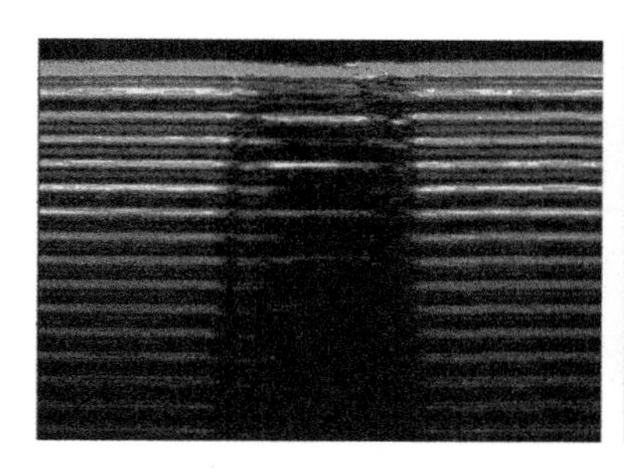 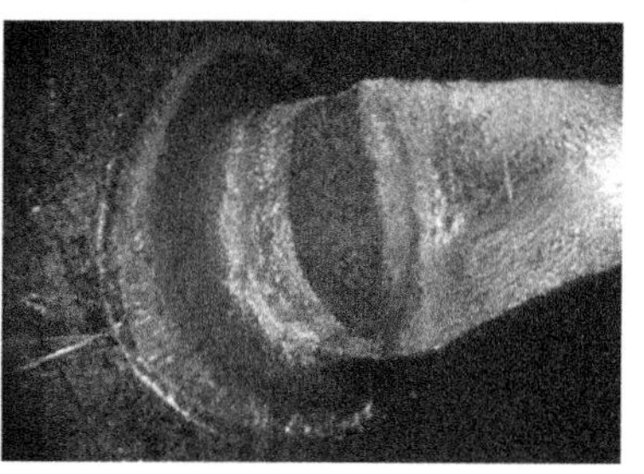

FIGURE 6.39
(See color insert.) B-scan and a peeled weld (current = 9500 A).

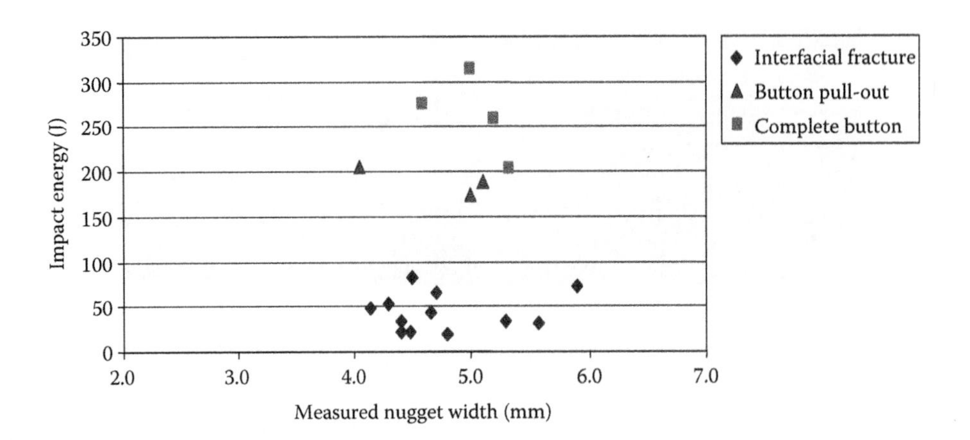

FIGURE 6.40
Relation between impact energy and nugget width.

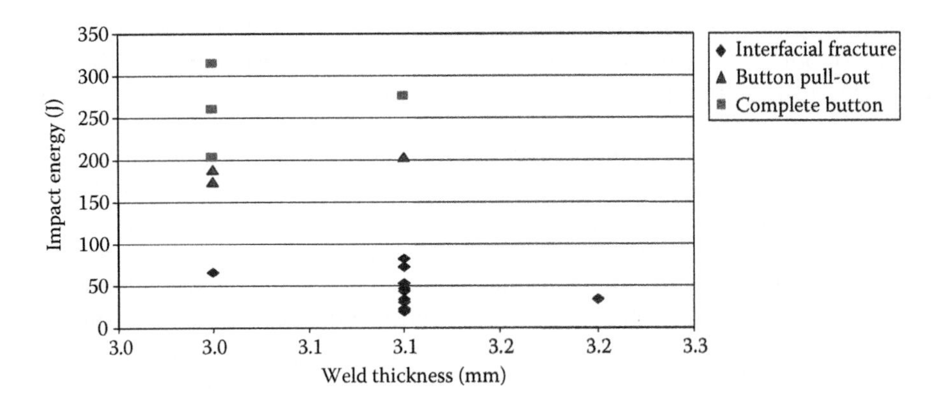

FIGURE 6.41
Relation between impact energy and weld thickness.

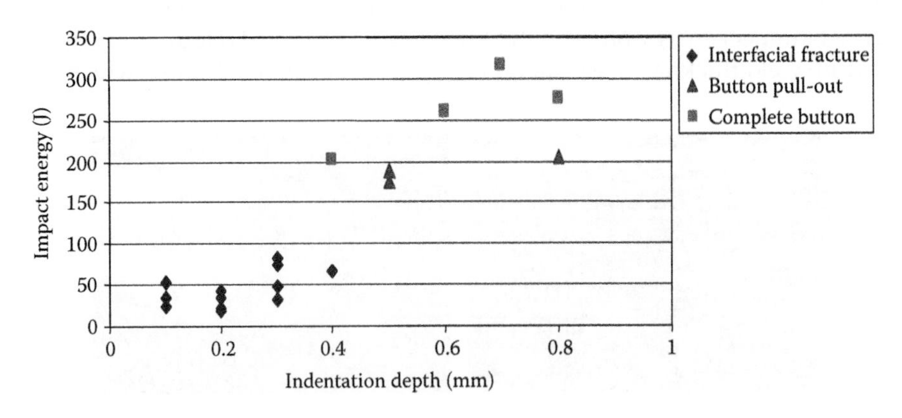

FIGURE 6.42
Relation between impact energy and indentation depth.

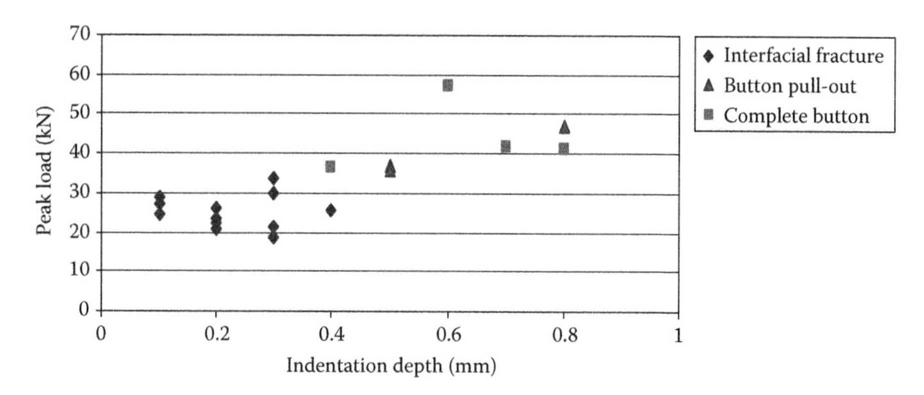

FIGURE 6.43
Relation between peak load and indentation depth.

for a certain adhesion strength. Again shown in this figure, the welds with complete button separation are stronger than those with other fracture modes. For this material, the electrode indentation has the closest link to a weld's impact strength. Figures 6.42 and 6.43 show that impact energy and peak load increase monotonically with indentation depth for most values of indentation. The impact energy peaks at an indentation of 0.7 mm, and the peak load at about 0.6 mm. A further increase in indentation depth actually weakens the weldment, resulting in a decreasing trend in both energy and peak load.

The conclusions are material dependent, that is, different materials have different correlations between weld attributes and weld performance. Such relations are essential in using nondestructive means for weld quality evaluation.

References

1. AWS A3.0:2001, *Standard Welding Terms and Definitions; Includes Terms for Adhesive Bonding, Brazing, Soldering, Thermal Cutting, and Thermal Spraying*, The American Welding Society (AWS) and the American National Standards Institute (ANSI), Miami, FL, 2001.
2. AWS D8.7: *Recommended Practices for Automotive Weld Quality—Resistance Spot Welding*, American Welding Society, Miami, FL, 2004.
3. Zuniga, S. M. and Sheppard, S. D., Determining the constitutive properties of the heat-affected zone in a resistance spot weld, *Modeling & Simulation in Materials Science & Engineering*, 3 (3), 391–416, 1995.
4. Gao, Z. and Zhang, K., Comparison of the fracture and fatigue properties of 16MnR steel weld metal, the HAZ and the base metal, *Journal of Materials Processing Technology*, 63 (1–3), 559–562, 1997.
5. *Procedure for Spot Welding of Uncoated and Coated Low Carbon and High Strength Steels, Resistance and Related Welding Processes*, International Institute of Welding, Doc III-1005-93, 1993.
6. Spinella, D. J., Using fuzzy logic to determine operating parameters for resistance spot welding of aluminum, Sheet Metal Welding Conference VI, Detroit, Michigan, 1994.
7. Newton, C. J., Browne, D. J., Thornton, M. C., Boomer, D. R., and Keay, B. F., The fundamentals of resistance spot welding aluminum, Sheet Metal Welding Conference VI, Detroit, MI, 1994.
8. Zhou, M., Hu, S. J., and Zhang, H., Critical specimen sizes for tensile–shear testing of steel sheets, *Welding Journal*, 78, 305-s, 1999.

9. Zhou, M., Relationship between spot weld attributes and weld performance, PhD Dissertation, University of Michigan, Ann Arbor, MI, 2000.
10. Keller, F. and Smith, D. W., Correlation of the strength and structure of spot welds in aluminum alloys, *Welding Journal*, 23 (1), 23-s–26-s, 1944.
11. McMaster, R. C. and Lindrall, F. C., The Interpretation of radiographs of spot welds in alclad 24S-T and 75S-T aluminum alloys, *Welding Journal*, 25 (8), 707-s–723-s, 1946.
12. Heuschkel, J., The expression of spot-weld properties, *Welding Journal*, 31 (10), 931-s–943-s, 1952.
13. Sawhill, J. M. and Baker, J. C., Spot weldability of high-strength sheet steels, *Welding Journal*, 59 (1), 19-s–30-s, 1980.
14. Thornton, P. M., Krause, A. R., and Davies, R. G., The aluminum spot weld, *Welding Journal*, 75 (3), 101-s–108-s, 1996.
15. Ewing, K. W., Cheresh, M., Thompson, R., and Kukuchek, P., Static and impact strengths of spot-welded HSLA and low carbon steel joints, SAE Paper 820281, 1982.
16. Koehler, J. R. and Owen, A. B., Computer experiments in design and analysis of experiments, In: Ghosh, S., Rao, C.R. (Eds.), *Handbook of Statistics*, Elsevier, Amsterdam, 261–308, 1996.
17. Ye, K. Q., Orthogonal column Latin hypercubes and their application in computer experiments, *Journal of the American Statistical Association*, 93 (444), 1430–1439, 1998.
18. AWS D8.9: *Recommended Practices for Test Methods for Evaluating the Resistance Spot Welding Behavior of Automotive Sheet Steel Materials*, American Welding Society, Miami, FL, draft, 2005.
19. Krautkramer, J. and Krautkramer, H., *Ultrasonic Testing of Materials*, Springer-Verlag, New York, NY, 1983.
20. Mansour, T. M., Ultrasonic inspection of spot welds in thin-gage steel, *Materials Evaluation*, 46, 650–658, 1988.
21. Raj, B., Subramanian, C. V., and Jayakumar, T., *Non-destructive Testing of Welds*, Narosa Publishing House, New Delhi, India, 2000.
22. *SWIS Operation Manual*, Applied Metrics, Fremont, CA, 2003.
23. Zhang, J., Ultrasonic evaluation of resistance spot weld quality, M.Sc. thesis, University of Toledo, Toledo, OH, 2003.
24. Mucciardi, A. N. and Gose, E. E., A comparison of seven techniques for choosing subsets of pattern recognition properties, *IEEE Transactions on Computers*, C-20, 1023, 1971.
25. Murthy, S. K., *Automatic Construction of Decision Trees from Data*, Siemens Corporate Research, Princeton, NJ, 1997.
26. Roye, W., *Ultrasonic Testing of Spot Welds in the Automotive Industry*, Special Issue no. SD 298, Krautkrämer GmbH & Co., Huerth, Germany, 1999.
27. Chertov, A. M., Maev, R. Gr., and Severin, F. M., Acoustic microscopy of internal structure of resistance spot welds. Available online at http://www.andrey-chertov.com/SAM%20for%20Industry%20Final%20Version%202006%2006.pdf. Accessed on Jan. 26, 2011.
28. Polrolniczak, H., *Ultrasonic Testing as a Means for Quality Assurance in Resistance Spot Welding*, A Special Issue no. SD 297, Krautkrämer, GmbH & Co., Huerth, Germany, 1999.
29. Shayan, A., Zhang, H., and Gan, Z., Quality test of AHSS steel spot welds using ultrasonic technique, SMWC XIV, Paper 3-2, 2010.
30. Saaty, T. L., *Fundamentals of Decision Making and Priority Theory with the Analytic Hierarchy Process*, RWS Publications, Pittsburgh, PA, 1994.
31. Saaty, T. L. and Vargas, L. G., *Models, Methods, Concepts, & Applications of the Analytic Hierarchy Process*, Kluwer Academic Publishers, Boston, 2001.
32. http://www.123ahp.com/OMetodi.aspx. Accessed on April 20, 2010.
33. Zhang, H., Jayatissa, A. H., and Gan, Z., Monitoring resistance spot welding using ultrasonic B-scan techniques, *EIT Conference, IEEE*, Illinois, May 20–22, 2010.
34. Karve, G., An impact tester and impact strength measurement of advanced high strength steel welds, M.Sc. thesis, University of Toledo, Toledo, OH, 2004.
35. Karve, G. and Zhang. H., Impact strength measurement of advanced high strength steel welds, in *Proceedings of Sheet Metal Welding Conference XI*, Sterling Heights, MI, Paper 5-3, 2004.

7

Expulsion in Resistance Spot Welding

A common phenomenon in resistance spot welding (RSW) is expulsion—the ejection of molten metal during welding. Expulsion, which can be observed frequently during spot welding, happens at either the faying surface or the electrode–workpiece interfaces, as shown in Figures 7.1 and 7.2, respectively. The latter may severely affect surface quality and electrode life, but not the strength of the weld if it is limited to the surface. On the other hand, expulsion at the faying surface may compromise a weld's quality, as it involves loss of liquid metal from the nugget during welding. The risk of expulsion is especially high in spot welding aluminum as well as magnesium alloys because of the very dynamic and unstable nature of the process, which is related to application of a high current in a short welding time, compared to welding steels. The causes of expulsion are both technical and human related. Expulsion is often used as a visual indicator of a correct welding process in steel welding. In order to achieve a weld size as large as possible to meet certain requirements, a prevalent practice is to use a large welding current, often close to or beyond expulsion limits. Expulsion limits are also often deliberately exceeded in production to reduce variations in weld quality caused by random factors. However, because of the loss of metal during expulsion, defects such as voids and porosity, which may reduce weld strength, are introduced to the nugget. In addition, expulsion has a negative influence on adhesive bonding, if it is used in conjunction with spot welding (so-called weld bonding), by damaging the adhesive layer.

The need to eliminate nonconforming welds in the sheet metal industry makes it necessary to avoid expulsion as much as possible in RSW. Prediction and control of expulsion are then of important practical interest.

In this chapter, the theories and models of expulsion are reviewed, and examples of some models are presented. The fundamental understanding of the expulsion phenomenon and its influence on weld quality presented in this chapter may lead to more research on expulsion, and developing effective means to control it.

7.1 Influence of Expulsion on Spot Weld Quality

In general, expulsion causes undesirable features to both the appearance and performance of a spot weld. It is often linked to excessive surface indentation, electrode wear, sheet distortion or separation, and defect formation. Because of heavy expulsion, a large cavity was formed after nugget solidification as a result of a deficit in volume, as shown in Figure 7.3.

Regarding the influence of expulsion on the performance characteristics of a spot weld, there are two opposing opinions. The first is that expulsion does not decrease weld performance and it is acceptable in limited ranges, which is represented by the works of Kimchi[1] and Karagoulis.[2] Another opinion is that expulsion has a detrimental effect on weld performance and appearance, and it should be suppressed, as suggested in the studies by

FIGURE 7.1
Expulsion traces at faying interface of a steel weldment. Arrows point at ejected and momentarily frozen liquid metals.

Newton et al.[3] and Hao et al.[4] This difference in opinions basically stems from the difference in materials tested. The first opinion was formed based on experiments on steels, whereas the latter one was based on aluminum alloys. However, in general, it is agreed that expulsion may induce some unfavorable features to the weldment.

To clarify the confusion, a series of experiments were conducted by Zhang[5] on the influence of expulsion on steel welding. In the first experiment, a fixed welding time (200 ms or 12 cycles) and electrode force (2.8 kN) were used. The welding current was altered in a small range:

FIGURE 7.2
Expulsion from surface of a sheet (interface between an electrode and a sheet) in welding an AA5754.

FIGURE 7.3
A cross section of a nugget in an AA5754 after heavy expulsion.

Case A: 7.7 kA, with very low expulsion probability
Case B: 7.8 kA, with moderate expulsion probability
Case C: 7.9 kA, with high expulsion probability
Case D: 8.0 kA, with very high expulsion probability

These cases were chosen based on a statistical study of expulsion, which is presented in detail in Sections 7.3.3 and 7.4.2. These cases represent the welding schedules from very low possibilities of expulsion to very high possibilities of expulsion. The small range of welding current was chosen to produce welds of similar size (and other physical attributes), with some having expulsion and others not. In the study, AKDQ steel of 0.78 mm gauge was used.

Figure 7.4 shows the measurement of weld attributes (average values and ranges of variation). Compared to the ones without expulsion, welds with expulsion have a slightly smaller nugget diameter, less penetration, and a smaller HAZ and total thickness of weldment, but slightly larger electrode indentation marks. The slight difference in the physical appearance might be because welding parameters used to create these welds were very close to each other. From the figure, it can be seen that the variations in the physical attributes are smaller, in general, for welds with expulsion compared to those without. This could be a reason for welding with expulsion on purpose in certain circumstances. A closer look at the microstructures of the nuggets and the HAZ reveals very little difference between welds with

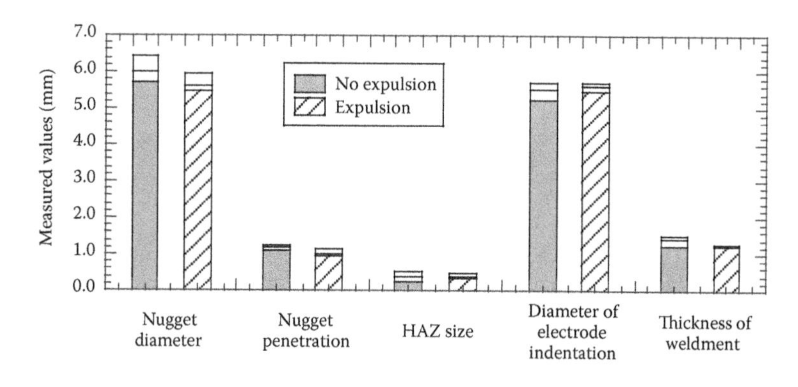

FIGURE 7.4
Physical attributes measured on sectioned specimens. For each measurement, welds with and without expulsion are compared, and ranges of variation are shown for minimum, average, and maximum values.

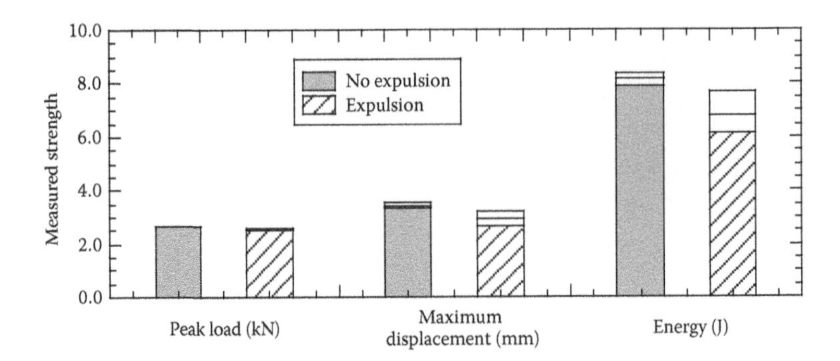

FIGURE 7.5
Influence of expulsion on strength measurements.

and without expulsion in untested specimens. An examination of the microstructures of tensile–shear-tested specimens also shows no significant difference. Columnar structures can be clearly seen inside the nugget, and precipitates are visible around the grain boundaries in the HAZ. Basically, no significant difference has been observed on the appearance of the welds. The only noticeable difference between the expelled and normal welds is that specimens with expulsion usually have larger distortion and sheet separation.

When the specimens were tensile–shear tested, the peak load showed very little difference between welds with and without expulsion (Figure 7.5). However, in general, welds with expulsion have lower maximum displacement, especially lower energy (these quantities are defined in Chapter 4). These differences would be neglected if only peak load was measured, as in many works on expulsion, where peak load was taken as the sole measure of weld strength. They can be clearly attributed to expulsion. Indeed, displacement and energy exhibit an important aspect of weld strength, that is, ductility (and fracture toughness) of a weldment. A quality weld should have sufficient load-carrying capacity (peak load), as well as certain ductility. Using peak load as the sole measure of weld strength is not only an incomplete description of weld quality; it may also be misleading, such as in the case of brittle welds. Another observation from Figure 7.5 is that the variations are larger when testing welds with expulsion, which is an implication of expulsion on the consistency of welding quality. An examination of tested specimens revealed that specimens with expulsion generally have fractures in the HAZ near the nugget (Figure 7.6a), whereas the ones without expulsion failed farther away from the

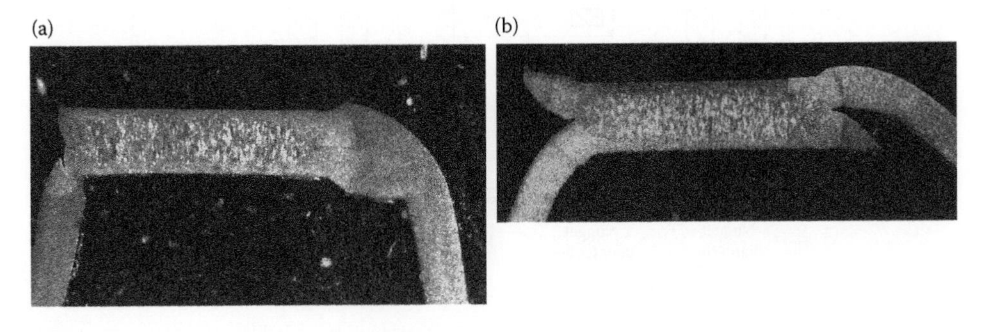

FIGURE 7.6
Tensile–shear-tested specimens (a) with and (b) without expulsion.

nugget (Figure 7.6b). Therefore, the difference in distortion may contribute the most to the difference in strength.

In another experiment, the effect of expulsion was investigated using schedules with larger ranges of electrode force, welding time, and welding current. The material used was 0.8-mm galvanized steel. To create various sizes of nuggets and HAZs, two electrode face diameters (6 and 10 mm) and several welding schedules were used. The electrode forces used were 600 and 1667 lb (2.67 and 7.4 kN), the weld current was in the range of 9968–23,232 A, and the weld time was between 6 and 24 cycles (100 and 400 ms).

Figure 7.7 shows the dependence of peak load and maximum displacement on average button size. Both peak load and displacement increase with button size. The trend is interrupted when expulsion happens. Compared with the welds of similar size without expulsion, the ones with expulsion show a significant drop in peak load (about 10%) and maximum displacement (about 30%). A similar trend can be observed in energy measurement (Figure 7.8). In this set of experiments, under the same welding conditions, welds with expulsion show slightly larger nugget and button sizes, but lower strength than their counterpart without expulsion.

From the analysis of the experimental results, the following conclusions can be drawn:

- Welds with expulsion may have lower strength than the ones of similar sizes but without expulsion.
- Button diameters correlate well with tensile–shear strength when there is no expulsion.

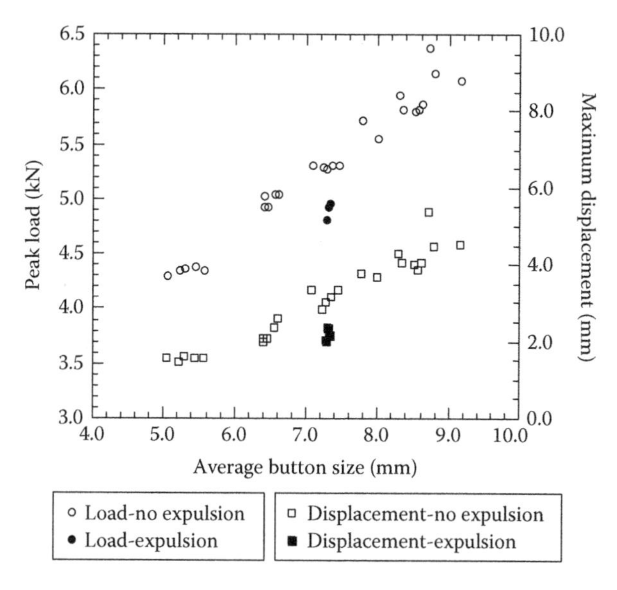

FIGURE 7.7
Dependence of peak load and maximum displacement on average button size.

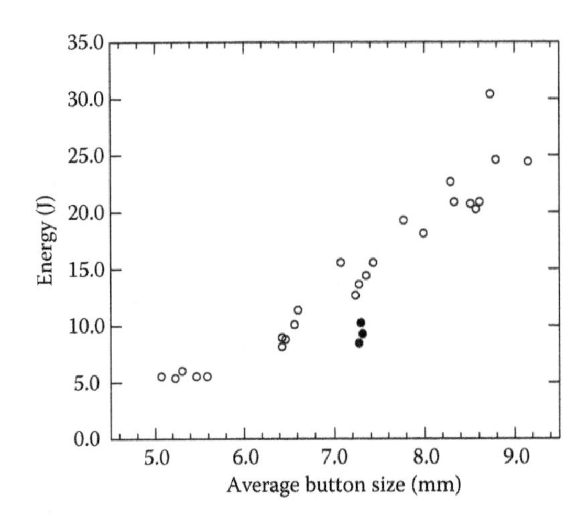

FIGURE 7.8
Dependence of energy on average button size.

7.2 Expulsion Process and Detection

To understand the expulsion phenomenon and to predict expulsion limits, several theories or hypotheses on the mechanisms of expulsion have been proposed. It was believed that the electrode force causes expulsion, as stated by Davies.[6] Expulsion happens when the molten metal of a weld nugget is squeezed out by the electrodes. However, this is contrary to observations by several researchers. It is worthwhile to mention the works by Dickinson et al.[7] and Wu.[8] Expulsion takes place, as stated in Dickinson et al.[7] when total useful energy applied to the weld exceeds a certain value defined as "critical expulsion energy," which is a function of physical properties or characteristics of the given material. However, this critical expulsion energy, as well as the total useful energy, is difficult to calculate in practice. Expulsion was linked by Wu[8] to excessive current densities, either by gross peak welding current or by highly localized/microscopic contact areas of faying surfaces with increased resistance by oxidation or contamination, at the early stage of spot welding. Although these studies provided certain insights to the understanding of the expulsion mechanism, quantitative analysis and application of these theories in practice are difficult. An analytical model of expulsion was developed by considering the balance of various forces acting on a weldment during welding.[9] Expulsion is believed to occur when the force from the nugget due to the internal pressure in a liquid nugget, caused by melting, liquid expansion, and other factors, exceeds the force from the electrodes.

As shown in the previous section, expulsion may decrease the strength of a spot weld; therefore, it should be avoided. As a frequently observed phenomenon in resistance welding, it involves interactions of all the processes during welding. For instance, the electrical current determines the heat input rate, which in turn influences the formation of the nugget, the temperature distribution in the weldment, and ultimately expulsion. Besides welding parameters, other factors, such as surface conditions, material strength (especially yield strength), loading, and thermal conditions, also influence expulsion. Because of its complex nature, it is difficult to predict or control expulsion. A welding process through the initiation and growth of a nugget, and eventually expulsion, was analyzed by dividing

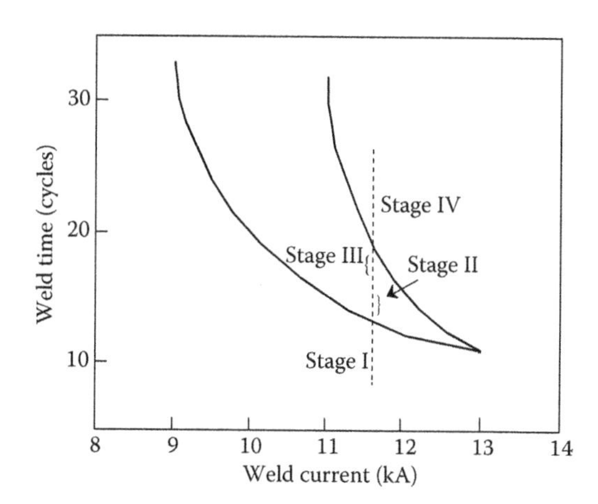

FIGURE 7.9
Welding stages for a particular current, superimposed on a lobe diagram for a high-strength low-alloy (HSLA) steel. (From Han, Z. et al., *Welding Journal*, 72, 209-s, 1993. With permission.)

it into principal stages as reflected in a typical lobe diagram (Figure 7.9): initiation of nugget (I), rapid nugget growth (II), steadily decreasing growth (III), and expulsion (IV).[7,10] Such diagrams, which are widely used in practice to determine welding schedules, only explore the expulsion phenomenon and suggest reasons for expulsion in terms of welding time and welding current. No fundamental understanding can be derived from them.

A few attempts have been made to detect expulsion by various techniques. According to published works, expulsion can be detected by measuring dynamic resistance,[7] acoustic emission,[11] electrode displacement,[12,13] and electrical signals.[4] Most of these measurements are difficult to make and often too expensive for use in production. The measurement proposed by Hao et al.[4] seemed promising. It described a robust method of monitoring expulsion in aluminum spot welding in high-volume production, based on electrical signals for both alternating-current (AC) and medium-frequency direct-current (MFDC) welding machines. All of these efforts are generally for expulsion detection only. Thus, corrections of welding parameters, if possible, can only be made on subsequent welds to avoid further expulsion after making the weld. Therefore, the effectiveness of these methods in eliminating/reducing expulsion is limited.

7.3 Expulsion Prediction and Prevention

The various techniques used to detect expulsion during spot welding provide important clues for understanding the expulsion phenomenon. For instance, the displacement between electrodes indicates how a weld nugget grows through thermal expansion of the weldment, and how electrodes collapse on the workpieces when expulsion happens. The displacement profile measured with a fiber-optic sensor on a 0.8-mm hot-dip galvanized (HDG) steel during welding is shown in Figure 7.10. Such observations may aid in understanding the fundamental process of expulsion, and in developing models for prediction and control of expulsion.

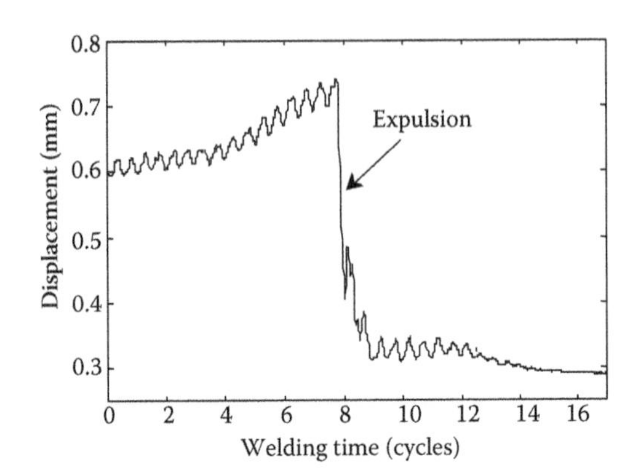

FIGURE 7.10
Displacement measured with a fiber-optic sensor on a 0.8-mm HDG steel. Weld cycle was based on a 60-Hz AC.

There have been a number of efforts in modeling the expulsion process, and most of them can be found in a review by Senkara et al.[9] Among them, four models provide reasonable explanation of the expulsion phenomenon and appear promising for practical use:

- Geometry comparison model
- Force balance model
- Liquid metal network model
- Statistical model

The following sections summarize the models and present examples on how these models are applied in practice.

7.3.1 Geometry Comparison Model

In a series of papers by researchers at Alcan Aluminum Company,[3,15,16] a model was proposed and verified based on the comparison of weld nugget size with the compressive zone size. In the model, the radius of a molten weld nugget is calculated [using a finite element model (FEM)] as a function of welding parameters. Expulsion takes place when the radius of the growing nugget exceeds that of the compressive force from the electrodes, $r_N > r_F$ in Figure 7.11. This model catches the characteristics of expulsion phenomenon; that is, the liquid metal is ejected when there is insufficient containment (by compressive force)

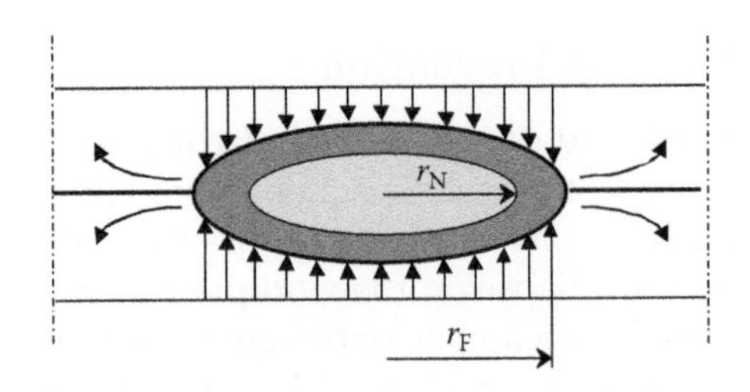

FIGURE 7.11
Mechanisms of expulsion of geometry comparison model.

through part of the solid periphery of the nugget. Fairly good agreement was obtained by the authors when the model prediction was compared with experimental results.[16]

However, because of limitations of the numerical simulation techniques and the small number of cases that can be simulated, this model is difficult to be used for accurately predicting expulsion in practice. Welding conditions and processes have to be idealized because of the limitations of numerical modeling, and the prediction relies solely on the comparison of geometric dimensions. Expulsion is usually predicted at a later stage of welding using this model, as a nugget needs time to grow to a certain size. However, there are observations that expulsion happens often at early stages of welding, when the size of the molten metal is considerably smaller than that of the compressive zone supplied by the electrodes. Expulsion results not only from improper welding time, current, and insufficient electrode force, but also from poor electrode conditions, electrode alignment, and workpiece conditions. In practice, cases in which the electrodes are aligned and the fit-up is perfect are rare. It has also been recognized that some materials tend to expel more than other materials. It is difficult to apply the geometry comparison model to such cases mainly because of the limitations of finite element modeling.

7.3.2 Force Balance Model

Based on the understanding of the physical processes during RSW, an expulsion model was proposed by analyzing the forces involved in welding.[9] An expulsion criterion was proposed by comparing the electrode force with the force from the liquid nugget, and expulsion occurs when the latter exceeds the former. An effective electrode force, instead of an applied/nominal electrode force, was used in the model. A methodology was proposed for determining the effective electrode force by analyzing loading conditions and locations of the nugget and electrodes. By thermodynamics analysis, the internal pressure in the liquid nugget caused by melting, liquid expansion, and other factors can be evaluated, and the force from the liquid nugget can be calculated by knowing the internal pressure and the dimensions of the nugget. Details of this model are presented in the following sections.

7.3.2.1 The Principle

Although there are many complicated reasons for expulsion, its basic process was described in the model as the interaction between the forces from the liquid nugget and its surrounding solid containment. Major forces acting on the weldment during welding are illustrated in Figure 7.12. They include the squeezing force provided by the electrodes ($F_{E,applied}$), the force from the liquid nugget (F_N) onto its solid containment, which is generated by the pressure (P) in the molten metal, and the compressive force (F_x) acting at the faying interface. There is a bonding provided by solid diffusion (corona bonding) at the faying surface that may provide some resistance to sheet separation. This force is usually much smaller than the others are and can be neglected in the analysis, as this model considers extreme expulsion conditions only.

Based on this understanding, a general model of expulsion was proposed. The criterion of expulsion can be stated as: Expulsion occurs when the force from the liquid nugget (F_N) onto its solid containment equals or exceeds the effective electrode force (F_E), that is, $F_N \geq F_E$.

These two forces and a schematic illustration of the model are shown in Figure 7.13. In practice, the applied electrode force is rarely aligned with the total force from the liquid

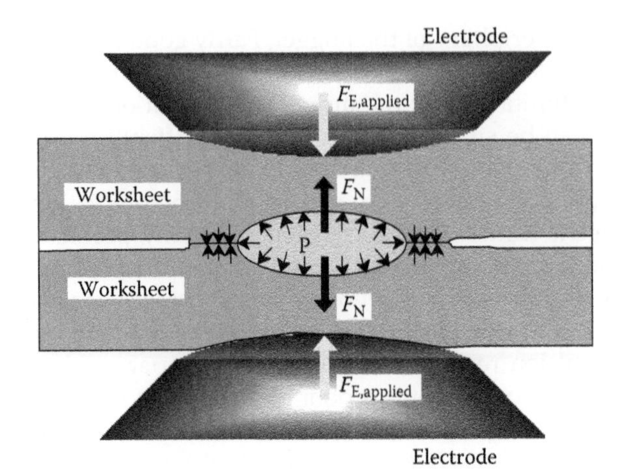

FIGURE 7.12
Forces acting on weldment during RSW under idealized conditions with aligned electrodes and perfect fit-up.

nugget because of imperfections in electrode alignment, part fit-up, and electrode geometry, such as changes induced by electrode wearing. Therefore, the applied electrode force, in many cases, is not the same as the one used to contain the liquid nugget from expulsion. The effective electrode force is introduced in this situation to accurately represent the force used to suppress the force from the liquid nugget.

7.3.2.2 Evaluation of Effective Electrode Force

An effective electrode force, which is usually a fraction of the total applied electrode force, is the one that actually balances the force from the liquid nugget. It can be estimated as follows.

The actual forces on one workpiece can be idealized, shown by the arrows in Figure 7.14, at the time of expulsion. $F_{E,applied}$ is the applied electrode force, F_N is the total force from

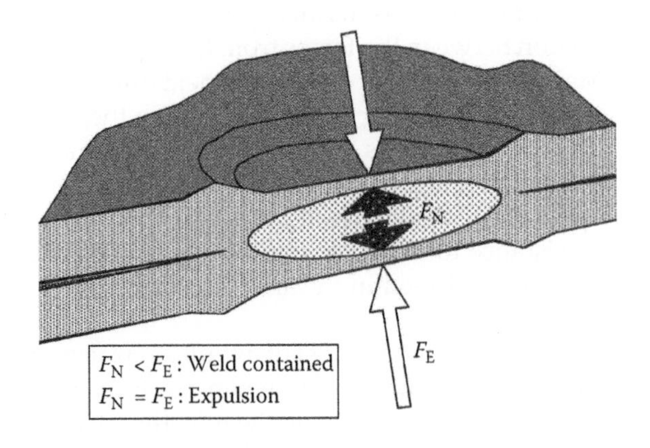

FIGURE 7.13
Schematic diagram of force balance. F_N is force from the nugget due to liquid pressure and F_E is effective electrode force.

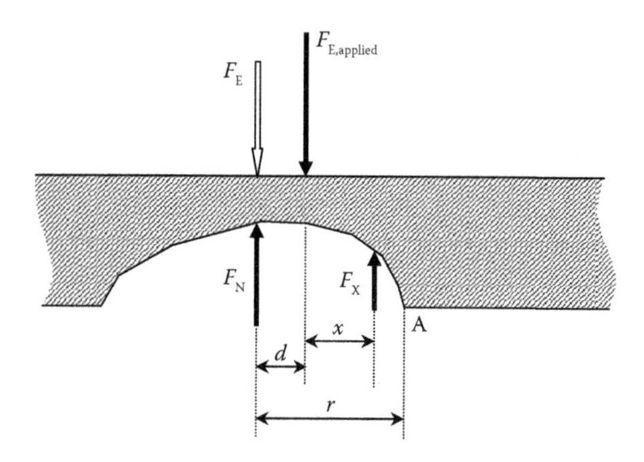

FIGURE 7.14
Schematic diagram of simplified forces and their locations on one workpiece at expulsion.

the liquid nugget against the solid containment, and F_x is a force imposed by the other workpiece.

In Figure 7.14, d is the distance between the total nugget force and the electrode force. r is the distance between F_N and the edge of the nugget. It is the nugget radius in the case of a round weld. x is the distance between force F_x and $F_{E,applied}$. Moment equilibrium with respect to the acting point of F_x produces the following relationship between $F_{E,applied}$ and F_N:

$$F_{E,applied}\, x = F_N(d + x) \tag{7.1}$$

Before the metal melts, $x = 0$ because $F_N = 0$, and $F_{E,applied}$ and F_x have to be collinear. As the liquid nugget grows, F_N gets larger ($F_N \propto$ the area of the nugget at the faying surface); thus, F_x gets smaller because $F_x + F_N = F_{E,applied}$, assuming $F_{E,applied}$ = constant. Meanwhile, x goes up, as can be derived from a moment equilibrium with respect to the acting point of $F_{E,applied}$: $F_N d = F_x x$ when assuming d = constant. Because the magnitude of F_N increases and that of F_x goes down, x has to get larger, or F_x gets farther away from the center of the nugget during nugget growth. It is reasonable to assume that when F_x moves across the right edge of the nugget (point A), the solid loses its containment of the nugget. Therefore, $x = r - d$ can be regarded as a critical condition for expulsion to happen. Therefore, the expulsion condition is

$$F_{E,applied}\,(r - d) = F_N r \tag{7.2}$$

If an equivalent force of magnitude

$$F_E = \frac{r - d}{r} F_{E,applied} \tag{7.3}$$

is applied in line with the force from the nugget, the workpiece is under equilibrium. Therefore, when there is a difference between the locations of the electrode force and the force from the nugget, the electrode force can be replaced by an equivalent, or effective electrode force in line with the force from the nugget. This effective electrode force is smaller than the applied electrode force; that is, only part of the electrode force is used in controlling expulsion.

The discrepancy d is usually created by asymmetric loadings, such as in the case of electrode misalignment (axial or angular misalignments), electrode wear, or improper workpiece fit-up. In Figure 7.12, the electrodes are aligned and the workpiece fit-up is perfect. The force provided by the electrodes is fully used against the nugget force such that d = 0 and $F_E = F_{E,applied}$. However, such an aligned and symmetric system is rarely seen, even in a laboratory setup. Figure 7.15 shows a case with angular misaligned electrodes. The nugget forms around the shortest electrical current path or the path with least electrical resistance, which is not the same as the one along which the total electrode force is applied because of the angular misalignment. As a result, an offset d is created between the applied electrode force and the force from the nugget. If the applied electrode force is not sufficient, expulsion may occur on one side of the nugget, as illustrated in Figure 7.15.

The sectioned AA5754 nugget shown in Figure 7.3 has an offset of 1.3 mm, and therefore $F_E \approx 0.7\, F_{E,applied}$, that is, only 70% of the applied electrode force was used to suppress expulsion. The location of the applied electrode force is estimated from the surface indentation, and that of the nugget force is through the geometric center of the nugget.

Electrode force can be applied through pneumatic or other mechanisms and, for simplicity, can be approximated as a constant during welding. The change of electrode force during welding is described in more detail in Chapter 8. The offset is not easy to determine, as it depends on a large number of random factors of actual welding processes. However, a guideline for selecting an electrode force/welding schedule can be developed by estimating the conditions of extreme cases.

The force from the liquid nugget can be calculated with the knowledge on its size and pressure. The following section is devoted to the calculations of pressures and forces in the molten metal.

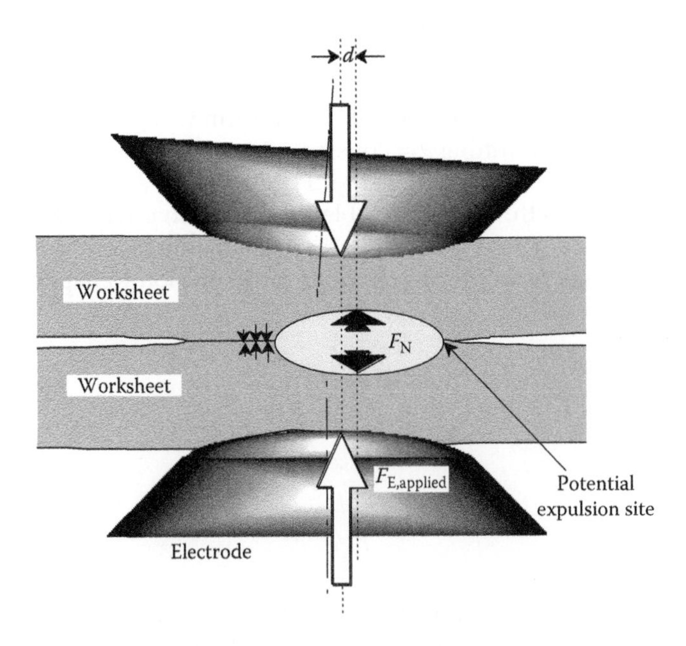

FIGURE 7.15
An offset between applied electrode force and that from the nugget, which is created by an angular misalignment of electrodes.

7.3.2.3 Pressures and Forces in Liquid Nugget

The pressure in a liquid nugget stems from the liquid volume expansion and the constraining imposed by its surrounding solid, and it is influenced by several factors. As an example, Figure 7.16 shows how the volume changes with temperature for pure iron and an Al–Mg alloy. The plots[9] are created using specific density data (in solid and liquid states, from the *Handbook of Chemistry and Physics*[17] and *Metals Handbook*[18] and coefficients of linear thermal expansion[19,20] for Al, Mg, and Fe. During welding, workpieces are heated from an initial temperature to the melting point and beyond. A volume expansion occurs during heating in the solid state, solid-to-liquid phase transformation, and heating in the liquid state. The volume change due to melting happens at the melting point for pure metals and between solidus and liquidus temperatures for alloys (except eutectic alloys). This volume change can be significant, as in the case of aluminum (about 7% volume change at its melting point). Further heating of the liquid generally produces a larger volume increase than thermal expansion in the solid state.

However, a free volume expansion of the liquid nugget during RSW is not possible due to its surrounding solid containment and the squeezing of electrodes. As a result, pressure in the nugget may be significant because of relatively low compressibility of liquid. It should be noted that if there is an expansion of surrounding solids around the liquid nugget, pressure in the nugget drops. Because the transformation of solid to liquid, which is the major contribution to internal pressure in the nugget, occurs at a relatively narrow temperature range, no significant volume change in the surrounding solid due to thermal expansion is expected. Therefore, the relief of pressure due to simultaneous solid expansion at melting can be neglected for simplicity. The solid farther from the molten metal in the workpieces, which is at a lower temperature, inhibits the expansion of the solids in the immediate vicinity of the nugget. This confinement, together with the squeezing imposed by the electrodes, makes it difficult for the liquid nugget to expand under the pressure from either melting or subsequent heating.

Another source of pressure in the liquid nugget is the pressure from metal vapors. Such pressure exists because at above the melting temperature, a closed system tends to reach liquid–vapor equilibrium according to general thermodynamics principles. This effect should be taken into account, especially if there are volatile elements in the workpieces,

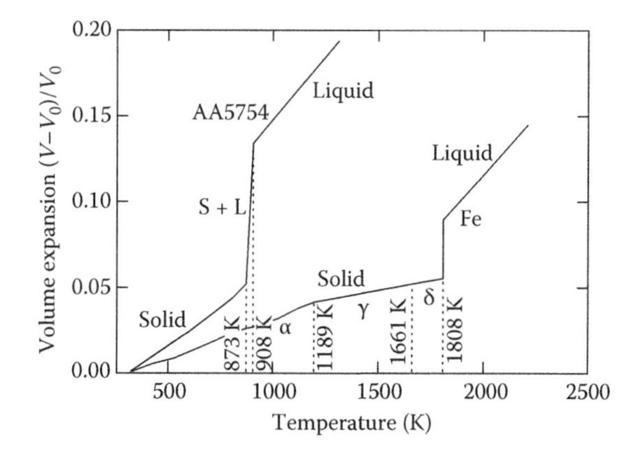

FIGURE 7.16
Calculated thermal expansion for an aluminum alloy 5754 and pure iron, in temperature ranges between room temperature and beyond melting point.

for example, Mg and Zn in certain series of aluminum alloys, Zn in coated steels, or Mg in magnesium alloys. The pressure from metal vapors may be significant even at a temperature below a metal's boiling point. In addition to metal vapor pressure, pressure from gases resulting from thermal decomposition of surface agents should also be considered. Examples of surface agents are lubricants on metal sheets, pretreatment agents, adhesives (in the case of weld bonding), and adsorbed moisture or gases. The pressure can be evaluated by considering the type and amount of gaseous products, their reactivity with the liquid alloy, and solubility in it.

Therefore, there are four major components of pressure in the liquid metal in RSW:

- Solid-to-liquid phase transformation (melting, P_{melt})
- Expansion in the liquid state ($P_{\text{exp.}}$)
- Vapors from the liquid metal (P_{vapor})
- Decomposition of surface agents ($P_{\text{lubr.}}$)

The total pressure in a liquid nugget is the summation of all these components:

$$P = P_{\text{melt}} + P_{\text{exp.}} + P_{\text{vapor}} + P_{\text{lubr}} \tag{7.4}$$

The material properties above their melting points should be known in order to calculate the total pressure. Because of a lack of certain data, an assumption has to be made that the property of an alloy Z_{A-B} of components A and B can be linearly interpolated, using the properties of its components Z_A and Z_B and their atomic fractions x_A and x_B in the alloy:

$$Z_{A-B} = Z_A x_A + Z_B x_B \tag{7.5}$$

The same assumption is also used for the remaining parts of this section. The calculation of the four pressure components is detailed below.

7.3.2.3.1 Pressure due to Melting

As a result of melting a certain portion of the metal surrounded by the solid phase, compression of the liquid takes place. The relationship between the volume V and pressure P in the liquid nugget at a given (absolute) temperature T can be described by the coefficient of compressibility κ[21]:

$$\kappa = -\frac{1}{V}\left(\frac{\partial V}{\partial P}\right)_T \tag{7.6}$$

Therefore, for a small decrement of volume, the increase in pressure is

$$\mathrm{d}P = -\frac{1}{\kappa V}\mathrm{d}V \tag{7.7}$$

Because the molten metal is not allowed to expand freely due to the containment of its solid surrounding and the electrode forces, the increase in pressure resulting from melting is approximately the same as that from compressing the liquid metal from the (free) liquid volume to its original volume before melting. This pressure can be obtained by integrating Equation 7.7:

$$\int_0^P \mathrm{d}P = -\int_{V_L}^{V_S} \frac{1}{\kappa V}\mathrm{d}V \tag{7.8}$$

where V_S and V_L are molar volumes of solid and liquid states, respectively, at melting temperature. Therefore, the pressure due to melting is

$$P_{\text{melt}} = \int_{V_S}^{V_L} \frac{1}{\kappa V}\mathrm{d}V = \frac{1}{\kappa}\ln V\bigg|_{V_S}^{V_L} = \frac{1}{\kappa}\ln\frac{V_L}{V_S} \tag{7.9}$$

Thus, a high volume change during melting results in a high-pressure contribution. Volume changes of several metals and alloys are shown in Figure 7.16 and Table 7.1.

7.3.2.3.2 Pressure due to Liquid Expansion

A quantitative relationship between pressure and temperature under a constant volume can be described by a thermal pressure coefficient β, defined in the following:

$$\beta = \frac{1}{P}\left(\frac{\partial P}{\partial T}\right)_V \tag{7.10}$$

Its value is unknown for most liquid metals. However, the partial derivative $\partial P/\partial T$ may be presented as the product of two partial derivatives:

$$\left(\frac{\partial P}{\partial T}\right)_V = -\left(\frac{\partial V}{\partial T}\right)_P \left(\frac{\partial P}{\partial V}\right)_T \tag{7.11}$$

TABLE 7.1

Selected Properties of Aluminum, Iron, and Main Components in Their Alloys

Property	Unit	Al	Fe	Cu	Mg	Zn
Solid density at melting point (ρ_S)	10^3 kg m^{-3}	2.55[a]	7.31[a]	8.32[b]	1.65[b]	6.84[b]
Liquid density at melting point (ρ_L)	10^3 kg m^{-3}	2.37[b]	7.07[b]	8.09[b]	1.58[b]	6.64[b]
Liquid density change rate ($-\mathrm{d}\rho_L/\mathrm{d}T$)	10^{-1} kg m^{-3} deg^{-1}	3.11[b]	6.34[b]	9.44[b]	2.60[b]	11.3[b]
Volume change due to melting $[(V_L - V_S)/V_S]$	%	7.06[a]	3.16[a]	4.68[a]	4.2[a]	2.9[a]
Coefficient of volume expansion in liquid at melting point (α)	10^{-4} deg^{-1}	1.31[a]	0.89[a]	1.17[a]	1.65[a]	1.70[a]

[a] Calculated values based on the data from *Handbook of Chemistry and Physics*,[17] *Metals Handbook*,[18] *ASM Metals Reference Book*,[19] and *Aluminum: Properties and Physical Metallurgy*.[20]

[b] Data from *Handbook of Chemistry and Physics*.[17]

By introducing a coefficient of volume thermal expansion, α, defined as

$$\alpha = \frac{1}{V}\left(\frac{\partial V}{\partial T}\right)_P \tag{7.12}$$

and using compressibility coefficient κ as defined in Equation 7.6, β can be expressed by variables whose values can be found in published metallurgical data sources:

$$\beta = \frac{1}{P}\frac{\alpha}{\kappa} \tag{7.13}$$

Hence, for a small increment of temperature, the increase in pressure is

$$\mathrm{d}P = \frac{\alpha}{\kappa}\mathrm{d}T \tag{7.14}$$

Integration of Equation 7.14 yields the contribution of pressure due to the expansion of the liquid nugget in the range from melting point T_m (ignoring the temperature range of melting for alloys) to a given temperature T at a constant volume in the following form:

$$P_{\text{exp.}} = \frac{\alpha}{\kappa}(T - T_{\text{melt}}) \tag{7.15}$$

In Equation 7.15, κ is a constant and α can be evaluated by considering the volume change during heating. Because the liquid density depends linearly on the temperature above the melting point, α can be expressed as

$$\alpha = -\frac{C}{\rho_{L,\text{melt}} + C(T - T_m)} \tag{7.16}$$

where C equals $-\mathrm{d}\rho_L/\mathrm{d}T$, and it is a constant that can be derived from the values listed in Table 7.1. α is listed in Table 7.1 for several metals.

7.3.2.3.3 Vapor Pressure

Vapor pressures of several commonly used metals are presented in Figure 7.17. Although pure metals are rarely welded in industrial practice, vapor pressures of alloys can be derived from those of their components.

The total vapor pressure over the liquid alloy, P_{vapor}, equals the sum of partial vapor pressures of particular components $\bar{p}_i$:

$$P_{\text{vapor}} = \sum_i \bar{p}_i \tag{7.17}$$

According to Raoult's law, $\bar{p}_i$ may be written as a product of thermodynamics activity a_i of given component i in the liquid solution and its vapor pressure in pure state p_i^o:

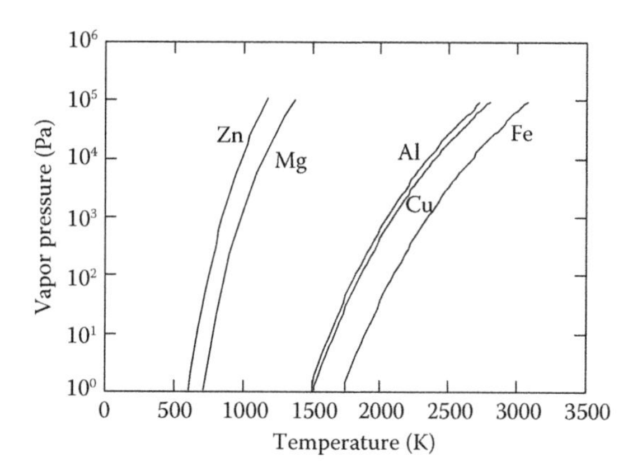

FIGURE 7.17
Plots of vapor pressures for several metals using data from *Handbook of Chemistry and Physics*.[17]

$$\bar{p}_i = a_i p_i^o \tag{7.18}$$

where a_i can be written as

$$a_i = x_i \gamma_i \tag{7.19}$$

x_i and γ_i are the molar ratio of component i in the solution and its coefficient of activity, respectively, at a given temperature. Unfortunately, data on a_i and γ_i and their temperature dependence for liquid metallic solutions are limited,[22] and in many cases, it is impossible to obtain a_i. However, it is possible to calculate γ_i using Gibbs–Duhem's equation of a multi-component metallic solution:

$$\sum_i x_i \mathrm{d} \ln \gamma_i = 0 \tag{7.20}$$

An especially useful solution to the above equation was developed by Krupkowski[23] for quasi-regular metallic solutions. According to his work, γ_i can be expressed as the product of two independent functions of temperature $w(T)$ and concentration $f(x_i)$:

$$\ln \gamma_i = w(T) f(x_i) \tag{7.21}$$

In two-component solutions (A–B), coefficients of activity can be expressed as

$$\ln \gamma_A = \left(\frac{a}{T} + b\right) x_B^m \tag{7.22}$$

and

$$\ln \gamma_B = \left(\frac{a}{T} + b\right)\left(x_B^m - \frac{m}{m-1} x_B^{m-1} + \frac{1}{m-1}\right) \tag{7.23}$$

where a, b, and m are constants that can be found in thermodynamics tables or calculated from existing data. Based on two-component solutions, the calculation of γ_i in a multicomponent solution is also possible by using the following equation:

$$\ln \gamma_i = \sum_{j=1,\,k=2}^{j=n-1,\,k=n} [\ln \gamma_i]_{j,k} \tag{7.24}$$

where $[\ln\gamma_i]_{j,k}$ is for the activity coefficient of i in a binary j–k solution.

7.3.2.3.4 Pressure due to Decomposition of Surface Agents

Before spot welding, other than the substrates, there are usually certain substances (e.g., lubricants, adhesives, and coatings) that exist at the faying interfaces. Additionally, thin layers of adsorbed gases and moisture may also exist on the surfaces. During spot welding, the heat decomposes surface agents and releases gases at the faying interfaces. Part of the gases may be trapped in the rapidly spreading liquid metal at the faying surface. Examples of such gases are H_2, H_2O, CO, CO_2, and C_xH_y chains. The type and amount of gases may vary in individual cases. However, the general steps of calculating their contributions to the total nugget pressure are as follows:

- Determine the types of particular gases in the area of interest and calculate their quantities (in terms of moles) under the standard condition (at 298 K).
- Subtract the amount of gases dissolved in liquid metal and those reacted with liquid metal. Gas–metal solubility as a function of temperature, as well as chemical affinity, should be considered.
- Calculate the pressure increase at the welding temperature (above T_m) with the knowledge of the total pressure contributed by other factors

In practice, the amount of specific gases that are released from the sheet surface during welding may be determined by a suitable chemical analysis. Thorough decomposition and zero gas solubility and reactivity can be assumed to obtain extreme values of pressure attributed to surface agents.

Therefore, the pressure from surface agent decomposition depends on the particular gas composition, solubility, and reactivity of the system. By using the ideal gas equation of state for the released gases, an equation for the pressure can be derived by thermodynamics consideration as

$$\kappa P_{\text{lubr.}} + \ln\left(V_{\text{nugget}} - \frac{nRT}{P_{\text{lubr.}} + P_{\text{L}}}\right) - \ln V_{\text{nugget}} = 0 \tag{7.25}$$

In Equation 7.25, $n = n_{\text{total}} - n_{\text{diss.}} - n_{\text{react.}}$, where n_{total} is the sum of the moles of the released gases under standard thermodynamics conditions, and $n_{\text{diss.}}$ and $n_{\text{react.}}$ are the moles of gases dissolving in and reacting with the liquid metal, respectively, at the welding temperature T. R is the universal gas constant and V_{nugget} is the volume of the liquid nugget. P_{L} is the total pressure due to melting, liquid expansion, and metal vapor pressure. Equation 7.25 can be solved numerically for particular systems.

7.3.2.3.5 Calculation of Total Pressure and Force from Liquid Nugget

The total force from the nugget can be estimated numerically (e.g., by using the finite element method). The liquid nugget can be considered as a simply connected domain. The minimum temperature is at the solid–liquid boundary, which is the melting temperature (*solidus* can be used for simplicity). Finite element simulation[24,25] shows that the maximum temperature is reached at the center of the nugget, and isotherms in the nugget are ellipsoidal in shape. Therefore, the temperature distribution in the liquid nugget can be approximated as layers (or shells), with the temperature being constant in each layer/shell. For any shell i reaching the melting point in the liquid nugget, the total pressure is

$$p_i = p_{i1} + p_{i2} + p_{i3} + p_{i4} \tag{7.26}$$

where partial pressures p_{i1-3} are as expressed in Equations 7.9, 7.15, and 7.17, and p_{i4} is the partial pressure due to decomposition of surface agents, which is calculated from Equation 7.25. Total pressure in the entire nugget is

$$P = \frac{1}{V_{\text{nugget}}} \int_{V_{\text{nugget}}} p\,\mathrm{d}V \approx \frac{1}{V_{\text{nugget}}} \sum_i p_i V_i = \sum p_i f_i \tag{7.27}$$

where p is the total pressure of a unit volume at a temperature, V_i is the volume of shell i in which the temperature is assumed constant, and f_i is the volume fraction of shell i. A detailed calculation of pressures is demonstrated in Section 7.4.1.1. Figure 7.18 shows the calculated total pressures of Al, Mg, and Fe.

Force from the liquid nugget onto its solid surroundings can be estimated once the pressure and the projected area in the direction of concern are known. Force in the direction of interest (here it is assumed to be the z-direction) on a segment of solid–liquid interface dS (as shown in Figure 7.29) is

$$\mathrm{d}F_z = P\mathrm{d}S\cos\gamma = P\mathrm{d}S_{xy} \tag{7.28}$$

where γ is the angle between the z-axis and the normal n of dS, and dS_{xy} is the projection of dS onto the x–y plane. The total force in the z-direction is

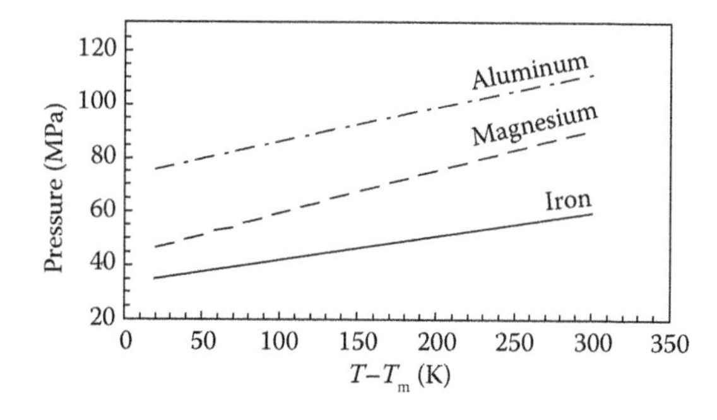

FIGURE 7.18
Calculated total pressure in liquid nuggets in a low-alloy steel, a Mg alloy, and an aluminum alloy (AA5754) as a function of overheating. T_m is melting temperature.

$$F_z = \int_S dF_z = P\int_S dS_{xy} = PS_{xy} \quad (7.29)$$

where S_{xy} is the total projection area of the nugget on the x–y plane. Therefore, the total force is independent of the particular shape of a nugget. It also implies that the location of the total force is the geometric center of S_{xy}.

This model provides a criterion for expulsion prediction and a procedure to analyze forces involved in RSW. In summary, this model provides

- A criterion of expulsion based on the interaction of forces acting on a weldment during resistance spot welding
- A detailed and systematic procedure to evaluate the pressure in the liquid nugget by thermodynamics analysis
- A procedure of calculating the force from the liquid nugget, from the pressure and the size of the nugget
- A method to evaluate the effective electrode force by knowing the offset between the applied electrode force and the force from the nugget

7.3.3 Expulsion through Molten Liquid Network in HAZ

In a study on resistance welding magnesium alloys AZ31 and AZ91, it was found that expulsion in welding AZ91D is significantly different from that in welding AZ31B or Al alloys.[26] In general, it is found that welding AZ31B is very similar to welding Al alloys, whereas welding AZ91D does not follow the force balance model presented in the previous section.

Interfacial expulsion in welding AZ91D was detected even when the electrode force was clearly sufficient to contain expulsion according to the force balance model. Although the overall appearance of an AZ91D weld, as shown in Figure 7.19, is very similar to that of an AZ31 or Al weld, with ejected metal debris on both sides of the nugget, and a large void and some cracks in the nugget, a closer look shows that the nugget is surrounded by a partial melting zone with a dense network of molten phases along the grain boundaries in the HAZ. By varying the electrode force level and changing the amount of heat input to create welds of various sizes, a series of experiments show that expulsion was almost inevitable in welding AZ91D. Even when the electrode force was raised to 9 kN in welding

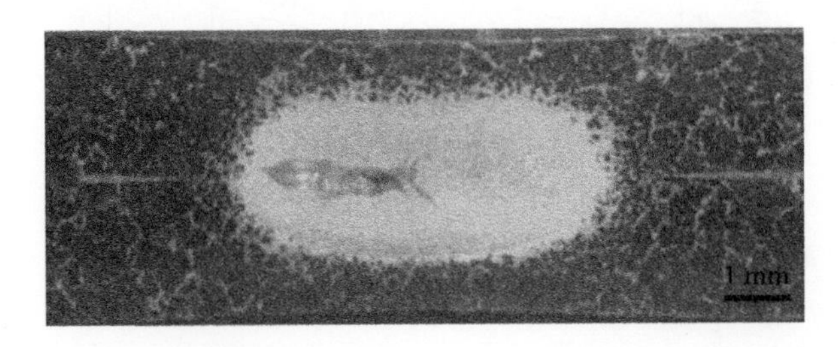

FIGURE 7.19
A 2.0-mm AZ91D weld with expulsion.

2.4 mm AZ91D, the severity of expulsion was only moderately reduced. In general, increasing electrode force is not as effective in controlling expulsion in welding AZ91D as in welding other metals. A new expulsion mechanism is proposed based on the observations in welding this alloy.

7.3.3.1 Expulsion Characteristics of AZ91D

It is common to observe a significant plastic deformation at the faying interface in the HAZ near the nugget in resistance spot welds of Al and other materials. One such example is an AZ31B weld shown in Figure 7.20a, in which the sheets are bound together at the faying interface next to the nugget. The workpieces near the weld nugget in the HAZ were heated to an elevated temperature below the melting point, and plastic deformation could easily occur when they were squeezed by the electrodes. Such a plastic deformation at elevated temperature promotes solid bonding, and a solid bonding ring formed through this process may effectively seal the liquid nugget from expulsion. However, an examination of the weld structures revealed that little plastic deformation of the sheets occurred at the interfaces in welding AZ91D. The weld in Figure 7.20b exhibits no plastic deformation and the original surfaces are virtually intact at the faying interface.

Another unique feature of AZ91D welds with expulsion is the existence of a dense network of grain boundaries in the HAZ surrounding the nugget, which possibly melts and then solidifies during welding. Figure 7.21 shows different views of a fairly small weld nugget surrounded by a massive network of dark grain boundaries in the HAZ, and some of them are cracks either open or filled with (solidified) liquid metal (Figure 7.21a). Although low-melting eutectics can be found at the grain boundaries in many material systems, those in AZ91 are different from most of others in their amount or volume fraction and the morphology. A comparison between AZ91 and AZ31 by Salman et al.[27] in their study on the corrosion properties of these two Mg alloys shows that the microstructure of AZ91 has a nearly continuous network of $Mg_{17}Al_{12}$ at its grain boundaries, whereas AZ31 contains merely discrete particles of the phase (Figure 7.22). A continuous network of the eutectics

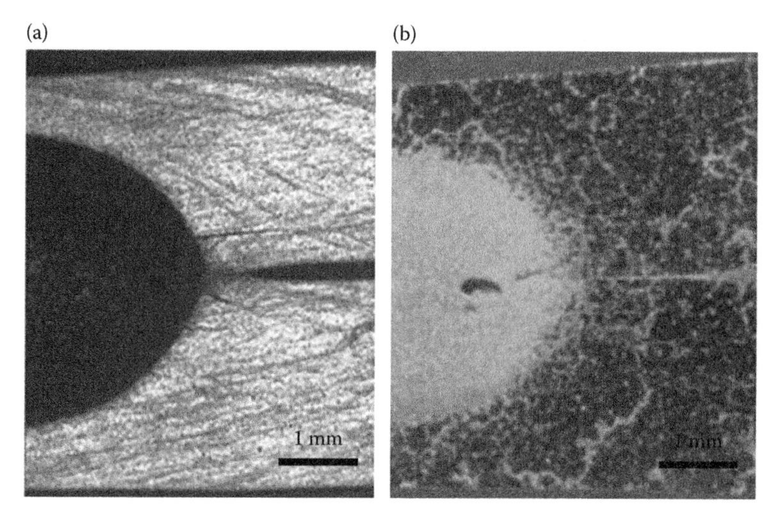

FIGURE 7.20
Plastic deformation in an AZ31B weld (a), and an AZ91D weld (b).

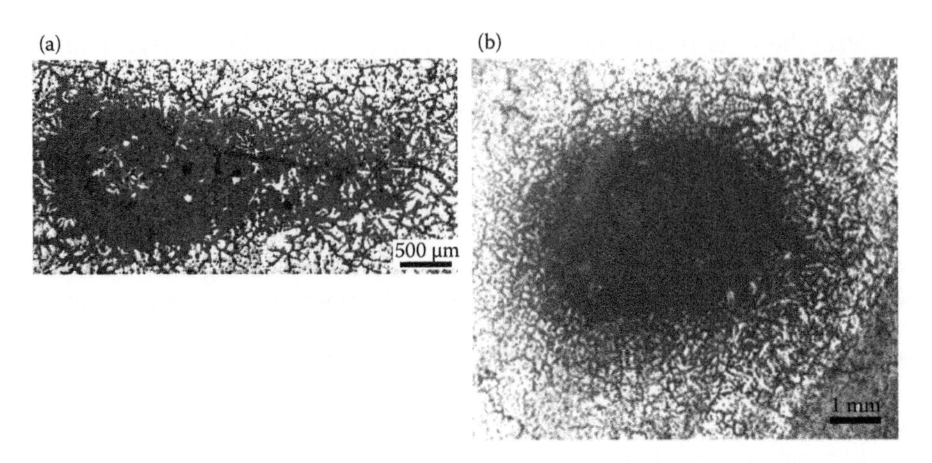

FIGURE 7.21
A section perpendicular (a) to, and a section parallel (b) to sheet surfaces of a small AZ91D weld.

may form along the grain boundaries only when the volume fraction is sufficiently large. The $Mg_{17}Al_{12}$ at the grain boundaries of these two alloys has a melting temperature of 437°C, more than 100°C lower than that of the alloys and, therefore, may melt during welding in the HAZ. Melting of such eutectics at grain boundaries was also observed in a study on electric arc welding of AZ91D.[28] A close look at the microstructure near the fusion line of an arc weld, as shown in Figure 7.23, confirms grain boundary melting, as well as cracking at the grain boundaries resulting from melting of the β-phase ($Mg_{17}Al_{12}$).

Grain boundary melting was observed in all AZ91D welds with expulsion in the current study. One example is shown in Figure 7.24 in which the HAZ has a high proportion of molten grain boundaries in the region next to the fusion line. As a result of melting and solidification, the grain boundaries in the HAZ are filled with fine equiaxed grains due to rapid cooling, similar to the observations of Munitz et al.[29]

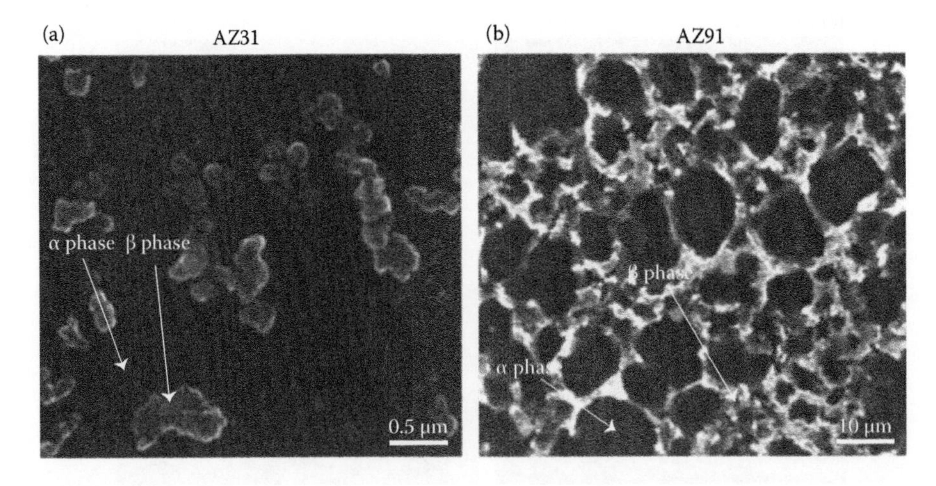

FIGURE 7.22
Microstructures of AZ31 (a) and AZ91 (b). (From Salman, S.A. et al., *Int. J. Corros.*, 2010, Article ID 412129, 2010. With permission.)

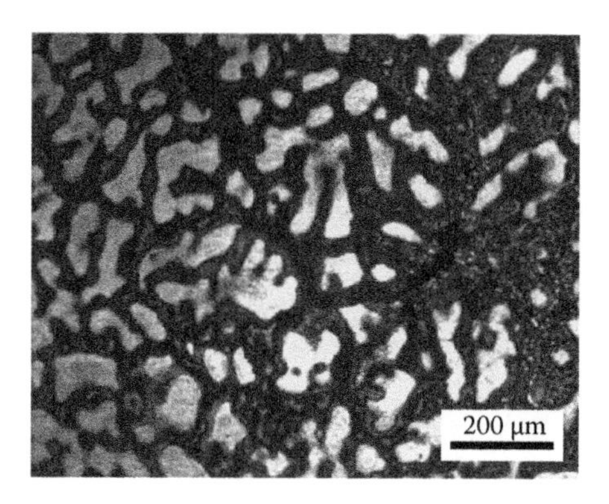

FIGURE 7.23
Microstructure of HAZ of an AZ91D arc weld. (From Luo, H., New joining techniques for magnesium alloy sheets, MS thesis, Institute of Metal Research, Chinese Academy of Sciences, June 2008. With permission.)

A continuous network of $Mg_{17}Al_{12}$ at grain boundaries in AZ91, when it melts, allows for the ejection of liquid metal from the nugget through the HAZ. Both the cross section of a small weld (Figure 7.21a) and a section along the faying interface of a weld (Figure 7.21b), made using the same welding parameters, show the existence of such a liquid network during welding. Expulsion may occur through the liquid network if the pressure in the nugget is high enough, and the network has a sufficient number of intersections at the faying interface as outlets. There are many junctions of the $Mg_{17}Al_{12}$ network with the faying interface near the nugget, and the number decreases along the interface away from the nugget, as seen in Figures 7.19 and 7.21. When there is such a liquid network, expulsion may occur even when the nugget is sealed at the faying interface by electrode squeezing, as the liquid metal could flow from the nugget through the web of molten grain boundaries, and reach the faying interfaces by circumventing the seal. An examination of the weld cross sections prove that this is what actually happened. Traces of ejected metals are observed in all directions around a weld shown in Figure 7.25a, and the debris of the ejected metal are clearly visible in the lower half of the figure covering the annular pressure ring, which is an evidence of expulsion through the faying interface. The upper half of the weld, however, appears to be sealed by electrode squeezing and the annular pressure ring is clear of residual metals. A large amount of ejected metal is found outside of the ring, possibly resulting from an expulsion through the web of molten grain boundaries,

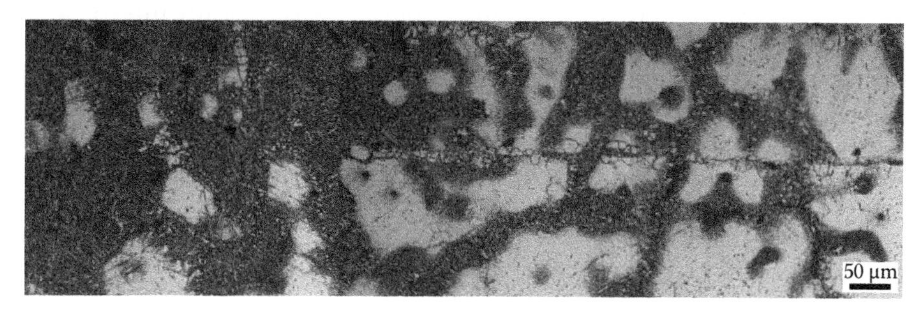

FIGURE 7.24
Microstructure of HAZ of an AZ91D spot weld.

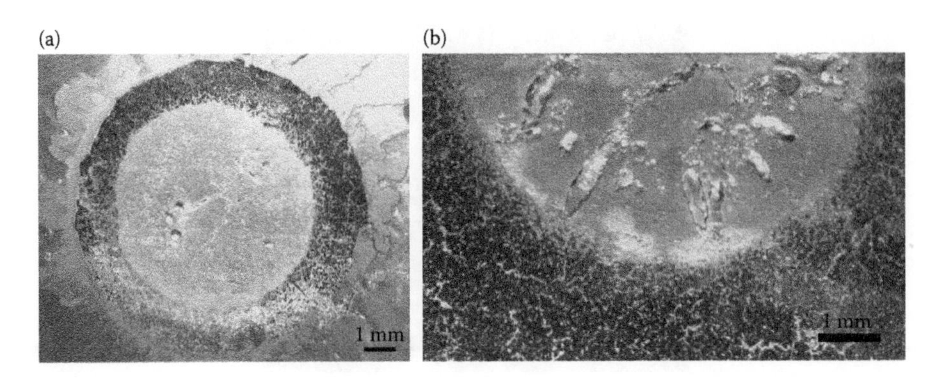

FIGURE 7.25
Interfacial expulsion in an AZ31D.

bypassing the portion of the faying interface sealed by the compressive ring. The large voids, distributed in the radial directions in the nugget shown in Figure 7.25b could be the result of such an expulsion.

7.3.3.2 Effect of Electrode Force

An effective way to deter expulsion is raising the electrode force level, as being recognized by many practitioners and explained in the force balance model of expulsion. However, it was found that increasing the electrode force was not effective in controlling expulsion in welding AZ91D, as it occurred as soon as the nugget grew to a certain size when the electrode force was set below 9 kN. When it was raised to 9 kN, the severity of expulsion was only moderately reduced.

The force balance model as described in the previous section works well in predicting expulsion by comparing the electrode force with that from the liquid nugget. According to the model, the expulsion is contained if the electrode force is bigger than that from the nugget. Whereas the electrode force can be measured directly, the force from a liquid nugget can only be approximated based on the pressure in the nugget. Depending on the chemistry, different metals have different liquid pressures and, therefore, welding these metals requires different electrode forces. If the force balance model is applicable to welding Mg, the electrode force needed would be smaller than for welding Al, as the liquid pressure is at least 20 MPa lower than in a comparable Al weld with the same amount of overheating. However, when welding AZ91D, expulsion was basically unavoidable when the electrode force was lower than 9 kN, even when the electrode force would be sufficient to suppress expulsion in similar-sized Al welds. For instance, the weld shown in Figure 7.21b has a diameter of approximately 5 mm. If a 200°C overheating (beyond the melting point) is assumed for this weld, the liquid pressure in the nugget will be about 75 MPa as estimated from Figure 7.18. The force from the liquid nugget, which is the same as the electrode force needed to contain the faying interface expulsion according to the force balance model, is 6 kN for this model. However, the electrode force used for making this weld was 9 kN, yet expulsion still occurred. Therefore, there must be mechanisms of expulsion that are not accounted for in the force balance model.

An implicit condition for applying the force balance model is the existence of a seal formed around the nugget at the faying interface in the HAZ. Making such a sealing requires both a reasonable plasticity of the sheets at the interface and a sufficiently high electrode force. Without a noticeable plastic deformation at the sheet surfaces in welding AZ91D, as discussed in the previous sections, it is difficult to create an effective seal even when the electrode force is high. The electrode force alone can only close a fraction of the apparent contact area and leave the rest open for the liquid metal to penetrate through and, therefore, the effect of electrode force is limited in controlling expulsion in welding this alloy.

7.3.3.3 **Expulsion through a Network of Liquid Grain Boundaries**

The fact that expulsion was almost inevitable in welding AZ91D even under large electrode force in the course of this investigation indicates that existing expulsion mechanisms are not effective in explaining the expulsion process in welding materials such as AZ91D.

As discussed in the previous sections, the force balance model of expulsion did not work well for welding AZ91D because of a lack of an effective sealing ring created by plastic deformation under electrode squeezing, and the existence of a liquid network made of molten eutectics at the grain boundaries. Their influence on the expulsion process in this alloy can be understood by approximating the evolution of various zones in a weldment during welding. For this purpose, one quarter of an AZ91D weldment was constructed (Figure 7.26) using actual corresponding sections from several welds. A liquid network of molten eutectics at grain boundaries around the nugget forms as soon as melting starts, as the temperature in the HAZ next to the nugget is just below the melting temperature for the alloy, but above that of the eutectics. The size of this liquid network, measured from the center of the weld, is r_{melt} (Figure 7.26), and it grows with the nugget (measured by r_{nugget}). The compressive zone generated by the applied electrode force at the faying interface surrounding the liquid nugget can be considered as a circular shape with radius r_{force} for simplicity. It is the region in which the compressive stresses are sufficiently high to seal either the liquid nugget or the intersections of the liquid eutectics network with the faying interface, if they fall within the range of this zone. Therefore, if an electrode force is too small to provide effective sealing, the compressive zone may not exist and $r_{force} = 0$, even when there are compressive stresses at the faying interface. Because $r_{melt} \geq r_{nugget}$ is always true, expulsion through the liquid network will occur before that directly through the faying interface and, therefore, expulsion happens when the liquid network outgrows the compressive zone, or $r_{melt} > r_{force}$.

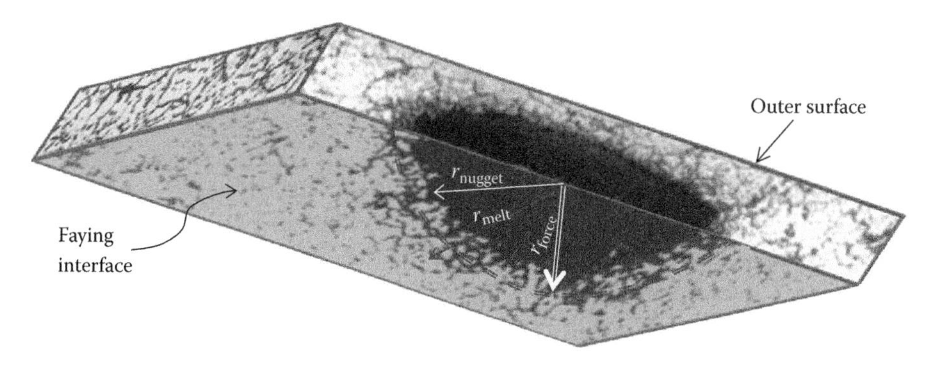

FIGURE 7.26
One quarter of an AZ91D weldment.

Therefore, expulsion in AZ91D depends on both the electrode force level and the welding stage, which determines the value of r_{melt}. Consider a growing weld nugget under a moderate electrode force. At the beginning, the partial melting zone is narrow, and it is enclosed by the compressive zone ($r_{melt} < r_{force}$). Expulsion is contained because the liquid cannot penetrate through the faying interface directly, nor through the molten liquid network. As welding proceeds, both the nugget and liquid network grow in size, and the liquid pressure and the resultant force from the nugget rise as well. Expulsion in AZ91D may occur if one of the following two conditions is met: when the force from the nugget reaches a level comparable to that of the electrode force; and when the liquid network grows beyond the compressive zone ($r_{melt} > r_{force}$). Therefore, expulsion in RSW AZ91D may occur either directly through the faying interface, or through the web of liquid grain boundaries. Whereas the former is common for all the materials, the latter is unique to materials such as AZ91D. The volume fraction of the low-melting phase ($Mg_{17}Al_{12}$ in this case) at the grain boundaries has to be large enough in order to make a connected liquid network that allows the liquid to go through.

Experiments were performed to verify the proposed mechanism of expulsion. When the electrode force was altered at 3, 5.4, and 9 kN, but the electrical current and welding time were kept unchanged, the nugget size decreased in both width and height, as shown in Figure 7.27. Electrode force is apparently beneficial to reducing the severity of expulsion, but not sufficient to prevent it from occurring, as expulsion was detected in all three welds including the one made using an electrode force of 9 kN. Another experiment was conducted to test the effect of compressive zone by using sheets clamped by two large washers, and a weld made under this confinement is shown in Figure 7.28. A significant reduction in expulsion was observed using such a confinement, which can largely be attributed to the extended range of the effective compression zone that is responsible for preventing expulsion through the liquid network. No expulsion was detected when an electrode force of 9 kN was used when welding with constraining washers.

Although electrode force alone is not sufficient in suppressing expulsion in AZ91D, a large electrode force helps to reduce the gap between the faying surfaces and enlarge the compressive zone. Therefore, a large electrode force is still beneficial in deterring expulsion as observed in this study. When the electrode force was raised to 9 kN, the proportion of welds with expulsion dropped to below 50%.

7.3.4 Statistical Modeling

Expulsion is influenced by many factors of electrical, mechanical, thermal, and metallurgical nature. The models presented in the previous sections generally deal with expulsion (or predict expulsion) with the knowledge of nugget size and other geometric factors. However, it is not always possible to obtain these quantities. Besides, random factors influencing expulsion, such as electrode alignment and workpiece fit-up, are not easy to include in the geometric comparison model and the force balance model. Welding current and time were not directly reflected in the models, which makes the use of these models inconvenient. Choosing correct welding schedules (usually electrode force, welding current, and welding time) is still the preferred way to control expulsion. With these considerations in mind, a model was proposed on expulsion prediction based on a statistical analysis.[30]

A common practice for determining a welding schedule is finding the limit of current for expulsion, with fixed electrode force and welding time. This is usually conducted in the form of lobe diagrams (an example is shown in Figure 7.29). Minimum acceptable weld sizes and expulsion limits, as functions of welding time and welding current, are

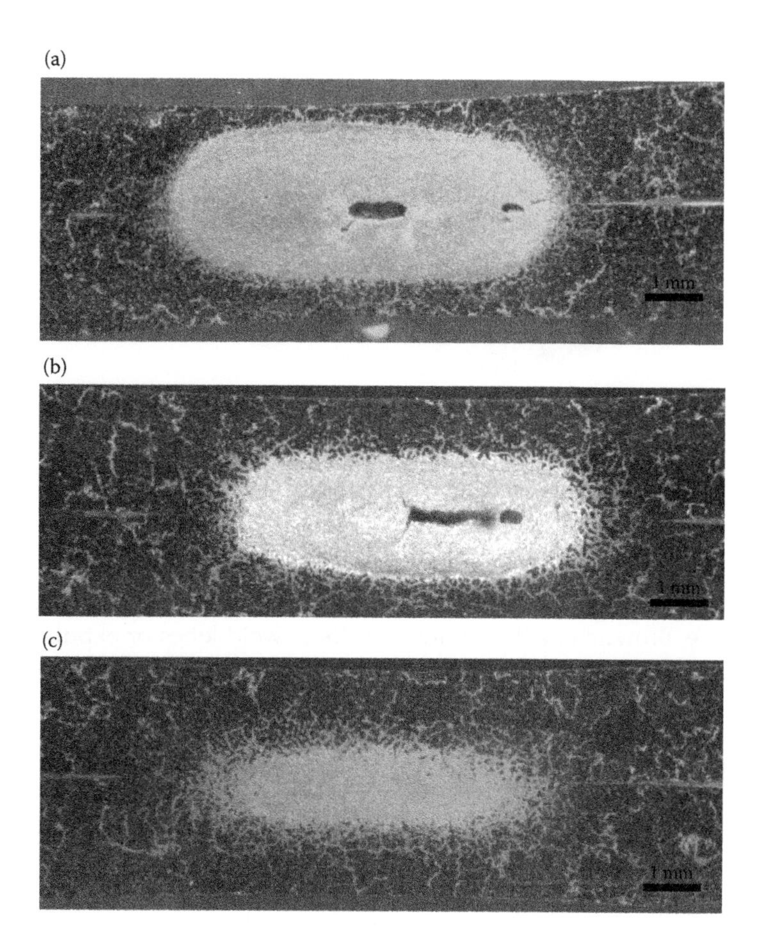

FIGURE 7.27
AZ91D welds made with electrode forces of (a) 3 kN, (b) 5.4 kN, and (c) 9 kN.

the boundaries in a lobe diagram. Although weld lobes provide a simplified description of the influence of complex interactions of mechanical, thermal, and electrical processes on expulsion limits, limited information can be obtained on the underlining physical processes of expulsion. As most lobe diagrams are created using (single) fixed electrode force, the influence of electrode force on expulsion is effectively neglected. This has been realized by many researchers and efforts have been made to reflect the influence of electrode

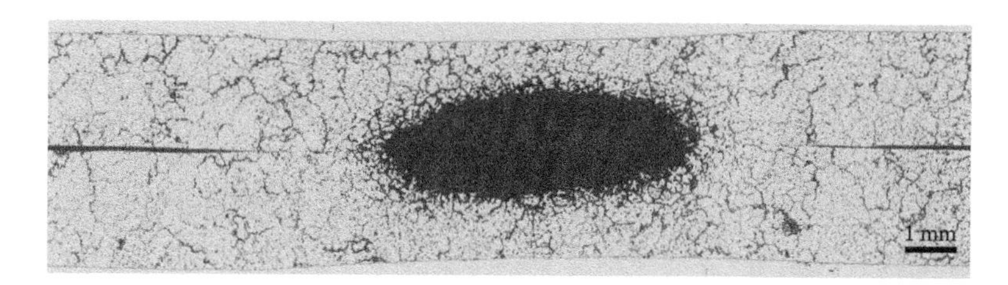

FIGURE 7.28
An AZ91D weld made with constraints.

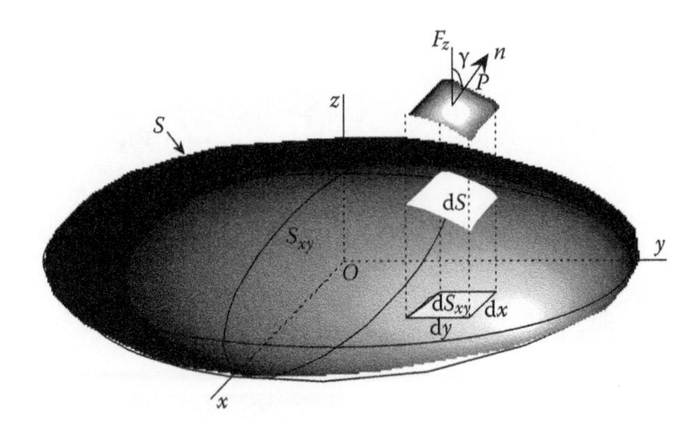

FIGURE 7.29
Schematic diagram of a liquid nugget. S is surface (boundary between solid and liquid) above x–y plane. dS is a small segment of S, and dS_{xy} is its projection onto x–y plane. Total projected area of S is S_{xy}.

force. For instance, Browne et al.[16,31] showed shifts of weld lobes or expulsion limits with electrode force in resistance welding an aluminum alloy, and their results are similar to those observed by Kaiser et al.[32] for welding low-carbon and HSLA steels. Karagoulis[2] also reported weld lobe shifting due to electrode misalignment. These results are consistent with the principles of the force balance model. There are also attempts by several researchers on treating electrode force in a similar way as welding current and time. In the work by Gould et al.,[33] three-dimensional weld lobes were created to describe the interactions of electrode force, welding current, and time in resistance seam welding. Only a limited number of electrode forces were used in this and other similar studies, partially because electrode force is believed to be less important than welding current and time, and partially because force is more difficult and tedious to change by using pneumatic cylinders. However, it is difficult to understand how welding parameters and their interactions influence the expulsion phenomenon from such analyses based on scattered data. In addition, the size of experiment matrix will be too large to handle if electrode force is taken as a variable in the same fashion as welding current and time.

In the work by Zhang et al.,[30] a statistical procedure was proposed to outline the expulsion limits by statistically analyzing experimental results. In this method, expulsion was not taken as an event happening at a single welding schedule as in previous works on expulsion limits; rather, its occurrence was treated as a probability that spans from no expulsion to 100% of welds having expulsion, considering the effects of random factors. Realizing the statistical nature of expulsion and using expulsion probability is useful in design and production where a certain percentage, rather than a definite number of conforming welds, is often more meaningful.

In this section, the outline of the statistical methodology for investigating expulsion limits is given. Examples using this method on welding an AKDQ steel, and AA5754 and AA6111 alloys are presented at the end of this chapter.

7.3.4.1 Modeling Procedure

The statistical model of expulsion includes two main parts: experiments and statistical analysis. As the model is derived based on the principle that expulsion is a phenomenon with probability of occurrence, the experiments need to be planned and conducted in accordance. Welding schedules can be selected around potential expulsion boundaries determined from prior knowledge, and adjustment on welding schedules needs to be constantly made during experiments to effectively capture the expulsion ranges. In the experiments, several electrode forces (instead of one single force) are used, and welding current and time are varied for each of the fixed electrode forces. The occurrence of expulsion is monitored and recorded in each welding test, and it is used to determine the welding parameters for the subsequent test. Expulsion can usually be recognized easily from abrupt changes in signals of dynamic resistance, electrode force, secondary voltage, or electrode displacement. Experienced visual judgment is fairly accurate as well. A certain number of replications (not less than five) are needed for each welding schedule to obtain an estimate of expulsion probability.

Once experimental data are collected, a statistical analysis can be performed to establish the analytical relationship between expulsion probabilities and welding schedules following the procedures outlined below. Statistical details can be found in Chapter 10.

The first step is to choose a proper statistical model. This model must be able to

- Explain and predict the frequency of occurrence of expulsion
- Identify important effects and estimate their magnitudes
- Describe the randomness of occurrence of expulsion

There are several statistical models that can be used to perform these tasks. A logistics model, which is ideal for dealing with continuous input and output variables of count data, is recommended.

In a logistics model, the common link function, which is used to describe the relationship between p_x (probability of getting expulsion) and x (welding schedule), is as follows:

$$\log(p_x/(1-p_x)) = f(x) \tag{7.30}$$

where $f(x)$ is a real function of x, and it can be approximated by the sum of polynomial terms of x. Substituting x by current (I), time (τ), and force (F), a cubic polynomial $f(I,\tau,F)$ can then be expressed as

$$\begin{aligned} f(I,\tau,F) \approx\ & \alpha_{000} + \alpha_{100}I + \alpha_{010}\tau + \alpha_{001}F + \alpha_{200}I^2 + \alpha_{020}\tau^2 + \alpha_{002}F^2 + \alpha_{110}I\tau \\ & + \alpha_{101}IF + \alpha_{011}\tau F + \alpha_{300}I^3 + \alpha_{030}\tau^3 + \alpha_{003}F^3 + \alpha_{210}I^2\tau \\ & + \alpha_{201}I^2F + \alpha_{021}\tau^2F + \alpha_{120}I\tau^2 + \alpha_{102}IF^2 + \alpha_{012}\tau F^2 + \alpha_{111}I\tau F \end{aligned} \tag{7.31}$$

where α_{ijk} values are the coefficients, usually called *parameters*, to be estimated using information from the experimental data. Third-order polynomial terms are used in Equation 7.31 for demonstration purposes only. The terms in the polynomial expansion are usually determined by the amount of data available and the levels of the variables. High-order terms are usually not preferred. (For details of logistics models, refer to the book by McCullagh and Nelder.[34])

Experimental data need to be transformed into a suitable form before performing statistical analysis. An orthogonal coding system can be used to translate polynomial vectors of x values into orthonormal vectors by the Gram–Schmidt process. Through a one-to-one transformation between linear combinations of polynomial terms of I, τ, and F values (current, time, and force in natural scale, respectively) and those of I_s, τ_s, and F_s values (current, time, and force in standardized scale, respectively), Equation 7.31 can be rewritten as follows:

$$\begin{aligned} f(I,\tau,F) \approx\ & \theta_{000} + \theta_{100}I_s + \theta_{010}\tau_s + \theta_{001}F_s + \theta_{200}I_s^2 + \theta_{020}\tau_s^2 + \theta_{002}F_s^2 + \theta_{110}I_s\tau_s \\ & + \theta_{101}I_sF_s + \theta_{011}\tau_sF_s + \theta_{300}I_s^3 + \theta_{030}\tau_s^3 + \theta_{003}F_s^3 + \theta_{210}I_s^2\tau_s \\ & + \theta_{201}I_s^2F_s + \theta_{021}\tau_s^2F_s + \theta_{120}I_s\tau_s^2 + \theta_{102}I_sF_s^2 + \theta_{012}\tau_sF_s^2 + \theta_{111}I_s\tau_sF_s \end{aligned} \quad (7.32)$$

where θ_{ijk} values are the coefficients with subscripts i, j, and k equal to 0, 1, 2, and 3 for constant, linear, quadratic, and cubic effects, respectively, in the order of the input variables.

The reason for using welding parameters in standardized scale is that the estimators of coefficients in a model formed by polynomial terms in the orthogonal coding system are more efficient and statistically independent. This makes a model selection procedure accurate, but at the expense of losing intuitive physical interpretation of the coefficients. The fitted model using the orthogonal coding system can be transformed back to be a function of I, τ, and F with more meaningful coefficients. Equation 7.32 is used to obtain a fitted model, and then it is transformed to obtain a model in the natural scale (Equation 7.31).

In the experiments, settings with low current and short time, as well as those with high current and long time, do not need to be used, because in these regions, expulsion either never happens (low settings) or always happens (high settings). Such information that can be obtained without conducting an actual experiment is called prior knowledge in statistics. Although there is no need to conduct experiments, information in such regions is needed when building the statistical model. It is easy to use pseudodata to represent the prior knowledge. A certain number of pseudo no-expulsion data on the low-current side and expulsion data on the high-current side can be added to the experimental data set without conducting experiments.

7.3.4.2 Statistical Analysis

After the data set is prepared, one can select an appropriate form of polynomials and their coefficients, or a model.

7.3.4.2.1 Model Selection

The polynomial terms are not equally influential; some are more important and have more influence on the output (probability of getting expulsion) than others. Insignificant effects can be screened out by means of model selection. Model selection also provides a balance between goodness-of-fit and generality, as described in Chapter 10.

The *Cp* criterion, which includes measurement of both goodness-of-fit and the number of effects, is applied to each subset of the full model. Then an appropriate model can be found by comparing the values of *Cp* for each submodel (subset of the full model).

7.3.4.2.2 *Identifying Influential Effects*

A model selected through previous steps usually contains many effects. Because of the collinearity between effects, some less important effects in the chosen model may be replaced by others and the new model still preserves the same goodness-of-fit.

The results of model selection can be used to identify important effects. Intuitively, if one effect has a strong influence on the response, it should appear in most of the good models. Therefore, the frequency of each effect appearing in most of the best models is used to identify influential effects.

7.3.4.2.3 *Estimating Magnitudes of Effects*

After choosing a statistical model by the model selection procedure described above, coefficients θ_{ijk}, magnitudes of effects of the model can be estimated. In the logistics model, the estimation is preceded by an iterative weighted least square procedure to get the maximum likelihood estimate of θ_{ijk} magnitudes.[34]

The selected model is then transformed back to the coding system of true scale (with true values of welding current, time, and force). By normalizing the welding parameters in $f(I,\tau,F)$, the influential effects in the true scale are identified. The fitted probability can be obtained by a simple transformation of $f(I,\tau,F)$ as

$$p_x = e^{f(I,\tau,F)}/(1 + e^{f(I,\tau,F)}) \tag{7.33}$$

After a statistical model is built, it usually needs to be judged by its closeness to the original data it is based on by using diagnostic methods, such as residual analysis, to see if there is any significant contradiction.

The statistical models built through the aforementioned procedure contain the influence of all welding parameters and random factors encountered during experiments, and they are important in understanding the complex phenomenon of expulsion. Although the established models are material/welding system dependent, so that they cannot be directly applied to other material systems, the methodology is generic.

In summary, the statistical model has the following characteristics:

- Contrary to traditional lobe diagrams, expulsion is treated as a dependent of both deterministic effects and random factors. Therefore, its boundary is expressed as a probabilistic range, rather than a line.
- All important welding parameters (i.e., welding current, time, and electrode force) are included in the model, unlike traditional lobe diagrams usually constructed using a fixed electrode force.
- Expulsion can be directly linked to welding parameters, which is an important advantage over other expulsion models.
- Logistics models can be developed in an orthogonal coding system containing linear, quadratic, cubic, and higher-order effects of welding current, force, and time, and their interactions. The models can be used in real-scale application after the coding systems are transformed back to natural scales.
- Influential welding parameters on expulsion can be determined using such models, and it provides guidelines for suppressing expulsion in practice.

7.3.5 Summary

Among existing models on expulsion, the geometry comparison model, force balance model, liquid network model, and statistical model have been verified and show the greatest promise in effectively controlling expulsion. As with any other models, they have both pros and cons:

- The geometric comparison model deals with expulsion without knowledge of the nugget formation and the influence of welding parameters. It captures the important geometric aspect of expulsion. However, it relies on finite element simulation to provide geometric dimensions, which seriously limits its use in practice.
- The force balance model provides a systematic procedure to calculate and analyze forces involved in RSW, and provides a criterion for expulsion. It reveals the physical processes involved in resistance welding, which leads to the understanding of the expulsion phenomenon, and it can be used to develop welding guidelines to suppress expulsion. However, it requires input from other sources for weld dimensions.
- The liquid network model describes a special expulsion phenomenon in certain materials when additional channels of liquid ejection exist, other than the faying interface. This model is particularly important in explaining certain observations of expulsion that cannot be explained otherwise. As the liquid network in the HAZ is responsible for such expulsion, the amount of the phases in the grain boundaries of the substrate must exceed certain values for this type of expulsion to occur. Therefore, this model is only applicable to certain material systems.
- The statistical model treats expulsion as a phenomenon with probability of occurrence, unlike previous works on this topic. Such a method is adequate in dealing with physical processes with uncertainty. Quantitative predictions or models on expulsion limits can be obtained by applying the generic statistical methodology on particular material/welding systems. The concept of expulsion probability provides a quantitative guideline for controlling resistance weld quality, by allowing the largest possible spot welds without expulsion.

These models describe the important aspects of the expulsion phenomenon. A proper use of these models makes it possible to not only monitor, but also control expulsion. As illustrated in Chapter 5, a feedback control system is possible by implementing the knowledge gained from the aforementioned models. Further study is needed in making the models applicable to various material systems and welding conditions.

7.4 Examples

In this section, examples of using the force balance model and the statistical model are presented in detail on several material systems.

7.4.1 Application of Force Balance Model

First, the force balance model is applied to RSW AA5754 alloy. This material is being used, on a limited scale, as a structural material in the automobile industry. In general, aluminum alloys are more sensitive to expulsion than steels, and the effect of electrode force is more significant, due to aluminum's high electrical and thermal conductivity, low melting temperature, and high thermal expansion (both in the solid and liquid states). The use of the force balance model, including calculation of forces from the liquid nugget, is demonstrated. The chemical composition of AA5754 is listed in Table 7.2.

7.4.1.1 Calculation of Pressures and Forces

The force from the liquid nugget can be calculated with knowledge of the pressure inside the nugget and the size of the nugget. All major components of pressure, that is, pressures due to solid-to-liquid transformation, liquid expansion, liquid–vapor pressure, and that due to surface agent decomposition, are dependent on temperature, except pressure due to melting. Therefore, it is important to know the temperature distribution in the liquid nugget. Once the temperature for a particular volume of liquid is known, its contribution to the pressure can be calculated by knowing the major pressure components and using the equations in Section 7.3.2.3. Numerical approximation is usually needed for calculating the total pressure and the forces. In general, the procedure is outlined as follows:

- Obtain material properties and temperature distribution in a liquid nugget.
- Divide the liquid nugget into segments and assume a constant temperature for each segment.
- Calculate pressure components in each segment.
- Sum up all pressure components in all segments and obtain the total pressure.
- Calculate forces in the directions of interest.

A temperature distribution in the liquid nugget is needed to calculate the pressure due to liquid expansion, vapor pressure, and pressure from surface agent decomposition. Temperature in the molten metal during welding is commonly perceived as nonuniform, as shown by Gould.[10] To the contrary, Alcini[36] claimed that temperature is uniform throughout the molten nugget. A rough approximation of possible temperature distribution in the nugget can be obtained from the results of a finite element simulation[25] as shown in Figure 7.30. Isotherms are approximated by ellipsoidal surfaces (or ellipses for two-dimensional cases). This is a nugget formed after 160 ms of welding for an AA5754 with a welding current (DC) of 28 kA and an electrode force of 7 kN. Fluid dynamics is not

TABLE 7.2

Chemical Composition (wt.%) of Commercial AA5754

Mg	Mn	Cu	Fe	Si	Ti	Cr	Zn
2.6–3.6	Max. 0.5	Max. 0.1	Max. 0.4	Max. 0.4	Max. 0.15	Max. 0.3	Max. 0.2

Source: *Automotive Sheet Specification*, Alcan Rolled Products Company, Farmington Hills, MI, 1994. With permission.

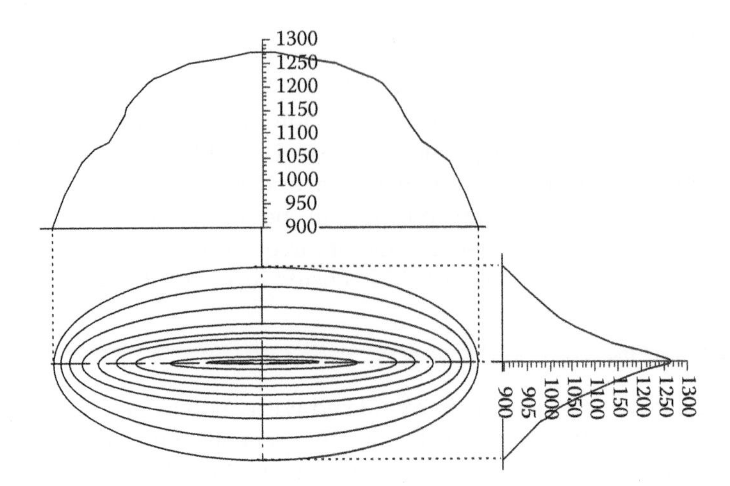

FIGURE 7.30
Temperature distribution predicted by an FEM.[25] Elliptic rings are isotherms in liquid nugget. Temperature is in degrees Kelvin (K).

considered in the calculation; therefore, this temperature distribution is an approximation of the real case. Temperature gradients are different along the width and height directions. For simplicity, the contribution from the liquid–vapor pressure and that from the surface agent decomposition are ignored, and the pressure components can be calculated based on the temperature distribution from the finite element simulation as

$$P_{\text{melt}} = 74.05 \text{ MPa}, P_{\text{exp.}} = 18.49 \text{ MPa}, P_{\text{vapor}} = 0, P_{\text{lubr.}} = 0$$

and the ratio of pressures is $P_{\text{melt}}/P_{\text{exp}}/P_{\text{vapor}}/P_{\text{lubr.}} = 80.02{:}19.98{:}0{:}0$ (%).

The calculated force due to the liquid pressure in the nugget height direction is 2.62 kN. To understand how the force from the nugget is related to temperature distribution and overheating (heating above the melting temperature), a comparison is made for three types of temperature distributions, as shown in Figure 7.31, with the same nugget size and

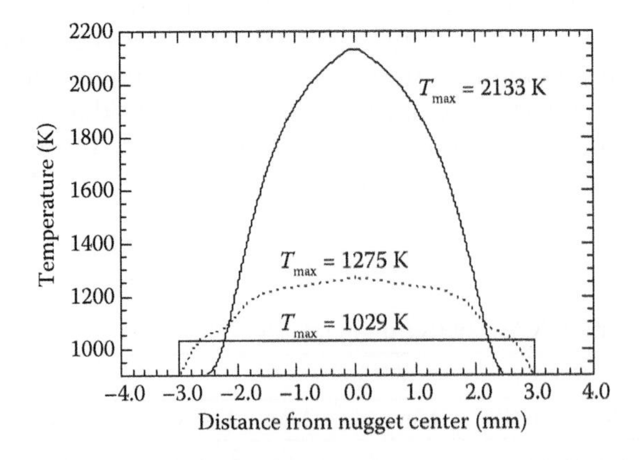

FIGURE 7.31
Temperature distributions in a nugget. Dashed line is taken from an FEM simulation,[25] and solid lines are assumed temperature distributions.

TABLE 7.3

A Comparison of Force, Pressure Ratio, and Heat for Various Temperature Distributions

Temperature Distribution	F_N (kN)	Pressure Ratio ($P_{melt}/P_{exp}/P_{vapor}/P_{lubr.}$)	Q (J)
FEM output	2.62	80.02:19.98:0:0	15.99
Bell-shaped	2.65	79.05:20.95:0:0	15.99
Uniform	2.61	80.09:19.91:0:0	16.02

same amount of heat in the nugget. The dashed line is the temperature distribution from the FEM, with a peak temperature of 1275 K. The other two lines are for a bell-shaped and a rectangular distribution, with peak temperatures of 2133 and 1029 K, respectively. The results are shown in Table 7.3.

From this table, it is easy to see that although temperature distributions and peak temperatures in Figure 7.31 are quite different, the total pressures in the nuggets are very similar because the overheating is similar for the same amount of total heat, and so are the forces from the liquid nugget. Therefore, a detailed and accurate temperature distribution is not of importance for force calculation. In fact, force from the nugget is directly linked to overheating. As shown in the cases in Table 7.3, the major contributions to the total pressure come from the volume change due to melting and from the expansion of liquid metal after melting. The latter is proportional to the overheating. Once the total overheating is obtained, average temperatures, which can be calculated by assuming a uniform temperature distribution, can be used for force calculation.

Figure 7.32 shows the convergence of the nugget force with the number of calculation shells. Each shell is made of ellipsoidal inner and outer surfaces. From all cases shown in the figure, 150 shells are sufficient for both force and overheating calculation. Figure 7.33 shows increases of force and overheating, with an average temperature in a nugget of 6.0 mm in diameter and 2.5 mm in height. There is approximately a 0.5-kN force increase for every 100 K rise in temperature.

The dependence of force from the nugget and overheating on the nugget size is shown in Figure 7.34. The height of the nugget is assumed to be half of its width, and a fixed average

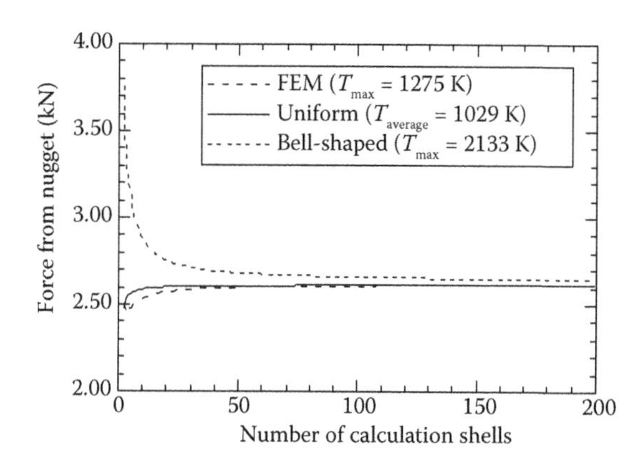

FIGURE 7.32
Force from nugget vs. number of shells used in calculation.

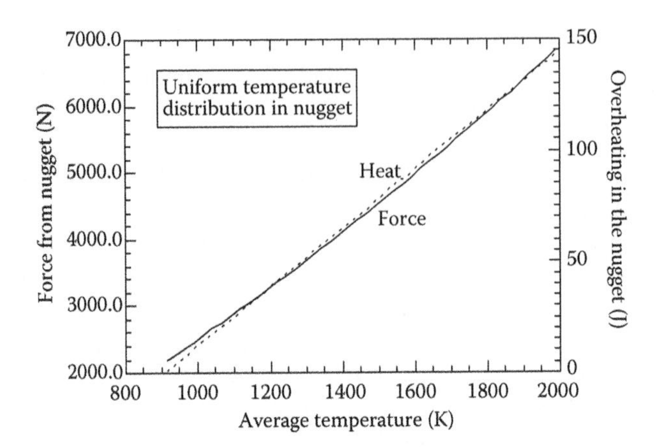

FIGURE 7.33
Nugget force and overheating of a nugget vs. average temperature in a nugget. A uniform temperature distribution is assumed. A constant nugget size of 6 mm in diameter and 2.5 mm of penetration are used.

temperature of 1275 K is used for all nugget sizes. Figure 7.34 shows an increase of force, as well as overheating, with nugget size. This is the force that has to be overcome by electrode forces to avoid expulsion. If the electrode force and the force from the nugget are not aligned, an effective electrode force, which is a fraction of the applied electrode force, must be large enough to balance the force from the nugget.

7.4.1.2 Experimental Verification

To verify the expulsion model, experiments were conducted on an aluminum alloy AA5754, the same material used for the calculation in the previous section.[9] It was chosen for verification of the model because aluminum alloys are generally more prone to expulsion than

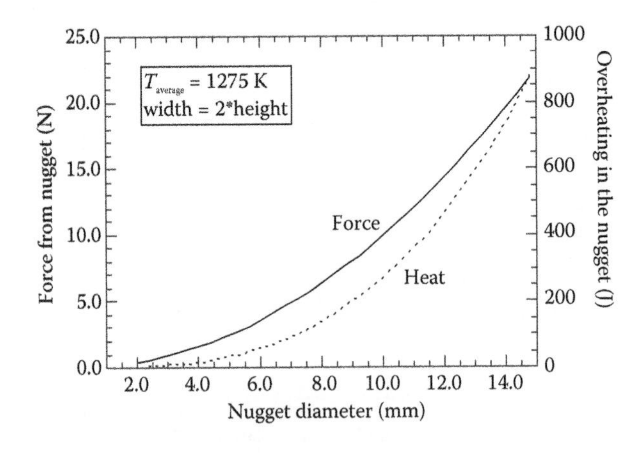

FIGURE 7.34
Dependence of nugget force and overheating on weld diameter. A uniform temperature distribution is assumed in nuggets.

steels. The sheets were supplied by Alcan Aluminum Company and treated by an Alcan surface treatment technique to ensure a repeatable surface condition. A pinch gun with an MFDC transformer was used in the experiments. The welding parameters were chosen to cover a wide range of possibilities. Electrode force was in the range of 2–9 kN, electrical current was between 20 and 35 kA, and welding time was varied between 67 (4 cycles) and 167 (10 cycles) ms.

During welding, the occurrence of expulsion was monitored. Welded samples were then cut, ground, polished, and etched following standard metallographic procedures. The nugget diameter, size of electrode indentation, and offsets between the nugget center and indentation center were measured by an optical microscope. Based on the results of the FEM, shown in the previous section, the average temperature for each weld when it was in a liquid state was estimated. An assumption was made that the average temperature in the liquid nugget is proportional to the total heat input after deducting the heat necessary to melt the nugget. This approximation avoids all thermal–electrical details during the welding process. However, the possible error caused by this assumption is small, as the contribution of pressure because of liquid expansion is usually less than 20% of the total pressure, as shown in the previous section. The force from a liquid nugget can then be calculated, using the weld dimensions and the average temperature, through the procedures outlined in the previous section. Equation 7.3 was then used to calculate the effective electrode force.

The results of calculated force from the nugget vs. effective electrode force are plotted in Figure 7.35. The diagonal line represents the equilibrium boundary between the two forces, and expulsion is expected when a point falls on or below this line according to the model. As shown in the figure, most expulsion points are below the boundary, some are near the line, and several points with expulsion are above the diagonal. There are a few possible explanations for this discrepancy. First, it is not always possible to accurately measure the size of nonsymmetrical nuggets by metallographic techniques. This might be the biggest contributor to the error, as the force from the liquid metal is proportional to the

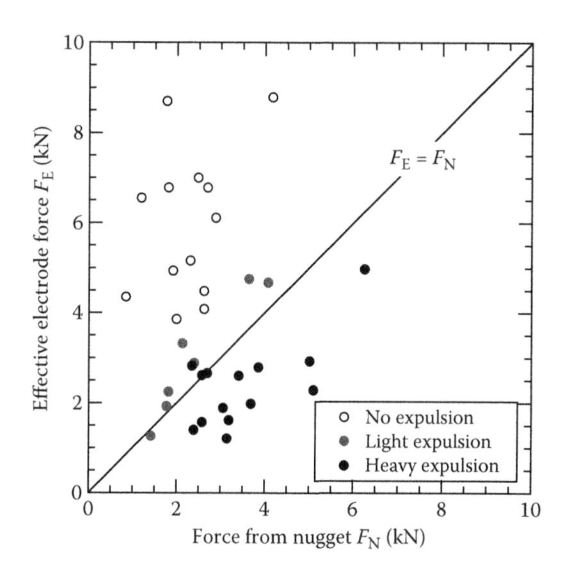

FIGURE 7.35
Comparison of model prediction and experimental observation.

size of the nugget. The offset between the applied electrode force and the force from the nugget is also difficult to determine accurately. Furthermore, RSW is very dynamic and many processes during welding are far from equilibrium, and random factors in welding are difficult to predict or control. Another contributing factor is that there are no well-accepted qualitative or quantitative definitions of expulsion and its severity. Considering these factors, the experiments fit well with the model's predictions.

Some welds had asymmetric nugget growth and asymmetric indentation marks (the depth of indentation marks on the surface is uneven across the surface), which are possibly the results of the characteristic closing (squeezing) of pinch guns and aluminum's sensitivity to contact resistance. This observation also confirms that it is important to distinguish the effective electrode force from the nominal electrode force; that is, they are different when there is asymmetric nugget growth.

7.4.1.3 Perspective Applications

The results of this model provide guidelines for the selection of electrode force. Figure 7.36 shows the levels of electrode force needed to suppress expulsion, with the uncertainty of welding conditions taken into account. An average temperature of 1275 K in the liquid nugget is assumed for an AA5754. In an ideal welding situation, that is, the electrodes are aligned and the fit-up is good, the offset between the applied electrode force and the force from the nugget is zero. This provides the minimum electrode force needed to contain the nugget, and is represented by the solid line as the expulsion–no-expulsion boundary. However, this kind of ideal welding conditions rarely exist, and the degree of nonconformity varies drastically in practice. If an offset of a quarter of a nugget's width is assumed, the applied electrode forces to contain the same nuggets are twice their minimum values, as shown by the dashed line in Figure 7.36. Assuming that this is the worst case, then if the electrode force is chosen in zone A, the probability of expulsion is low. In zone C, the risk of expulsion is high. Whether it happens in zone B depends on the welding condition. In experiments conducted to verify the model, offsets of 0.01 to 0.38 × (nugget width) have been observed. The same set of data was also plotted in the figure. There is no expulsion for the data points falling in zone A, and expulsion for all in zone C. All possibilities (no

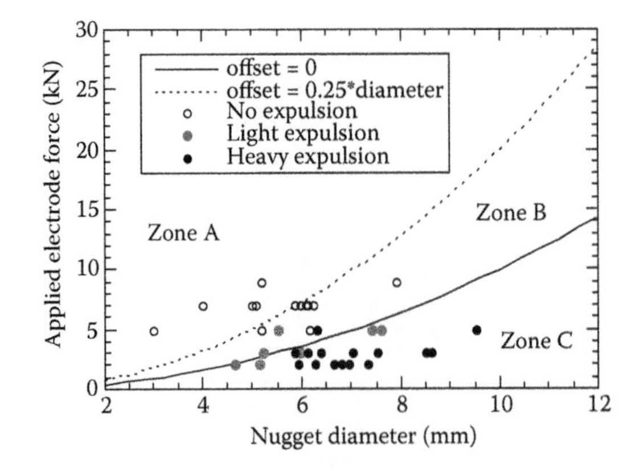

FIGURE 7.36
Estimated electrode force to suppress expulsion in an AA5754 alloy. Solid dots and circles are data from the experiment, as shown in Figure 7.35.

expulsion, light expulsion, and heavy expulsion) are observed in zone B. Therefore, the experimental results have basically confirmed the model prediction. In addition to expulsion, other factors, such as the influence of the electrode force on electrode indentation and contact resistance, should also be considered when choosing an electrode force.

Hence, it is possible to use this model to control expulsion, although its use requires knowledge of the applied electrode force, the force from the liquid nugget, and the offset. It is important to know the material properties, especially those at elevated temperatures, in the application of the model. Some of them are difficult to calculate accurately, but can be approximated using existing databases.

7.4.2 Examples of the Use of Statistical Model

In the work by Zhang et al.,[30] both steel and aluminum alloys were used to verify the statistical model. The experiments were planned, conducted, and analyzed using the procedures outlined in Section 7.3.4.

7.4.2.1 Experiments

7.4.2.1.1 Steel

A bare AKDQ steel of 1.2 mm gauge was used. Its chemical composition is shown in Table 7.4. A single-phase AC pedestal welder was used for welding. The experiments contained a large number of combinations of welding current, time, and electrode force. Welding current (root mean square value, RMS) ranges from 6.5 to 13.9 kA, welding time from 133 to 400 ms (8–24 cycles), and electrode force from 2.7 to 5.3 kN (600–1200 lb). There were a total of 76 welding parameter combinations (runs) with 10 replicates each.

7.4.2.1.2 Aluminum Alloys

Aluminum sheets of AA5754 (2.0 mm) and AA6111 (1.0 mm) were supplied by Alcan Aluminum Company and treated by an Alcan surface treatment technique to ensure a repeatable surface condition. AA5754 is currently used for structural components, and AA6111 is used for enclosures in selected automobile models. The chemical compositions provided by the producer are listed in Table 7.2 (AA5754) and Table 7.5 (AA6111). A pinch gun with an MFDC transformer was used in the experiments. The welding parameters were chosen to cover a wide range of possibilities. For AA5754, electrode force = 2 to 9 kN (450–2000 lb), electrical current = 20 to 35 kA, and welding time = 67 to 167 ms (4–10 cycles). For AA6111, electrode force = 2 to 6 kN (450–1350 lb), electrical current = 5 to 40 kA, and welding time = 17 to 84 ms (1–5 cycles). The experiments on AA5754 had 35 runs with five replicates each, and those on AA6111 had 132 runs with five replicates each.

Expulsion was monitored during welding. The occurrence of expulsion was calculated in terms of the percentage of welds with expulsion for the replicates at a fixed welding

TABLE 7.4

Chemical Composition (wt.%) of the AKDQ Steel Tested

C	Mn	P	S	Si	Cu	Ni	Cr	Mo	Sn	Al	Ti
0.035	0.210	0.006	0.011	0.007	0.020	0.009	0.033	0.006	0.004	0.037	0.001

Source: Provided by National Steel Corp., Livonia, MI. With permission.

TABLE 7.5

Chemical Composition (wt.%) of Commercial AA6111-T4

Mg	Mn	Cu	Fe	Si	Ti	Cr	Zn
0.5–1.0	0.15–0.45	0.5–0.9	<0.4	0.7–1.1	<0.10	<0.10	<0.15

Source: Kaiser, J.G. et al., *Welding J.*, 61, 167-s, 1982. With permission.

schedule. The experimental results were then used in the statistical analysis and modeling of expulsion limits for these materials.

The model selection criteria described in Section 7.3.3.2 are then used to choose the best statistical models for these materials. Many models can be chosen based on the *Cp* criterion with a small *Cp* value, but they may not be able to reflect the underlying physical process. Physical consideration has to be made when choosing the best model. For instance, both of the models for the steel shown in Figure 7.37 have small *Cp* values, but the trend shown in (a) is slightly contrary to, whereas the one in (b) is consistent with practical experience. Therefore, the model shown in Figure 7.37b is a better choice.

Following the procedures outlined in Section 7.3.3.2, the coefficients θ_{ijk}, magnitudes of effects of the models, were estimated by an iterative weighted least square procedure to get the maximum likelihood estimate of θ_{ijk}. After transforming from the standardized coding systems back to the coding systems of natural scales (with true values of welding current, kA; time, ms; and force, kN), the probabilities of expulsion for the steel and aluminum alloys can be expressed explicitly as follows.

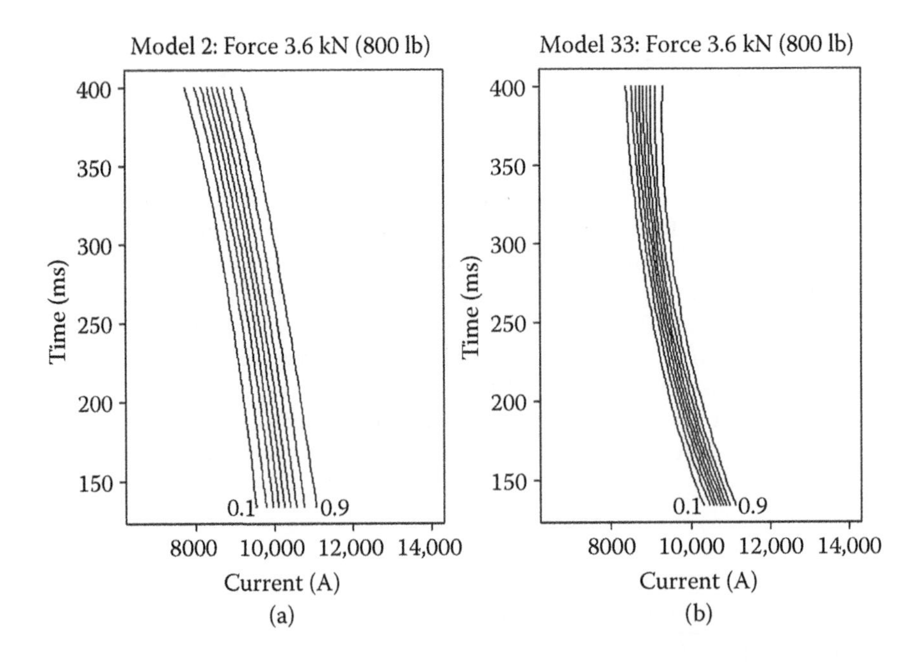

FIGURE 7.37
Contours of two models with similar small *Cp* values. Probabilites of expulsion between 0.1 and 0.9 are shown in plots.

7.4.2.1.2.1 AKDQ Steel

$$
\begin{aligned}
\ln\left(p_x/(1-p_x)\right) \approx\ & (-7.6449\times 10^{2}) + (1.6731824\times 10^{2})I \\
& + (7.12636\times 10^{-1})\tau + (9.7174\times 10^{1})F \\
& + (-1.54168327\times 10^{1})I^2 + (-1.49\times 10^{-5})\tau^2 \\
& + (-4.234\times 10^{1})F^2 + (6.251982\times 10^{-1})I^3 \\
& + (1.4202468)F^3 + (-1.540455\times 10^{-1})I\tau \\
& + (8.088965)IF + (6.08688\times 10^{-2})\tau F \\
& + (7.5306\times 10^{-3})I^2\tau + (-1.4449971)I^2 F \\
& + (-5.12\times 10^{-5})\tau^2 F + (2.6919807)IF^2 \\
& \equiv f(I,\tau,F)
\end{aligned} \tag{7.34}
$$

The influential effects in the true scale are identified as (in the order of importance) I^2, I^3, I, I^2F, IF^2, $I\tau$, $I^2\tau$, F^2, and IF.

7.4.2.1.2.2 AA5754

$$
\begin{aligned}
\ln\left(p_x/(1-p_x)\right) \approx\ & (-2.372\times 10^{1}) + (2.172)I \\
& + (3.9\times 10^{-3})\tau + (6.56\times 10^{-1})F \\
& + (-2.79\times 10^{-2})I^2 + (-1.4\times 10^{-3})I\tau \\
& + (-1.796\times 10^{-1})IF + (-8.0\times 10^{-4})\tau F \\
& + (2.7\times 10^{-3})I^2F + (3.0\times 10^{-4})I\tau F \\
& \equiv f(I,\tau,F)
\end{aligned} \tag{7.35}
$$

The influential effects are identified as I, F, IF, I^2, I^2F, and τF.

7.4.2.1.2.3 AA6111

$$
\begin{aligned}
\ln\left(p_x/(1-p_x)\right) \approx\ & (-1.394\times 10^{1}) + (1.15\times 10^{1})I \\
& + (9.81\times 10^{-2})\tau + (-3.15\times 10^{-1})F \\
& + (-1.69\times 10^{-2})I^2 + (6.17\times 10^{-2})F^2 \\
& + (-6.3\times 10^{-3})I\tau + (-1.37\times 10^{-1})IF \\
& + (1.0\times 10^{-4})I^2\tau + (2.8\times 10^{-3})I^2F \\
& \equiv f(I,\tau,F)
\end{aligned} \tag{7.36}
$$

The influential effects are I, IF, I^2F, $I\tau$, I^2, and $I^2\tau$. It is noteworthy that models for AA5754 and AA6111 have very similar influential effects, which reveals the similarity in welding aluminum alloys.

Residual analyses on these models showed reasonable agreement between the observed and fitted values.

7.4.2.2 Discussion

The statistical models shown in Equations 7.34 through 7.36 present the probability of expulsion as a continuous function of welding time, current, and force. There are a few observations that can be made by examining the results predicted by these statistical models.

Figure 7.38 shows a comparison of influencing factors in steel and aluminum welding. The influence of a factor is represented by its relative value (percentage) in its own group of influencing factors. Factors of values less than 5% are not presented. As shown in the figure, welding current-related effects are most influential in both steel and aluminum welding, largely because Joule heating is the basis of resistance welding. Electrode force-related effects are the second most influential in determining expulsion. That is especially true in steel welding. It is also interesting to see that welding time has the smallest influence.

Three-dimensional surface plots of the fitted p_x as a function of welding current and time are presented in Figure 7.39 derived from the models. They can be regarded as three-dimensional lobe diagrams of expulsion. There are two plateaus in the surface plots for all three models. One is on the low-setting side with zero probability of expulsion, and the other on the high-setting side with a probability of 1. Between these two, there is a transition zone, in which the probability of expulsion changes continuously from 0 to 1. Therefore, an expulsion limit is not presented as a boundary or a line—as is done in most research works on this subject; rather, it is presented as a range. Although weld lobes have been widely used for selecting welding parameters, there has not been a well-accepted procedure to determine expulsion boundaries. This is partially because there is no clear line between no-expulsion and expulsion. Because the occurrence of expulsion appears random in the transition region, it is reasonable and practical to treat expulsion statistically with occurrence probability.

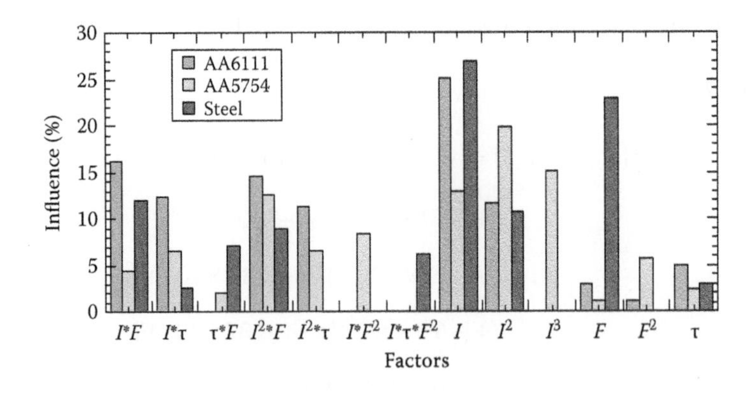

FIGURE 7.38
A comparison of influencing factors in models.

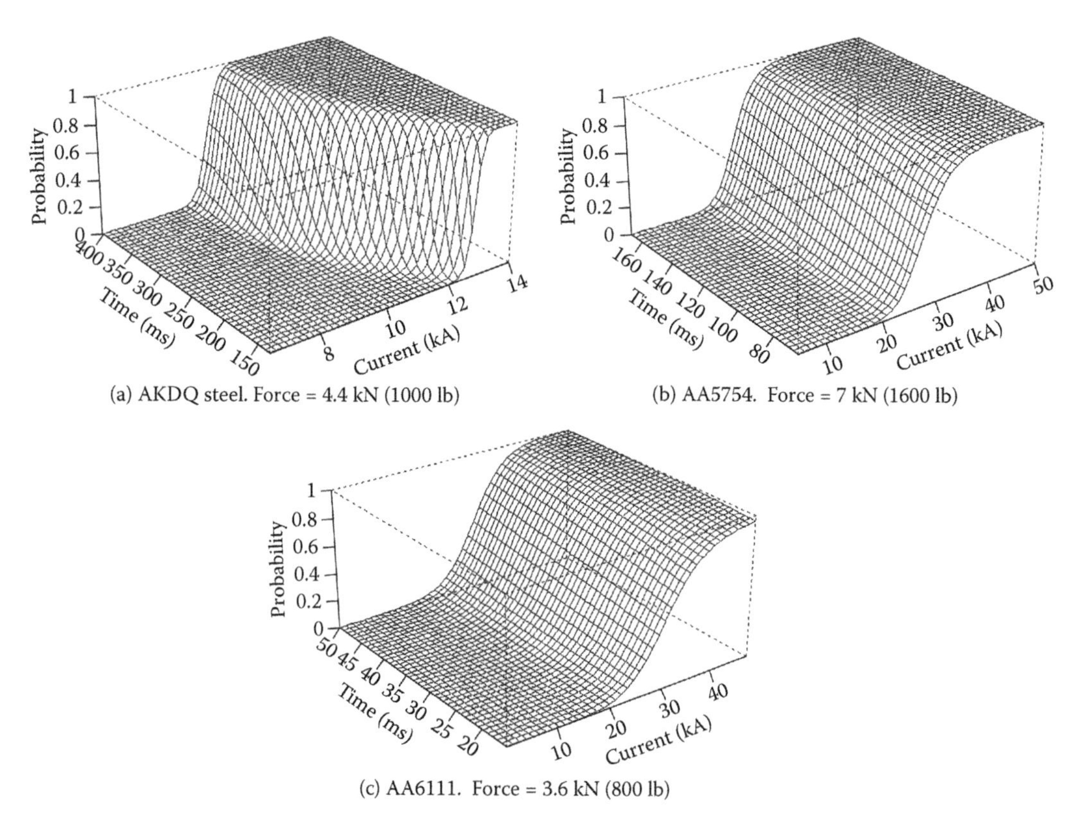

(a) AKDQ steel. Force = 4.4 kN (1000 lb)

(b) AA5754. Force = 7 kN (1600 lb)

(c) AA6111. Force = 3.6 kN (800 lb)

FIGURE 7.39
Surface plots of expulsion probability at fixed forces.

The shape of expulsion probability surfaces for the steel (Figure 7.39a) is different from those of aluminum alloys (Figure 7.39b and c), as shown in the different transition regions in the plots. Expulsion limits are dependent on welding time in steel welding, the same as was observed experimentally by other researchers, such as Kaiser et al.[32]; however, welding time is less influential in aluminum welding, as shown by Browne et al.[31] An obvious difference between expulsion limits of the steel and aluminum alloys is that expulsion boundaries (transition zones from no-expulsion to all-expulsion) of aluminum alloys are generally wider than those for the steel, which means that welding aluminum alloys has a larger uncertainty in terms of expulsion in the ranges selected for the experiments. The transition appears smoother for AA6111 than for AA5754. These are more evident from the contour plots of expulsion probabilities of 0.05 and 0.95 in Figure 7.40 created using the same models. In the figure, expulsion boundaries move to the right side as electrode force increases. This phenomenon is similar to that observed by Kaiser et al.[32] and is consistent with the principles of the force balance model, in which expulsion is directly linked to the effective electrode force.

In the statistical study of aluminum welding experiments, the authors also found that expulsion behavior depends strongly on electrode conditions. Because of the high affinity of copper for aluminum, electrode surfaces can be easily contaminated by aluminum pickup, and the copper–aluminum alloy (or α-phase bronze) was observed to form on the electrode face. As it has a lower yield strength and can be easily squeezed out or deposited

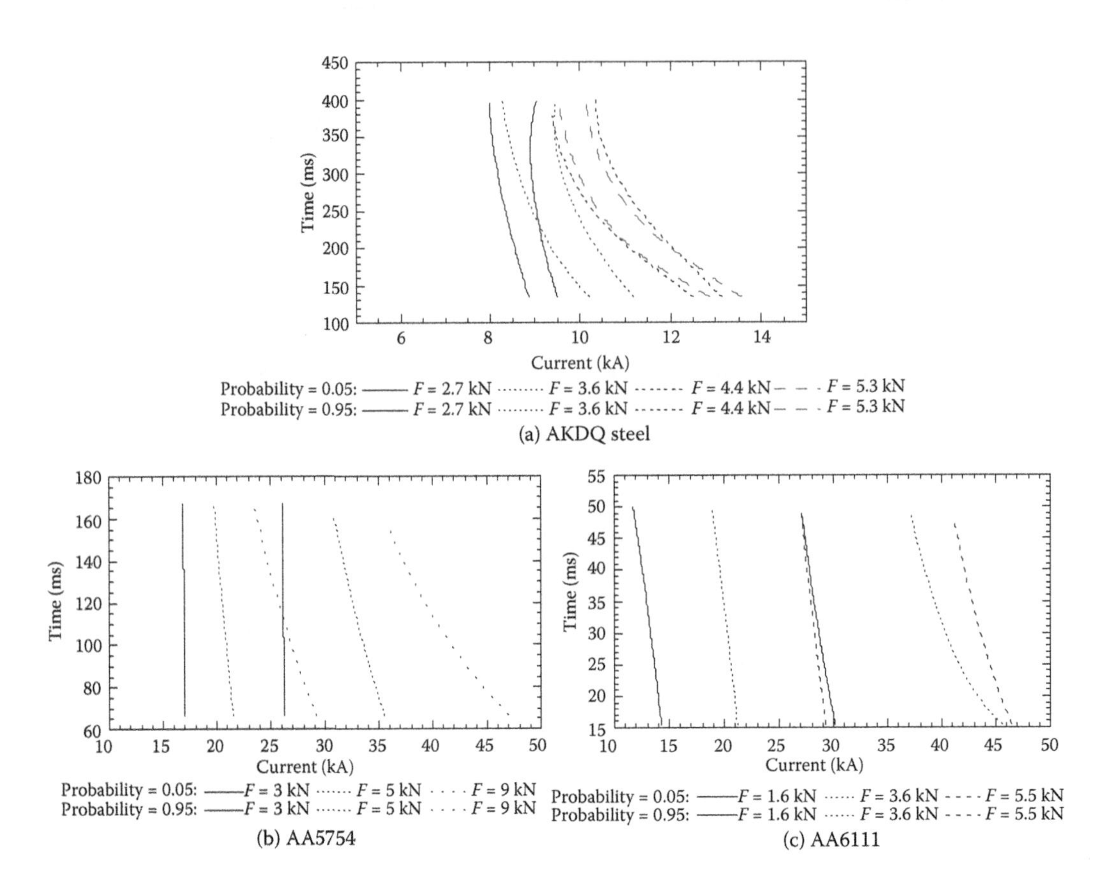

FIGURE 7.40
Contour plots of expulsion limits at various electrode forces.

onto the weld surface during welding, the electrode face changed constantly between welding, producing certain random effects on the welds created, and expulsion occurrence. All these factors contribute to the uncertainty, or wide ranges, of expulsion probability in aluminum welding.

The effect of electrode force on expulsion limits is significantly larger for Al welding than for steel welding. This can be attributed to the fact that steel welding is dominated by bulk electrical resistance, whereas surface resistance is dominant in Al welding. Electrode force or pressure has little, if any, influence on bulk resistance, but has a significant effect on surface resistance. It is directly responsible for breaking aluminum oxide (Al_2O_3) layers on the surfaces of workpieces. Therefore, contact resistance, which provides most of the overall resistance in Al welding, is a strong function of electrode force. Figure 7.41 shows how total resistance changes with electrode force in welding AA6111, using dynamic resistance signals recorded during welding experiments. It clearly shows that for the same welding time and current, a larger electrode force results in lower total dynamic resistance in aluminum alloys by lowering the surface resistance. Similar conclusions can be drawn from the works by Auhl and Patrick[37] and Patrick and Spinella[38] on the influence of surface characteristics on aluminum welding.

It is also interesting to see the dependence of expulsion probability on welding force and current (Figure 7.42), as opposed to its dependence on welding current and time (Figure 7.39). In general, expulsion depends on both electrode force and welding current. Treating electrode force as a continuous variable in the models enables a better understanding of its influence in welding. An increase in electrode force reduces the chance of expulsion. However, there is a strong interaction between the electrode force and welding time for the steel in terms of influence on expulsion, and a weaker interaction for Al alloys, as shown by the shape changes in surfaces in Figure 7.42. Electrode force becomes less important for the steel welding when welding time is long. This is primarily because the weld nugget size becomes influential, in addition to the electrode force, in controlling expulsion when the nugget size is close to the size of the electrode face after sufficient heating. The trends in surfaces of expulsion probability for AA5754 and AA6111 (Figure 7.42c to f) are similar to those in the steel. Both electrode force and welding current affect expulsion probabilities with less magnitude than in the case of steel welding. Generally, the influence of welding time is very limited in Al welding, consistent with the observations from Figure 7.39. An examination of displacement signals shows that expulsion during Al welding often occurs at an early stage, and therefore, total preset welding time has little influence.

Generally, welding time has far less significance in influencing expulsion than the other parameters. A closer look at how expulsion boundaries depend on welding current and electrode force provides certain insights on the expulsion process. Figure 7.43 shows expulsion probabilities of 0.05 and 0.95 as functions of welding current and electrode force. For a fixed welding time, the electrode force needed to contain expulsion increases with current. However, this increase is not constant and it is drastically different in welding steel and aluminum alloys.

According to the characteristics of the curves in the figures, the effect of electrode force can be categorized into three types, in the form of various current zones as marked in the figures. Take Figure 7.43a as an example. In zone I, when current is low, the influence of random factors such as asperity and workpiece fit-up is significant. Therefore, a small variation in current may induce significant changes in expulsion probability. However, the influence of random factors or variations can be suppressed by increasing electrode force. In zone II, the dependence of expulsion on force and current is more predictable. A window exists in this zone for welding with a high degree of certainty of expulsion occurrence. In this region, the force balance expulsion model proposed by Senkara et al.[9] should dominate. When welding current further increases, the influence of electrode force on

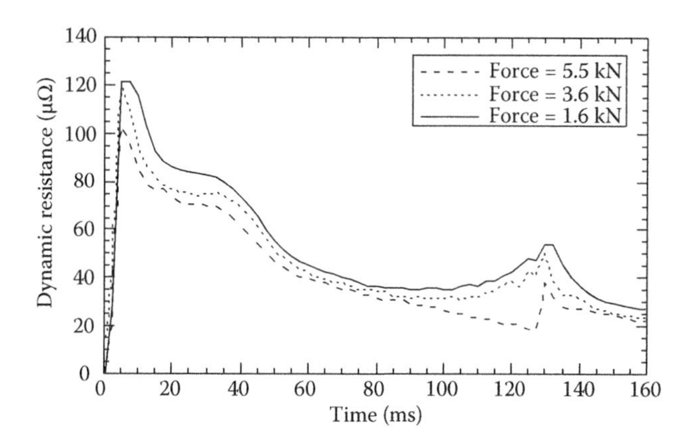

FIGURE 7.41
Dynamic resistance vs. welding time for AA6111. Welding time and current were the same, whereas electrode force was altered.

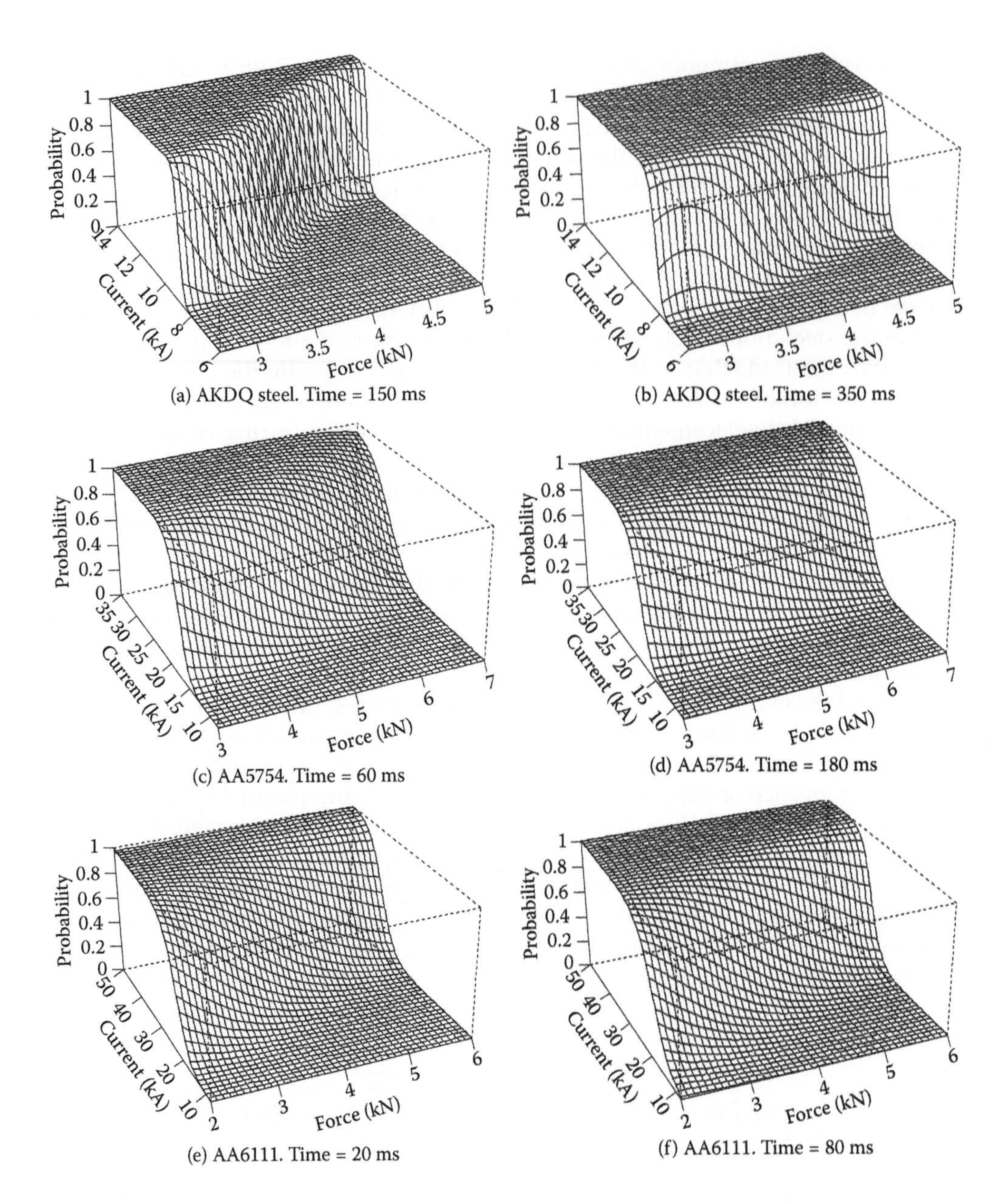

FIGURE 7.42
Expulsion probability vs. welding current and electrode force.

expulsion diminishes, as in zone III. This is primarily due to the interaction between electrodes and the nugget during welding. A large current (for a fixed welding time) produces a large weld nugget, and the electrode force will not be able to contain the liquid nugget, no matter how big the force is, once the edge of the nugget grows beyond the containment of the electrode. This may happen when the nugget size is comparable to the compressive zone by the electrode, or when part of the nugget is beyond the compressive region created by the electrode force. Therefore, the fundamental concept of the geometry comparison model is consistent with the expulsion behavior in zone III. In general, the welding of aluminum alloys has a larger uncertainty of expulsion than the welding of steel as they possess significantly narrower zone II, as shown in Figure 7.43.

In spite of a large amount of research efforts, the mechanisms of expulsion and the influence of expulsion on weld quality have not been well understood. The statistical model appears to be especially appropriate in this situation as it directly links expulsion to welding schedules without a detailed, quantitative understanding of the physical process, and some insights of the expulsion phenomenon can be obtained. It shows clearly that welding steel has different characteristics from welding aluminum alloys. Adjusting electrode force is generally an effective way to control expulsion, which is consistent with the basis of the force balance model. When the nugget grows to a size close to that of the electrode face, increasing electrode force is no longer effective, and the geometry comparison model is more applicable.

As a routinely observed phenomenon in RSW, expulsion is not desirable because of its adverse effects on weld strength, electrode wear, weld appearance, weld quality, weld performance, and others. Although welds with expulsion, especially steel welds, can

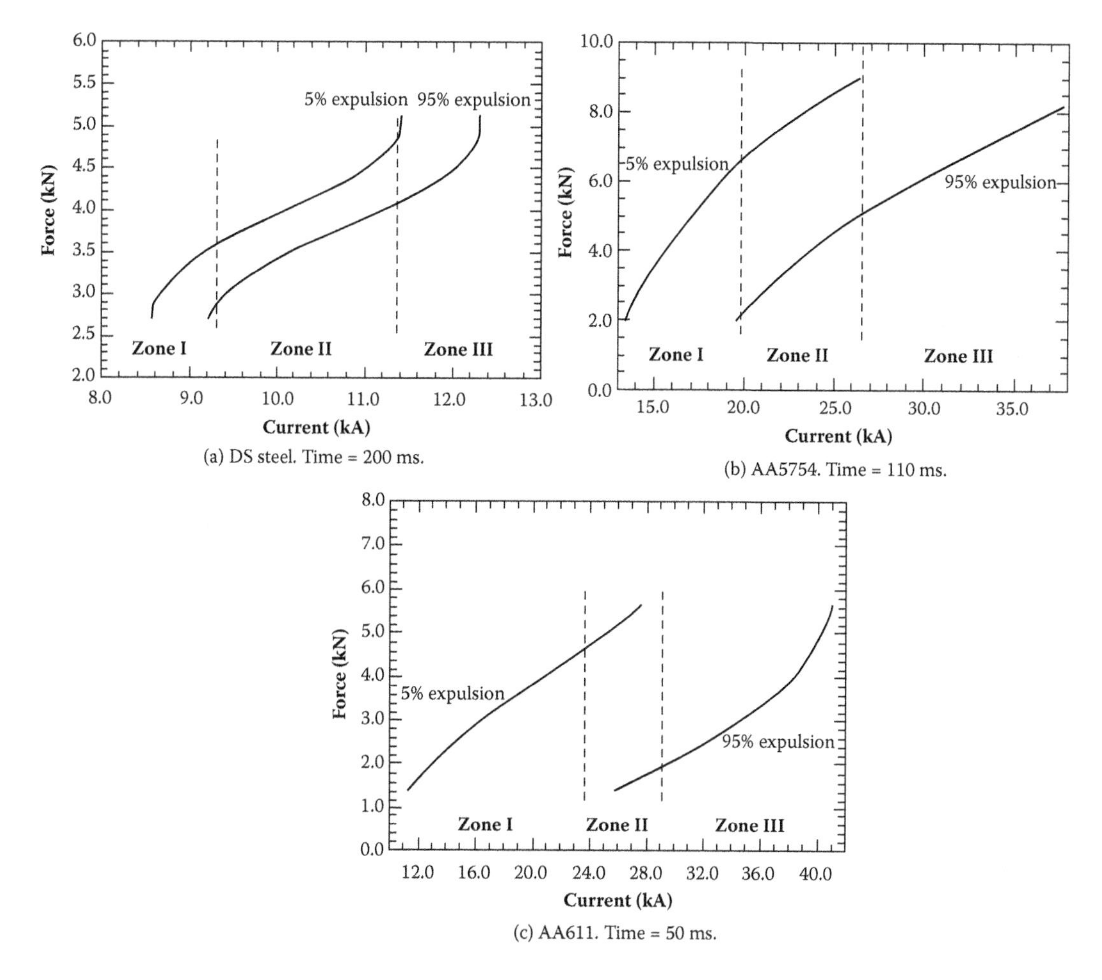

FIGURE 7.43
Electrode force vs. welding current for expulsion limits.

often meet or exceed industrial requirements, it does not justify the practice of welding deliberately at or over the expulsion limits. Expulsion should be suppressed to achieve high weld quality and optimal electrode life, which are especially important in aluminum and magnesium welding. Based on the discussions in this chapter, it is possible to develop an on-line expulsion prediction and control algorithm for use in welding practice.

References

1. Kimchi, M., Spot weld properties when welding with expulsion—A comparative study, *Welding Journal*, 63, 58-s, 1984.
2. Karagoulis, M. J., Control of materials processing variables in production resistance spot welding, in Proceedings of AWS Sheet Metal Welding Conference V, Detroit, MI, Paper No. B5, 1992.
3. Newton, C. J., Browne, D. J., Thornton, M. C., Boomer, D. R., and Keay, B. F., The fundamentals of resistance spot welding aluminum, in Proceedings of AWS Sheet Metal Welding Conference VI, Detroit, MI, Paper No. E2, 1994.
4. Hao, M., Osman, K. A., Boomer, D. R., and Newton, C. J., Developments in characterization of resistance spot welding of aluminum, *Welding Journal*, 75, 1-s, 1996.
5. Zhang, H., Expulsion and its influence on weld quality, *Welding Journal*, 78, 373-s, 1999.
6. Davies, A. C., *The Science and Practice of Welding, Vol. 2, The Practice of Welding*, 10th edition, Cambridge University Press, United Kingdom, 1993.
7. Dickinson, D. W., Franklin, J. E., and Stanya, A., Characterization of spot welding behavior by dynamic electrical parameter monitoring, *Welding Journal*, 59, 170-s, 1980.
8. Wu, K. C., The mechanism of expulsion in weldbonding of anodized aluminum, *Welding Journal*, 56, 238-s, 1977.
9. Senkara, J., Zhang, H., and Hu, S. J., Expulsion prediction in resistance spot welding, *Welding Journal*, 83, 123-s, 2004.
10. Gould, J. E., An examination of nugget development during spot welding using both experimental and analytical techniques, *Welding Journal*, 66, 1-s, 1987.
11. Vahaviolos, S. J., Carlos, M. F., and Slykhouse, S. J., Adaptive spot-weld feedback control loop via acoustic emission, *Materials Evaluation*, 39, 1057, 1981.
12. Kilian, M. and Hutchenrenther, A., Monitoring and control of electrode indentation, in Proceedings of AWS Sheet Metal Welding Conference VI, Detroit, MI, Paper No. C4, 1994.
13. Hao, M., Osman, K. A., Boomer, D. R., Newton, C. J., and Sheasby, P. G., On-line nugget expulsion detection for aluminum spot welding and weld bonding, SAE Paper No. 960172, 1996.
14. Han, Z., Indacochea, J. E., Chen, C. H., and Bhat, S., Weld nugget development and integrity in resistance spot welding of high-strength cold-rolled sheet steels, *Welding Journal*, 72, 209-s, 1993.
15. Browne, D. J., Chandler, H. W., Evans, J. T., and Wen, J., Computer simulation of resistance spot welding in aluminum-part I, *Welding Journal*, 74, 339-s, 1995.
16. Browne, D. J., Chandler, H. W., Evans, J. T., James, P. S., Wen, J., and Newton, C. J., Computer simulation of resistance spot welding in aluminum—Part II, *Welding Journal*, 74, 417-s, 1995.
17. Lide, D. R., editor, *Handbook of Chemistry and Physics*, 74th edition, CRC Press, Boca Raton, FL, 1993–1994.
18. *Metals Handbook*, Vol. 1, 8th edition, ASM, Metals Park, OH, 1977.
19. *ASM Metals Reference Book*, 2nd edition, ASM, Metals Park, OH, 1984.
20. Hatch, J. E., *Aluminum: Properties and Physical Metallurgy*, ASM, Metals Park, OH, 1984.
21. Prigogine, I. and Defay, R., *Chemical Thermodynamics*, Longmans, London, 1967, 156.

22. Hultgren, R., Orr, R. L., Anderson, P. D., and Kelley, K. K., *Selected Values of Thermodynamic Properties of Metals and Alloys*, Wiley, London, 1974.
23. Krupkowski, A., *Basic Problems in Theory of Metallurgical Processes*, Polish Science Publications, Warsaw (in Polish), 1974.
24. Tsai, C. L., Jammal, O. A., Papritan, C., and Dickinson, D. W., Modeling of resistance spot welding nugget growth, *Welding Journal*, 71, 47-s, 1992.
25. Zhang, H., Huang, Y., and Hu, S. J., Nugget growth in spot welding of steel and aluminum, in Proceedings of AWS Sheet Metal Welding Conference VII, Troy, MI, Paper No. B3, 1996.
26. Luo, H., Hao, C., Zhang, J., Gan, Z., Chen, H., and Zhang, H., Characteristics of resistance welding magnesium alloys AZ31 and AZ91, *Welding Journal*, in print, July 2011.
27. Salman, S. A., Ichino, R., and Okido, M., A comparative electrochemical study of AZ31 and AZ91 magnesium alloy, *International Journal of Corrosion*, 2010, Article ID 412129, 2010.
28. Luo, H., New joining techniques for magnesium alloy sheets, MS thesis, Institute of Metal Research, Chinese Academy of Sciences, June, 2008.
29. Munitz, A., Kohn, G., and Cotler, C., Resistance spot welding of Mg–AM50 and Mg–AZ91D alloys, Magnesium Technology, ed. Kaplan, H. I., TMS (The Minerals, Metals & Materials Society), Warrendale, PA, 2002.
30. Zhang, H., Hu, J. S., Senkara, J., and Cheng, S., Statistical analysis of expulsion limits in resistance spot welding, *Transactions of ASME—Journal of Manufacturing Science and Engineering*, 122, 501, 2000.
31. Browne, D. I., Newton, C. I., and Boomer, D. R., Optimization and validation of a model to predict the spot weldability parameter lobes for aluminum automotive body sheet, in Proceedings of International Body Engineering Conference IBEC'95, Advanced Technologies and Processes Section, Detroit, MI, 1995, 100.
32. Kaiser, J. G., Dunn, G. J., and Eagar, T. W., The effect of electrical resistance on nugget formation during spot welding, *Welding Journal*, 61, 167-s, 1982.
33. Gould, J. E., Kimchi, M., Leffel, C. A., and Dickinson, D. W., Resistance seam weldability of coated steels, Part I, Weldability envelopes, Edison Welding Institute Research Report, No. MR9112, Columbus, OH, 1991.
34. McCullagh, P. and Nelder, J. A., *Generalized Linear Models*, 2nd edition, Chapman & Hall, London, 1989, 21.
35. *Automotive Sheet Specification*, Alcan Rolled Products Comp., Farmington Hills, MI, 1994.
36. Alcini, W. V., Experimental measurement of liquid nugget heat convection in spot welding, *Welding Journal*, 69, 177-s, 1990.
37. Auhl, J. R. and Patrick, E. P., A fresh look at resistance spot welding of aluminum automotive components, SAE Paper No. 940160, 1994.
38. Patrick, E. P. and Spinella, D. J., The effects of surface characteristics on the resistance spot weldability of aluminum sheet, in Proceedings of AWS Sheet Metal Welding Conference VII, Troy, MI, Paper No. B4, 1996.

8

Influence of Mechanical Characteristics of Welding Machines

8.1 Introduction

Very often two sets of welded specimens created using exactly the same welding parameters on identical workpieces, but different welders, possess significantly different geometric characteristics and strength when tested. The only logical explanation to this observation is that welders make the difference. A resistance welder consists of two distinctly different, yet closely related systems: electrical and mechanical, as illustrated in Figure 8.1. The characteristics of the mechanical system, such as machine stiffness, friction, and mass, play an important role in the functionality and performance of a welding machine, and subsequently influence the welding process and weld quality.

According to published literature, research on the influences of welding machines began in the 1970s. Early work focused on the differences of machine types without attempting to understand the mechanisms. For example, Ganowski and Williams[1] investigated the influence of machine types on electrode life in welding zinc-coated steels. Kolder and Bosman[2] studied the influence of equipment on the weld lobe diagrams of high-strength low-alloy (HSLA) steel using five different welders. Satoh et al.[3,4] concluded that machine type was an important factor in weld performance based on their experiments on four types of welders. Hahn et al.[5] found that large displacement of electrodes resulted in variance in electrode contact and a decrease in weld quality. Williams et al.[6] discovered that an increase in throat depth and electrode stroke decreased electrode life. Similarly, Howe[7] found that electrode deflection significantly influences electrode life.

Researchers have also addressed the effects of individual machine characteristics on various aspects of the welding process.[8–11] Dorn and Xu[8,9] showed that the stiffness of a lower (stationary) arm had an effect on the oscillation and mean value of electrode force at electrode touching. The papers published by Tang et al.[10,11] emphasized the use of various sensors in monitoring the effects of welder mechanical characteristics. Another study by Wang[12] explored further the use of process signals to characterize and differentiate welders. These studies are critical in understanding the mechanisms of machine characteristics, and are the basis of optimal welder design.

In this chapter, the important mechanical characteristics of welders are discussed, focusing on their possible influence on weld quality.

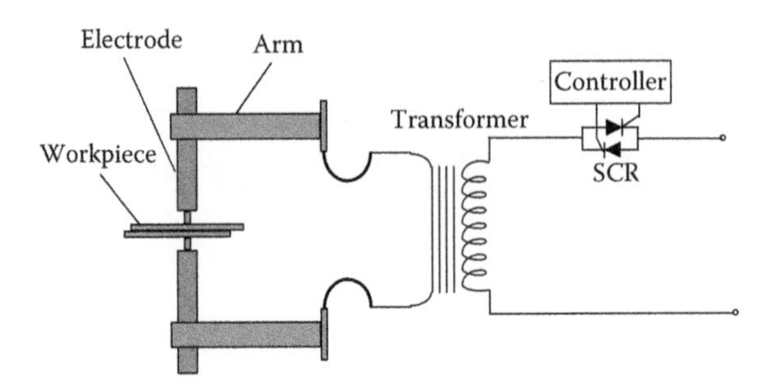

FIGURE 8.1
A schematic drawing of a spot welding system.

8.2 Mechanical Characteristics of Typical Spot Welders

Resistance welders are often classified according to their operating mechanisms, such as how the welder arms move relative to the workpieces. However, mainly because most welders are custom-made for particular applications, there are considerable differences in terminology and definitions for welders. Nevertheless, drawings of five typical welders are shown in Figure 8.2 to illustrate the main differences among resistance welders.

Pneumatic cylinders are the most common driving devices used in welders for electrode closing and releasing, although servomotor-driven electric guns are becoming popular.

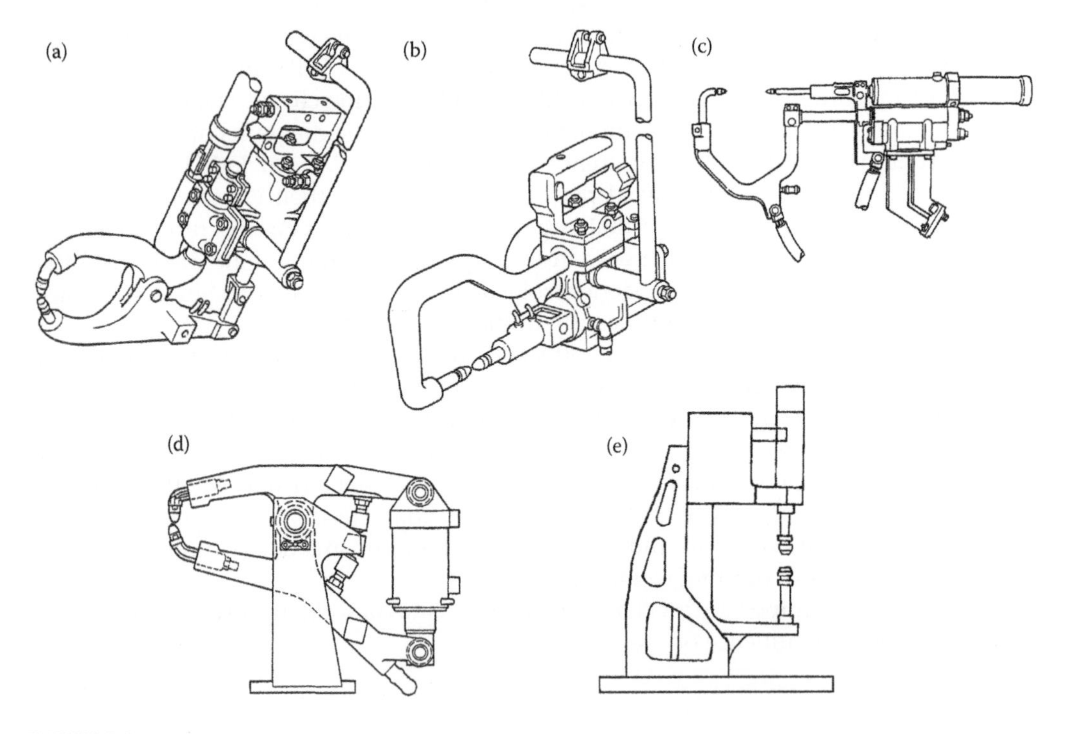

FIGURE 8.2
Schematics of typical welders: (a) rocker-type gun, (b) straight-acting gun, (c) equalizing gun, (d) pinch gun, and (e) pedestal welder.

The moving part of a welder consists of the assembly from the piston head to the electrode. The combined effect of friction inside the cylinder and through the guideway may contribute to the motion of the electrodes both at touching and during welding (follow-up).

As a result of rapid pressure buildup in a cylinder, the piston, together with the electrode and very often a shank, is pushed out. The magnitude of the force depends directly on the pressure and cylinder size. However, it is limited by the rigidity or stiffness of the welder arms—both the moving and stationary arms. Such limitation can be gauged by the maximum allowable deflection of the welder arms to avoid excessive electrode misalignment due to deflection. A very common practice in welder or weld gun design is that one arm is much stiffer than the other is. Therefore, often only the less stiff one needs to be considered.

During welding, electrodes move relatively to each other due to expansion and contraction of the workpieces. Although this movement is extremely small (in the magnitude of 0.1 mm or less), the resultant electrode force change may have a significant impact on the weld formation. The influence of moving mass may be considerable during the initial contact with the workpieces (touching).

These factors and other machine mechanical system-related issues are discussed in the following sections. They can be better understood by examining the detectable signals in a complete welding cycle. They include electrode force, electrode displacement, electric current, and voltage. A detailed examination of the force and displacement signals can provide insights not possible through other means. A possible outline of a data acquisition system as shown in Figure 5.2 can be used for signal collection.

Typical signals for electric voltage, current, electrode force, and displacement are shown in Figure 5.3. Depending on the purpose, either one or three linear variable differential transducer (LVDT) sensors are used to monitor the relative movement (displacement) of the electrodes in a welding cycle.

A displacement sensor can be used to record motion along its axial direction, and calculate the axial stiffness of the welder arms. However, for some welding systems, the lateral stiffness in a direction perpendicular to the gun tips is also a concern, as angular misalignment may be induced in that direction, especially under large electrode forces. In this case, a fixture can be designed to measure both axial and lateral stiffness, as shown in Figure 8.3. The fixture may hold three LVDT sensors, and the collected axial

FIGURE 8.3
A C-gun and fixture for displacement sensors.

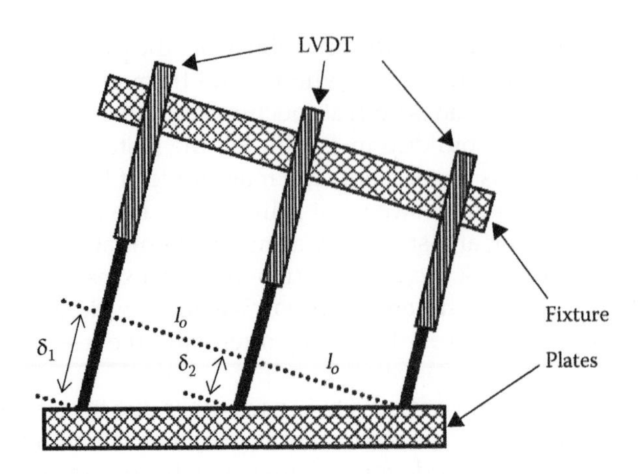

FIGURE 8.4
Dimensions used for calculating axial and lateral deflections.

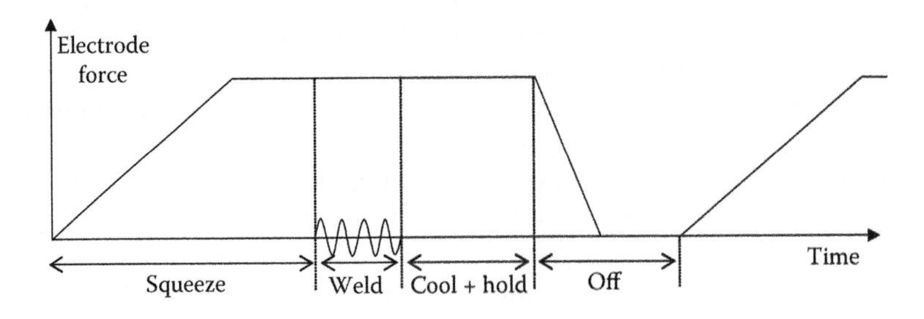

FIGURE 8.5
Stages in a typical weld cycle.

displacement signals are used to calculate displacements in both directions, as illustrated in Figure 8.4.

The signals of electrode force, displacement, and electric current for a complete weld cycle are usually needed to characterize the overall performance of a welder. A typical welding cycle consists of squeeze, weld, cool and hold, and off stages, as sketched in Figure 8.5.

These stages are usually selected based on experience/guidelines and controlled through an electronic controller. The mechanical system interacts with the controller of a welder, and its true responses often deviate from the designated ones, depending on both the mechanical and control systems, as illustrated in Section 8.6.

In the following sections, possible influences of the mechanical factors of a welder are discussed using signals collected during welding, and weld quality is used as a measure of the influence.

8.3 Influence of Machine Stiffness

Stiffness is usually measured on a structure for its ability to resist deflection under loading. In resistance welding, the interest is the amount of relative displacement of the electrodes under the applied electrode force. Its influence is reflected in many aspects of a welding process. In this section, the differences resulting from different machine stiffness values in electrode force, electrode displacement, and weld formation process are illustrated using experimental observations. A method of estimating the stiffness value proposed in a recent work is also included.

8.3.1 Effect on Electrode Force

The effect of stiffness is directly reflected in the electrode force signals collected during welding. The work by Tang et al.[10] shows a detailed experimental work on the influence of stiffness. In their experiments, only the stiffness of the lower structure of a pedestal welder was considered, because the upper structure was moving and much stiffer. The welder was modified to have two levels of stiffness by adjusting the spring stiffness between the lower (stationary) electrode and its support structure on the pedestal welder (Figure 8.6). The values of stiffness were 8.8 and 52.5 kN/mm. As shown in Figure 8.7, although the two cases have very similar touching behavior and reach the preset force level through almost identical paths, the electrode force responds differently with different machine stiffness when an electric current is applied. The increment of the force under lower stiffness is 133 N (30 lb), whereas it is 334 N (75 lb) under higher stiffness, that is, greater stiffness results in a greater change in electrode force. Different machine stiffness provides different electrode forces and, therefore, different degrees of constraint to nugget growth. The nugget expansion is more difficult under higher stiffness, as it causes a greater reaction force from the electrodes.

8.3.2 Effect on Electrode Displacement

Electrode displacement may be the best indicator of nugget initiation and growth, and of other welding process characteristics. Although the amount of electrode movement during welding varies depending on the stiffness of the welder, similar displacement characteristics have been observed when welding with different machine stiffnesses using the

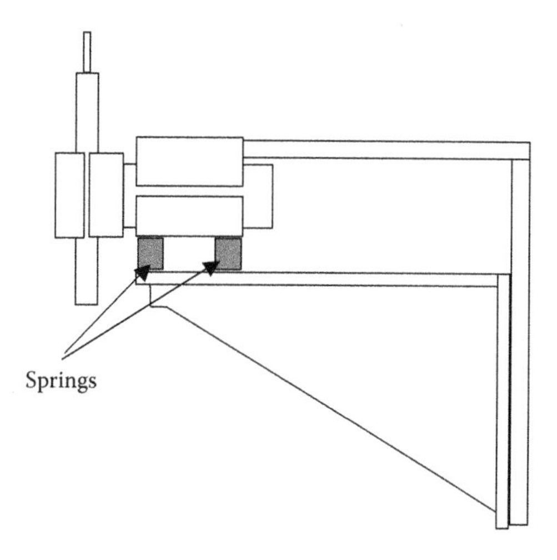

FIGURE 8.6
Modification of machine stiffness on lower arm of a pedestal welder. (From Babu, S., Web site: http://mjndeweb.ms.ornl.gov/Babu/default.html, 2004. With permission.)

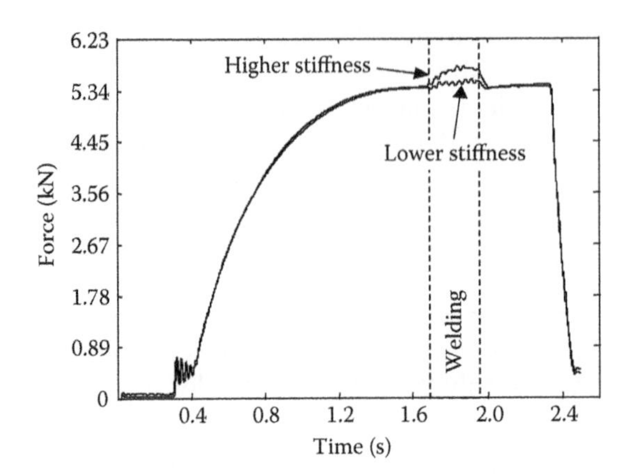

FIGURE 8.7
Electrode forces under different machine stiffnesses.

same pedestal welder as mentioned above. As shown in Figure 8.8, the amount of expansion is similar, that is, $d_1 \approx d_2$ for nuggets to grow, but with the stiffer machine, it takes longer for expulsion to happen. This is because a stiffer machine provides a larger constraint or force, which delays the expulsion according to Zhang.[13] In the case of Figure 8.8, welding time was set at 16 cycles, and expulsion occurred at the 5th cycle (0.083 s) in the weld stage when the machine stiffness was low, whereas expulsion happened at the 13th cycle (0.217 s) when stiffness was high. If welding time had been set at 10 cycles, the expulsion would not have occurred under the higher stiffness, but still could have occurred under the lower stiffness. Therefore, stiffer machine delays expulsion and widens the operation window. One extreme case of low rigidity of a welder is as in the experiments by Howe[14] in which the bolts used to mount the electrode assembly to the shaft of the head of a welder got loose after repeated use. The electrode life was doubled because of the skidding and deflection of an electrode that effectively made more of the electrode tip surface available for welding.

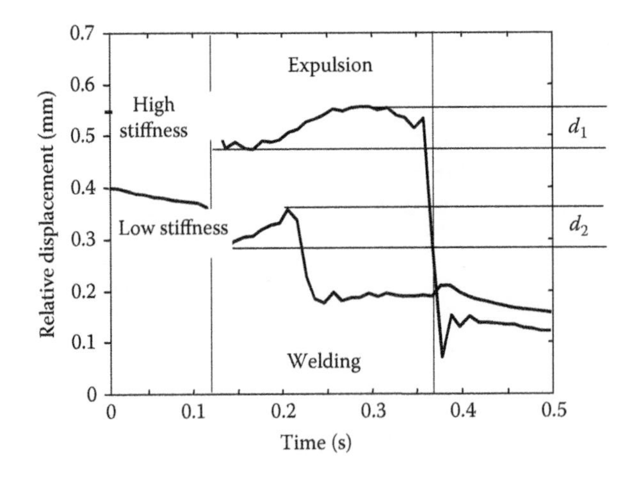

FIGURE 8.8
Displacement comparison under different machine stiffnesses.

8.3.3 Effect on Electrode Touching Behavior

The stiffness of the welder also directly affects the strike of electrodes on workpieces when they touch. Such impact determines the magnitude of the instantaneous force and the duration of electrode force oscillation. Therefore, the electrode touching behavior is directly linked to the electrode life, as the mechanical impact between the moving electrode and the workpiece may significantly deform the electrode face. Dorn and Xu[8,9] showed that the stiffness of the lower arm of their welder had an effect on the oscillation and mean value of electrode force at electrode touching. Another possible influence of touching is related to the fact that it is desired that the welding current start as early as possible to save cycle time—often before the electrodes are stabilized in practice. The consequence of doing so is to weld with insufficient electrode force, the primary reason for expulsion, as discussed in Chapter 7. If the weld current is applied when there is a gap, no matter how small it is, between the electrodes and workpieces due to the oscillation of welder arms, electric arcs may be created. Whether there is such oscillation depends on the machine, the workpiece material, and the fit-up conditions. The oscillation can be clearly observed from the electrode displacement or force signals, whereas the latter is more apparent, as shown in Figure 8.9. The excessive heat generated at the electrode face because of arcing significantly reduces the electrode life through alloying with the workpiece or coating materials.

8.3.4 Effect on Weld Formation

The stiffness has a direct effect on electrode force, which in turn affects the welding process. Therefore, it is natural to link the stiffness of a welder to a weld formation process. Experiments such as those conducted by Tang et al.[11] have shown clearly the relationship between weld formation and machine stiffness.

8.3.4.1 Expulsion

As shown in Figure 8.7, the actual electrode force can be significantly different during welding for machines with different stiffnesses. This difference may alter the welding

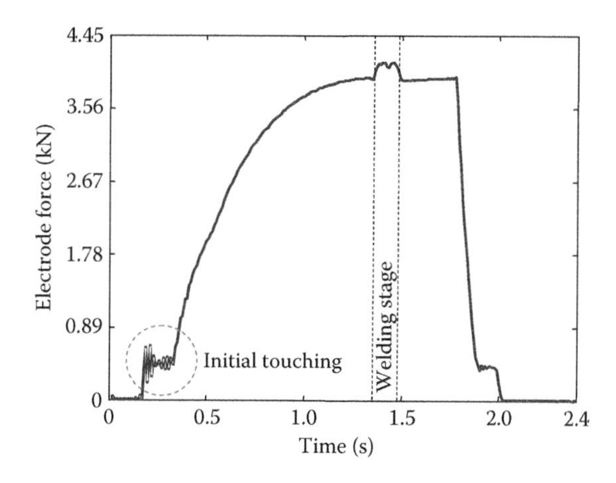

FIGURE 8.9
Force profile during initial touching and welding. (From Babu, S., Web site: http://mjndeweb.ms.ornl.gov/Babu/default.html, 2004. With permission.)

TABLE 8.1
Influence of Stiffness on Expulsion Limits

Material	Limit (kA)		Increase of Expulsion Current (kA)
	Low Stiffness	High Stiffness	
0.8-mm bare steel	7.0	8.0	1.0
0.8-mm galvanized steel	8.3	8.7	0.4
1.7-mm bare steel	6.9	7.1	0.2

process in terms of the occurrence of expulsion and forging effect (on nugget structure). From Figure 8.8, it can be clearly seen that expulsion can be delayed by increasing machine stiffness.

It was reported that expulsion limits (in terms of welding current) increase with weld gun frame stiffness using a modified C-gun. This is because a high stiffness machine frame imposes high constraining forces on the workpieces and, therefore, expulsion is less likely to happen, as explained in Chapter 7. The increase was more significant for thin-gauge sheets. The amount of increase in expulsion limit, in terms of expulsion current, is shown in Table 8.1. Because higher expulsion limits allow higher welding current, potentially larger welds can be made without expulsion.

8.3.5 Effect on Weld Strength

A comparative experiment was conducted to illustrate the effect of stiffness on weld quality. The stiffness of a pedestal welder was altered between original stiffness and higher stiffness. A slight increase in tensile–shear strength with machine stiffness was observed (Figure 8.10). However, the improvement is not significant because only about a 3% difference exists and the data ranges overlap.

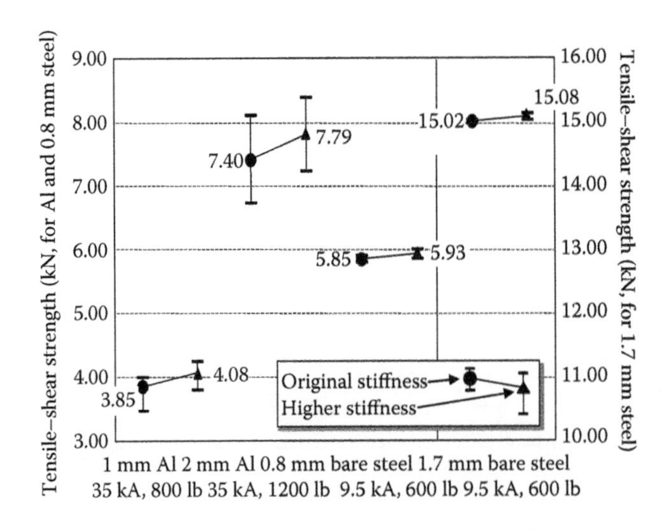

FIGURE 8.10
Influences of stiffness on weld strength. (From Khan, J. A. et al., in *Proceedings of AWS Sheet Metal Welding Conference IX*, Sterling Heights, MI, Paper No. 5-1, 2000. With permission.)

8.3.6 Effect on Electrode Alignment

Ideally, electrodes should be aligned during the resistance spot welding (RSW) process because a misalignment induces unfavorable features to the process and weld quality. Misalignments, either axial or angular, may cause irregularly shaped welds and reduce the weld size because they result in asymmetrical distribution of pressure and current. A welder's frame stiffness has an obvious influence on electrode alignment, as electrodes on a less stiff machine tend to have large misalignment, usually both axial and angular, as shown in Figure 8.11. Therefore, a stiff welder frame is generally preferred. However, the appropriate level of stiffness should be determined because extremely high stiffness is neither economical nor necessary.

In general, high machine stiffness is preferred because it ensures good electrode alignment, provides large forging force, and raises expulsion limits. Therefore, high stiffness is recommended for the structure design of RSW machines.

8.3.7 Stiffness and Damping Ratio Estimation

Because of the complexity of a resistance spot welder, it is not possible to obtain closed-form expressions for calculating the stiffness and other characteristics of a welder. However, attempts have been made to develop an experimental procedure for convenient yet accurate measurement of machine stiffness and damping ratio. For a dynamic welding system, the change in electrode force can be expressed, through a general force equilibrium consideration, as

$$\Delta F_e = M\frac{d^2x}{dt^2} + \rho\frac{dx}{dt} + K\Delta x + A\Delta P \tag{8.1}$$

where x is the relative displacement between the electrodes, Δx is the change in x, ΔF_e is the electrode force change, which can be measured by a force sensor, M is the mass of the moving part, ρ is the damping ratio, K is the stiffness along the electrode arms, A is the inner cross section area of the pneumatic cylinder, and ΔP is the pressure fluctuation due to the displacement x. The influence of friction from the moving parts is ignored for simplicity. Because of the complex processes involved in welding, the coefficients in Equation

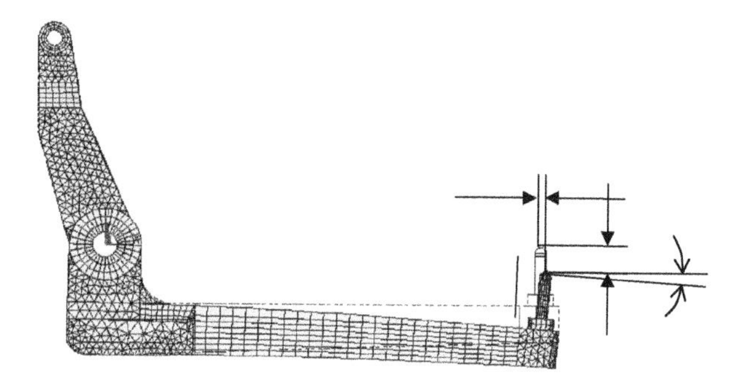

FIGURE 8.11
Deflection of a welder's arm under electrode force. (From Babu, S., Web site: http://mjndeweb.ms.ornl.gov/Babu/default.html, 2004. With permission.)

8.1 cannot be assumed constant. However, it is reasonable to regard them as constants in a squeeze stage when only a mechanical process is involved. The equation can be further simplified by recognizing the relatively small effect of moving mass M, because of a very small acceleration after the electrodes touch the worksheets, as reported in the work by Tang et al.[11] The pressure in the cylinder can be assumed constant for a large cylinder with reasonable capacity of pressure follow-up. With such effects accounted for, Equation 8.1 can be reduced to

$$\Delta F_e = \rho \frac{dx}{dt} + K\Delta x \tag{8.2}$$

In this equation, ΔF_e and x can be measured using the sensors, and dx/dt can be derived from the displacement signal. The two unknowns, ρ and K, can then be estimated using sets of ΔF_e and x (or Δx) values.

The experiments were conducted using two C-guns (C-gun I and C-gun II) and a pedestal welder, which have different stiffnesses and damping ratios. Although the two C-guns had similar mechanical characteristics, they were equipped with different types of electronic controllers, which might produce different responses in squeeze and cool periods during a welding cycle.

In practice, the raw signals collected contain certain noises. If the welding stage is not of concern, then simple filtration can usually produce sufficiently clean force and displacement signals for the calculation of ρ and K. A sample plot for the signals collected on C-gun I is presented in Figure 8.12.

Although the signals are not polluted by an electromagnetic field in a squeeze stage, a treatment of the collected signals is usually needed for calculating the damping ratio and stiffness. Because the differences of displacement and their derivatives are used as the denominator in calculation, large errors could result from small fluctuations in data. To overcome this, a fitted curve, instead of the original data, was used. Such a curve fitting should be done for the period of interest only, and it should be done for both force and

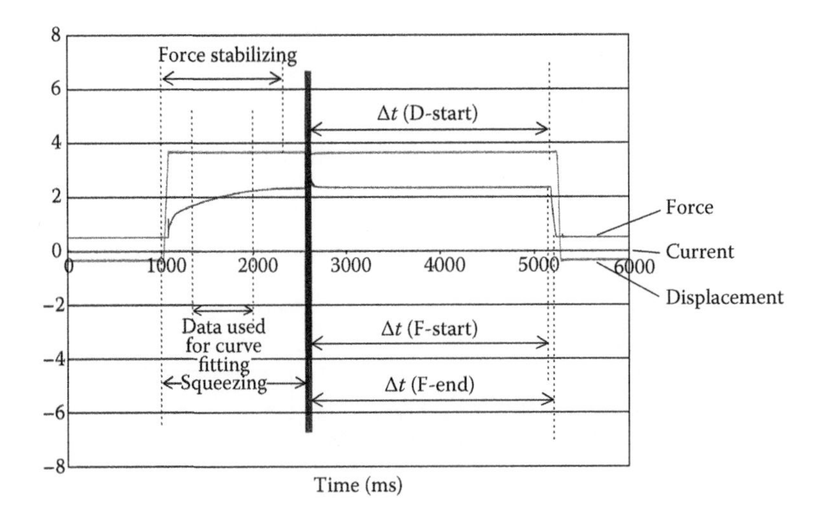

FIGURE 8.12
(See color insert.) Signals obtained for a typical welding cycle using DAQ system.

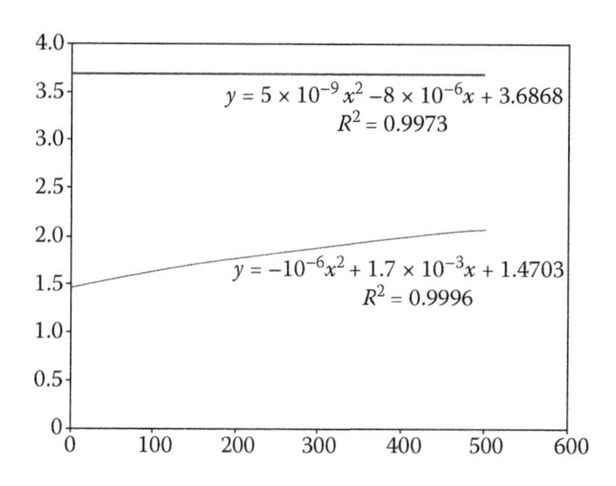

FIGURE 8.13
Curve fitting for force and displacement signals.

displacement signals. Fitted force and displacement curves are shown in Figure 8.13, taken from the portion of interest as indicated in Figure 8.12. Using the data treated following the aforementioned procedure, the stiffness and damping ratio were calculated on C-gun I, as presented in Table 8.2.

The calculated values for the stiffness and damping ratio are fairly constant, as expected for a squeeze stage. This shows that the model (Equation 8.2) is adequate for estimating ρ and K.

TABLE 8.2
Calculated Stiffness and Damping Ratio for C-Gun I

Force (kN)	Displacement (mm)	Velocity (mm/s)	Stiffness (kN/mm)	Damping Ratio (kN s/mm)
5.54091	1.8367998	–	–	–
5.54092	1.8374992	0.019982	–	–
5.54093	1.8381982	0.019970	17.33853474	1.318514441
5.54094	1.8388968	0.019958	17.33853474	1.319913624
5.54095	1.8395950	0.019946	17.33853474	1.321314143
5.54096	1.8402928	0.019934	17.33754601	1.322942507
5.54097	1.8409902	0.019922	17.33858821	1.324124114
5.54098	1.8416872	0.019910	17.33853474	1.325523742
5.54099	1.8423838	0.019898	17.33853474	1.326929629
5.54100	1.8430800	0.019886	17.33853474	1.328336866
5.54101	1.8437758	0.019874	17.33848191	1.329740551
5.54102	1.8444712	0.019862	17.33858821	1.331160358
5.54103	1.8451662	0.019850	17.33848191	1.332561778
5.54104	1.8458608	0.019838	17.33858821	1.333984317
5.54105	1.8465550	0.019826	17.33858757	1.335398277
5.54106	1.8472488	0.019814	17.33842843	1.336842551
5.54107	1.8479422	0.019802	17.33858821	1.338221543
5.54108	1.8486352	0.019790	17.33853474	1.339615844
5.54109	1.8493278	0.019778	17.33848191	1.341563246
5.54110	1.8500200	0.019766	17.33858821	1.342421879

Near the end of a squeeze stage, if the preset force level is reached, both the force and displacement will remain as constants until the electric current is applied. During this period, the electrodes are virtually stationary, and dx/dt can be considered zero. As a result, Equation 8.2 can be further simplified as

$$\Delta F_e = K\Delta x \quad (8.3)$$

Therefore, it is possible to calculate stiffness K directly from the force and displacement signals. In order to do this, schedules of different force levels were used, and corresponding displacement signals were collected. Deriving from Equation 8.3, the stiffness can be expressed by the ratio of difference in force to that in displacement between two sets of data:

$$K = \frac{\Delta F_e}{\Delta x} \quad (8.4)$$

The results are shown in Table 8.3 for C-gun I. The stiffness value calculated using Equation 8.4 is slightly lower than that obtained using Equation 8.2 (shown in Table 8.2). This could be attributed to several factors. Although the force measurement is fairly repeatable, the displacement measurement may change from run to run. Because of the impact of electrodes during touching, the electrode faces may be slightly plastically deformed (in the micron scale). As a result, the measured displacement value may increase from run to run, and the system may appear more compliant than it is. Another possible explanation is that there might be slight relative movement between the LVDT sensors and the fixture at the beginning and during electrode touching. These factors had a minimal effect when Equation 8.2 was used to calculate K, as the signals collected in the same weld cycle, instead of between two weld cycles, were used in the calculation. The method using Equation 8.2 appears to be more accurate for calculating the stiffness, in addition to the benefit of obtaining the damping ratio at the same time. Nonetheless, Equation 8.3 (or 8.4) provides a convenient procedure for estimating the stiffness of a welder.

This procedure was also applied to a pedestal welder, which was expected to have a noticeably different stiffness from the C-guns. The force and displacement signals for the pedestal welder are similar to those shown in Figure 8.12. Similar portions of force and displacement were selected for calculation, as presented in Figure 8.14 with fitted curves.

TABLE 8.3

Stiffness Calculation Using Equation 8.4

Force (kN)	ΔF (kN)	Displacement (mm)	ΔD (mm)	Stiffness (kN/mm)
3.6662	–	42.26516	–	–
4.0244	0.3582	42.29054	0.02538	14.11
5.0468	1.3806	42.39051	0.12535	11.01
5.4118	1.7456	42.41589	0.15073	11.58
6.1732	2.5070	42.46572	0.20056	12.50
6.7072	3.0410	42.51586	0.25070	12.13

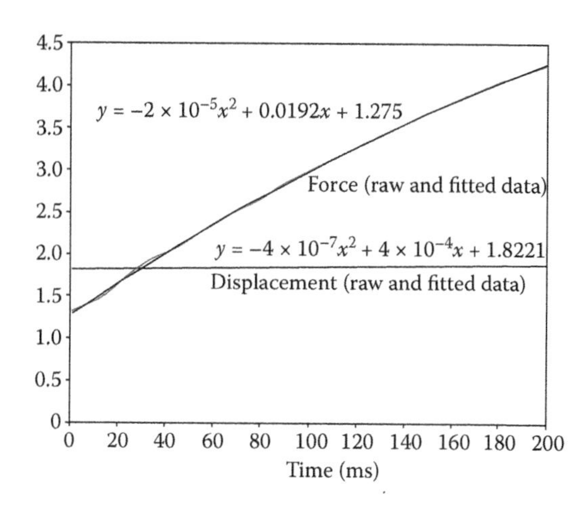

FIGURE 8.14
Curve fitting for pedestal welder.

During this time, force is less volatile than that at the beginning of touching, and displacement has a visible increase, which is needed for calculation. The calculated stiffness and damping ratio using Equation 8.2 for the pedestal welder are around 50 kN/mm and 1.0 kN s/mm, respectively, as listed in Table 8.4.

Another C-gun (II) was also used to compare with C-gun I. These two guns had similar mechanical characteristics, but different controllers. For this C-gun, in addition to the stiffness in the axial direction, lateral stiffness was also measured using all three LVDT sensors in the fixture shown in Figure 8.3. The electrode force was resolved into components parallel and perpendicular to the axis of the arms, and their corresponding displacement components were calculated according to the diagram shown in Figure 8.4. The calculated stiffness values are axial stiffness = 13.26 kN/mm and lateral stiffness = 0.025 kN/mm. The measurement for C-gun II was similar to that for C-gun I.

TABLE 8.4
Calculated Stiffness and Damping Ratio for Pedestal Welder

Force (kN)	Displacement (mm)	Velocity (mm/s)	Stiffness (kN/mm)	Damping Ratio (kN s/mm)
1.29418	1.822500	–	–	–
1.31332	1.822898	0.3988	–	–
1.33242	1.823296	0.3980	50.00000007	1.000000035
1.35148	1.823694	0.3972	49.99999993	0.999999965
1.37050	1.824090	0.3964	50.00000007	1.000000035
1.38948	1.824486	0.3956	49.99999990	1.000000035
1.40842	1.824880	0.3948	50.00000010	1.000000035
1.42732	1.825274	0.3940	49.99999993	0.999999965
1.44618	1.825668	0.3932	50.00000007	1.000000035
1.46500	1.826060	0.3924	49.99999993	0.999999965

8.4 Influence of Friction

The influence of machine friction has been noticed by some researchers. Satoh et al.[4] found that friction had effects on nugget diameter and sheet separation. They also noticed that weld expansion occurred mainly in a direction perpendicular to the electrode axis if the friction effects were significant. Dorn and Xu[8] concluded that the increase of friction improved electrode touching behavior. They further found that an increase in friction reduced the tension–shear strength and torsion moment of welds.

In general, friction exists between two contact surfaces only when there is a relative movement or a moving tendency. The moving parts and possible sources of friction can be identified by visual inspection of a welder. Improper gun installation or maintenance may cause excessive friction. In general, the total friction force contains two parts: that due to the internal friction inside the cylinder and that from the guideway. However, friction measurement of a particular welder is not straightforward. Neither static nor kinetic friction is constant in any pneumatic driving system. The values measured under the no-weld condition can only serve as an indicator of friction during welding. For this reason, quantitative measurement (and comparison) is not only difficult to make, but unnecessary. In order to study the effect of friction while keeping other factors unchanged, a device can be designed to add additional friction to a welder. In the work by Tang et al.,[11] the friction of the welder was varied by using a specially designed device, as shown in Figure 8.15. The device was mounted between the upper and lower structures of the pedestal welder. It can provide about 0.36 kN (80 lb) friction force when there is a slight movement between the structures. The static friction force of the device was about 0.45 kN (100 lb). Because there is a movement between electrodes during the weld stage, the effect of static friction was not considered in their study.

In their experiments, two different friction situations were considered: the original setup and one with an additional 0.36 kN (80 lb) of friction. This device creates significant static friction, as evidenced in Figure 8.16. However, increasing preset cylinder pressure can compensate such static friction. The dynamic friction is the one that influences the electrode follow-up and, therefore, the weld formation.

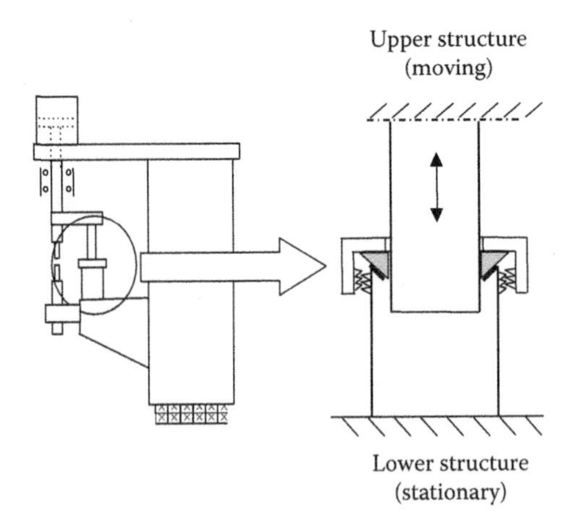

FIGURE 8.15
Modification of machine friction on a pedestal welder.

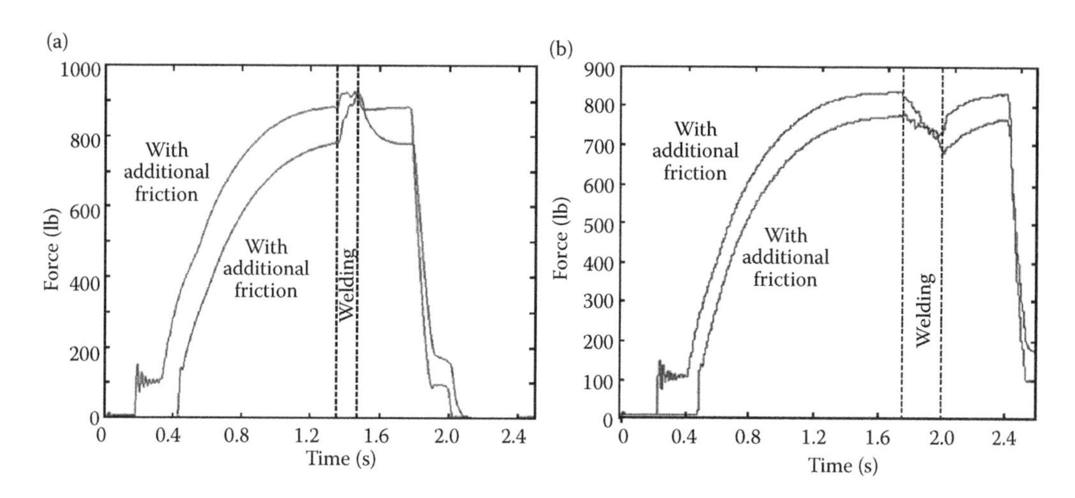

FIGURE 8.16
Effect of friction on electrode force: (a) steel welding and (b) aluminum welding.

8.4.1 Effect on Electrode Force

The friction in RSW machines contributes significantly to total electrode force. Under the same preset value, the actual force with larger friction is smaller than that with smaller friction (Figure 8.16). When large friction exists, electrode movement is sluggish and cannot promptly follow the contraction. Thus, it is most likely that internal discontinuities, such as porosity, appear in the nugget. The desired situation should be that the electrodes move freely in the welding and holding stages. During electrodes touching workpieces, the additional friction reduces force oscillation because the machine with greater friction has a stronger damping capacity. In addition, the touching is delayed. The total force is smaller before welding when friction is greater because the friction opposes and cancels out some of the cylinder force. During welding, however, the friction force applies toward the nugget and adds more to the total force because the nugget expands and pushes the electrodes away. Thus, the force in the case with additional friction increases more significantly than that without additional friction. Friction can be considered the main source of force change during welding. In general, it is proportional to the normal force on contact surfaces. The normal force, furthermore, is in proportion to the preset force because of the bending moment of machine structures and the imperfect alignment of electrodes. In other words, friction force is proportional to the preset force. It is expected, therefore, that the force change is more significant when the preset force is greater.

8.4.2 Effect on Electrode Displacement

The displacement signal provides explicit information on weld indentation. An example of 2 mm aluminum welding under 3.56 kN force is shown in Figure 8.17. The electrodes extrude less into the workpieces with larger friction. The difference in displacement is about 0.1 mm as electrodes were retracted. This 0.1-mm reduction in indentation is a 25% improvement considering the original 0.4-mm indentation. This observation agrees with the indentation measurements. In addition, the initial impact of electrodes onto workpieces in a process similar to forging, is also reduced by friction, which may be beneficial to electrode life.

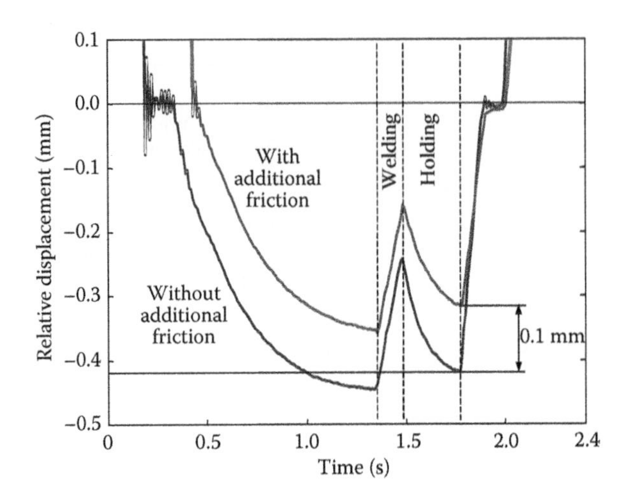

FIGURE 8.17
Comparison of electrode displacement with different friction.

8.4.3 Effect on Microstructure

In order to gain further understanding, the welded specimens were sectioned and examined through standard metallographic techniques. Typical cross sections are shown in Figures 8.18 through 8.20. The differences in the cross sections under different friction conditions are clearly visible and the friction always opposes electrode movement, which makes electrodes difficult to follow nugget contraction during the hold stage. This helps the creation of internal discontinuities in welds.

For 0.8-mm steel (Figure 8.18), incomplete fusion near the faying surface was observed when the machine had additional friction. This could be the reason for the reduction of weld strength. For 1.7-mm steel (Figure 8.19), it is obvious that the weld under greater friction has shrinkage porosity. A similar situation was observed in the weld of 2 mm aluminum (Figure 8.20). However, the internal porosity may not affect the tensile–shear strength of the welds.

8.4.4 Effect on Tensile–Shear Strength

In most cases, the additional friction only promotes the formation of porosity, not other types of defects in a weldment. As a result, the tensile–shear strength of the welds does

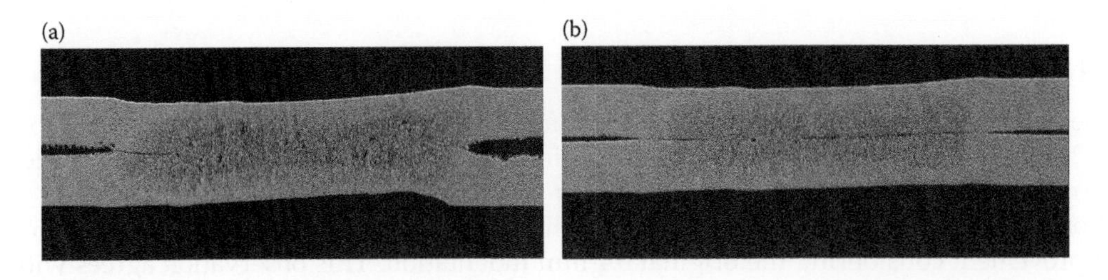

FIGURE 8.18
Weld cross sections with different frictions (0.8 mm steel). (a) Without additional friction. (b) With additional friction.

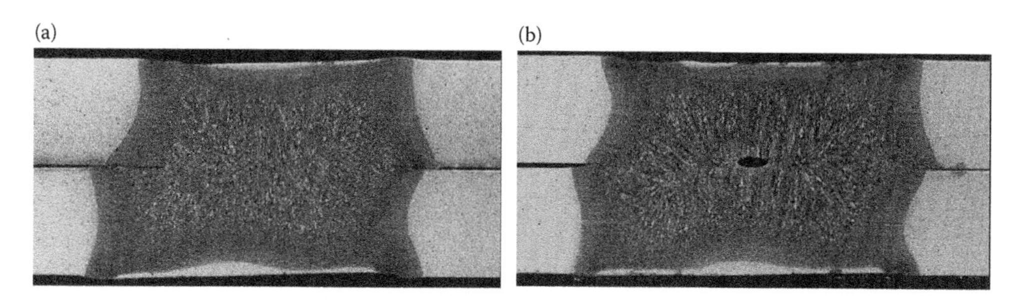

FIGURE 8.19
Weld cross sections with different frictions (1.7 mm steel). (a) Without additional friction. (b) With additional friction.

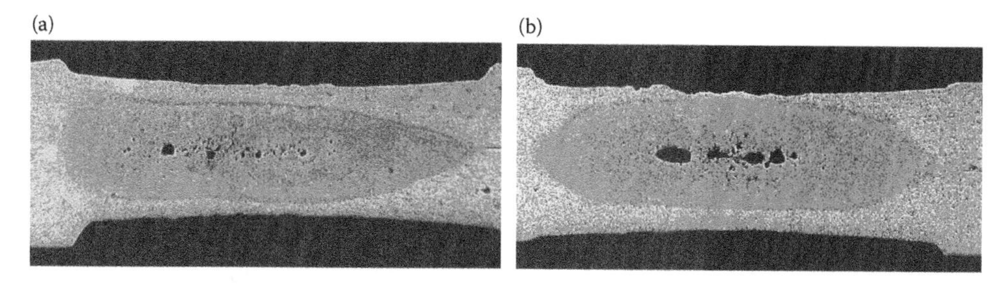

FIGURE 8.20
Weld cross sections with different frictions (2 mm aluminum). (a) Without additional friction. (b) With additional friction.

not suffer much (refer to Figures 8.21 and 8.22). The reduction in strength is not statistically significant for the steels and aluminum alloys tested.

These figures also show the comparisons of the joint strength under different conditions. It can be concluded, based on the comparisons, that friction is unfavorable for both steel and aluminum welding. However, some of strength reduction may not be statistically

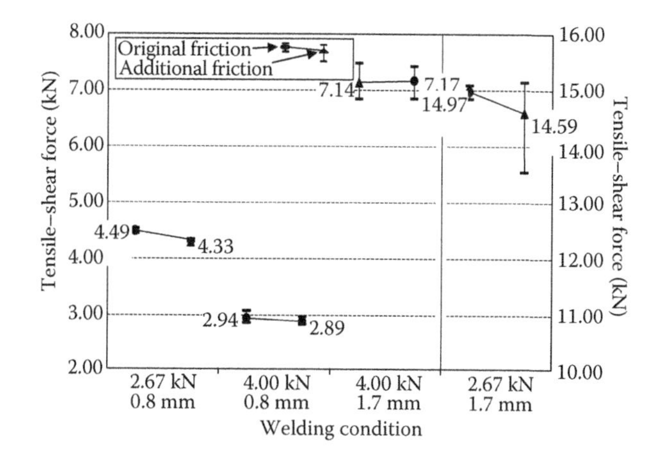

FIGURE 8.21
Influence of friction on weld strength (steel).

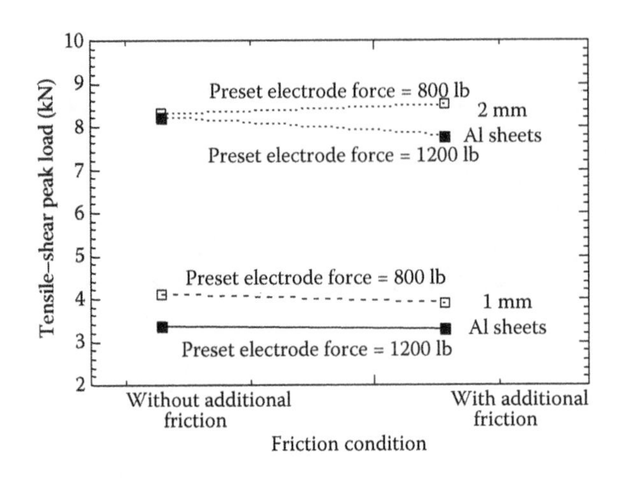

FIGURE 8.22
Influence of friction on tensile strength (aluminum).

significant because the data ranges overlap. In general, the influence of friction varies with welding conditions.

In summary, machine friction influences welding process and weld quality. The friction always opposes electrode movement and makes it difficult for electrodes to follow nugget expansion during heating and contraction during cooling. The latter effect may help the creation of internal discontinuities in welds. Therefore, machine friction is in general unfavorable to weld quality. Friction should be kept as small as possible for this reason. There are several practical ways to minimize friction. For instance, the moving parts of RSW machines should be supported by a roller guide, such as using ball screws, rather than using a sliding mechanism.

8.5 Influence of Moving Mass

The moving mass of RSW machines has been found to be less important to weld quality than stiffness and friction. Satoh et al.[4] did not find much influence from moving mass on weld nugget formation. However, they stated that an optimal weight of the moving part existed for electrode life in relation to the natural frequency of an RSW machine. Dorn and Xu[9] observed that the moving mass affected vibration at low friction with a rigid lower arm. However, they did not detect any clear influence of the mass on weld quality. Theoretical attempts have been made by Gould and Dale[15] and by Tang et al.[11] on the dynamic behavior of moving parts of a welder.

8.5.1 A Dynamic Force Analysis

The insignificance of the effect of the moving mass is generally expected because of the very small amount of motion of electrodes during welding. The effect of mass, in the form of dynamic force, can be significant only when weld volume thermally expands with a large acceleration. The dynamic force can be obtained if moving mass and the acceleration of electrode movement are known. The force (F) can be calculated by $F = M\frac{d^2x}{dt^2}$, where M is the moving mass and $\frac{d^2x}{dt^2}$ is acceleration, which can be approximated by the following differential equation:

$$\frac{d^2x}{dt^2} = \frac{x(t+\Delta t) - 2x(t) + x(t-\Delta t)}{(\Delta t)^2} \tag{8.5}$$

where $x(t)$ is the electrode displacement during welding at the instant t and Δt is the sampling interval.

Once the electrode displacement profile is recorded for a welding process, the acceleration can be estimated using Equation 8.5. For a 1.7-mm steel welding using 6.8 kA current and 2.67 kN (600 lb) force, as an example, the calculated acceleration of weld expansion is shown in Figure 8.23. The results show that the largest acceleration occurred during the first cycle in the weld stage, but its magnitude is very small (0.23 m/s^2). Afterward, the acceleration is nearly zero. If the moving mass were 40 kg, then the maximum dynamic force would be 9 N (2 lb). Its effect can be ignored compared with the applied electrode force.

In addition to its small magnitude, the dynamic force happens only at the very beginning in the welding stage; therefore, it has little effect on weld formation. In order to gain a better understanding of the effect of moving mass and verify the findings drawn from Figure 8.23, a pedestal welder was modified to isolate the effect of moving mass (Figure 8.24). The original moving mass was estimated to be about 40 kg. A 20-kg weight was added to the upper structure of the welder in the experiment. As a result, the moving mass of the welder was increased from 40 to 60 kg. Visible differences can be observed in electrode force (Figure 8.25) only at initial touching and when electrode leave the workpieces. The difference during welding can be ignored.

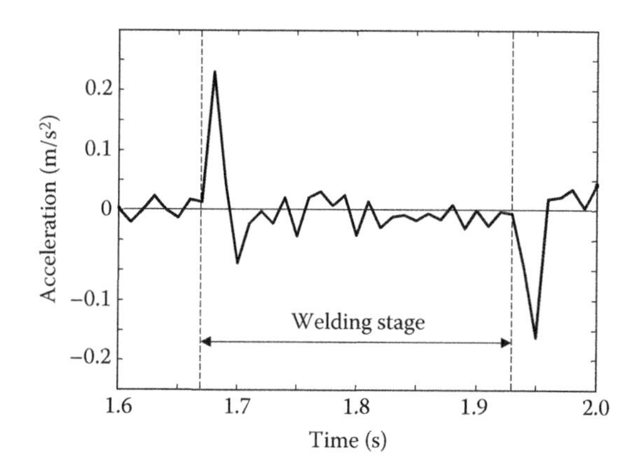

FIGURE 8.23
Acceleration of electrodes along the electrode direction.

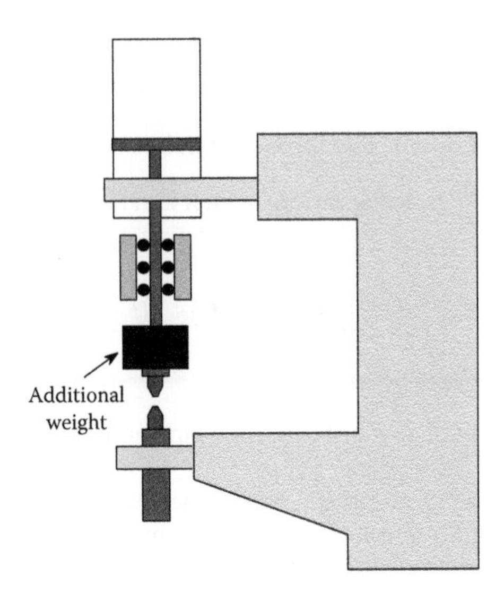

FIGURE 8.24
Modification of machine moving mass.

In general, the effect of moving mass on electrode force is negligible once the electrode force is stabilized after initial touching, as can be observed from Figure 8.25. Therefore, the influence of moving mass on weld quality is expected to be insignificant. However, the influence may be significant during the touching of electrodes, as shown in the figure, where both the duration and magnitude of fluctuation of electrode force are affected by the moving mass. As a result, excessive electrode deformation may result from the large fluctuation at the beginning, and in turn, moving mass may influence the electrode life.

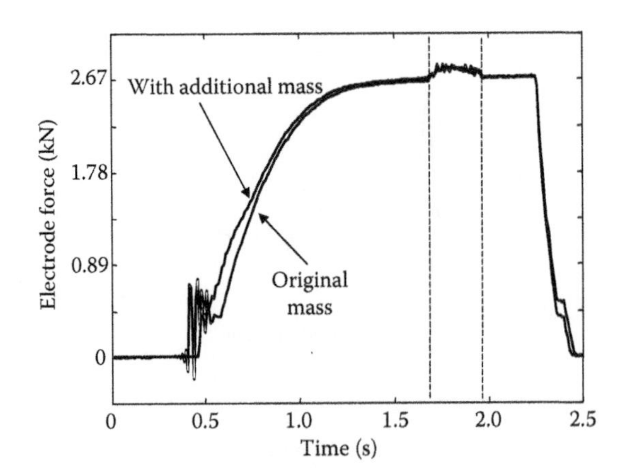

FIGURE 8.25
(See color insert.) Effect of moving mass on electrode force.

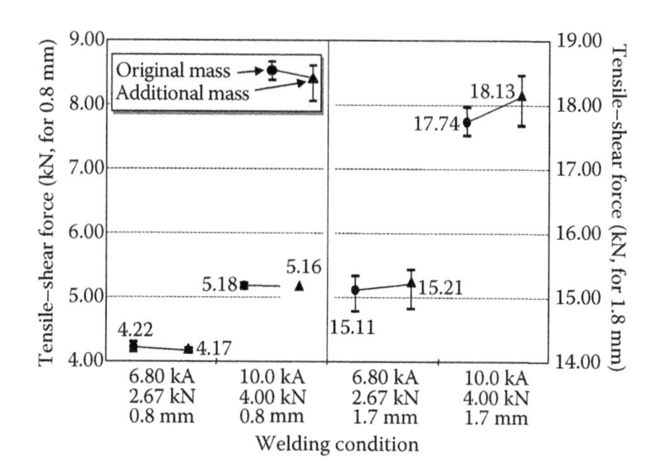

FIGURE 8.26
Influence of moving mass on weld strength for steel.

8.5.2 Effect on Weld Quality

Various experiments were conducted by altering the mass of the upper structure of a pedestal welder in the work by Tang et al.[11] The tensile–shear strength and welding expulsion limits do not show any significant difference for welds made under different moving masses. Strength comparisons of welds made with and without additional moving masses are shown in Figures 8.26 and 8.27 for steel and aluminum sheets. Therefore, it is safe to conclude that the moving mass has no effect on weld quality.

Therefore, the mass or weight should be minimized to reduce the impact at touching for improved electrode life and to improve gun portability for energy saving and ergonomic benefits.

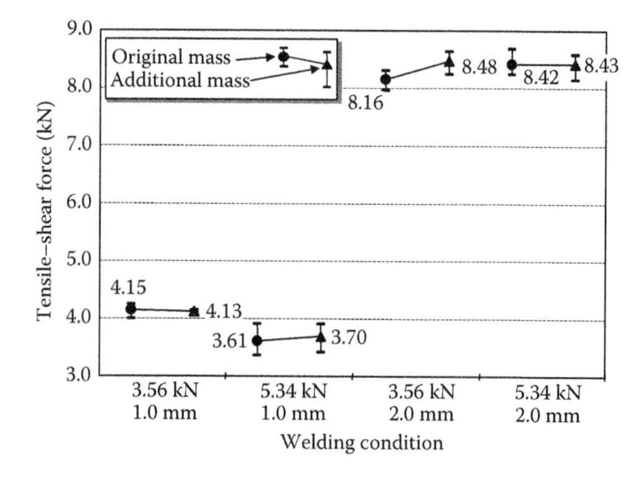

FIGURE 8.27
Influence of moving mass on weld strength for aluminum.

8.6 Follow-Up in a Welding Cycle

Using the application of electric current as a landmark, a welding cycle can be divided into a touching stage and a follow-up stage. Such classification may help better understand the effect of a mechanical system on welding. In this section, the possible mechanical and pneumatic (assuming an air cylinder is used for driving the electrodes) processes during and after the welding current is applied are analyzed.

8.6.1 Thermal Expansion

A direct consequence of Joule heating during the application of welding current is thermal expansion of a weldment. However, such expansion is constrained by the arms of a welder. The degree of constraining depends on the stiffness of the machine. A stiffer machine allows less thermal expansion than a more compliant one. As a result, the electrode force is higher with a stiffer machine, as shown in Figure 8.7. This phenomenon can be better understood by estimating the expansion of the weldment and the constraining exerted onto it by the electrodes. The thermal expansion of a weldment includes both liquid and solid expansion. As shown in Chapter 7, the melting, or solid-to-liquid phase transformation, contributes the most to the total thermal expansion of the weldment. Therefore, the electrode force needed to counter the expansion can be approximated using the pressure–volume relation as shown in Equation (7.9):

$$P = \frac{1}{\kappa} \ln \frac{V_L}{V_S} \tag{8.6}$$

In this equation, κ is a constant and P is the pressure needed to compress a liquid of volume V_L to a solid of volume V_S (the same volume as that in solid state before melting). If the liquid is not compressed to V_S, but some intermediate volume V ($V_S \le V \le V_L$), then the pressure needed is

$$P = \frac{1}{\kappa} \ln \frac{V_L}{V} \tag{8.7}$$

Assuming the projection area A of the liquid nugget normal to the axis of the electrodes does not change during the compression, and only the height changes as a result of the volume change, then the force needed to compress the liquid metal from V_L to V is

$$F = PA = \frac{1}{\kappa} A \ln \frac{V_L}{V} = \frac{1}{\kappa} A \ln \frac{H_L}{H} \tag{8.8}$$

where H_L and H are the heights of the liquid metal at pressure-free state and under the pressure P, as defined by $H_L = V_L/A$ and $H = V/A$, respectively. This force is illustrated in Figure 8.28 as a function of the height of the liquid metal. In this figure, $H_S = V_S/A$; (H_{high}, F_{high}) and (H_{low}, F_{low}) are the liquid metal heights and electrode forces corresponding to high and low welder stiffness, respectively.

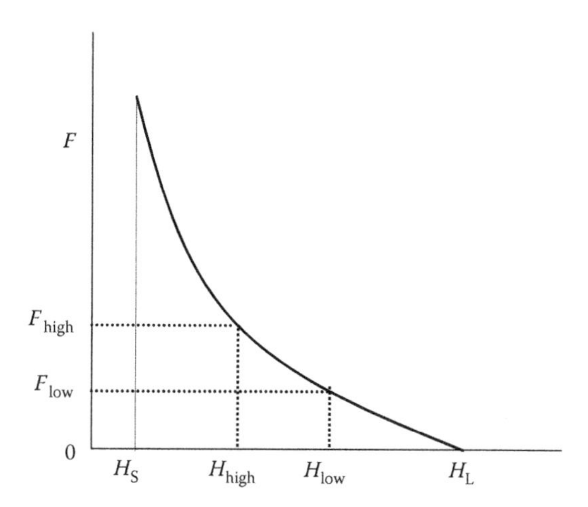

FIGURE 8.28
A schematic of relationship between force exerted onto a liquid and height of liquid metal.

Because a welder with high stiffness allows small expansion during welding, the force exerted onto the liquid by the electrode through the welder frames is high, as shown in Figure 8.28, compared with a welder of low stiffness. This approximation aims to provide a rough rationale for the relationship between stiffness and electrode force during welding. It should not be used for quantitative calculation or prediction as it is oversimplified and idealized. Other factors, such as material softening, may contribute significantly to the overall behavior of electrode movement during welding.

8.6.2 Effect of a Pneumatic Cylinder

The closing of electrodes in a spot weld can be driven in several ways, based on pneumatics (air cylinder), hydraulics, and electricity (servomotor), or a hybrid form (combination) of two of the above mechanisms. Among them, air cylinders are used predominately for economical reasons, and they are used as a model system in the following analysis. In this section, simplified analytical expressions of force and pressure change during welding are derived, and corresponding experimental results are presented.

As discussed in previous sections, both the distance between the electrodes and the electrode force change during welding are directly influenced by the thermal–mechanical behavior of the weldment, and they are also functions of the machine mechanical system through the influence of friction and other factors related to the moving system.

Friction of the moving parts is directly linked to the follow-up of electrodes during welding. Although friction reduces the preset value of electrode forces, it increases the electrode force significantly during welding, because of the sluggishness of the moving parts. However, this increase is delayed, so it may not serve the purpose of forging and expulsion prevention. Force forging, an important process in deterring the formation of internal discontinuities, can be fully utilized when the electrodes are free to move in. For this purpose, the self-locking mechanism that prohibits the retraction of the pneumatic cylinder after the preset force value is reached may promptly provide the needed electrode forces.

Because the acceleration of the electrode motion during welding is quite low, the influence of moving mass is usually negligible. However, the cylinder size and line pressure, if a pneumatic cylinder is used to drive the moving arm, or the speed of a servomotor, for the case of an electric gun, also contribute to the agility of the electrodes during welding.

8.6.2.1 Theoretical Analysis

The influence of a moving system on the follow-up can be better understood through an analysis of the forces in the system using a simplified configuration and assumptions on working conditions.[16]

The air cylinder and the weld head of a resistance welder can be simplified as a cylinder, a piston, a guideway, and an electrode and its holder, as shown in Figure 8.29a. During welding, various forces act on the moving system. They are F_p, force due to the air pressure on the piston head; F_{f1}, force due to friction between the piston head and the cylinder wall; F_{f2}, friction in the guideway; and F_e, the reaction from the weldment, or the electrode force (Figure 8.29b).

A resultant force due to friction, $F_f = F_{f1} + F_{f2}$, can be used because the loading state of particular locations is not of concern and the moving assembly can be assumed a rigid body. Newton's second law of motion yields

$$F_e - F_f - F_p = ma \tag{8.9}$$

where m is the total mass of a moving assembly and a is acceleration. a is usually fairly small, as discussed in Section 8.5, so for simplicity, ma can be ignored. Therefore,

$$F_e - F_f - F_p = 0, \text{ or } F_e = F_f + F_p \tag{8.10}$$

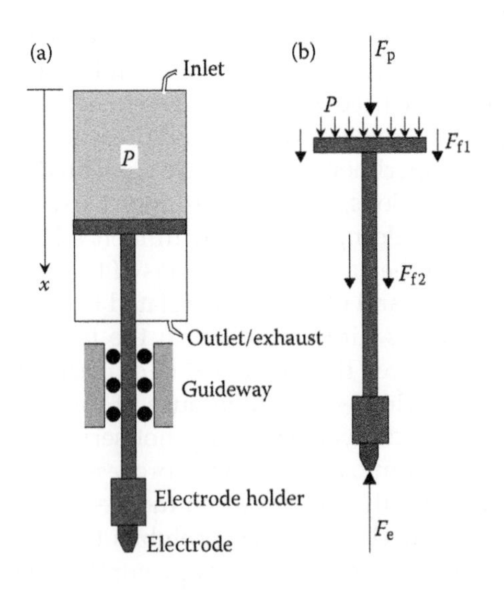

FIGURE 8.29
(a) A schematic of an air cylinder and upper electrode assembly. (b) Forces acting on moving components.

Any change in electrode force can be reflected by changes in pressure and in friction:

$$dF_e = dF_f + dF_p \tag{8.11}$$

Assume that welding or the application of electric current starts after the electrodes are stabilized, that is, the initial friction is zero. Then the change in friction force is a kinetic friction and can be assumed a constant in value:

$$dF_f = \left(-\frac{dx}{|dx|}\right)\mu_k N \tag{8.12}$$

where μ_k is the kinetic friction coefficient and N is the total force at the contact surfaces between moving and stationary parts, in the direction normal to the surfaces. μ_k and N are assumed constant. dx is the change in length, or displacement of the moving electrode. Thus, $-dx/|dx|$ indicates the opposite direction of motion. Friction force is usually non-constant, especially when the sliding system (guideway) is not properly installed. Bending caused by misalignment may contribute considerably to the normal force N and change the value of N, depending on the severity of bending.

The force change due to pressure can be approximated utilizing the ideal gas law, which describes the relation between pressure, P; volume, V; molar number of the gas, n; and temperature, T:

$$PV = nRT \tag{8.13}$$

where R is the gas constant. If one considers the state of air in the cylinder only, as marked in Figure 8.29, then its temperature can be assumed constant as the heat transfer in the air cylinder can be ignored because of the extremely short period of welding. In general, the pressured volume of a cylinder changes because of the oscillation of electrodes; so does the pressure and the amount, or the molar number of air.

The instantaneous volume of the pressure chamber can be expressed by the length x and cross section area A of the chamber, $V = xA$. Substituting it into Equation 8.13:

$$xPA = nRT \tag{8.14}$$

Differentiating it yields

$$x dP + P dx = \frac{RT}{A} dn \tag{8.15}$$

or

$$dP = \frac{RT}{xA} dn - P\frac{dx}{x} \tag{8.16}$$

It can be further simplified as

$$dP = P\left(\frac{dn}{n} - \frac{dV}{V}\right) \tag{8.17}$$

Equation 8.17 indicates that a change in pressure has contributions of two parts: that due to an increase in the amount of air and the decrease due to volume expansion. Using Equation 8.17, the change in force onto the piston head resulting from air pressure, $F_p = AP$, can be expressed as

$$dF_p = AdP = F_p\left(\frac{dn}{n} - \frac{dV}{V}\right) \tag{8.18}$$

Therefore, the change in electrode force is

$$dF_e = F_p\frac{dn}{n} - F_p\frac{dV}{V} - \mu_k N\frac{dV}{|dV|} \tag{8.19}$$

Note that $dV/|dV| = dx/|dx|$. Equation 8.19 is the general quantitative expression of force change during welding. It shows that change in electrode force is approximately proportional to the electrode force value, or preset electrode force. As electrodes close onto the workpieces, $dV > 0$, so the change in electrode force dF_e is determined by whether the air supply can react promptly, or the flow rate dn/n is high enough to offset the decrease due to cylinder volume expansion and friction.

A special case is when expulsion occurs. Because it happens in a very short period (usually in a few milliseconds), airflow in and out of the cylinder can be neglected, or $dn/n \approx 0$. The electrodes move in onto the workpieces during expulsion because part of the liquid metal is ejected from the nugget, so $dV > 0$ and

$$dF_e = -F_p\frac{dV}{V} - \mu_k N\frac{dV}{|dV|} < 0 \tag{8.20}$$

A drop in electrode force then results. The change in air cylinder pressure due to expulsion can be estimated using the ideal gas law. Because $dn \approx 0$, so $P_{after}V_{after} \approx P_{before}V_{before}$ and

$$\begin{aligned}
\Delta P &= P_{after} - P_{before} \\
&= \frac{P_{before}V_{before}}{V_{after}} - P_{before} \\
&= P_{before}\frac{V_{before} - V_{after}}{V_{after}} \\
&= -P_{before}\frac{\Delta V}{V_{after}}
\end{aligned} \tag{8.21}$$

This expression can be further simplified as

$$\begin{aligned}\Delta P &= -P_{\text{before}} \frac{\Delta x}{x_{\text{after}}} \\ &= -P_{\text{before}} \frac{\Delta x}{x_{\text{before}} + \Delta x} \\ &\approx -P_{\text{before}} \frac{\Delta x}{x_{\text{before}}}\end{aligned} \tag{8.22}$$

It is therefore evident that the sudden movement of the electrode caused by the expulsion causes the sudden pressure drop in the air cylinder. The magnitude of pressure change would reflect the extent of the expulsion. The sensitivity of the pressure change to the expulsion is proportional to the pressure level in the air cylinder, and inversely proportional to the initial height of the upper air chamber.

An air cylinder usually has controlled pressure on both sides of the piston head: one in the inlet chamber, and the other on the outlet (exhaust) side. The pressure discussed above is the differential of the two pressures, with consideration of the area covered by the piston rod on the exhaust side.

8.6.2.2 Experiment Results

An experiment was conducted on a Taylor Winfield 30 kVA press-type machine at the Edison Welding Institute.[17] Figure 8.30 shows the experimental setup. The two displacement transducers mounted near the electrodes measure the displacement between the upper electrode and the machine frame, and the displacement between the two electrodes, respectively. The displacement traces were used to confirm the event of the expulsion. Two pressure transducers of 100 lb/in^2 in range and 0.2% in accuracy were mounted at the inlet

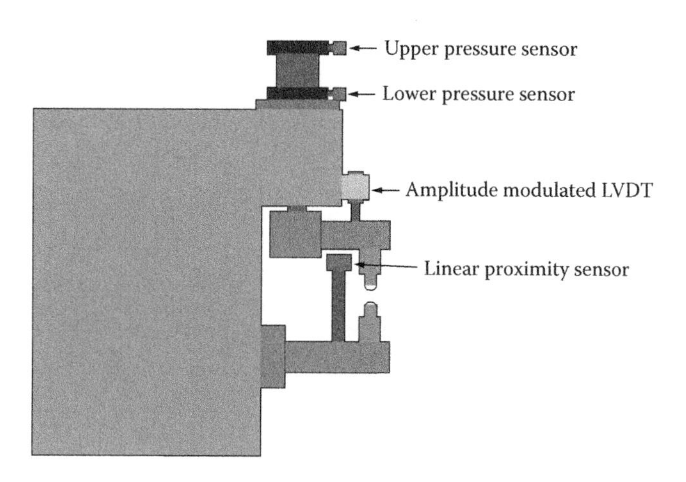

FIGURE 8.30
Experimental setup for monitoring cylinder pressure and displacement on a pedestal welder.

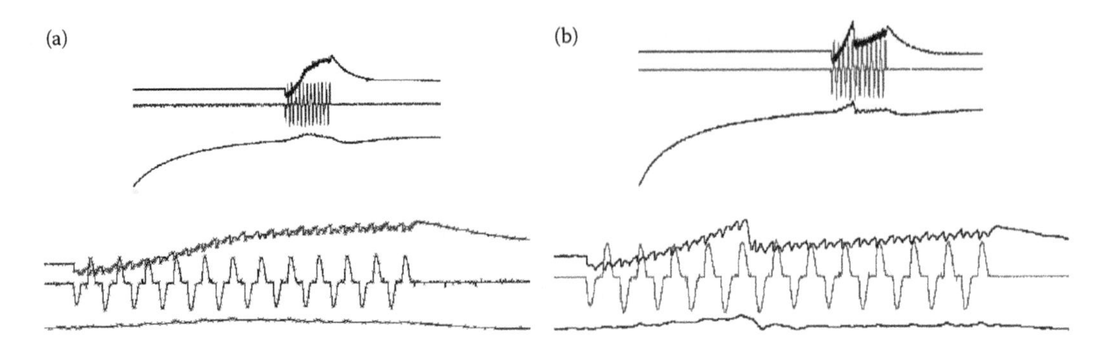

FIGURE 8.31
Displacement and pressure variations during welding (a) without expulsion and (b) with expulsion. Lower part is an enlarged view of welding period.

and outlet ports of the air cylinder, to monitor the pressure variation during the welding cycle. The secondary welding current was also collected.

Figure 8.31a and b shows examples of displacement and pressure variations during welding, without and with the occurrence of weld expulsion, respectively. The secondary welding current trace provides the time reference for the other signals on the plots. When there is no expulsion (Figure 8.31a), the pressure increases gradually during the welding cycle, corresponding to thermal expansion of the weldment due to heating, and gradually decreases after the welding current is shut off. On the other hand, there is a sudden drop in pressure associated with the occurrence of expulsion (Figure 8.31b). The pressure signals presented in the figures were directly taken from the sensors without postprocessing of the signals. They had very little electrical interference from the secondary welding current.

Table 8.5 summarizes the pressure changes in the air cylinder during expulsion for different cylinder sizes, preset pressure levels, and sheet thicknesses, collected during the experiments. The pressure varied from 0.2 to 1.5 lb/in^2, which is measurable using standard pressure transducers.

A comparison can be made between the calculation based on the theoretical analysis in the previous sections, and an actual case using the setup shown in Figure 8.30. The measurement from the experiment is as follows: $x_0 = 25.4$ mm, $P_0 = 80$ lb/in^2, $\Delta x = 0.15$ mm, and $\Delta P = 0.3$ lb/in^2. Using Equation 8.22, a change of 0.47 lb/in^2 is obtained, which is reasonably close to the test data.

TABLE 8.5
Summary of Pressure Decrease Associated with Expulsion

Material Thickness (in)	Cylinder Diameter (in)	Pressure (upper/lower) (lb/in^2)	Expulsion (yes/no)	ΔP During Weld (lb/in^2)
0.034	4	80/15	Yes	1.190
0.034	4	50/15	Yes	0.300
0.047	4	50/15	Yes	0.300
0.047	4	80/15	Yes	1.480
0.034	3	80/15	Yes	0.210
0.047	3	80/15	Yes	0.196

These testing results suggest that the sudden pressure drop during welding could be used for automated weld expulsion monitoring on production lines. It would also lead to the design of simple and effective control feedback systems for eliminating expulsion.

In conclusion, both the experimental data and the theoretical analysis reveal that there is a direct correlation between the sudden motion of electrode and the sudden change in pressure in air cylinder when expulsion happens in welding. The magnitude of pressure change can be used as an indication for the extent of expulsion. The level of pressure change depends on the pressure level in the cylinder and the height of the upper air chamber. The pressure changes range from 0.2 to 1.5 lb/in^2 for the machine used, which can be readily measured with out-of-shelf pressure transducers.

Measuring the pressure signal appears to be a viable method for monitoring the electrode motion during expulsion, thus providing an attractive alternative for expulsion detection and feedback control. There are several advantages of using pressure signal for expulsion detection, because of the location of the pressure transducers. It is less intrusive than the displacement transducers in the weld area. It also experiences much less electrical interference from the welding current loop, which often causes great difficulties in force and displacement measurement in RSW. It appears that the expulsion detection algorithms using the pressure changes would be rather simple and straightforward to implement in a production environment.

The follow-up process is also a strong function of electrode geometry. Flat-faced electrodes have a relatively stable contact area with the workpieces, and the contact area between dome-faced electrodes and sheets changes constantly during welding due to thermal softening. Therefore, the signals such as electrode displacement, force, and dynamic resistance have different characteristics for welding using different electrodes. Caution must be exercised before applying findings from one welding process to others.

8.7 Squeeze Time and Hold Time Measurement

The measured signals of electrode force, displacement, and electric current can be used to analyze mechanical properties of a welder, as discussed in Section 8.3. One application is to quantify the differences between setup parameters and the true responses of a machine. The setup values, often input through a keypad on an electronic controller, have been regarded as the true responses obtained during a welding cycle with negligible errors. However, a mechanical system may not respond as desired because it usually takes a certain amount of time for a mechanical system to react after it receives a command from a controller. For instance, a finite amount of time is needed for a pneumatic cylinder to reach the preset pressure after the valves are opened by an electronic actuation. This may have an influence on both the squeeze stage and the hold period of a weld cycle. If the preset value of electrode force is not reached before the current is applied, excessive heat could be generated at the contact surfaces because of high contact resistivity, and in some cases, an electric arc could form between two metals. If, on the other hand, the electrodes are retracted at a different instant from the designated schedule, from the workpieces after the current is shut off, the just-formed weld could be undercooled or overcooled. This could be a serious problem for high-strength steels. Therefore, it is of practical interest to understand the extent of differences that may exist between setup values and the true responses for a welder. It is reasonable to assume that such differences are machine or system dependent. As large-scale production often strives for a balance between high-quality welds and short process time, such study is essential in understanding the welding system and in predicting the weld quality once a relationship is established between the (true) process parameters and weld quality.

In a study by Wang,[12] the difference between the setup squeeze time and the measured squeeze time was calculated using the signals. Three sensors (i.e., force, displacement, and electric current sensors) were used for the study. The electric current sensors were used to identify the end of the squeeze period (as the start of the current) and the start of the cooling period (as the end of the current). The measured squeeze time is defined as the period starting from the moment at which electrodes start to move, characterized by the start of increase in displacement value, as shown in Figure 8.12. The time ends when electric current is applied. Another time to be considered during squeezing is the time to stabilize the electrode force, when the electrode force reaches a steady level (or the preset value). These times are defined in Figure 8.12. The comparison between setup squeeze time and measured squeeze time, as well as the time needed for stabilizing the electrode force, is presented in Table 8.6 for C-gun I and in Table 8.7 for C-gun II.

Table 8.6 shows that there is an approximately constant difference between real (measured) squeeze time and the setup time. A positive value (~4 cycles) means that the measured time is longer than the setup, which is desirable because electric current would not be applied, or welding would not occur prematurely. In contrast, the difference changes from positive to negative when the setup squeeze time increases for C-gun II, as shown in Table 8.7. For this welder, welding could start before the electrode force reaches the desired level and, therefore, additional squeeze time should be used. The electronic controller probably should take the main responsibility for such discrepancies between welders. The time needed to reach a stable electrode force appears fairly constant for each welder. C-gun I needs about 30 cycles and C-gun II needs about 21 cycles to reach the preset force level. These are the minimum squeeze times needed for welding. Shorter times may result in unconformable welds or excessive electrode wear because of overheating of the contact interfaces.

The hold time response can also be analyzed using the same signals. The hold time starts at the end of the electric current application, but the end of the hold period can be defined in several ways. The start of electrode force drop (F-start) can be considered the end of electrode cooling of the weldment. The moment when the electrode force drops to zero (F-end) and that when electrodes start to leave the workpieces (D-start, so the electrodes are not in contact with the workpieces) can also be defined as the end of the cooling period. These periods are illustrated in Figure 8.12. Tables 8.8 and 8.9 list the various hold

TABLE 8.6

Setup and Measured Squeeze Time for C-Gun I

Setup (cycle)	Measured (ms)	Measured (cycle)	Stable (ms)	Stable (cycle)	Difference in Squeeze (ms)	Difference in Squeeze (cycle)
35	518.0	31.1	486.8	29.2	65.10	3.9
35	519.0	31.1	502.0	30.1	64.10	3.8
37	548.0	32.9	483.0	29.0	68.42	4.1
38	559.0	33.6	483.0	29.0	74.08	4.4
39	572.5	34.4	477.0	28.6	77.24	4.6
39	579.0	34.8	440.0	26.4	70.74	4.2
99	1573.0	94.4	491.0	29.5	76.34	4.6

TABLE 8.7

Setup and Measured Squeeze Time for C-Gun II

Setup (cycle)	Measured (ms)	Measured (cycle)	Stable (ms)	Stable (cycle)	Difference in Squeeze (ms)	Difference in Squeeze (cycle)
1	425.0	25.5	342.8	20.6	408.34	24.5
2	461.2	27.6	362.7	21.8	427.88	25.7
5	557.6	33.5	359.1	21.6	474.30	28.5
10	173.4	10.4	346.7	20.8	6.80	0.41
20	260.6	15.6	335.5	20.1	–72.60	–4.4
30	429.2	25.8	329.1	19.8	–70.60	–4.2
35	505.8	30.3	320.7	19.2	–77.30	–4.6
40	590.3	35.4	333.6	20.0	–76.10	–4.6
45	674.3	40.5	341.1	20.5	–75.40	–4.5
50	755.6	45.3	335.5	20.1	–77.40	–4.6
60	920.6	55.2	335.5	20.1	–79.00	–4.7
70	1091.0	65.5	362.6	21.8	–75.20	–4.5
80	1263.0	75.8	361.2	21.7	–69.80	–4.2
90	1422.0	85.3	357.2	21.5	–77.40	–4.6
99	1573.5	94.4	343.4	20.7	–75.84	–4.6

times (expressed as the difference between measured and setup hold times) measured according to different definitions on C-guns I and II.

The hold time characterized by the start of electrode motion seems a reasonable choice, as the physical separation of the electrodes from the workpieces marks the end of cooling. As shown in Tables 8.8 and 8.9, the C-guns have fairly consistent larger hold times than the setup values. C-gun I has about 5 to 7 cycles extra hold time than the preset value, and C-gun II has about 5 cycles more than the setup. As the cooling rate determines the microstructure and, therefore, the mechanical properties of a weldment after electric current is shut off, the discrepancies between true and setup hold times are of certain importance. Because hold time effect is different for mild steel, HSLA steel, and advanced high-strength steel, such an effect deserves a systematic study and the deviation of real hold time from its setup value should be accounted for.

TABLE 8.8

Setup and Measured Hold Time for C-Gun I

Setup (cycle)	Δt (F-start, cycle)	Δt (F-end, cycle)	Δt (D-start, cycle)
0	3.12	6.72	6.88
1	2.60	5.95	6.35
9	2.29	5.64	5.96
9	2.32	5.56	5.83
12	2.26	5.62	5.96
40	2.32	5.22	5.70
50	2.69	5.55	6.22
100	2.34	6.08	6.12
150	2.52	6.66	6.24
180	3.66	7.92	7.38
198	2.46	6.78	6.12

TABLE 8.9

Setup and Measured Hold Time for C-Gun II

Setup (cycle)	Δt (F-start, cycle)	Δt (F-end, cycle)	Δt (D-start, cycle)
1	2.1	4.7	5.5
10	1.4	4.7	4.9
20	1.4	4.6	5.1
30	1.5	4.7	5.1
40	1.5	4.6	5.2
60	1.3	4.7	5.3
80	1.5	4.8	5.2
99	1.6	4.9	5.3

The C-guns used in the study showed differences in responses, and it can be expected that welders of different types will be different. Therefore, the numeric values of the times obtained for the C-guns cannot be generalized, but the methodology is applicable.

8.8 Other Factors

Based on the discussions in previous sections, it can be concluded that stiffness is probably the most critical mechanical property of a welder. It has a direct impact on electrode alignment and electrode force and, therefore, a weld's quality. These two factors and the influence of workpiece materials are further explored in this section.

8.8.1 Electrode Alignment and Workpiece Stack-Up

It is desirable to have electrodes aligned to make electrode faces parallel and to have them overlapped during welding. The purpose of the alignment is to create a contact as uniform as possible for conducting electric current and generating heat for nugget formation. The deviation from the ideal positions, either axially or angularly, is called misalignment. These conditions are illustrated by Figure 8.32.

An approximate relationship can be established between the contact area and the electrode axial alignment due to structure deflection based on a geometric consideration. An axial misalignment reduces the (overlapped) contact area between the sheets, as shown in Figure 8.33. The actual contact area (C_r), in percentage of electrode face area, can be approximated by the following equation:

$$C_r = \frac{2r^2 \arccos\left(\dfrac{\delta}{2r}\right) - 2\delta\sqrt{r^2 - \left(\dfrac{\delta}{2}\right)^2}}{\pi r^2} \tag{8.23}$$

where r is the radius of electrode face and δ is the axial misalignment. It can be seen from the equation that the reduction in contact area is strongly dependent on the axial misalignment.

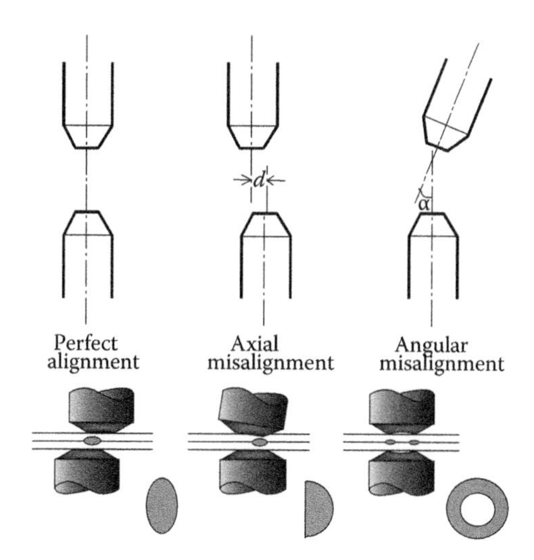

FIGURE 8.32
Perfectly aligned, axial, and angular misaligned electrodes, and possible nugget shapes due to abnormal electrode conditions.

Another consequence of misalignment is the change induced in contact pressure. Assuming a misalignment of 0.75 mm axially and 0.28° in angle, a finite element model produces an asymmetric pressure distribution at the faying surface under 2.67 kN (600 lb) of electrode force, as shown in Figure 8.34. The average pressure with perfect alignment on the faying surface is 83.0 MPa. Under ideal alignment, high pressure occurs around the electrode edge, which plays a role in constraining the molten nugget and preventing possible welding expulsion. When the electrodes misalign, the pressure distributes asymmetrically. Obviously, this asymmetrical pressure distribution is unfavorable in terms of expulsion prevention[13] and electrode life.

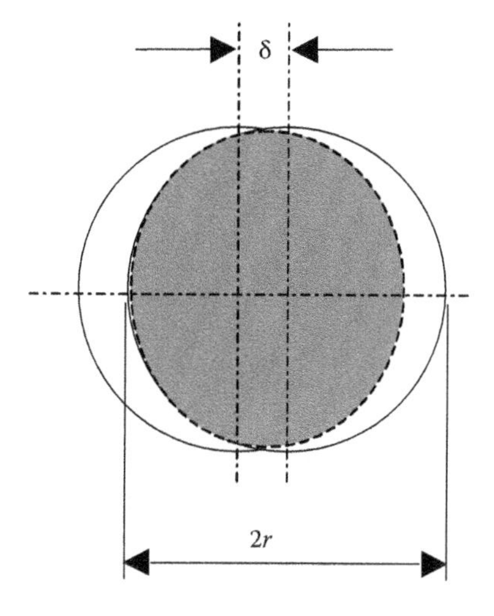

FIGURE 8.33
Geometric model for calculating contact area due to axial misalignment.

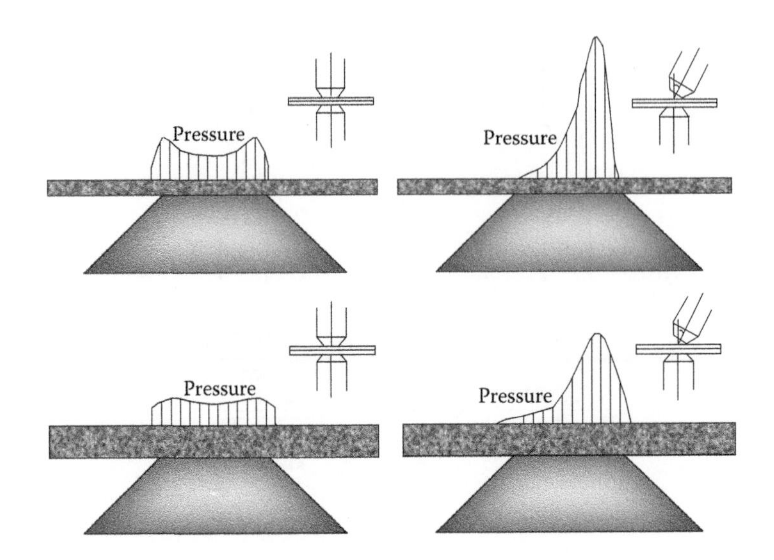

FIGURE 8.34
Pressure distribution at faying surface for perfectly aligned (left) and misaligned (right) electrodes in welding 0.8 mm (upper) and 1.7 mm (lower) steel sheets.

In addition, the analysis also shows that sheet thickness is a factor on pressure distribution. Thicker sheets tend to have a more uniform distribution of pressure at the faying surface.

Although this calculation did not consider thermal–mechanical interaction occurring during welding, it points out the possible influence of machine stiffness on electrode alignment and possibly on weld quality. In practice, a certain amount of angular misalignment can be compensated by the fact that electrodes are usually slightly worn after a few welds, and by using dome-shaped electrodes. In general, establishing a tolerance of axial and angular misalignments for RSW machine design needs further finite element analysis and experimental study.

Electrode alignment is directly related to the stiffness of welding machine frames and the workpiece stack-up. Aligned electrodes alone cannot guarantee a good alignment, as both workpiece stack-up and the relative position of the electrodes with respect to the workpieces determine the final alignment. Some possible stack-ups are shown in Figure 8.35.

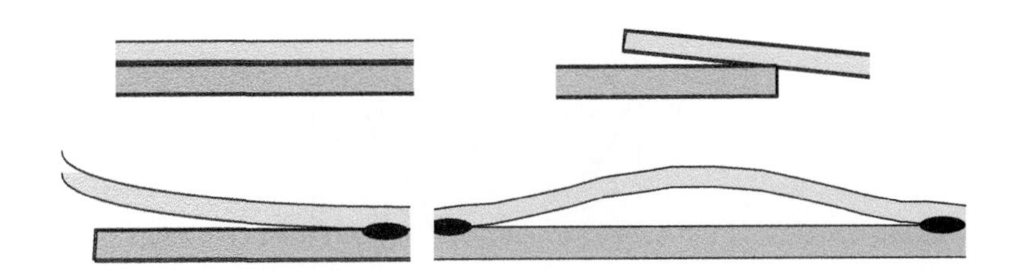

FIGURE 8.35
Typical workpiece stack-ups.

A large number of stack-up possibilities exist, considering possible combinations of electrode alignments with workpiece stack-ups, such as using the cases shown in Figures 8.32 and 8.35. Therefore, electrode alignment is case dependent and should be treated with the consideration of both welder setup and configuration of workpieces. The imprint method using carbon paper to record the relative locations of electrode faces is a convenient way to measure axial electrode alignment. However, it provides very little information on the condition of angular alignment. Caution should be taken when this method is used on gun-type welders with one or two rotating arms to close electrodes (squeezing). Workpiece stack-up (both total thickness and fit-up) may significantly affect both axial and angular alignments. The effect of workpiece fit-up depends on the sheet thickness. As shown in an experimental study, little practical effect of moderately poor fit-up was found when welding light-gauge, low-strength steel.[18]

Besides the installation of electrodes and their holders, stiffness is probably the most influential factor affecting electrode alignment. It dictates the deflection of welder arms and, therefore, the alignment. Such deflection tends to increase with applied electrode forces. Thus, alignment should be determined under the largest electrode force possible in the working ranges.

Because the alignment conditions of electrodes directly affect the electrical contact and current distribution, they are the single most important factor in determining the nugget location, as discussed in Chapter 7. Therefore, good alignments tend to generate less offset and, therefore, more applied electrode forces are used to contain the liquid nugget to avoid expulsion. Electrode misalignment is also responsible for irregularly shaped nugget formation and excessive electrode wear due to uneven distribution of contact pressure between the electrodes and the workpieces, as shown in Figure 8.34. For example, Howe[7] found that electrode deflection could significantly change electrode life, which may result from the misalignment of electrodes created by electrode deflection.

The effect of electrode misalignment can be alleviated a bit if domed instead of flat electrodes are used. For the same reason, slightly worn electrodes may perform better than new electrodes because they are more forgiving in terms of angular alignment.

8.8.2 Electrode Force

As discussed in previous sections, when a mechanical characteristic of a machine, such as stiffness or friction, changes, it usually directly results in a change in electrode force. Therefore, a change in electrode force not only affects the weld quality, but also indicates a welder's capacity and the possible constraining that the electrodes can impose onto a weldment. Understanding the trend of electrode force change should be beneficial to weld quality improvement and welder design.

In the work by Tang et al.,[10] experiments were conducted under several preset forces with the other parameters held constant, in order to find the influence of the preset force on the force characteristics during the weld stage. Force profiles monitored are shown in Figure 8.36. A general observation is that the force increase becomes greater with a larger preset force. For example, the increment is about 338 N (76 lb) for a 5339 N (1200 lb) preset force and is 160 N (36 lb) for a 3500 N (800 lb) preset force. The force changes (ΔF) and the change ratio ($\Delta F/F$) are shown in Figure 8.37.

Note that electrode force does not always increase during welding. As shown in Figure 8.16, the force actually decreases when welding an aluminum alloy.

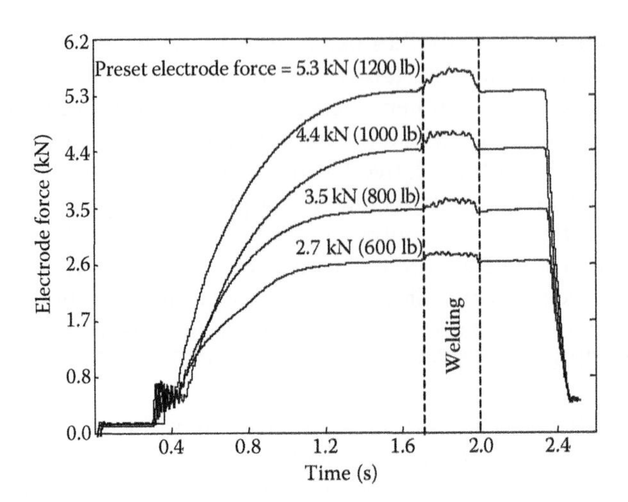

FIGURE 8.36
Force profiles vs. preset forces.

8.8.3 Materials

Besides electrical properties, the thermal–mechanical behavior of workpieces greatly influences welding processes. For instance, during heating, the weldment expands due to both solid and liquid expansion. Such expansion is constrained by the electrodes, and different materials behave differently under such constraint according to their unique thermal–mechanical characteristics. Some differences in behavior are clearly demonstrated by comparing the force profiles of steel and aluminum welding. The electrode force in steel welding has a larger value than the preset force during the application of current, whereas aluminum welding exhibits a drop in electrode force during welding. This phenomenon stems from the differences in mechanical and thermal behavior of these two materials at elevated temperatures, and whether the force exerted on the electrodes by the weldment increases or drops largely depends on the competition of thermal expansion and softening. A comparison of yield

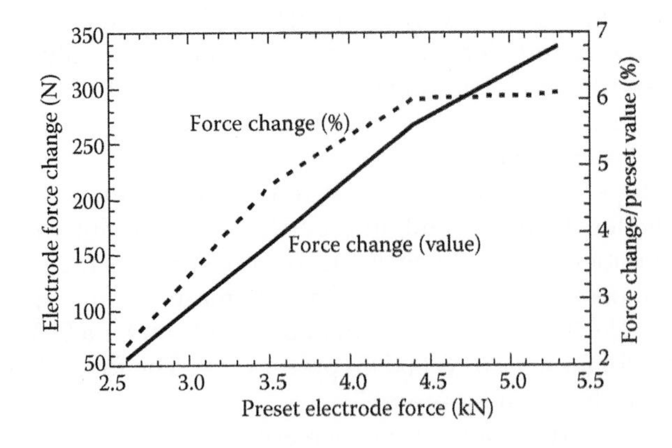

FIGURE 8.37
Force change under various preset forces.

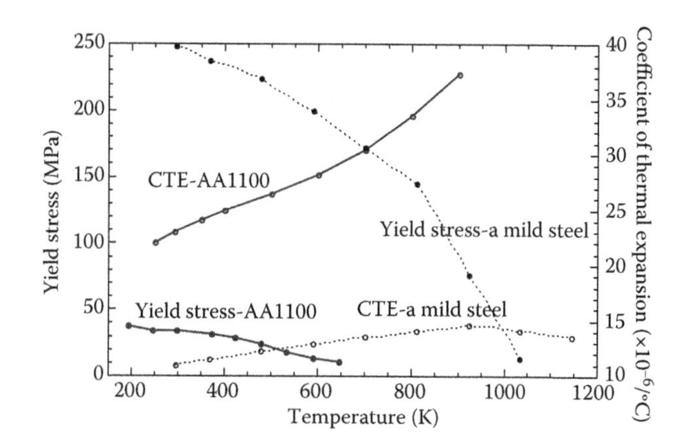

FIGURE 8.38
Comparison of material properties between a steel and an aluminum alloy.

strength and coefficient of thermal expansion between steel and an aluminum alloy is shown in Figure 8.38. The expansion of a weldment (including both liquid and solid phases) in a weldment pushes the electrodes apart and, therefore, exerts additional force onto the electrodes. On the other hand, the softening of materials at elevated temperatures makes it easy for the electrodes to extrude into the weldment and results in a drop in electrode force. The separation of sheets during welding may also contribute to the force decrease. In steel welding, the effect of thermal expansion overcomes that of softening and, therefore, the total electrode force increases. On the contrary, the electrode force decreases during aluminum welding because the softening has a larger effect. Note that the trend of electrode force change reverses around the middle of welding for both steel (decreasing) and aluminum (increasing). This can also be explained by the competing processes between thermal expansion and softening, with the larger influence of temperature when melting starts or the melting pool grows to a certain size.

The thermal–mechanical effect due to heating is also reflected by the change of electrode force during welding as a function of welding current, as shown in Figure 8.39. The

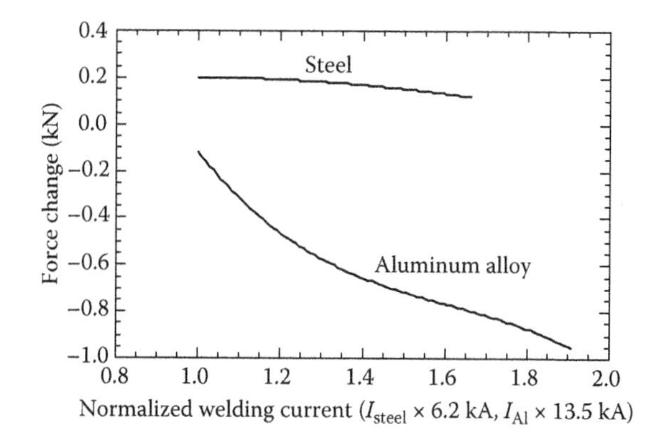

FIGURE 8.39
Dependence of electrode force change on welding current.

maximum electrode force is larger than the preset value during welding for steel, and it is smaller for the aluminum. However, the increase in the case of steel welding diminishes as the welding current increases due to increased heating/softening. The decrease in electrode force for aluminum welding becomes more severe with welding current for the same reason.

Because of the differences in material properties, aluminum alloys act much differently from steel during welding. Therefore, the welding schedules for aluminum are different from those of steel. For example, if the electrode face is 6 mm in diameter for steel welding, the pressure on the workpiece surface is 126 MPa, which is about 36% of steel yield stress at room temperature. Comparatively, the pressure is about 133 MPa for aluminum welding with a 7-mm-diameter electrode, which is about 116% of aluminum yield stress. Even at room temperature, the aluminum alloy yields under preset electrode force. The different properties of aluminum alloys from those of steels determine that new welding machines, instead of those used for steel welding, are needed for aluminum welding.

References

1. Ganowski, F. J. and Williams, N. T., Advances in resistance spot and seam welding of zinc coated steel strip, *Sheet Metal Industries*, 49, 692, 1972.
2. Kolder, M. W., and Bosman, A. W. M., Influence of the welding equipment on the weldability lobe of an HSLA-steel, IIW Doc. No. III-796-84, 1984.
3. Satoh, T., Katayama, J., and Okumura, S., Effects of mechanical properties of spot welding machine on electrode life on electrode life for mild steel, IIW Doc. No. III-912-88, 1988.
4. Satoh, T., Katayama, J., and Nakano, T., Effect of mechanical properties of spot welding machine on spot weld quality, IIW Doc. No. III-912-88, 1988.
5. Hahn, O, Budde, L., and Hanitzsch, D., Investigations on the influence of the mechanical properties of spot welding tongs on the welding process, *Welding and Cutting*, 42, 6, 1990.
6. Williams, N. T., Chilvers, K., and Wood, K., The relationship between machine dynamics of pedestal spot welding machines and electrode life, IIW Doc. No. III-994-92, 1992.
7. Howe, P., The effect of spot welding machine characteristics on electrode life behavior on two welders, in *Proceedings of AWS Sheet Metal Welding Conference VII*, Detroit, MI, Paper No. A3, 1996.
8. Dorn, L. and Xu, P., Influence of the mechanical properties of resistance welding machines on the quality of spot welding, *Welding and Cutting*, 45, 12, 1993.
9. Dorn, L. and Xu, P., Relationship between static and dynamic machine properties in resistance spot welding, *Welding and Cutting*, 44, 19, 1992.
10. Tang, H., Hou, W., Hu, J. S., and Zhang, H., Force characteristics of resistance spot welding of steels, *Welding Journal*, 79, 175-s, 2000.
11. Tang, H., Hou, W., Hu, S. J., and Zhang, H. Influence of machine mechanical characteristics on RSW process and weld quality, *Welding Journal*, 82 (5), 116-s–124-s, 2003.
12. Wang, Y., Mechanical characterization of resistance welding machines, MS Thesis, The University of Toledo, 2005.
13. Zhang, H., Expulsion and its influence on weld quality, *Welding Journal*, 78, 373s, 1999.
14. Howe, P., The effect of spot welding machine characteristics on electrode life behavior, SMWC VII, Paper 3, 1996.
15. Gould, J. E. and Dale, W. N., Theoretical analysis of weld head motion, in *Proceedings of AWS Sheet Metal Welding Conference VII*, Detroit, MI, 1994.

16. Gould, J. E. Feng, Z., Chou, J., and Kimchi, M., Analytical models for the mechanical response of a resistance spot welding machine, CRP Report SR9902, Edison Welding Institute, Columbus, OH, 1999.
17. NIST—ATP Intelligent Resistance Welding Quarterly Progress Report, No. 304, Ann Arbor, MI, 1998.
18. Natale, T. V. and Pickett, K., The effect of workpiece fit-up and electrode composition on the resistance spot welding behavior of hot-dip galvanized sheet steel, SMWC IV, Paper 9, 1990.

9

Numerical Simulation in Resistance Spot Welding

9.1 Introduction

Unlike other welding processes, resistance spot welding (RSW) is difficult to directly monitor on the weld nugget development, because melting and solidification processes primarily take place between the workpieces. A common practice is to control the input, such as welding parameters, and monitor the output, such as the attributes of a weld and process signals. However, little can be learned about the nugget formation process from the input and output. Complexity rises because of the interacting electrical, mechanical, thermal, and metallurgical processes. Numerical simulation (e.g., finite element analysis) is a powerful tool in this situation. Detailed thermal profile, stress and strain distributions, as well as distortion at various stages can be revealed by numerical simulations. Welding process parameters, such as electrode force, welding current, and welding machine stiffness, can be easily altered using the finite element method to study their influence. Performing a similar study experimentally would be extremely difficult if not impossible.

Finite element simulation of RSW has been attempted by many researchers, and most of the work has been on resistance welding steel. For instance, a commercial finite element simulation package, ANSYS™,[1] was used for simulating RSW of steel by sequentially coupling electrical–thermal and thermal–mechanical processes.[2,3] A small portion of the existing work is on aluminum welding. Computational models were also proposed to study effects of electrode geometry, electrode wear, and thermoplastic constitutive relationship.

An accurate prediction of weld structure and properties requires a precise simulation of the interactive mechanical–thermal–electrical process of RSW. This is not always possible because there is no commercial software package available to handle fully coupled aforementioned processes. In addition to the lack of a capable program, the scarce source of material properties, especially their temperature dependence, available for numerical simulation has significantly hindered the progress in this regard. For instance, the large variation of contact resistance during welding cannot be accurately accounted for because of the lack of data and process randomness. The improvement in simulation precision has been made in recent years, but it is generally limited by the advances in numerical simulation techniques and computation software and hardware. More reliable material data are needed that can be only obtained from material research.

In this chapter, the fundamentals of finite element simulation of the RSW process are reviewed. Their applications in nugget growth modeling and microstructure evolution are then presented.

9.1.1 Comparison between Finite Difference and Finite Element Methods

As two major numerical methods for solving engineering problems, both the finite difference method (FDM) and finite element method (FEM) have been used in RSW process simulation. The FDM was used almost exclusively in early works, whereas the FEM took over in the more recent efforts of simulating the RSW process. These two methods are different in terms of discretization, handling of boundaries, problem formulation, and simulation accuracy. In this section, they are compared in order to give the reader a simplified view of similarities and differences between these two methods.

9.1.1.1 Discretization

Both methods require the discretization of a structure, object, or region being analyzed. However, the way of discretization is fundamentally different. Such a difference can be clearly seen in the two-dimensional rectangular elements used in these two methods, as shown in Figure 9.1.

In a finite difference analysis, the object or region being analyzed is divided into a finite number of lumps, whereas in the finite element analysis, it is divided into a finite number of elements. Thus, in a finite difference approach, each lump is assumed to have a constant value of a pertinent field variable. For example, in a thermal analysis, the field variable temperature is assumed constant over a lump and the entire lump becomes isothermal. This indicates that the node for a lump is not associated with the corners of the lump, but rather with its geometric center or centroid, as shown in the figure. In the case of finite element analysis, the nodes are associated with the corners of an element and may have different values of a field variable, such as different temperatures at the corners. A field variable may vary over an element in a prescribed manner, depending on the interpolation scheme (functions) used. Higher-order interpolating polynomials can be used to form higher-order elements in finite element modeling.

9.1.1.2 Geometry

In a finite difference analysis, discretization assumes that the nodes are equally spaced. Thus, if rectangular lumps are used in an analysis and the problem requires nodes to be placed at boundaries (to impose prescribed boundary conditions), surface lumps that are essentially half the size of the interior lumps, and corner lumps of one fourth the size of their interior counterparts are needed, as shown in Figure 9.2. On the other hand, finite element modeling may define elements with nodes on the boundaries (Figure 9.2b).

For the same reason, discretizing complex geometries such as curvatures on surfaces into lumps in the FDM requires the use of jagged effective boundaries, as shown in Figure 9.3a

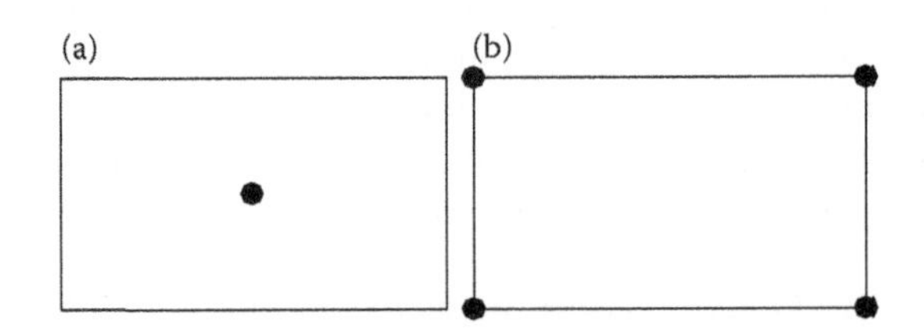

FIGURE 9.1
(a) A lump used in finite difference analysis. (b) An element used in finite element analysis.

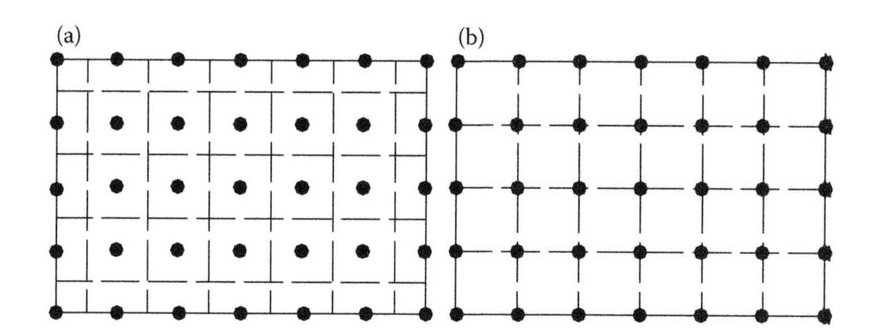

FIGURE 9.2
Discretization of a plate into (a) lumps for a finite difference analysis and (b) elements for a finite element analysis, for a rectangular domain.

by dashed lines. In finite element modeling, however, curved boundaries can be closely simulated using various types of elements, such as triangular (for plane problems) or tetrahedral (for three-dimensional problems) elements. The elements may be constructed to have a required shape of boundaries and have their nodes lie exactly on the prescribed boundaries.

Reducing the size of lumps or elements usually makes the mesh closer to the actual configuration. Yet there is a trade-off between accuracy and computation cost, and it often requires writing special finite difference equations for the boundary nodes that explicitly include the effect of curved boundaries in the FDM. However, in finite element analysis, various elements such as triangular or quadrilateral elements can be used to handle the complex geometries. This is the major advantage of finite element analysis over the FDM. Variable spacing of nodes and various shapes of elements can also be routinely handled by the FEM.

9.1.1.3 Formulation

In the FDM, the governing equations are written for each node. In the heat transfer formulation, for example, an energy balance is made on each lump. In FEM, the direct energy balance approach can be used, but there are some other approaches (e.g., virtual work, variational method, and weighted residual method) that are more powerful and flexible for engineering applications.

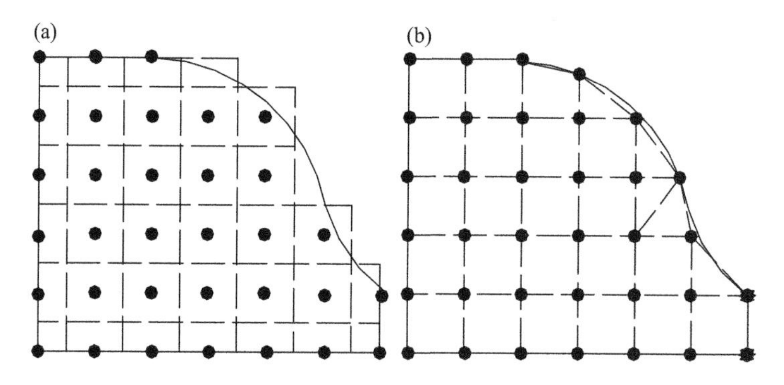

FIGURE 9.3
Irregularly shaped boundaries discretized into (a) lumps and (b) triangular and quadrilateral elements.

9.1.1.4 Accuracy and Others

It is difficult to make a general statement on accuracy, as a simulation's accuracy generally depends on many factors. The FEM seems to be more accurate when curved boundaries are present. It is also difficult to compare the execution times between the FEM and the FDM. In general, the FEM has a longer execution time. However, the FDM usually requires more preparation than the FEM.

9.1.2 Methods of RSW Process Simulation

The majority of early RSW simulations were based on FDMs. In the last decade or so, commercial finite element packages, such as ANSYS[1] and ABAQUS™,[4] have demonstrated their versatility in the simulation of RSW process, in particular by incorporating more realistic process parameters and conditions. However, FDMs remain an effective tool to some fundamental physics and mechanics associated with resistance welding process. The early work of RSW simulation can be traced back to the 1960s. For instance, temperature distribution of an RSW process was first determined numerically by Bentley et al.[5] Rice and Funk[6] created a simplified one-dimensional heat transfer model using the FDM to predict the temperature history. Moreover, using a one-dimensional finite difference model, Gould[7] compared simulation results with measured nugget growth in experiments.

These models possess a common inherent inability to deal with radial heat transfer because of the assumptions made, which inhibits the calculation of nugget growth in the radial direction. Also, these models could not account for nonuniform current density distribution. The use of two-dimensional models may overcome these shortcomings. Axisymmetric heat transfer models for analyzing the RSW process were developed later[8,9] using FDMs, which addressed some of these concerns.

The thermomechanical aspects of the spot welding process have generally been ignored in the finite difference models because of the limitations of the method. Consequently, contact areas at the electrode–sheet and sheet–sheet interfaces had to be assumed. For instance, in the two-dimensional model by Cho and Cho,[9] the diameter of the sheet–sheet contact area was assumed to be twice as much as that of the electrode tip surface. Huh and Kang[10] developed a three-dimensional thermoelectrical finite element model. Using a commercial FEM package, ANSYS,[1] temperature histories were obtained in the model for the electrode–sheet assembly, and another commercial program, ABAQUS,[4] was used to model the thermomechanical process with emphasis on finite strain thermoplasticity effects. Wei and coworkers[11–13] reported a model considering transport of mass, momentum, energy species, and magnetic field intensity. These models generally represent significant improvement over the early works. However, their focus was on the heat transfer response only; thermomechanical and thermoelectrical aspects of an RSW process were overlooked.

The early simulation works were generally on RSW steels. As the use of aluminum alloy sheets has increased rapidly in recent years in the automotive body assembly, research on RSW aluminum, including numerical simulation, has attracted more attention. The basic physical principles remain the same for resistance spot welding steels and aluminum alloys, so do the fundamental equations for simulation. However, there are major differences in the physical processes between welding aluminum alloys and steels, mainly due to their differences in mechanical, thermal, electrical, and metallurgical properties. For instance, because of its lower electrical resistance, aluminum requires higher welding currents, which can be three to four times more than welding steel. Some parameters/properties, such as contact resistance, may not carry much weight in (uncoated) steel welding but are important in aluminum welding. Such differences make aluminum welding quite different from steel welding and warrant different treatment in numerical simulations.

There are a number of attempts in numerical simulation of resistance spot welding aluminum alloys. The effort of RSW aluminum simulation can be represented by the work of Browne et al.[14–16] Along with the simulation of nugget growth, they predicted the boundaries of minimum nugget size and expulsion. A criterion of expulsion was proposed based on the modeling, as depicted in Chapter 7 as the geometry comparison model. Khan et al.[17–19] also developed a model of aluminum welding process to predict the nugget development. The model included the thermal contact resistance of the workpiece–electrode interface, the effect of the interface friction coefficient, the contact resistance of the workpiece–electrode interface and faying surface, the contact resistance variation on contact pressure and temperature, and the convection in the mushy and melting zones.

Although the FEM offers tremendous flexibility in modeling welding processes, general commercial packages do not provide a simulation of completely coupled mechanical, thermal, and electrical processes. Therefore, various attempts have been made using available commercial software packages to simulate the coupling in the welding process. The history of welding process simulation can be characterized by the extent of coupling in the simulation at different periods of time. An uncoupled or loosely coupled simulation may calculate the temperature history from the electrical–thermal process using a program such as ANSYS, and it is then exported as an input to the thermal–mechanical simulation using the same or different software, such as ABAQUS. The contact areas at the electrode–sheet and sheet–sheet interfaces have to be determined, as they have a direct impact on current density. Theories were developed to predict the change in contact areas during welding. Because the processes are basically decoupled, the interactions among the processes are not accounted for, and errors could be significant. An improvement is sequentially coupled analysis, as in the work carried out by a number of researchers. The basic scheme is outlined in a paper by Zhang et al.[20] Instead of completing the temperature history of the entire welding period before it is exported to the thermal–mechanical module, either in the same or different software, the electrical–thermal modeling is interrupted after a small increment in welding time, usually less than a quarter of a cycle. In the electrical–thermal module, electric current is applied while mechanical load is kept constant. A new temperature distribution is obtained based on previous temperature field and additional heat generated. This new temperature field is then imposed onto the thermal–mechanical coupling system, which provides updated information on geometry, contact area, etc. This information is then passed to the electrical–thermal module for the next increment of heating. By repeating this procedure, nugget formation process, that is, heating, melting, and cooling of the workpieces, is modeled. Time increments need to be carefully chosen to catch details of the process, yet avoid an unreasonably long execution time.

An ideal and realistic model for simulating RSW processes should include a thorough heat transfer analysis, electrical field analysis, thermo–elastic–plastic analysis, actual variation of contact resistance, phase change, and temperature-dependent material properties. The most difficult aspect of RSW simulation is probably dealing with melting and solidification of metals. The work by Feng et al.[21] and Li et al.[22] showed the possibility of predicting the structure and properties for a heat-affected zone (HAZ) and solidified nugget, and residual stresses in a weldment. Their works are reviewed at the end of this chapter.

9.2 Coupled Electrical–Thermal–Mechanical Analysis

Because RSW is a very specialized process, currently available general-purpose engineering FEM software packages do not have the ability to simulate the electrical–thermal–mechanical interactions needed in RSW. The difficulties in formulating and coding such a process also inhibit the development of a customized program. In this section, the general formulations of a coupled RSW process are presented for three-dimensional and two-dimensional problems.

9.2.1 A General (Three-Dimensional) Finite Element Model

The electrical, thermal, and mechanical processes can be formulated separately, and then they are linked to each other during coding based on the mutual dependence of the process variables. However, it is convenient to consider the electrical and thermal processes together, as they are directly related. Then coupling with the mechanical process can be realized by considering the thermal effects in a stress–strain analysis. The basic steps of formulating the electrical–thermal process are as follows:

1. First, the electrical potential is obtained for the entire domain and scaled according to the given electric current. The electric field is used to calculate the energy dissipation due to the electrical resistance of materials.
2. Then the energy dissipation from the electrical analysis is used as heat generation for calculating the temperature distribution using the heat conduction equation.
3. All material properties are updated element-wise according to the calculated temperature, which will be used in the stress–strain analysis.

In this analysis, the contact resistance at electrode–sheet and sheet–sheet interfaces plays an important role. It can be treated using artificial interface elements in which the material properties for electrical and thermal analyses are artificially imposed for physically reasonable simulation. This procedure simulates the welding process with the consideration of variation of process parameters such as the electric current, contact resistance, and the electrical and thermal properties of electrode and sheet materials.

9.2.2 Formulation of Electrical Process

The electrical potential can be expressed by a Laplace equation. Assuming that there is no internal current source, the governing equation can be written as

$$\frac{\partial^2\phi}{\partial x^2}+\frac{\partial^2\phi}{\partial y^2}+\frac{\partial^2\phi}{\partial z^2}=0 \tag{9.1}$$

where $\phi(x, y, z, t)$ is the electrical potential, which is a function of coordinates (x, y, z) and time t. Equation 9.1 is equivalent to the following:

$$\frac{\partial}{\partial x}\left(\frac{\partial\phi}{\partial x}\right)+\frac{\partial}{\partial y}\left(\frac{\partial\phi}{\partial y}\right)+\frac{\partial}{\partial z}\left(\frac{\partial\phi}{\partial z}\right)=0 \tag{9.2}$$

Through the Galerkin approach, this equation can be rewritten as

$$\int \bar{\phi} \frac{\partial}{\partial x}\left(\frac{\partial \phi}{\partial x}\right) + \int \bar{\phi} \frac{\partial}{\partial y}\left(\frac{\partial \phi}{\partial y}\right) + \int \bar{\phi} \frac{\partial}{\partial z}\left(\frac{\partial \phi}{\partial z}\right) = 0 \tag{9.3}$$

where $\bar{\phi}$ is a weighting function. Solving it by integrating by parts and using Ohm's law, the equation becomes

$$\int_V \frac{\partial \bar{\phi}}{\partial x} \cdot \sigma \cdot \frac{\partial \phi}{\partial x} \mathrm{d}V = \int_S \bar{\phi} \cdot j \mathrm{d}S \tag{9.4}$$

where $\sigma(\theta, f)$ is electrical conductivity matrix; $\theta = \theta(x, y, z, t)$ is the temperature, which is a function of both time and coordinates; f is for predefined field variables; and $j = -\boldsymbol{j} \cdot \boldsymbol{n}$ is the current density entering the control volume across S.

9.2.3 Formulation of Heat Transfer Process

The general three-dimensional differential equation governing heat conduction with an internal heat source can be expressed as

$$\frac{\partial}{\partial x}\left(k\frac{\partial \theta}{\partial x}\right) + \frac{\partial}{\partial y}\left(k\frac{\partial \theta}{\partial y}\right) + \frac{\partial}{\partial z}\left(k\frac{\partial \theta}{\partial z}\right) + \dot{Q} = \rho\frac{\partial U}{\partial t} \tag{9.5}$$

where ρ is the material's density, k is the thermal conductivity, $\dot{Q}$ is the internal heat generation rate per unit volume, and U is the internal energy. The variational form of the above equation using Galerkin's approach is

$$\int_V \rho \dot{U} \bar{\theta} \mathrm{d}V + \int_V \frac{\partial \bar{\theta}}{\partial x} k \frac{\partial \theta}{\partial x} \mathrm{d}V = \int_V \partial \theta r \mathrm{d}V + \int_S \bar{\theta} q \mathrm{d}S \tag{9.6}$$

where x represents the vector form of coordinates, k is the thermal conductivity matrix, q is the heat flux per unit area of the body flowing into the body, and r is the heat generated within the body due to Joule heating.

As the workpiece experiences a large temperature span during the RSW process, from room temperature to above melting point, it is natural to consider the temperature-dependent electrical and thermal properties in simulation. However, such dependence is not always available, especially in elevated temperature ranges. Extrapolation schemes are typically used to estimate the properties in high temperature ranges from those available in lower temperature ranges. It is observed that such temperature dependence may not offer drastic improvement in terms of accuracy for nugget growth modeling.

9.2.4 Boundary Conditions

The total surface area S can be divided into S_p and S_i. S_p is the surface on which the boundary conditions can be prescribed, and S_i represents the surface that interacts with nearby surfaces of other bodies. The rate of electrical energy dissipated by letting current flow through a conductor is represented by P_{ec}. Then, Equations 9.4 and 9.6 can be rewritten as

$$\int_V \frac{\partial\bar{\phi}}{\partial x}\cdot\sigma\cdot\frac{\partial\phi}{\partial x}\mathrm{d}V=\int_{S_p}\bar{\phi}\cdot j\,\mathrm{d}S+\int_{S_i}\bar{\phi}\cdot j\,\mathrm{d}S\begin{pmatrix} a_{11} & \cdots & a_{1n} \\ \vdots & \ddots & \vdots \\ a_{m1} & \cdots & a_{mn} \end{pmatrix} \tag{9.7}$$

and

$$\int_V \rho U\bar{\theta}\,\mathrm{d}V+\int_V \frac{\partial\bar{\theta}}{\partial x}\cdot k\cdot\frac{\partial\theta}{\partial x}\mathrm{d}V=\int_V \bar{\theta}\cdot\eta_V P_{ec}\cdot\mathrm{d}V+\int_{S_p}\bar{\theta}q\,\mathrm{d}S+\int_{S_i}\bar{\theta}(q_c+q_r+q_{ec})\,\mathrm{d}S \tag{9.8}$$

where η_V is the energy conversion factor for calculating the amount of electrical energy released as internal heat, q_c is the heat conduction, q_r is the radiation, and q_{ec} is the heat generated at the interfaces.

9.2.5 Formulation of Thermomechanical Analysis

The localized temperature gradients due to heat generation and dissipation, in addition to the mechanical loading through electrodes, induce a thermomechanical response in a weldment. The equilibrium conditions can be established using the theorem of virtual work, which states that a virtual change of the internal strain energy must be offset by an identical change in external virtual work due to the applied loads:

$$\delta U=\int_V \tau\cdot\delta\varepsilon\cdot\mathrm{d}V=\int_S T\cdot\delta u\cdot\mathrm{d}S=\delta V \tag{9.9}$$

A realistic temperature dependence of a material's mechanical properties plays an important role in simulation accuracy. As for the temperature dependence of electrothermal properties, there is also a serious lack of experimental data, especially at elevated temperatures. The use of realistic thermomechanical properties is expected to significantly improve simulation accuracy.

In early simulation works, assumptions of small strain and often linear elastic material response were usually made. For instance, small-strain plastic deformation was assumed in the work by Browne et al.[14,15] Ideally, finite strain effects should be accounted for in order to establish accurate contact during nugget formation, as well as to obtain the information on the development of residual stresses and springback. More sophisticated simulations can be performed as computing hardware and software advance.

9.2.6 Simulation of Melting and Solidification

When temperature reaches the melting point during heating, the solid starts to melt and a nugget starts to form. It is difficult to directly treat the melting as a metallurgical process in simulation. To simulate melting, a common practice is assigning rapid changes in material properties (e.g., Young's modulus, specific heat) at temperatures close to the solidus or liquidus. The effect of latent heat is considered by an increase in specific heat, as given in the following equation:

$$\bar{C} = C + \frac{L}{T_L - T_S} \tag{9.10}$$

when ($T_S < T < T_L$). L is latent heat, T_S is solidus temperature, and T_L is liquidus temperature. The solidification process can be treated in the same manner.

Upon melting, the metal basically loses its ability of bearing any load. Therefore, its strength drops from a finite number to zero. A temperature dependence of mechanical properties, such as Young's modulus is commonly created, which has a low value (usually less than 10% of its value at room temperature) at temperatures beyond melting. A smooth transition between the variable's value at solid state and that at melting is usually needed to avoid a divergence problem in simulation. Once melting occurs, the strain definition, particularly plastic strain in the molten nugget, ceases to exist within the context of solid mechanics, on which a commercial program such as ANSYS or ABAQUS is based. Simulation of just-solidified metal also deserves special attention, as solidification processes cannot be modeled using commercial FEM packages. As a molten metal solidifies, a prior zero-straining history must be imposed in the finite element modeling. User-specified subroutines, such as UMAT in ABAQUS, are generally needed to simulate the property changes in melting and solidification processes.

9.2.7 Finite Element Formulation

In a finite element analysis, a domain is divided into elements such as isoparametric elements. The use of isoparametric triangular or rectangular elements in two-dimensional problems is illustrated in Figure 9.3b. Isoparametric and higher-order elements are generally available in commercial codes, which can be used for two-dimensional or three-dimensional simulation of the resistance welding process.

Consider a two-dimensional, four-node isoparametric element. An unknown variable, taking the electrical potential (ϕ) as an example, in an element can be expressed as a sum of the product of nodal values (ϕ_i) of the electrical potential and shape functions:

$$\phi = \phi_1 \cdot N_1 + \phi_2 \cdot N_2 + \phi_3 \cdot N_3 + \phi_4 \cdot N_4 \tag{9.11}$$

where N_1, N_2, N_3, and N_4 are shape functions for a four-node element.

Similar equations can be written for temperature and strain/stress. The equations can be solved by finite element transient analysis. For this generally, a Crank–Nicolson scheme, which gives a stable solution, can be used.

For finite element simulation, Equation 9.7 is first solved to calculate heat generation from electrical potential. The heat generation calculated is then substituted in Equation 9.8, which is solved for temperature distribution at a specific time. According to the temperature distribution obtained, all electrothermal properties are updated element-wise for the next time increment. Using the temperature distribution, the thermomechanical model is solved for strain and stress, which are also updated for the next time increment.

9.2.8 Two-Dimensional Finite Element Modeling

Although practical problems are three-dimensional in nature, many of them can be reduced to two-dimensional with sufficient accuracy and a significantly simplified procedure. The general principles depicted in the last section remain the same, and their use in formulating a two-dimensional problem can be described in the following:

- First calculate the electric potential.
- The electric field obtained is then used to calculate the energy dissipation due to the electric resistance of materials, that is, due to joule heating.
- After obtaining the temperature distribution through a heat transfer analysis, the thermomechanical model is then used to calculate the stress distribution, the deformation of the workpieces and electrodes, and the change in interface contact.

9.2.8.1 Formulation for Electrical Analysis

In two-dimensional problems, the governing equation of the electrical potential distribution is expressed by the Laplace equation:

$$\frac{\partial^2 \phi}{\partial x^2} + \frac{\partial^2 \phi}{\partial y^2} = 0 \tag{9.12}$$

where $\phi = \phi(x, y, t)$ is the electrical potential as a function of coordinates and time. The finite element modulation of this equation can be formulated in the same way as in the three-dimensional analysis, and there are two kinds of boundary conditions to be specified in the electrical analysis:

$$\phi = \phi_0 \tag{9.13}$$

on the boundaries in contact with the power supply with known potential ϕ_0, and

$$\frac{\partial \phi}{\partial n} = 0 \tag{9.14}$$

on the free boundaries, where n denotes the normal direction of a boundary.

After determining the potential distribution, the current density can then be calculated as

$$J_x = -\frac{1}{\rho}\frac{\partial \phi}{\partial x} \text{ and } J_y = -\frac{1}{\rho}\frac{\partial \phi}{\partial y} \tag{9.15}$$

where J is the current density and ρ is the electrical resistivity.

The heat generation rate per unit volume is calculated using the formula

$$\dot{Q} = \rho \cdot J^2 \tag{9.16}$$

9.2.8.2 Formulation for Thermal Analysis

The governing differential equation for two-dimensional transient heat conduction with an internal heat source is

$$\frac{\partial}{\partial x}\left(k\frac{\partial T}{\partial x}\right)+\frac{\partial}{\partial y}\left(k\frac{\partial T}{\partial y}\right)+\dot{Q}=\gamma\cdot C\frac{\partial T}{\partial t} \tag{9.17}$$

where T is the temperature as a function of coordinates and time, k is the thermal conductivity, γ is the mass density, and C is the heat capacity per unit mass.

The three types of boundary conditions involved are

$$T = T_0 \tag{9.18}$$

on the boundaries with specified temperatures,

$$-k\frac{\partial T}{\partial n}=0 \tag{9.19}$$

on the lines of symmetry, and

$$-k\frac{\partial T}{\partial n}=h(T-T_e) \tag{9.20}$$

on the free surfaces taking the convective heat exchange into account. h is the convection heat transfer rate of the surrounding air and T_e is the ambient temperature.

9.2.8.3 Finite Element Formulation

The domain is divided into a set of finite elements, such as triangular or rectangular elements. The solution of the entire domain can be obtained by analyzing each element and then solving the assemblage of the elemental equations over the entire domain. Assuming that rectangular isoparametric elements are used in the electrothermal analysis, the unknown electrical potential and temperature can be expressed as

$$\phi=\phi_1\cdot N_1+\phi_2\cdot N_2+\phi_3\cdot N_3+\phi_4\cdot N_4 \tag{9.21}$$

and

$$T=T_1\cdot N_1+T_2\cdot N_2+T_3\cdot N_3+T_4\cdot N_4 \tag{9.22}$$

with shape functions the same as those in Equations 9.11.

Applying Galerkin's approach and solving by integration by parts, the set of equations for the potential distribution can be written as

$$[A][\phi]=[F] \tag{9.23}$$

where $[A]$ is electrical conductance in the form of an $n \times n$ matrix, where n is the total number of nodal points in the model. $[F]$ is a boundary condition vector and $[\phi]$ is a vector of unknown potential values.

Similarly for the temperature distribution, the set of equations can be written as

$$[A][T]+[C]\left[\frac{\partial T}{\partial t}\right]=[F] \tag{9.24}$$

where $[A]$ is an $n \times n$ matrix of thermal conductance, $[C]$ is the heat capacity matrix, $[F]$ is the vector of heat source and boundary condition, and $[T]$ is the vector of unknown temperature values.

The Crank–Nicolson scheme can be used to solve the above transient problem:

$$\{[C] + 0.5\Delta t[A]\}[T]_{t+\Delta t} = \{[C] - 0.5\Delta t[A]\}[T]_t + \Delta t[F]_t \tag{9.25}$$

where $[A]$ and $[C]$ are $n \times n$ matrices, and $[T]$ and $[F]$ are $n \times l$ matrices. By solving the above equations, electrical and temperature distributions can be obtained.

9.2.9 Axisymmetric Problems

A special case of two-dimensional problems is axisymmetric models. The two-dimensional approach described in previous sections can be directly applied; however, cylindrical coordinates have to be used in these models. Therefore, the equations, such as those for electrical potential and heat transfer, have the same physical meanings, but different forms.

9.3 Simulation of Contact Properties and Contact Area

Correct modeling of contact properties, such as electro- and thermal conductivity, and change of contact area, as a function of solution variables such as temperature and pressure, are critical in RSW simulation. In early simulation works, the contact resistance was simulated by assigning equivalent contact resistance properties using two-dimensional solid elements. In the models by Browne et al.,[14–16] contact areas at the electrode–sheet and sheet–sheet interfaces were updated at selected time increments. The majority of modeling works on the RSW were conducted by prescribing contact resistance values on a trial-and-error basis to achieve optimal fit between the predictions and experimental measurements. A consistent methodology remains to be developed so that contact properties as a function of solution variables can be readily incorporated and the contact areas can be updated without human intervention.

Because it is difficult to directly measure the contact resistivity, contact resistance is usually dealt with in simulations instead. In addition, the overall resistance can be divided into static resistance and dynamic resistance for a better understanding.

Static resistance results from a thin film on a surface formed by lubricant, dirt, oxides, etc., and its effect is more pronounced in the initial stage of a welding process. It can be measured using a search current of relatively small magnitude, which generates insufficient heat near the contact. For this, a direct-current (DC) bridge is used, which may provide a current, for measurement purposes, of the order of 1 A. Experiments revealed that measured static resistance values can be mostly attributed to the surface condition, or film effect. Besides surface condition, it also depends on current level, temperature, and applied force. Electrical contacts under such conditions are normally characterized as quasi-metallic. The surface film breaks down under the mechanical loading and intensive heating at the initial stage of welding, and they make the contact purely metallic. However, the measured values are of little use as they are usually not consistent and cannot be directly used in numerical analysis. As a matter of fact, the breakdown of surface film takes very little time, and the resistance measured in this period has very little implication on the resistance in the rest of the welding process. The contact resistance before the film breakdown can be generally ignored because of its complex nature and inconsistency, and because this resistance lasts only for a small fraction of the first welding cycle. Existing models mostly focus on the resistance after the surface film breakdown, that is, on the dynamic resistance.

In general, dynamic resistance can be measured as the ratio of the instantaneous voltage to the current passing through a purely resistive conductor at the same instant in time during actual welding operations at selected moments, as described in Chapter 5. The dynamic resistance values are more representative during an actual welding process. However, there are many factors that would affect the dynamic resistance values (e.g., electrode geometry, welding current, sheet thickness, electrode force), and their effects are difficult to be individually quantified.

Static and dynamic resistances are not pure material properties; they depend on welding conditions and welding parameters, and such dependence can be measured experimentally. The relationship between electrode force and static resistance is shown in Figure 9.4.[23,24] The static resistance decreases almost linearly when electrode force increases (Figure 9.4a). Similarly, dynamic resistance obtained under various applied forces is inversely proportional to the force, or the larger the force, the smaller the resistance (Figure 9.4b). This phenomenon has also been observed by other researchers, such as Babu,[25] as shown in Figure 9.5.

However, the overall resistance measured in experiments does not help much in RSW simulation, although it can be used to compare, or calibrate, an FEM model. Detailed

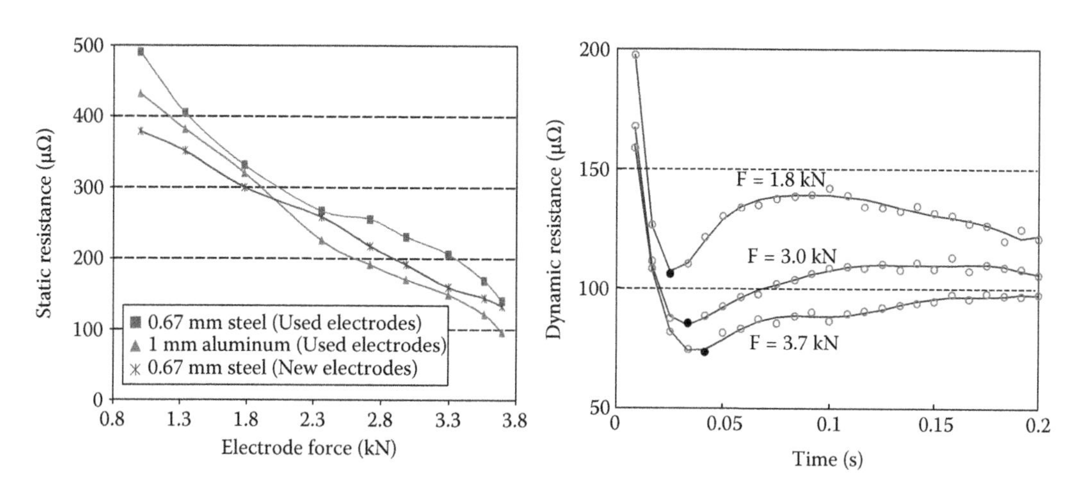

FIGURE 9.4
Static and dynamic resistance changes with electrode force.

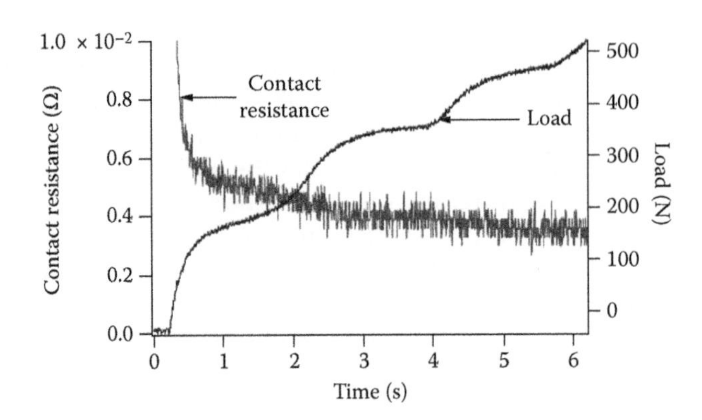

FIGURE 9.5
Dynamic resistance change with electrode pressure on a DF140 steel.

information of the dependence on temperature and pressure for both contact and bulk resistivity is needed. Bulk resistivity is in general a weak function of pressure, but a strong function of temperature. It can be determined fairly accurately by experiments. Contact resistivity, on the other hand, is a very strong function of the surface and loading conditions, and a carefully conducted experiment can only provide an averaged value of this quantity. Therefore, various models have been developed to describe the dependence of contact resistivity or, in many cases, contact resistance on pressure, temperature, and other conditions, for use in FEM simulation.

The contact resistivity ρ at an interface can be calculated according to Bay and Wanheim's model[26] taking into account the plastic deformation of surface asperities to determine the real contact area between rough surfaces:

$$\rho_{\text{contact}} = 3\left(\frac{\sigma_{\text{soft}}}{\sigma_{\text{n}}}\right)\left(\frac{\rho_1+\rho_2}{2}\right)+\rho_{\text{contaminant}} \tag{9.26}$$

In this equation, σ_{soft} is the flow stress of the softer metal, σ_{n} is the contact (normal) pressure at the interface, ρ_1 and ρ_2 are bulk resistivities of the metals, and $\rho_{\text{contaminant}}$ is the resistivity of surface agents, such as oxides, water vapor, and grease. It is necessary to include a term for surface contaminant, as it exists on all surfaces and it may contribute significantly to the overall contact resistance for certain materials during certain periods of time.

Similarly, the contact thermal conductivity k_{contact} at the interfaces can be expressed as

$$k_{\text{contact}} = \frac{1}{3}\left(\frac{\sigma_{\text{n}}}{\sigma_{\text{soft}}}\right)\left(\frac{k_1+k_2}{2}\right) \tag{9.27}$$

A similar model developed by Vogler and Sheppard[27] considers the surface asperities, which provide the electric contact. The model links contact resistance to both surface and bulk properties of the sheets in contact:

$$R_{\text{CA}} = (\rho_1+\rho_2)\left(\frac{1}{4\eta a}+\frac{3\pi}{32\eta l}\right) \tag{9.28}$$

In this model, ρ_1 and ρ_2 are bulk resistivities of contacting members (steel), η is the number density of asperities in actual contact, a is the average radius of contacting asperities, and $2l$ is the average in-plane distance between asperities. In order to use this model for resistance welding simulation, quantities related to surface asperities need to be characterized as functions of temperature and pressure.

A model developed by Kohlrausch[28] describes the relationship between the voltage drop across a metallic contact interface and its resistance. Bowden and Williamson[29] and Greenwood and Williamson[30] also supported this model. According to this model, the voltage drop across the contact interface can be estimated as

$$V^2 = 4L(T_s^2 - T_o^2) \tag{9.29}$$

where V is the voltage drop across the contact interface, T_s is the maximum temperature at the interface, T_o is the bulk temperature, and L is the Lorentz constant. The value of the Lorentz constant for most metals is 2.4×10^{-8} $(V/K)^2$.

This model is valid for metallic contacts, and contact members obey the Wiedemann–Franz–Lorentz law:

$$k\rho = LT \tag{9.30}$$

where k is thermal conductivity in W/m K, ρ is electrical resistivity in Ω/m, and T is temperature in K.

Thus, contact electrical conductance can be expressed as a gap conductance and written as

$$\sigma_g = \frac{1}{\rho h} = \frac{1}{R_c A_c} \tag{9.31}$$

or

$$\sigma_g = \frac{1}{2\pi r_c^2 \sqrt{L(T_s^2 - T_o^2)}} \tag{9.32}$$

The above formulation is based on the assumption that the contact area (represented by radius r_c) is under intimate metallic contact. It shows that there is a pressure dependence of the contact gap conductance (through r_c). Research shows that the number of contact asperities increases almost proportionally with the increase of contact pressure before the entire interface is in metallic contact.

The true contact area is generally only a fraction of the macroscopic apparent contact area. In the absence of the contamination film on the surface, the true contact area is the same as the load-bearing area. The true contact area is expected to increase with loading. It is apparent that the contact resistance is inversely proportional to the true contact area. Thus, resistance is roughly inversely proportional to the contact pressure. However, such dependence is material dependent. For example, in the case of mild steels, contact resistance follows the inversely proportional relation with respect to pressure before reaching a plateau value beyond a threshold pressure. This implies that the gap conductance should be roughly proportional to the contact pressure before reaching the value indicated by the above equation. Thus, if the gap interface pressure is greater than the threshold pressure, electrical conductance is calculated by the above equation. It can be obtained through linear interpolation if the gap interface pressure is less than the threshold pressure.

Therefore, the voltage drop across the interface at any bulk temperature (T_0) below T_s can be used to calculate the equivalent contact resistance/resistivity as a function of temperature by dividing the voltage drop by the welding current. At temperatures above the solidus temperature T_s, bulk electrical resistivity values for that material can be used for the contact elements in simulation.

In a simulation, the contact radius r_c and contact gap pressure obtained from the previous thermal–mechanical analysis are extracted and used as input to the electrical–thermal analysis module through the option of field variables. A user interface subroutine can then be coded to calculate gap conductance based on the nodal temperature, nodal pressure, and total contact area.

9.4 Simulation of Other Factors

In addition to contact resistivity, there are other factors that need to be considered in numerical simulation of an RSW process. For instance, zinc-coated steel sheets have drastically different conductivity from bare steels, and such effects should be included in a numeric model. The effects of zinc coating and electric current profile on simulation are briefly described in the following.

9.4.1 Effect of Zinc Coating

Typically, for galvanized coating, the zinc at the sheet–sheet interface melts shortly after the electric current is applied (in about one cycle of welding time). The molten zinc is contained in the middle of the contact area because the periphery of the mechanical contact is still in solid contact at this stage. After a few cycles, the zinc coating at the periphery of the mechanical contact area at the sheet–sheet interface is melted. The applied electrode force thus pushes a certain amount of molten zinc to the periphery of the mechanical contact area until solid-to-solid contact is reestablished at the sheet–sheet interface. The extruded molten zinc enlarges the effective electric contact area and consequently reduces the contact resistance and current density. A few more cycles of heating will make the temperature at the periphery of the mechanical contact area high enough to boil the zinc. As a result, the amount of molten zinc at the periphery of the mechanical contact area is reduced, and the effective electric contact area turns close to the mechanical contact area. In general, the free zinc on the surface affects both the contact resistivity and contact area, and accurately modeling its effects is a difficult task.

This analysis is valid for galvanized coating only. It cannot be generalized to galvanealled coating, as it alters the total resistance as a solid phase during welding.

9.4.2 Effect of Electric Current Profile

Many models in resistance welding simulation used root mean square (RMS) electric current values for alternating current (AC). However, using an RMS current value in a simulation does not differentiate an AC welding from a DC welding. Although using RMS provides significant simplicity in modeling, it misses the important heating–cooling cycles pertaining to an AC profile. A typical AC profile is shown in Figure 9.6. Rather than reaching the preset value immediately after electric current is turned on, it takes a few cycles (~3 cycles) to reach its preset current and the number of cycles needed depends on the transformer and weld control. When the current approaches zero (at the crosses with time axis) in any half of a cycle, less heat is generated, and the weld is effectively under cooling from the base metal and water-cooled electrodes. Realistic simulations should consider the actual current value at any moment to include the effective heating–cooling cycles. The time increment chosen in simulating the coupling process as described in Section 9.1 should reflect the cycling nature of AC. In order to do so, the largest time increment should not be bigger than 1/8 of a cycle.

9.5 Modeling of Microstructure Evolution

Efforts have been made to simulate microstructures of a resistance spot weldment. This is essential in determining the mechanical properties of a weldment and, therefore, its performance. Through a numerical simulation of a welding process, the thermal history can be determined in the fusion zone and surrounding solids. Knowing the thermal history and relevant metallurgical information about the alloy being welded, the microstructure throughout the entire weldment can be predicted. The microstructure gradient in the weldment results in a mechanical property gradient, which in turn determines the response of a weldment to an external loading. Various models, mostly using the FEM as a platform, were developed to simulate the formation of microstructures in a spot weldment by modeling the phase transformations during heating, cooling, and tempering processes. Simulation in this regard can be divided into modeling microstructure development in the fusion zone and in the HAZ. In steel welding, this includes the phase transformations between austenite and ferrite, the precipitation of carbides in martensite, and the dissolution of carbides and nitrides.

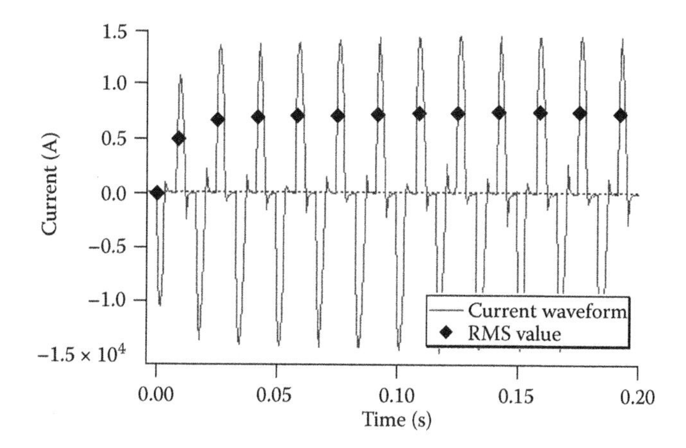

FIGURE 9.6
A typical alternating electric current profile in resistance welding.[25]

In the model by Ion et al.,[31] the carbon equivalence value and the cooling time were used to determine the volume fractions of bainite, martensite, and ferrite–pearlite mixtures. Another model by Watt et al.[32] works better for high-carbon steels, using transformation rate equations developed based on high-carbon steel time–temperature transformation (TTT) diagrams. Bhadeshia and Svensson[33] developed a phenomenological model for sequential decomposition of austenite to various ferrite morphologies. This model is applicable for a wide range of cooling rates and for low-carbon steels. For example, the predictions of TTT and continuous-cooling transformation (CCT) diagrams based on this model are quite applicable to low-carbon steels. However, this model is incapable of describing simultaneous transformation of austenite to different ferrite phases. The model developed by Jones and Bhadeshia[34] is more refined and describes the simultaneous formation of idiomorphic and allotriomorphic ferrite based on overall transformation kinetic equations. It is possible to extend their work to other ferrite morphologies and reverse transformations.

9.5.1 Effect of Cooling Rate

Once the electric current is turned off, the thermoelectrical analysis becomes a purely heat transfer analysis during the so-called hold time in RSW. The legitimacy of electrode holding is to maintain certain forging pressure during the solidification of the molten nugget to eliminate porosity and other defects. A hold time is basically determined by the welding process requirements, and it directly affects the cooling process as heat dissipates rapidly through the water-cooled electrodes.

Besides hold time, there are other factors that affect the cooling process. For instance, the presence of a zinc annulus surrounding the electrode tip also plays a significant role in accelerating the cooling process. It has been estimated that the cooling rate at the center of a nugget when the electrodes are in contact with the workpieces, that is, during hold time, is very high—in the magnitude of several thousand degrees Celsius per second at temperatures above 500°C. It decreases with temperature, but the cooling rate still remains around 1000°C/s, even at temperatures around 500°C. Cooling rates in such magnitude are extremely difficult to analyze by physical simulations in a well-controlled manner.

In the following sections, the mechanisms of solid-state phase transformations are discussed first, using steel welding as an example. The simulation of weld nugget structures then follows, which has many similarities to the HAZ simulation once the nugget is solidified.

9.5.2 Microstructure Evolution in HAZ

Unlike in the nugget, the peak temperature in the HAZ is below the melting temperature during welding. Therefore, the HAZ experiences solid-state transformations only, and as a result, the HAZ has different structures and mechanical properties from either the nugget or the base metal. As discussed in other chapters, this area is susceptible to development of microdefects due to the gradient of (often weakened) mechanical properties, stress concentration/residual stresses, and metallurgical processes involved during welding, such as liquid metal embrittlement. Therefore, understanding the microstructure evaluation and mechanical properties in the HAZ is of considerable importance in predicting a weld's quality and performance.

The simulation/calculation of solid-state phase transformations can be conducted based on the fundamental metallurgical principles, as outlined in Chapter 1. The calculation is material dependent, and welding steel is analyzed in this section to illustrate the basic procedures.

A thermal history of the HAZ is needed in order to predict the microstructure evolution, which is usually provided by a separately performed finite element analysis. The major functions of a model for simulating microstructure evolution in the HAZ include:

1. Predicting equilibrium phase composition and transformation characteristics
2. Reaction kinetics modeling of austenite decomposition upon cooling
3. Kinetic modeling of austenite grain growth

The input to such a model is thermal history and material composition. The history of microstructure evolution and the resultant room temperature hardness distribution are usually the output. As the HAZ experiences nonequilibrium cooling, temperatures of phase transformations cannot be obtained from equilibrium phase diagrams; rather, they should be computed based on thermodynamics considerations.

During heating, cementite usually dissolves and ferrite transforms to austenite, and some grain growth is also expected in steel. The thermal cycle and steel composition primarily determine the microstructural evolution in the HAZ of carbon steels. The basic metallurgical processes involve heating and cooling a piece of steel between the room temperature and the melting temperature, and they are outlined in Chapter 1.

To simplify the problem, simulation of microstructural evolution in a weldment may start with the decomposition of austenite into other phases during cooling. In general, phase transformations during cooling are controlled by inclusion, austenite grain size, hardenability, and cooling rate. The austenite-to-ferrite decomposition in steels can be either reconstructive or displacive during cooling, as summarized in Figure 9.7.

Several approaches have been developed to predict carbon steel microstructures. In the work by Ion et al.,[31] carbon equivalence was used to relate the volume fraction of martensite and bainite to time for cooling from 800°C to 500°C (Figure 9.8). It has been fairly successful in general, but the model is not very satisfactory when the cooling rate is low. Another approach is to use equations based on TTT/CCT diagrams, as illustrated in the work by Kirkaldy and Venugopolan[35] and Watt et al.[32] (Figure 9.9). However, the approach is not applicable to welding low-carbon steels. More sophisticated models were developed later for predicting solid-phase transformations in steels, such as the phenomenological models of austenite decomposition to various ferrite phases by Bhadeshia and Svensson[33] and the simultaneous transformation kinetics equations derived by Jones and Bhadeshia.[34]

9.5.3 Simulation of Microstructure of a Nugget

Weld solidification microstructure is controlled by temperature gradient and crystal growth rate. In general, weld pool shape, cooling rate, and composition of the weld affect the microstructure, and variation in weld microstructure depends on temperature gradient (G), growth rate (R), and combinations of them ($G{\cdot}R$ or G/R).

Equilibrium solidification can be described by either

$$C_S^* = kC_o(1-f_S)^{k-1} \quad \text{(Scheil's equation)} \tag{9.33}$$

or

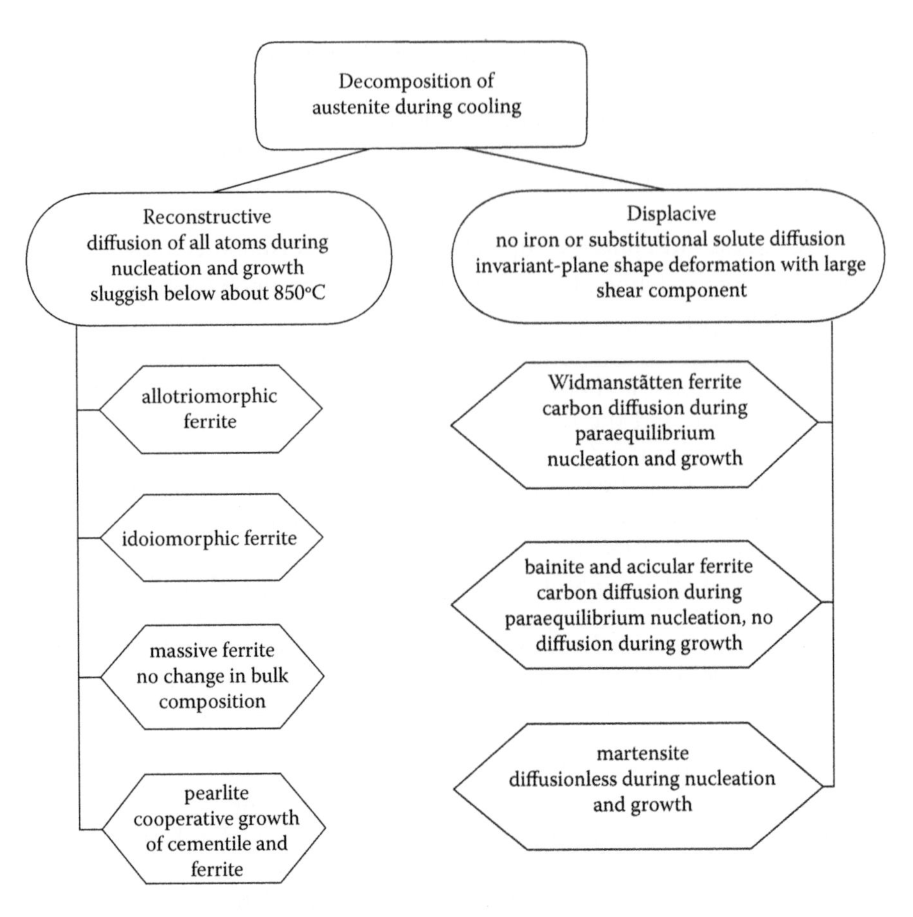

FIGURE 9.7
Decomposition of austenite during the cooling cycle of a steel welding process.[25]

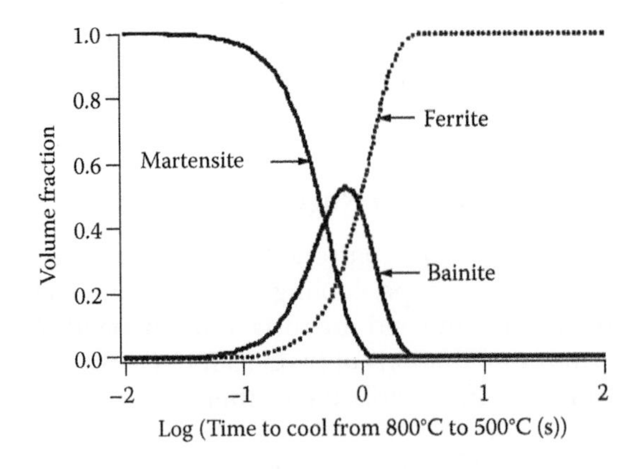

FIGURE 9.8
Volume fractions of various phases predicted using equations based on carbon equivalence values and cooling time.

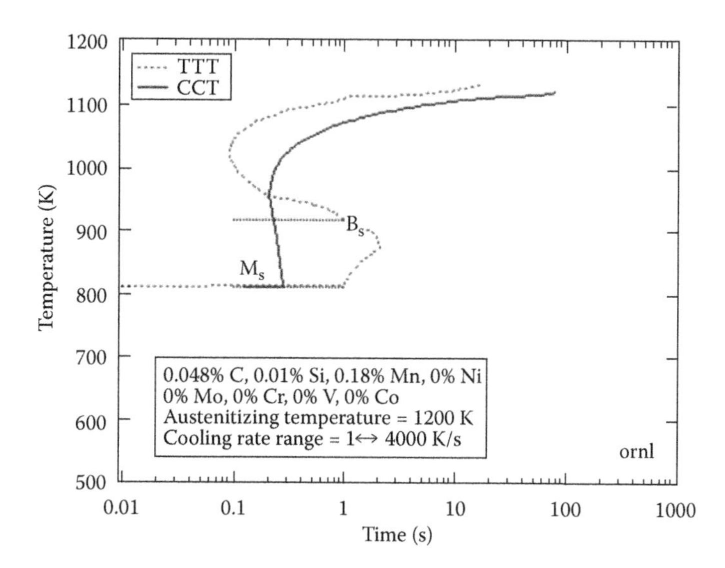

FIGURE 9.9
TTT/CCT diagram calculations for predicting the formation of bainite and martensite at cooling rates of >3000°C/s.[25]

$$C_S^* = kC_o\left(1 - \frac{f_S}{1+\alpha k}\right)^{k-1} \quad \text{(Brody and Fleming´s equation)} \tag{9.34}$$

However, the solidification process in a resistance spot weld nugget is far from equilibrium. Significant modification is needed for these equations in order to predict the microstructure of a nugget. Some attempts have been made to describe nucleation and growth during solidification under rapid cooling in RSW. In general, studies of weld solidification require a detailed analysis following Babu's reasoning[25]:

- Crystallography. Epitaxial growth of grains is predominant in liquid metals. Depending on the composition, heterogeneous nucleation on inclusions may occur.
- Macro- and microsegregation. Fluid flow, such as that caused by temperature gradient, and cooling rate affect segregation. There is a need to predict multicomponent partitioning and kinetics of solidification.
- Phase selection. Upon large undercooling, austenite may outgrow ferrite.
- Nonequilibrium partitioning. At high interface velocities, solute trapping occurs and the partitioning coefficient may approach unity.

The undercooling in a liquid nugget contains three parts[36]:

$$\Delta T = \Delta T_d \text{ (constitutional)} + \Delta T\gamma \text{ (capillarity)} + \Delta T_k \text{ (kinetics)} \tag{9.35}$$

Each part can be analyzed separately and then the overall undercooling can be estimated. In a closed system such as a molten spot weld nugget, fluid flow may result from convection

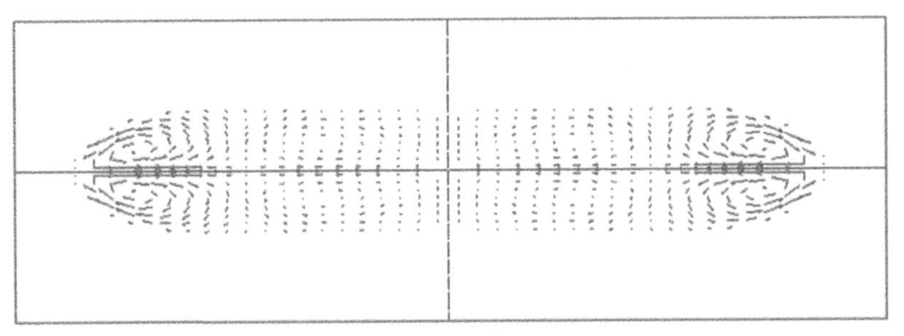

Radial velocity: -2.294×10^{-5} to 1.479×10^{-5}m/s
Axial velocity: -9.126×10^{-6} to 4.247×10^{-6}m/s

FIGURE 9.10
Velocity distribution in an aluminum nugget and its surrounding mushy zone.[18]

due to magnetic stirring and temperature gradient, and it has effects on the solute distribution. Using an axisymmetric model of coupled thermal–electrical–mechanical analysis, Khan et al.[18] predicted convection in a liquid nugget during resistance spot welding aluminum. Convection effects resulted from the interactions between phases in the porous mushy zone, and the buoyancy forces from the temperature difference were determined to be insignificant in the weld nugget formation. As shown in Figure 9.10 of the velocity vectors in the mushy zone and molten zone, the maximum liquid velocity obtained is about 2×10^{-5} m/s, which is too low to produce a significant convection effect.

9.5.4 An Example of Simulating Microstructure Evolution in a Spot Weldment

The process model developed by Babu[25] can be used to predict the thermal history, microstructure, and resulting mechanical properties. Details of the calculation and modeling can be found in Refs. 21 and 25. Based on the commercial FEM package ABAQUS, their model predicted the temperature profiles of a carbon steel (Fe—0.05 wt.% C—0.1 wt.% Si—1.0 wt.% Mn) at different locations in a weld nugget (Figure 9.11). In the figure, Node 1 corresponds to a location close to the center of the weld nugget, which experiences melting during heating, and nodes 14 and 270 are in the HAZ. The resultant microstructures

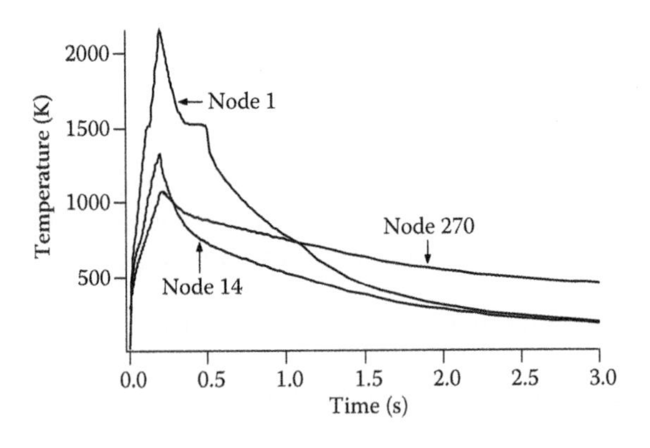

FIGURE 9.11
Temperature profiles at various locations in a weldment.[25]

TABLE 9.1

Phases and Hardness Generated in the Nugget and HAZ

Location	Ferrite (vol%)	Bainite (vol%)	Martensite (vol%)	Hardness (HV)
Node 1	0.01	0.02	0.97	245
Node 14	0.00	0.01	0.99	258
Node 270	0.50	0.00	0.50	182

Source: Babu, S., Web site: http://mjndeweb.ms.ornl.gov/Babu/default.html, 2004. With permission.

(phases) and properties (hardness) are listed in Table 9.1, and the hardness distribution in the weldment is shown in Figure 9.12.

9.6 Examples of Numerical Simulation of RSW Processes

9.6.1 Case Study I: Effect of Electrode Face Geometry

In a work by Babu,[25] the effects of electrode geometry (using new and used electrodes) and the process coupling were studied. As shown in Figure 9.13, new or unconditioned electrodes have well-defined edges, whereas conditioned ones are slightly worn with a rounded edge. Such differences in electrode surface profiles result in different contact pressure distributions on the worksheets at the electrode–sheet interface, therefore resulting in different contact resistance during welding. In addition, the pressure distribution at the sheet–sheet interface is also influenced by the electrode geometry, although to a smaller extent, and it affects the expulsion behavior, as explained in Chapter 7.

The simulated nugget growth using conditioned electrodes (as shown in Figure 9.13) is shown in Figure 9.14. In the simulation, instead of RMS values, actual current waveform was used with an increment of 1/8 of a welding cycle. It shows that heating is concentrated near the faying interface at the beginning of welding (Figure 9.14a). The temperature approaches the melting point of the steel after 2.125 cycles of welding time. Melting starts at the faying interface with further heating (Figure 9.14b, after 3.125 cycles), and the nugget

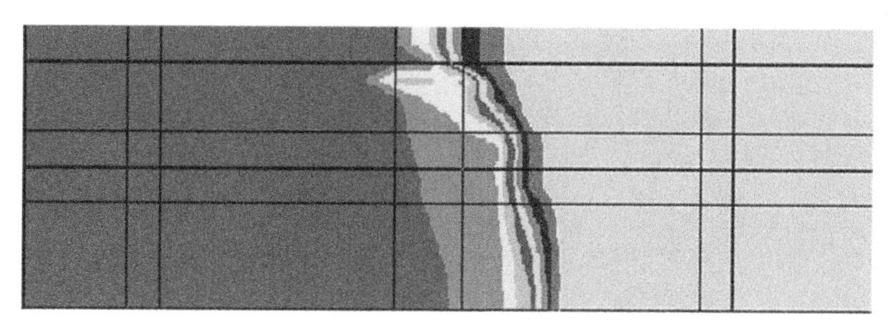

FIGURE 9.12
(See color insert.) Hardness gradients predicted in weldment.[25]

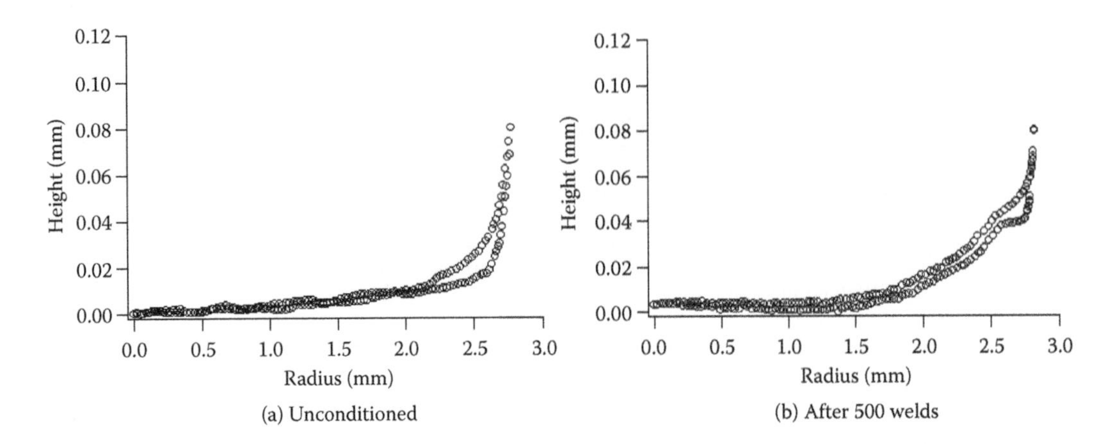

FIGURE 9.13
Measured surface profiles (both sides from center to edge) of unconditioned and conditioned electrodes used in simulation.

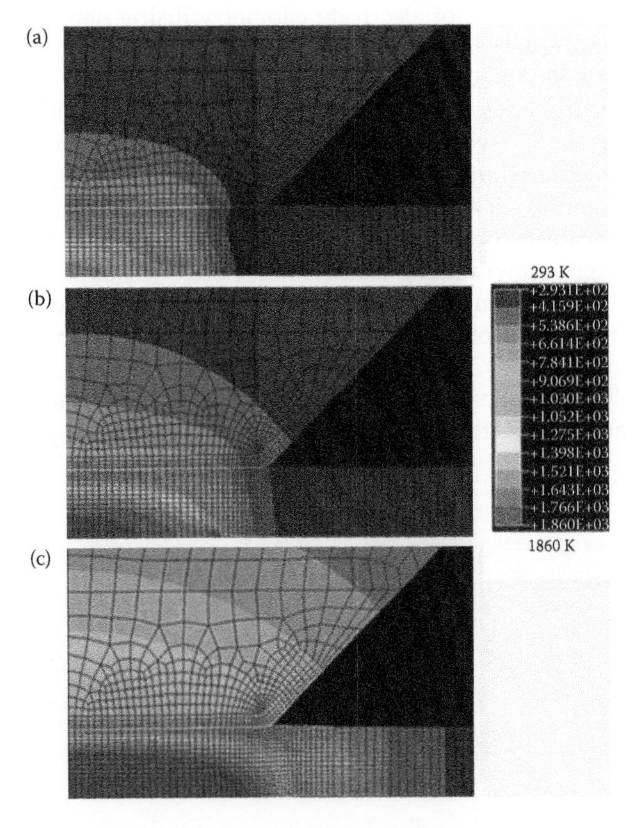

FIGURE 9.14
(See color insert.) Simulation results of nugget formation using conditioned electrodes.

grows to its full size after 12 cycles (Figure 9.14c). The results were compared with experiments and good agreement was achieved.

9.6.2 Case Study II: Differences between Using Coupled and Uncoupled Algorithms

It is especially interesting to see the difference in nugget growth simulated using different algorithms in dealing with the interaction among electrical, thermal, and mechanical processes. Using the same finite element mesh, but different coupling algorithms, Babu[25] clearly showed that the coupling has a significant effect on the nugget growth in modeling a mild steel welding. In an uncoupled modeling, temperature (as a function of time) was calculated for the entire welding cycle at once in the electrothermal module. The contact resistance was assumed as a function of temperature only, and contact area had to be assumed beforehand, instead of calculated. Therefore, the effect of pressure on contact resistance at the interfaces, which was calculated in the thermomechanical module, was not accounted for. As a result, unrealistic excessive heating is observed, as shown in Figure 9.15a. The nugget size after 12 welding cycles [part (d) in the uncoupled model] is significantly larger than that obtained through the coupled modeling algorithm [part (c) in the coupled model], especially in penetration. In the coupled modeling, the contact area and

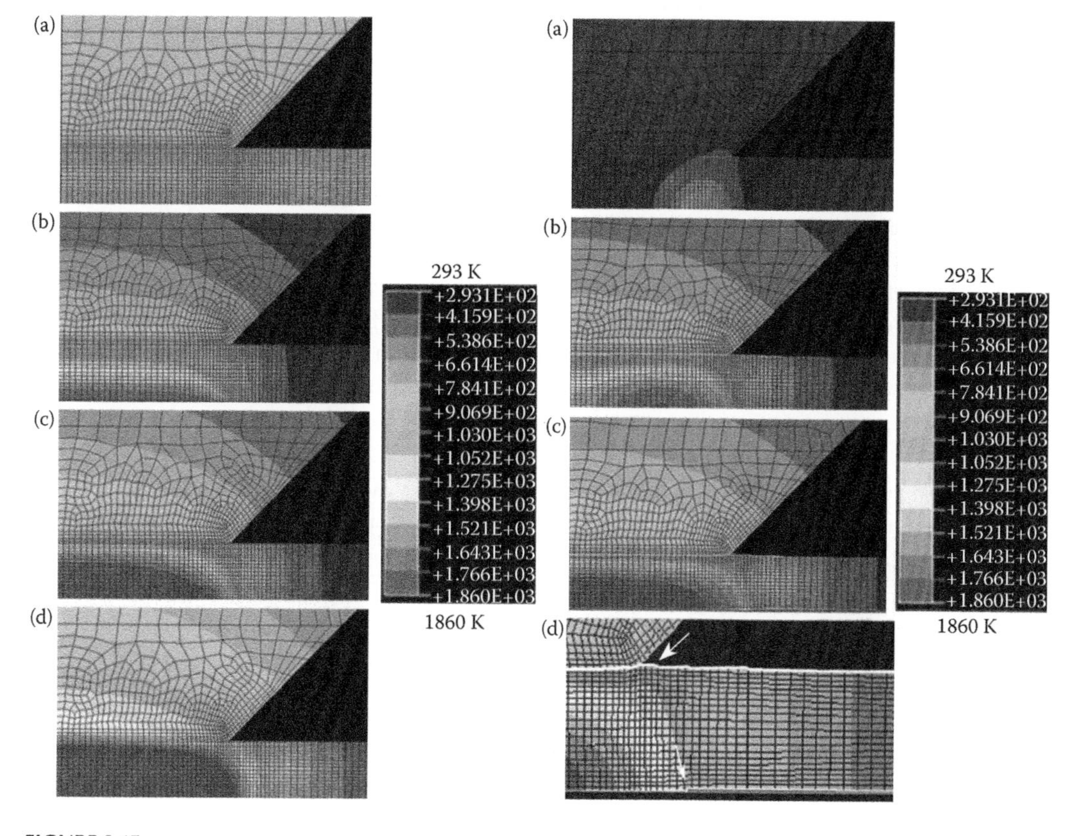

FIGURE 9.15
(See color insert.) Nugget growth simulation using uncoupled and coupled electrical–thermal–mechanical algorithms.

contact resistance, as functions of both contact pressure and temperature, were updated during the simulation. The effect of instantaneous deformation of the weldment during welding can be directly reflected using the coupling algorithm, as shown in the figure, which is crucial in obtaining realistic simulation results. In fact, fairly good agreement has been obtained when compared with experimental results, using the coupling algorithm for both unconditioned (new) electrodes and conditioned electrodes, as shown in Figure 9.16.

9.6.3 Case Study III: Effect of Electrode Axial Misalignment

Using a FEM model based on incrementally coupled electrical–thermal–mechanical processes, a study was conducted on the electrode axial misalignment effect in steel spot welding (Li and Sun[37]). The welding conditions are as follows:

0.8 mm bare steel sheet
A-nose electrode (6.4-mm face diameter)
8 kA welding current (RMS)
450 lb squeeze force
50% offset misalignment

A two-dimensional plane model was used in this study. It was found that a two-dimensional plane model generates higher temperature and larger deformation than three-dimensional modeling, because of less heat conduction and less restraining of the liquid nugget from its surrounding solid than in a three-dimensional model. However, using a two-dimensional model considerably reduces the modeling and computing time. Therefore, it was used to qualitatively illustrate the effect of electrode misalignment. In the model, equivalent current and load were calculated based on the distributed current density and load. Equivalent contact resistance was also calculated based on the plane model. Deformed geometry and contact area changes were updated every half cycle.

The misalignment reduces the contact area of the lower electrode–sheet interface and causes a severe stress concentration at the offset contact. The misalignment also directly affects the heating and nugget formation process. Unlike in the case of perfect alignment, the heating has an asymmetric temperature distribution in the weldment (Figure 9.17).

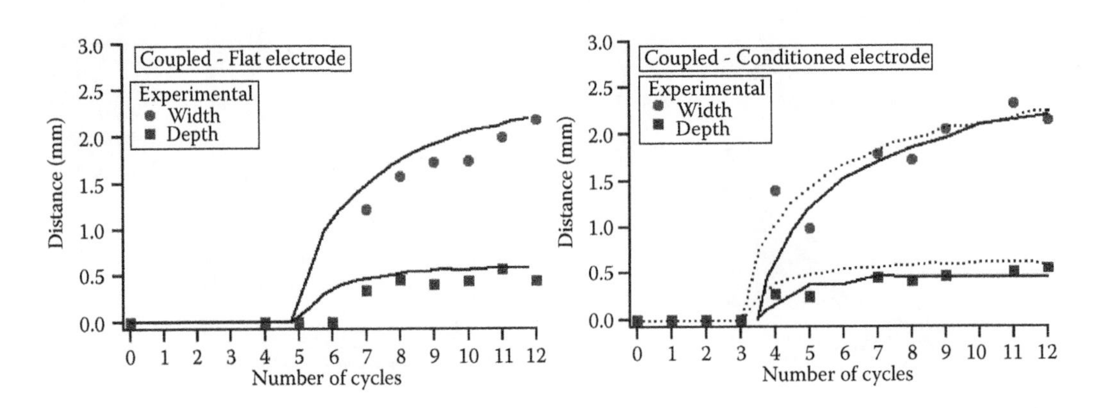

FIGURE 9.16
Comparison between simulation and experiments using coupled modeling algorithm.

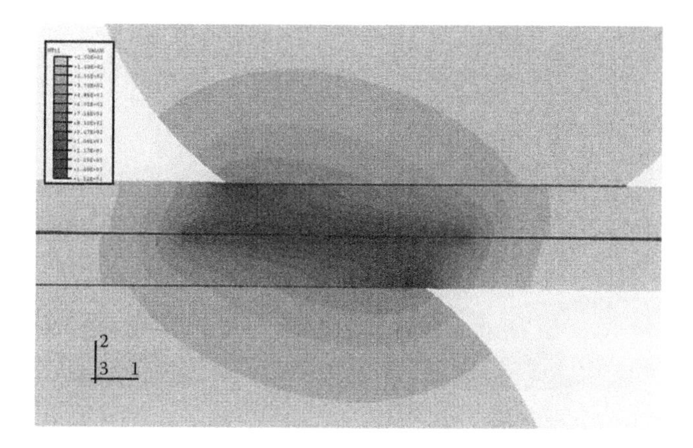

FIGURE 9.17
Temperature contours at end of 1.5th cycle.

Initial melting is first observed at the faying interface near the electrode face edges after 1.5 cycles. In a modeling of welding using the same material and process parameters, but under perfectly aligned conditions, melting does not occur until the third or fourth cycle. Therefore, melting occurs earlier when there is a misalignment because of smaller contact area or a more concentrated heating. As a result, the nugget formed has an irregular shape.

9.6.4 Case Study IV: Effect of Angular Misalignment of Domed Electrodes

In a study by Sun and Li,[23] standard dome-shaped Alcan electrodes were used to investigate the effect of angular misalignment in welding 2.0 mm aluminum alloy AA5754. The welding conditions used are as follows:

26 kA welding current
5° angular misalignment
60 cycles of squeeze time
10 cycles of weld time
30 cycles of hold time

An incrementally coupled two-dimensional plane model was used. Equivalent current, contact resistivity, and load were calculated based on the plane model. Deformed geometry and contact area changes were updated every half cycle.

Figures 9.18 and 9.19 show the temperature contours after 2 and 5.5 cycles of welding, respectively. Initial melting was first observed at the faying interface at the end of the second cycle. At this time, the shape of the molten zone is similar to that of a perfectly aligned welding case. After 5.5 cycles, however, the molten zone begins to tilt and the tip of the upper electrode is embedded in the worksheet. Moreover, expulsion is expected at the faying interface because the radius of the pressure distribution on the faying interface is smaller than the radius of the molten zone, using the geometry comparison model for expulsion, as shown in Chapter 7.

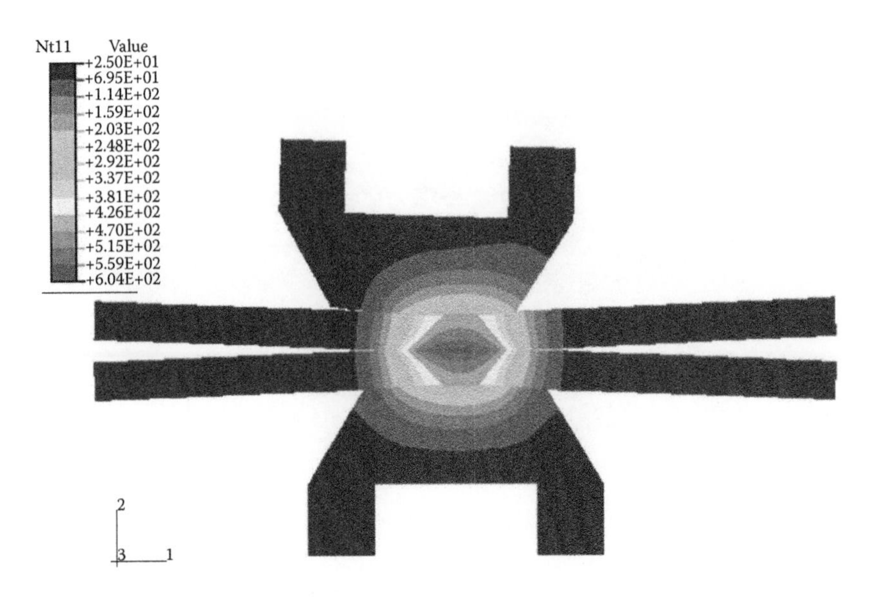

FIGURE 9.18
(See color insert.) Temperature distribution after second cycle.

Figure 9.20 shows the molten zone size and shape at the end of the eighth cycle. By this time, melting has already occurred at the electrode–sheet interface, and this is detrimental to electrode life. As a comparison, Figure 9.21 shows the nugget size at the end of the eighth cycle under perfectly aligned conditions using the same welding parameters. In the aligned case, the molten zone is symmetric with respect to the electrode centerline, and no melting through or expulsion is expected.

In general, the technical difficulties in numerical simulation of RSW lie on coupling the physical processes involved in welding. Each of the electrical, thermal, mechanical,

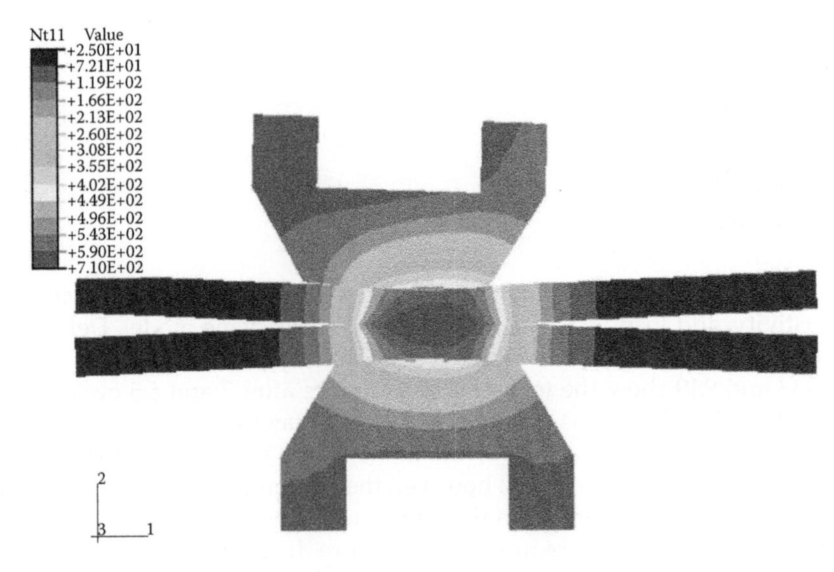

FIGURE 9.19
(See color insert.) Temperature distribution after 5.5th cycle.

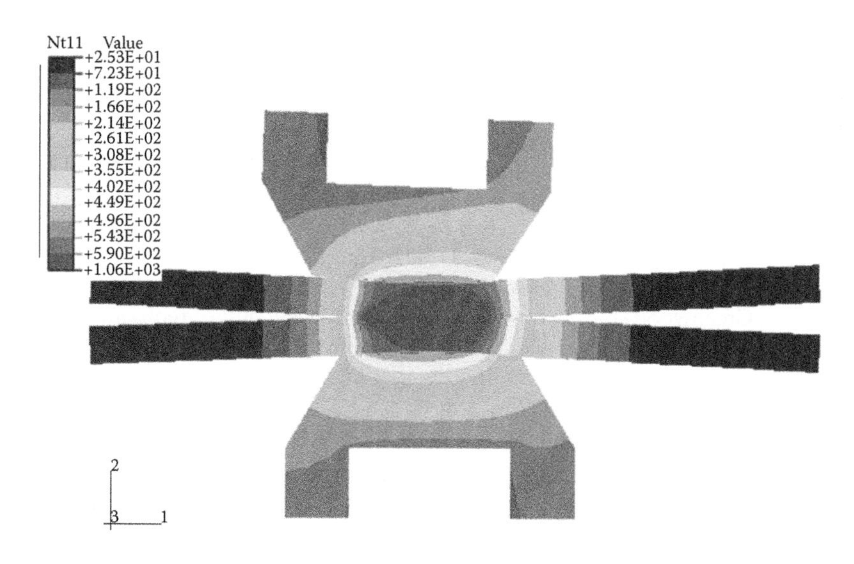

FIGURE 9.20
(See color insert.) Temperature distribution after 8 cycles.

and metallurgical processes has been well understood and analytically expressed in their respective fields. Combining these into a reliable software package requires coordinated efforts of all disciplines, including numerical simulation. Besides simulation of nugget formation and microstructure evolution, numerical simulation has also been used in other aspects of RSW (e.g., strength characterization, testing of specimen size determination), as illustrated in other chapters. In addition, a general lack of reliable material data, such as the temperature dependence of electrical conductivity of a material (often an alloy than a pure metal) imposes a significant challenge to the accuracy of numerical simulation of the RSW process.

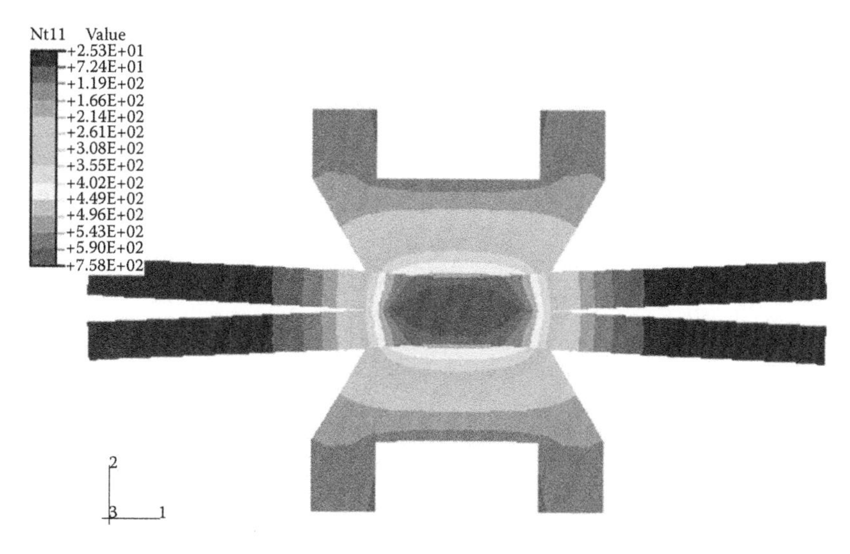

FIGURE 9.21
(See color insert.) Temperature distribution after 8 cycles with perfectly aligned electrodes.

References

1. *ANSYS*, Ansys Inc., Canonsburg, PA, 1999.
2. Nied, H. A., The finite element modeling of the resistance welding process, *Welding Journal*, 70, 339-s, 1991.
3. Tsai, C. L., Jammal, O. A., Papritan, C., and Dickinson D. W., Modeling of resistance spot welding nugget growth, *Welding Journal*, 71, 47-s, 1992.
4. *ABAQUS*, Hibbitt, Karlsson & Sorensen Inc., Pawtucket, RI, 1988.
5. Bentley, K. P., Greenwood, J. A., Knowlson, P. M., and Backer, R. G., Temperature distribution in spot welds, *British Welding Journal*, 10, 613–619, 1963.
6. Rice, W. and Funk, E. S., An analytical investigation of temperature distributions during resistance welding, *Welding Journal*, 46, 175-s, 1967.
7. Gould, J. E., An examination of nugget development during spot welding, using both experimental and analytical techniques, *Welding Journal*, 66, 1-s, 1987.
8. Tsai, C. L., Jammal, O. A., Papritan, C., and Dickinson, D. W., Modeling of resistance spot welding nugget growth, *Welding Journal*, 71, 47-s, 1992.
9. Cho, H. S. and Cho, Y. J., A study of the thermal behavior in resistance spot welding, *Welding Journal*, 68, 236-s, 1989.
10. Huh, H. and Kang, W. J., Electrothermal analysis of electric resistance spot welding process by a 3-D finite element method, *Journal of Materials Processing Technology*, 63, 672, 1997.
11. Wei, P. S. and Ho, C. Y., Axisymmetric nugget growth during resistance spot welding, *Journal of Heat Transfer*, 112, 309, 1990.
12. Wei, P. S. and Yeh, F. B., Factors affecting nugget growth with mushy-zone phase change during resistance spot welding, *Journal of Heat Transfer*, 113, 643, 1991.
13. Wei, P. S., Wang, S. C., and Lin, M. S., Transport phenomena during resistance spot welding, *Journal of Heat Transfer*, 118, 762, 1996.
14. Browne, D. I., Newton, C. I., and Boomer, D. R., Optimization and validation of a model to predict the spot weldability parameter lobes for aluminum automotive body sheet, in *Proceedings of International Body Engineering Conference*, IBEC'95, Advanced Technologies and Processes Section, Detroit, MI, 1995, 100.
15. Browne, D. J., Chandler, H. W., Evans, J. T., Wen, J., and Newton, C. J., Computer simulation of resistance spot welding in aluminum: Part I, *Welding Journal*, 74, 339-s, 1995.
16. Browne, D. J., Chandler, H. W., Evans, J. T., Wen, J., and Newton, C. J., Computer simulation of resistance spot welding in aluminum: Part II, *Welding Journal*, 74, 418-s, 1995.
17. Khan, J. A., Xu, L. J., and Chao, Y. J., Prediction of nugget development during resistance spot welding using a coupled thermal–electrical–mechanical model, *Journal of Science and Technology of Welding and Joining*, 4, 201, 1999.
18. Khan, J. A., Chao, Y. J., and Xu, L. J., Modeling and simulation of resistance spot welding process for Al-alloy, in *Proceedings of AWS Sheet Metal Welding Conference IX*, Sterling Heights, MI, Paper No. 5-1, 2000.
19. Khan, J. A., Xu, L. J., Chao, Y. J., and Broach, K., Numerical simulation of resistance spot welding process, *Numerical Heat Transfer Part A—Applications*, 37, 425–446, 2000.
20. Zhang, H., Huang, Y. J., and Hu, S. J., Nugget growth in resistance spot welding of aluminum alloys, in *Proceedings of Sheet Metal Welding Conference VII*, Detroit, MI, Paper No. B3, 1996.
21. Feng, Z., Babu, S. S., Santella, M. L., Riemer, B. W., and Gould, J. E., An incrementally coupled electrical–thermal–mechanical model for resistance spot welding, in *Proceedings of the 5th International Conference on Trends in Welding Research*, ASM International, Pine Mountain, GA, edited, 1998, 599.
22. Li, M. V., Dong, P., and Kimchi, M., Analysis of microstructure evolution and residual stress development in resistance spot welds of high strength steels, in *Proceedings of Sheet Metal Welding Conference VII*, Detroit, MI, Paper No. 5-6, 1998.

23. NIST—ATP Intelligent Resistance Welding Quarterly Progress Report, No. 202, Ann Arbor, MI, 1997.
24. NIST—ATP Intelligent Resistance Welding Quarterly Progress Report, No. 203, Ann Arbor, MI, 1997.
25. Babu, S., Web site: http://mjndeweb.ms.ornl.gov/Babu/default.html. Accessed in 2004.
26. Bay, N. and Wanheim, T., Real area of contact between a rough tool and a smooth workpiece at high normal pressures, *Wear*, 38, 225–234, 1976.
27. Vogler, M. and Sheppard, S., Electrical contact resistance under high loads and elevated temperatures, *Welding Journal*, 231-s–238-s, June, 1993.
28. Kohlrausch, F., *An Introduction to Physical Measurements: With Appendices on Absolute Electrical Measurements* (translated from the 2nd German edition by T. H. Waller and H. R. Proctor), D. Appleton and Company, New York, 1891.
29. Bowden, F. P. and Williamson, J. B. P., Electrical conduction in solids. I. Influence of the passage of current on the contact between solids, *Proceedings of Royal Society of London*, 1, 246, 1958.
30. Greenwood, J. A. and Williamson, J. B. P., Electrical conduction in solids: II. Theory of temperature-dependent conductors, *Proceedings of Royal Society of London*, 13, 246, 1958.
31. Ion, J. C., Easterling, K. E., and Ashby, M. F., A second report on diagrams of microstructure and hardness for heat-affected zones in welds, *Acta Metallurgica*, 32, 1949, 1984.
32. Watt, D. F., Coon, L., Bibby, M., Goldak, J., and Henwood, C., An algorithm for modelling microstructural development in weld heat affected zones (Part A). Reaction kinetics. *Acta Metallurgica*, 36, 3029, 1988.
33. Bhadeshia, H. and Svensson, L. E., Modelling the evolution of microstructure in steel weld metals, in *Mathematical Modeling of Weld Phenomena*, edited by H. Cerjak and K. E. Easterling, Institute of Materials, London, UK, 1993, 109.
34. Jones, S. J. and Bhadeshia, H., Kinetics of the simultaneous austenite into several transformation products, *Acta Metallurgica*, 45, 2911, 1997.
35. Kirkaldy, J. S. and Venugopolan, D., *Phase Transformations in Ferrous Alloys*, edited by A. R. Marder and J. I. Goldstein, AIME, Warrendale, PA, 1984, 125.
36. Brooks, J. A., Yang, N. C. Y., and Krafcik, J. S., On the origin of ferrite morphologies of primary ferrite solidified austenitic stainless steel welds, *Recent Trends in Welding Science and Technology*, edited by S. A. David and J. M. Vitek, ASM International, Materials Park, OH, 1992, 73–80.
37. NIST–ATP Intelligent Resistance Welding Quarterly Progress Report, No. 201, Ann Arbor, MI, 1997.

10

Statistical Design, Analysis, and Inference in Resistance Welding Research

10.1 Introduction

A significant amount of research has been devoted to finding the dependence of weld quality on welding parameters in published works. However, a relationship obtained experimentally is very often influenced by random factors, such as variations in sheet coating thickness and composition, line voltage, and cooling water temperature. The large number of possible variables also poses a challenge to quantitative study in resistance welding. It is often difficult to isolate the effects of a single variable. Even if a one-variable-at-a-time condition is realized, the results may render limited useful information because such a situation rarely exists in practice.

To understand and predict the outcome of a physical process, the conventional (and also prevalent) approach is to choose a set of governing equations with appropriate boundary/initial conditions (based on the processes involved), and then derive the solution through analytical or numerical means. Such a procedure is called deduction; an example of which is structural analysis of solids. Once the loading condition is given and the mechanical behavior of a solid is understood, in the form of a set of equations such as stress and strain relations, a numerical modeling may usually produce very accurate results. In the case of resistance welding, such a method proves to be ineffective. The difficulty lies in the complex physical processes (electrical, thermal, mechanical, and metallurgical) involved, and the interactions of these processes during welding. In addition, a lack of reliable material data of temperature dependence makes accurate and quantitative predictions impossible. Therefore, many studies in resistance welding research draw the effect of a variable or variables through the analysis of experimental results (data), without attempting to understand the physical processes involved. This is the so-called statistical approach.

The statistical approach reverses the deduction procedure. It takes real data that have arisen from practical situations (perhaps fortuitously or through a designed experiment) and uses the data to validate a specific model, to make rational guesses or estimates of the numerical values of relevant parameters, or even to suggest a model in the first place. This reverse procedure is called induction. This reverse, inductive process is possible only because the language of probability theory is available to form the deductive link. A statistical approach starts from a systematic collection of data that provides relevant information about the process. The observed data are then used to obtain a numerical estimate of the unknown parameters in a statistical model. Its aim is to enable inferences, conclusions, or predictions to be drawn about the model from the information provided by the sample data (or perhaps by other information), or to construct procedures to aid in making decisions relevant to the practical situation. A statistical model is quite different from a physical model obtained through a deduction procedure. In statistical analysis, a physical process is replaced by a model, and the process is deduced from the characteristic behavior of data. Hence, a statistical model is an approximation of the true physical process. As the idealization of a physical process, a statistical model usually contains two parts: (1) a polynomial form that describes/approximates the relationship between input and output, and (2) a probabilistic part that reflects random variation. What constitutes relevant information, the implied differences in the descriptive and decision-making functions of statistical theory, and examples of application of statistical theory to resistance welding research are presented in this chapter.

10.2 Basic Concepts and Procedures

Statistics is a science that works with numerical observations, usually called data. The methodology of systematic study of data developed in statistics has now been widely applied in many disciplines. Data, in statistics, are regarded as observations generated from an underlying system that involves uncertainty. As nothing in the physical world is certain, conclusions drawn based on data usually contain a certain degree of uncertainty. Statistics develops strategies and methodologies to separate systematic patterns from variation and uncertainty. It acquires an understanding of the systematic patterns by means of data study. In pursuit of this aim, statistics divides the study of data into three steps: (1) data collection, (2) statistical modeling and data analysis, and (3) inference or decision making. The objective of data collection is to produce good data, which is representative, in order to draw correct information. Statistical design offers principles and methodologies to generate representative data. In modeling and analysis, a statistical model constructed in the probabilistic language is often used to represent the system of interest. A model is a simplified representation of the system, but should capture the prominent features of the system. Data analysis, often presented in graphs and numerical summary based on the statistical model, reveals important information from data. It is a practical art of processing and presenting data in an understandable manner. Graphical and numerical tools and various strategies can be used for exploring/mining information from data. The inference or decision making utilizes the results from data analysis to answer the questions of interest to investigators. Probability is used to accompany the conclusions drawn from data by a formal statement of how confident the investigator is about its correctness. Each of the three steps is further elaborated in the following sections.

10.2.1 Data Collection

A statistical analysis starts with data collection, and how data are collected directly affects the validity of the inference drawn from the analysis based on such data. Before numbers can be used, they must be observed, collected, or produced. Observational data are those recorded without intentionally interfering with the system being observed, such as data recorded at a welding station. The use of observational data requires judgment and caution. The observational data may only present one aspect of the many facets of a fact, and inference based on such data may be misleading when it is generalized. Therefore, in order to see how a system responds to a change in input, the change must be actually induced. In contrast to the observational approach, a more active method to collect data is by experimentation, which has been widely applied in engineering/industrial studies. The conditions can be designed and controlled in experiments, and therefore, the relationship between input and output variables can be systematically established. Experimentation can be used to study the effects of specific variables of interest, either individually or in combinations, rather than simply observe the natural change of a system. A design is the plan or outline of an experiment made according to the characteristics of a system and the statistical treatments to be applied. The choice of a plan is crucial to the success of an experiment. A poor design may capture insufficient information, which may not be enough to answer questions of interest. On the other hand, a comprehensive experiment may offer a large amount of information, but at a greater cost. Furthermore, some of the information may not be related; hence, it is useless. An efficient design needs to have a balance between the amount of information and the expense. For instance, if the weld quality is monitored at a welding station on a production line as a function of welding process parameters such as electrode force and cooling water flow rate, process improvement can be made, but no optimization can be made beyond the existing capacity of the welder and its controller. Even the improvement is limited because the ranges of the variables are usually not wide enough and it is difficult to consider the interaction of the parameters.

10.2.2 Statistical Modeling and Data Analysis

A statistical model can be regarded as the realization of an underlying physical system, especially the probabilistic mechanism that governs the system. Statistical models for data from experiments often contain a deterministic component that describes the relationship between input and output variables. The only unspecified part in a statistical model is the unknown values of some parameters that can then be estimated by the use of data. Every statistical model is constructed based on some assumptions that simplify the complexity of the system. The main concern in a statistical modeling is to construct an adequate model whose assumptions are appropriate. Data analysis is the process of understanding the information contained in data. In data analysis, the true system is replaced by the statistical model, and any description based on the result of data analysis about the model is then regarded as a substitute for the system. Data are organized, summarized, and numerically calculated so that information can be extracted and presented for human use. The first step in data analysis is often to organize and display data in graphs, as they are intuitive and provide a preliminary understanding of data. Human eyes and mind can capture the information from a graph much easier than from a series of numbers. A more advanced analysis of data is usually based on a statistical model. For example, model-fitting techniques can estimate the parameters in a model and move from a general model form to a specific form; testing examines if some specific hypotheses expressed as a function of parameters are acceptable or should be rejected; validation of models, such as residual analysis, checks if the structure and assumptions in the model are adequate. The statistical model plays an important and indispensable role in these analyses.

10.2.3 Inference and Decision Making

In statistics, inference is dealt with using the information obtained from data analysis to answer questions of interest to the investigators. Decision making extends the inference to suggesting rules by incorporating assessments of consequences. In this step, not only statistical results but also available process knowledge should be considered to increase the precision of inference and decision making.

Even if all the steps and principles noted above are carefully followed and implemented, there is still no guarantee that correct information and inference can be obtained. For instance, the data may contain faulty information, or the experiment may have been interrupted unexpectedly. There are many factors that may cause an experiment to fail, such as variations that are not recognized when planning experiments, like day-to-day, batch-to-batch, or operator-to-operator effects; unstable conditions of a process that are not identified in experiments; or measurement error or even typos when recording data. Even though the collected data contain correct information, the statistical analysis may misinterpret the information just because some prior knowledge about the underlying system is not considered. An inappropriate statistical model or inference can cause these types of faults. Many of these situations can be prevented if statisticians and investigators/experimenters work closely with each other. The application of statistical tools is more effective when appropriate subject matter knowledge is taken into account. Interdisciplinary collaboration and exchange of knowledge are critical for an experiment or a data analysis to be successful. Both types of information, the prior knowledge about the underlying system and that gathered from empirical approaches such as data from experiments, should be considered in an investigation. Knowledge of the physical process should be used to confirm the conclusions of a statistical analysis. If the physical knowledge does not provide statistical results, a reasonable explanation, or support, the whole statistical procedure needs to be examined to see if any mistakes or deviations have been made. Note that the statistical technique is an important and effective means to help investigators in gathering information, rather than a replacement for the natural skill of the investigators. Experimenters and statisticians should collaborate and rely on each other for the success of an investigation. The focus of this chapter is to demonstrate the procedures of statistical analyses in resistance welding research. For this reason, some definitions and terms are explained first.

In general, the objective of an experiment is to study how the changes of input variables influence the output variables, and build a relationship between input and output variables. In designed experiments, or design of experiments, the input variables, called *factors*, are usually simultaneously (rather than one factor at a time) varied in a planned manner. The settings of factors are referred to as *levels* and a combination of factor levels is called a *run*. The output variable(s), called *response*, is the experimental outcome that interests the investigator. In physical experiments, responses are usually random and they may be either continuous or discrete. A continuous response takes any conceivable values within a range. For example, the diameter, thickness, and strength of a spot weld are continuous responses. A discrete response takes values that range over categories, such as expulsion, which either occurs or not, or it may be categorized as no expulsion, light expulsion, or heavy expulsion. On the other hand, if a quantitative signal is used to detect the occurrence of expulsion and the severity of expulsion, using a displacement sensor, as discussed in Section 7.3, expulsion becomes a continuous response. In addition to physical experiments, experiments performed on a computer code, which is a simulation of the physical system, have become more popular in recent years. A computer simulation experiment offers a different type of response from the conventional physical experiment. Unlike in physical experiments, the response is deterministic, rather than random. Observations with the same input are always identical. This characteristic, that is, a lack of randomness, calls for different techniques for the experiment design and analysis. In Section 10.3, the applications of statistical design, modeling and data analysis, and inference are explained in detail with real data sets as examples for experiments with continuous response. Experiments with discrete responses and computer simulation experiments are presented in Sections 10.4 and 10.5.

10.3 Experiment with Continuous Response

There are many examples in resistance spot welding (RSW) with continuous responses as output. The button-size measurement of welds created from similar welding schedules usually appears continuous, instead of discrete. This is the result of complicated physical responses of the system to the welding settings, in addition to the large number of random factors involved during welding, testing, and measurement. Other such continuous responses are electrode indentation depth, heat-affected zone (HAZ) size, tensile–shear strength, etc. Therefore, statistical experiment design and analysis should take such characteristics of resistance welding into consideration.

10.3.1 Statistical Design

10.3.1.1 Factorial Designs

For a scientific investigation in which the interest is to *simultaneously* study the effects of multiple factors, factorial designs are the most effective and most commonly used.[1,2] In a full factorial design, all the combinations of factor levels are performed. For example, a full factorial design of k factors, each with two levels, would require 2^k runs. It can be used to study the linear main effect of each factor and any i-factor interactions, where $i \leq k$. When k is large, a full factorial design may require a large number of runs (e.g., there are 128 runs if $k = 7$), which is often too expensive to perform in practice. For economic reasons, fractional factorial designs (FFDs), which consist of appropriately chosen subsets of the full factorial designs, are more suitable. For example, a 2^{k-p} FFD, which is a subset defined by *p defining words*, can be used to study k factors by using only 2^{k-p} runs. However, there is a trade-off in such a design. The reduction on the run size causes the effect aliasing in FFD. When two effects are aliased, they cannot be distinguished from each other. Based on the principle that lower-order effects are more likely to be important than higher-order effects, some criteria were developed for choosing optimal FFDs. Two popular criteria, *maximum resolution* and *minimum aberration*, were proposed under the justification of the principle. These two criteria work well in preventing aliasing between lower-order effects. For a complete list of minimum aberration FFDs and their aberrations, see the work of Wu and Hamada[2] (Appendix 4A for two-level designs and Appendix 5A for three-level designs) and Montgomery.[1] A key property of FFD is that all the estimable effects are orthogonal. The orthogonality property promises that the estimation of an effect is independent of other effects. From the perspective of statistical analysis, this property makes the analysis simpler and easier. Some graphical tools, such as half-normal plots, can be used efficiently for analysis of data from an FFD.

10.3.1.2 Orthogonal Arrays

The FFD designs are constructed by defining relations among factors. Designs that do not possess this property are usually called nonregular designs. A class of nonregular designs that are often used in applications is orthogonal arrays (OAs). In an OA, for instance, of strength 2 for any two columns, all possible combinations of factor levels appear equally often. OAs have some advantages over FFDs. FFDs usually have a run size that is a power of either 2 or 3 (as many experiments have two or three levels). The OAs are more flexible on the run size. For example, there are 12-run, 18-run, 20-run, 24-run, 36-run, 48-run, and many other run-size OAs. This is because in OAs, the factors may have different levels, whereas in an FFD, the factors are required to be of the same level. A collection of useful OAs is given in Appendices 7A–7C of Wu and Hamada.[2] Although OAs have this advantage of flexibility over FFDs, their statistical analysis is usually more complicated. Unlike the estimable effects in the FFDs, which are orthogonal, the effects in OAs usually have an aliasing pattern that is more complex. This is referred to as complex aliasing, which means that the correlation between an effect and other effects is usually nonzero. Advanced statistical tools for analyzing experimental data with complex aliasing can be found in the work by Wu and Hamada[2] (Chapter 8).

10.3.1.3 Second-Order Designs

Second-order designs are often used in the study of response surface, in which they are applied by a sequential experimentation strategy together with first-order designs such as two-level FFDs, as noted in previous sections. A second-order design allows all the linear effects, quadratic effects, and linear-by-linear interactions in the second-order model to be estimated. Central composite designs are the most commonly used second-order designs. They consist of three parts: cube points (factorial points), center points, and star points (axial points). When the experimental region is near or within the optimum region, a second-order design can be performed to fit a second-order model. Canonical analysis can then be used to classify the response surface.

10.3.1.4 Robust Parameter Designs

Robust parameter design is a methodology for reducing the variation of a system. There are two types of factors in such a design: control factors and noise factors. The control factors are input variables whose values will remain fixed at the optimum levels for the best performance. The noise factors are variables that are uncontrollable or for which influences are difficult to measure in the normal process or under normal conditions. However, the settings of noise factors are varied systematically in the experiments to represent their variation in a normal process. Through robust parameter designs, the optimum setting of control factors can be chosen to make them less sensitive to the variation caused by noise factors. FFDs or OAs are often used as designs for robust parameter experiments. There are two formats for an experiment design: cross arrays and single arrays. In cross arrays, location–dispersion modeling can be applied to analyze data, whereas for single arrays, the location–dispersion modeling approach is not applicable. The response modeling approach should be used instead. Comparisons between these two formats and their corresponding analysis strategies are given by Wu and Hamada[2] (Chapter 10).

10.3.1.5 Nested Designs

One fundamental assumption in factorial designs is that any of the levels of one factor can be paired with any level of another factor. In other words, all the possible combinations of factor levels are meaningful in the undertaken process. This is not always possible for many experiments including welding. Consider an experiment for understanding the expulsion phenomenon in welding a 1.2-mm AKDQ steel and a 1.0-mm aluminum alloy AA6111 as described in Chapter 7. If the factors are taken as material, welding current, and welding time, a complete factorial design (i.e., using the mathematical combinations of the levels of these factors) does not make much sense. For instance, the welding current for the

AKDQ steel ranges from 6.5 to 13.9 kA, and that for the AA6111 is from 5 to 40 kA. In addition, the welding times for the AKDQ steel and AA6111 are from 133 to 400 ms and from 17 to 84 ms, respectively. Therefore, the welding time and current values for one material cannot be used for another material, or they cannot be "crossed" between the steel and aluminum, and a design other than a complete factorial design is needed. Nested designs are developed to deal with situations such as this. If the levels of one factor, such as welding current, are similar to but not identical to those of another factor, such as material in the aforementioned example problem, then a nested design is warranted.

A three-stage nested design for the aforementioned hypothetical experiment is illustrated in Figure 10.1. This design is exactly the same as that in Montgomery.[1] As the values (levels) of welding current (factor B) depend on the material (levels of factor A), factor B is called "nested" within A, and the welding times (levels of factor C) for different welding currents are usually different; C is "nested" within B. The model for analysis of variance is different from that of a complete factorial design, and one fact is that it does not evaluate the interactions between a factor and those nested within it. In this sample problem, there will be no interactions between materials, welding current, and welding time, as such interactions are meaningless.

Many experiments in welding need nested designs and ignoring the need for nesting may yield erroneous inferences.

10.3.1.6 Use of Blocks

The precision of an experiment may be reduced by systematic sources of variation, which may not be of primary interest to investigators. Examples of systematic variations are day-to-day or week-to-week effect, operator-to-operator effect, and batch-to-batch or lot-to-lot effect. Blocking is a useful way to reduce the influence of such variations and improve the efficiency of experiments by arranging homogeneous experimental runs into groups. Not to block is a common mistake in dealing with engineering problems, including welding. If an experiment involves using coated steels, the coating thickness may not be uniform for different specimens if they are cut from different locations across the width a sheet blank. If such variation is not identified and isolated, it may mess up the experiment/analysis by exaggerating the variability of the responses if coating has a considerable influence in the response. For a comprehensive discussion on the role of blocking in experimental design, see Wu and Hamada.[2] For optimal arrangement of 2^{n-k} (3^{n-k}) design in 2^p (3^p) blocks, see Appendices 3A, 4B, and 5B of the same book.

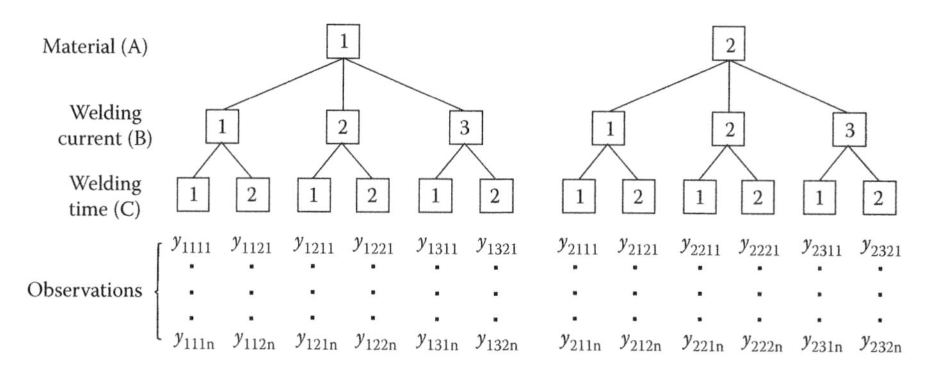

FIGURE 10.1
A three-stage nested design.

10.3.2 Analysis and Modeling

10.3.2.1 Use of Graphs

Graphs are a powerful tool in providing a visual summary of data. Various informative graphs are often seen in daily life, such as a pie chart that shows the fractions of monthly expenses, or an x–y plot that shows the dependence of weld size on welding current. Graphs convey an overall visual impression that is much stronger than a numerical summary of data. In statistical data analysis, graphs are an effective way to offer an immediate vivid perception of information in data. For many data sets, appropriate graphs are often enough to reveal most of the information. Some common types of graphs for experimental data are listed as follows:

1. *Main effect plots*: A main effect plot is drawn by linking the averages of observations at each level of a factor by a line. The shape of the line shows the influence of the level changes. For example, for two-level factors, a steep line is related to a significant main effect, whereas a flat line is related to an insignificant one. The main effect plots of all factors are often placed together in one graph for a comparison of their relative magnitudes.
2. *Interaction plots*: Interactions are effects that are defined by two or more factors. They reflect the joint effects of factors on the response. Interactions are usually defined as the component-wise product of the corresponding main effects. For example, the linear-by-linear interactions of two factors A and B are defined as the component-wise products of the linear main effects of A and B, respectively. Interaction plots are used to graphically show the joint influences. They display the averages of observations at each level combination of the two factors and connect the averages by lines. Those averages from the same level of one of the factors are connected by a line. The number of lines in the plot is the same as the number of levels of this factor. By comparing the shapes of these lines, more insight into the interactions can be obtained: lines of similar shape and trend indicate no interaction, whereas different ones imply the existence of interactions.
3. *Scatter plots*: These are used to explore the relationship between two variables by representing each paired observation of the two variables as a point in plane. They are often seen in residual analysis.
4. *Box plots*: Box plots are a tool for bringing up both the location and variation information from data sets. For a single variable, a box plot displays the minimum, 25th percentile, median, 75th percentile, and maximum of its data. From a box plot, immediate visuals are the center, the spread, and the overall range of the variable. They are particularly useful for comparing location and variation changes between different groups of data, by viewing box plots of the groups side by side.
5. *Normal and half-normal plots*: These are graphical tools for judging effect significance. They contain the (absolute) effect estimates against their corresponding quantiles of the standard normal distribution. Effects that are not significant form a cluster near the origin (0, 0). Through the cluster, a straight line can be drawn. Significant effects tend to lie away from the straight line. Both plots are methods based on visual judgment and are particularly useful for experiments without replicates.

Appropriate precaution should be taken when making conclusions solely based on graphs. For instance, the "steepness" in a main effect plot may appear significantly different if the scale bar (range) of the response is altered. More rigorous statistical analyses are needed in addition to the visual means that are convenient, yet often illusive and arbitrary.

10.3.2.2 Multiple Regression Model

Multiple regression models are the most widely used statistical tool for modeling experimental data with continuous responses. Such a model interprets the response as a linear combination of effects and experimental error. A multiple regression model relates a continuous response y to p corresponding effects, $x_1, x_2, \ldots, x_p$, as follows:

$$y = \beta_0 + \beta_1 x_1 + \cdots + \beta_p x_p + \varepsilon \tag{10.1}$$

where ε is the error term that represents the unpredicted experimental error. It is a random variable and usually assumed to be normally distributed with zero mean and variance σ^2. By taking expectation on both sides of Equation 10.1, the structural part of the model is

$$E(y) = \beta_0 + \beta_1 x_1 + \cdots + \beta_p x_p + E(\varepsilon) = \beta_0 + \beta_1 x_1 + \cdots + \beta_p x_p \tag{10.2}$$

because $E(\varepsilon) = 0$.

Hence, $E(y)$ is linear in parameters $\beta_0, \beta_1, \ldots, \beta_p$. Each of the parameters represents the magnitude of an effect and is regarded as an unknown constant. The estimation of parameters can be done using the information from data. If n observations are collected in an experiment, the model developed from them takes the form

$$y_i = \beta_0 + \beta_1 x_{1i} + \cdots + \beta_p x_{pi} + \varepsilon_i,\ i = 1, \ldots, n \tag{10.3}$$

where y_i is the ith value of the response and $x_{1i}, \ldots, x_{pi}$ are the corresponding values of the p effects. These n equations can be written in matrix form as

$$\mathbf{y} = \mathbf{X}\boldsymbol{\beta} + \varepsilon \tag{10.4}$$

where $\mathbf{y} = (y_1, \ldots, y_n)^T$, $\beta = (\beta_0, \beta_1, \ldots, \beta_p)^T$, $\varepsilon = (\varepsilon_1, \ldots, \varepsilon_n)^T$, and $\mathbf{X}$ is an $n \times (p + 1)$ matrix with the vector $(1, x_{1i}, \ldots, x_{pi})$ in its ith row, $i = 1, \ldots, n$. By using the matrix notation, the estimate of the unknown parameters, β, can be easily derived. A commonly used estimation technique for β is the least square estimate, which minimizes the sum of the square of the difference between the observed response and predicted response:

$$(\mathbf{y} - \mathbf{X}\boldsymbol{\beta})^T(\mathbf{y} - \mathbf{X}\boldsymbol{\beta}) \tag{10.5}$$

The matrix form of the least square estimate of β can be expressed as follows:

$$(\mathbf{X}^T\mathbf{X})^{-1}\mathbf{X}^T\mathbf{y} \tag{10.6}$$

After the estimation, standard statistical hypothesis tests can be performed to test whether the estimates are statistically significant. More details about multiple regression are given by Draper and Smith.[3]

10.3.2.2.1 Coded Effects

In the analysis of experimental data, it is usually convenient and computationally efficient to use coded effects. The coded effects are converted from factors in natural scale and related to the response through an equation such as Equation 10.1. A coded effect usually contains a physical interpretation of factors. For a factor with two levels, its (linear) main effect is usually represented as a vector in which the low and high levels are coded as –1 and +1, respectively. For a quantitative factor with three levels, its linear main effect is a vector in which the low, median, and high levels are coded as –1, 0, and 1, respectively, and the quadratic main effect as 1, –2, and 1, respectively. For a categorical three-level factor, there are various ways of main effect coding. For example, (–1, 1, 0) or (–1, 0, 1) can be used for main effect coding of categories 1, 2, and 3, respectively. The first three-digit vector represents the difference between categories 1 and 2, and the second one between categories 1 and 3. These two coded effects are appropriate to use when the main interest is the comparison of the first category with other categories. A more detailed discussion is given by Wu and Hamada[2] (Section 5.6).

10.3.2.2.2 Use of Dummy Variables

Dummy variables are very useful in categorizing data into two groups, such as in the case of treatment or control. Typically, two values, such as 0 and 1 (or –1 and 1), are assigned to two different categories for generating a dummy variable. A dummy variable can be included in the multiple regression models as an effect for estimating the difference between the two groups and testing if the difference is significant. An illustration for the use of dummy variables can be found in Example 10.3. When there are more than two groups in the data, several dummy variables may be required to quantitatively identify these groups.

10.3.2.2.3 Model Selection Techniques

The objective of model selection is to obtain a best regression equation, which reaches a balance between goodness of fit and generality. On the one hand, to make the regression equation useful for prediction, the model should contain as few effects as possible to reduce bias. The generality increases when bias is reduced. On the other hand, to keep the variance of the prediction small, the model should include as many effects as possible. This results in better goodness of fit. The two extremes are compromised, or a balance can be achieved by the techniques of model selection. There are various model selection procedures that have been proposed. The commonly used ones include all subset selections based on Akaike's Information Criterion, adjusted R^2, C_p, and stepwise procedures such as forward and backward selections. When there are replicates in an experiment, Draper and Smith[3] suggested an all-subset selection procedure based on the widely used Mallow's C_p criterion. The formula for the C_p criterion is[4]

$$C_p = (\text{RSS}_p/\hat{\sigma}^2) + 2p - n \tag{10.7}$$

where p is the number of effects in the model; n is the total number of settings; RSS_p is the residual sum of squares calculated under this submodel, which contains p effects; and $\hat{\sigma}^2$ is the estimation of variance under the full model. The first part of the formula ($\text{RSS}_p/\hat{\sigma}^2$) can be viewed as a measure of goodness of fit. $\hat{\sigma}^2$ is a constant over all submodels. RSS_p is small if the fitted surface is close to y_x/n_x at experimental setting x. The rest of the formula, $2p - n$, is small if the number of effects in submodels is small. Hence, a small C_p value is preferred. If C_p is graphed with p, Mallow recommends choosing the model where C_p first approaches p.

When there are no replicates in the experiment, the C_p criterion cannot be used because it is unable to estimate the error variance. To deal with such situations, Hamada and Wu[5] proposed a stepwise regression type of model selection method based on the concept of effect sparsity and effect heredity. Effect sparsity states that not all the effects are likely to be significant. Effect heredity means that an interaction effect is likely to be significant only if one of the involved main effects is significant. Although Hamada and Wu's method can identify most significant effects, it does not give a stopping criterion for the stepwise regression.

Some model selection techniques have been developed to resolve complex aliasing, noted in Section 10.3.1, such as iterative forward selection and the Bayesian model selection based on the effect heredity principle. For details, see Wu and Hamada[2] (Sections 8.4 and 8.5).

10.3.2.3 Residual Analysis

Residual analysis is often used to detect problems in a data set or a fitted model, and assess if the model is adequate. Residuals are defined as the difference between the observed responses and the fitted responses. Several assumptions are needed in a regression model (Equation 10.1). For example, the errors are assumed to be independent and normally distributed with constant variance. Residual analysis can be used to assess if these assumptions are satisfied in the fitted model. Furthermore, it can be used to answer questions such as if all the important effects are captured, or if there exist extreme or unreasonable values (such as outliers) in the data set. The residual analysis is usually performed by visual check of the scatter plots, such as the residuals against fitted response, residuals against factor levels, or residuals against time index.

10.3.2.4 Location–Dispersion Modeling for Variance Reduction

When the experiment is performed with several replicates or a robust parameter experiment is conducted with cross arrays, location–dispersion modeling can be performed in the data analysis for the purpose of variance reduction. In this approach, two regression models, one for location and another for dispersion, are built separately. The location and dispersion models are built for understanding the average yield and the variation of the response, respectively. The average yield usually influences the performance of products, and a small variation can result in products that have stable quality. The location modeling is done by relating the average of (noise) replicates to the factorial effects through a model such as that shown in Equation 10.1. For the dispersion, its model is formed by replacing the left-hand side of Equation 10.1 by $\log(s_i^2)$, where log denotes natural logarithm and s_i^2 is the sample variance of (noise) replicates at the i-th (control) factor setting. A main reason for taking the log transformation is that it maps $(0, \infty)$, which is the range of sample variance, to $(-\infty, \infty)$, which is the range of the linear combination on the right-hand side of Equation 10.1. The techniques presented in these sections can be utilized to estimate parameters and test their significance. Model selection techniques can be applied to develop fitted models individually for location and dispersion. From the fitted location and dispersion models, a two-step procedure (see Wu and Hamada[2]; Section 10.5) can be applied to obtain an optimum setting that minimizes dispersion and brings the location to a desired target. An illustration of location–dispersion modeling is given in Example 10.6.

10.3.3 Inference and Decision Making

10.3.3.1 Factor Screening

A factor screening experiment is usually performed in the preliminary stage of a study. In the beginning of a study of a system, there is usually only very limited knowledge about the system, whereas there may be quite a lot of potentially important factors. A large experiment with all factors and treatment combinations considered will be costly and very often impractical. Furthermore, it is often the case that only a few of these factors are truly important. The aim of factor screening is to distinguish the vital few factors from the trivial many so that investigators can focus on the most important factors in the follow-up experiments. The run size of a factor screening experiment is usually relatively small when compared with those experiments whose objective is to develop a sophisticated model. A rough rule of choosing experimental plans for factor screening experiments is that the run size can be close to the number of factors under study. The procedure of factor screening is illustrated by the following example.

Example 10.1: Identifying Influential Factors in Abnormal Welding Conditions

Li et al.[6] performed an experiment to study the effects of abnormal conditions in welding processes. Six types of abnormal conditions were considered in the study: axial misalignment (denoted as Ax), angular misalignment (An), edge-weld (Eg), fit-up (Ft), cooling water (Cl), and electrode wear (Wr). The six factors were chosen to be two levels, coded as –1 for normal condition and +1 for abnormal condition. The response of interest was button diameter. For a screening experiment with six factors, a 12-run Plackett–Burman design (see Wu and Hamada[2]; Chapter 7) is a good choice. However, in addition to factor screening, the authors would also like to study the main effects and two-factor interactions of these factors in the study. An experiment with more runs was designed. Because electrode wear is a hard-to-change factor, a split-plot experiment was used. The whole experiment was divided into two batches, for new and worn electrodes, respectively. Within each batch, a 2_{IV}^{5-1} (16 runs) fractional factorial matrix was used for the other five factors, that is, Ax, An, Eg, Ft, and Cl. Thus, there were a total of 32 runs. At least five replicates were made for each run. The conventional data analysis for factor screening is to first estimate the main effects of factors and then identify as important those factors whose main effects are significant. The effects of interactions are usually not considered in the conventional analysis for factor screening. Box and Meyer[7] proposed an alternative Bayesian analysis method that allows for the interactions, as well as main effects, to be investigated. In the Bayesian analysis, the marginal posterior probabilities that a factor is active are used as an index for the identification of important factors. The marginal posterior probabilities for the data are given in Figure 10.2. The result quite clearly indicates that the two abnormal conditions, edge-weld and axial misalignment, are much more influential on the button diameter than the other three factors (angular misalignment, fit-up, and electrode wear). The findings are consistent with experience.

10.3.3.2 Treatment Comparison

A treatment is a setting of input parameters such as a welding schedule. Experiments for treatment comparison are often conducted when there are several treatments to be applied. The investigators are generally interested in questions such as if there is any difference between treatments, which treatments are similar, and which treatments are better than others. The following example shows how RSW schedules are studied in order to obtain welds of good quality and small variation.

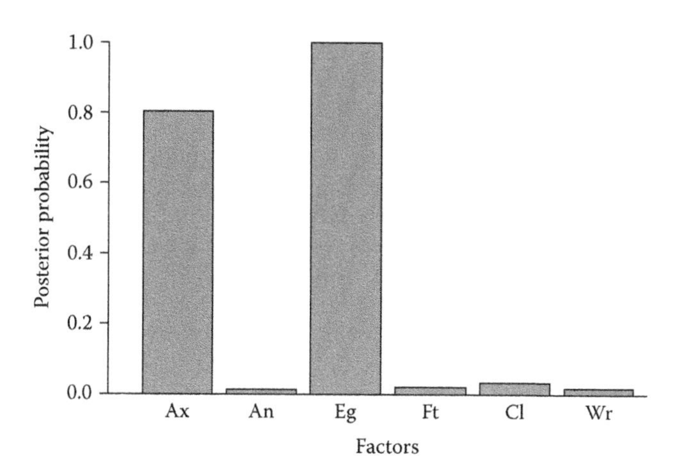

FIGURE 10.2
Marginal posterior probabilities that factors are active.

Example 10.2: A Comparison of Welding Schedules and Tests

To understand the sensitivity of weld quality and quality variation to welding schedules, and to correlate the measurement of the tensile–shear test, peel test, and impact test, an experiment was performed on a 0.8-mm galvanized interstitial free (IF) steel. Five different treatments or settings, each of which is a combination of current (in amperes) and time (in cycles), were used: (1) current = 11,000, time = 6; (2) current = 10,000, time = 10; (3) current = 9750, time = 14; (4) current = 9500, time = 16; and (5) current = 9000, time = 18. Electrode force was fixed at 2.8 kN. Each treatment was performed with 30 replicates. In the measurement step, each of the three types of testing, namely, tensile–shear, peel, and impact, was conducted on 10 of the 30 replicates. Several responses, such as weld diameter, displacement, peak force, maximum load, and energy, were then measured. The investigators were interested in the difference between the five treatments and between methods of measurement (testing). Only the analysis of diameter is presented here. The data analysis method for treatment comparison is called analysis of variance (ANOVA), which examines whether the treatments for comparison generate the same (or similar) outcomes. ANOVA was applied on data from tensile–shear, peel, and impact tests separately. The F values and p values are presented in Table 10.1. An F value in ANOVA is the statistic for evaluating whether the treatment means are equal. The calculated F value is compared with that obtained from tables such as Appendix D of Wu and Hamada's book.[2] A large F value implies a significant difference between the treatments. A p value is the probability of obtaining a value that is significantly different from the observed F value. Because all the p values are quite small (much smaller than 0.05), it is safe to conclude that the five treatments have significantly different effects on the response. A similar conclusion can be drawn based on the F value. However, ANOVA only tests the hypothesis that these treatments have the same effects.

TABLE 10.1

F Values and p Values, ANOVA

	***F* Value**	***p* Value**
Tensile–shear	25.0815	7.284151E–011
Peel	27.9803	1.371314E–011
Impact	27.9327	1.090872E–011

After the hypothesis is rejected, a more interesting question is how they are different. An advanced analysis based on a linear model is required to answer such a question. Before a formal analysis is performed, a graphical analysis using, for example, box plots, as shown in Figure 10.3, is helpful. From left to right, the box plots for tensile–shear (TS), peel (P), and then impact (I) tests, as well as for each treatment (T), 1 to 5, are drawn.

Although the plots offer valuable information and intuitive understanding of data, the conclusion, based on a graph alone, is usually rough or not sufficiently accurate. Some effects that describe the differences between treatments and methods are then introduced and linear models are usually used for quantitative analysis. A model selection technique based on the C_p criterion can be applied to develop a model in which all the effects are significant. The model obtained for the example problem is as follows:

$$\begin{aligned} D = 6.06 + (-0.14)\cdot \text{M2} + (-0.07)\cdot \text{M3} \\ + (0.07)\cdot \text{T3} + (-0.16)\cdot \text{T5} + (-0.04)\cdot \text{M2T3} \\ + (-0.02)\cdot \text{M2T4} + (0.04)\cdot \text{M2T5} \end{aligned} \tag{10.8}$$

where D is the weld diameter after testing; M is for the testing method, that is, TS (= M1) (tensile–shear), P (=M2) (peel), or I (=M3) (impact); and T (1 through 5) is for treatment. The constant (6.06) is the average diameter of welds of all treatments (schedules) measured by tensile–shear tests (M1).

The significant effects identified in the model have strong support from the box plots. The effect M2 describes the averaged difference between the second method (peel) and the first method (tensile–shear). From its coefficient, −0.14, it is known that the outcome of peel is lower than tensile–shear by an average of 0.14 mm. From the box plots, we can see that the group of box plots for peel (the middle five plots) uniformly shifts down when compared with the left five box plots, which are for tensile–shear. The other effects can also be related to the box plots. M3 describes the difference between the average diameter of impact and those of tensile–shear. The significance of T3 and T5 is that the third treatment generates a larger diameter by an average of 0.07 mm, whereas the fifth treatment generates a smaller diameter by an average of 0.16 mm, compared to the

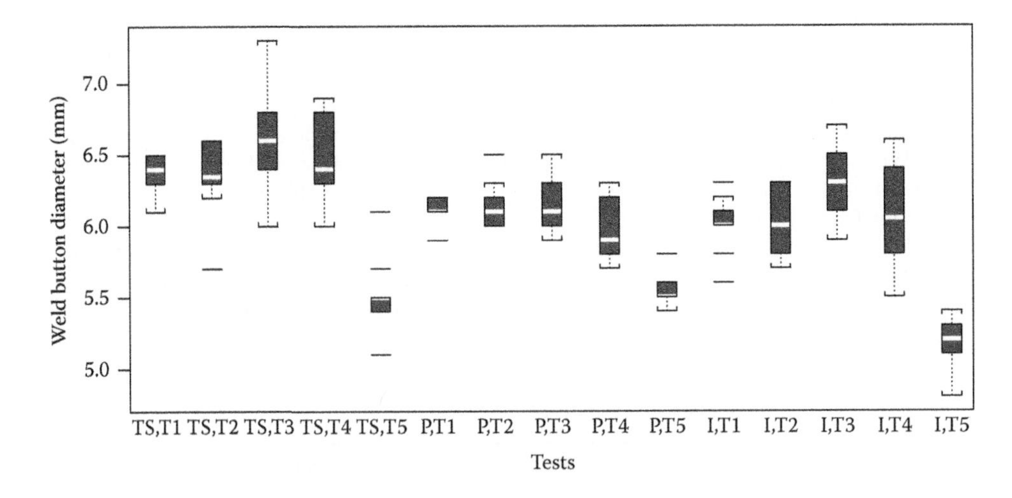

FIGURE 10.3
Box plots of weld diameter for different testing methods and treatments.

first treatment. The three effects, M2T3, M2T4, and M2T5, are interactions that show that in the second testing method (peel), the yields of the third, fourth, and fifth treatments are lower or higher when compared with the performance of the third, fourth, and fifth treatments in other methods. As reflected in the box plots, the trend from left to right of the five box plots in peel has less variation than the trends in tensile–shear and impact. It reveals that the peel test generates a more uniform outcome with less variation. In this analysis, an intuitive understanding is obtained from the plots and the numerical analysis extracts detailed information. A proper mix of graphical and numerical analyses can make information (data) easier to understand and conclusions more accurate.

10.3.3.3 Combination of Experiments

Experiments may have to be conducted on different subsystems for convenience or for practical reasons. Data gathered from each subsystem can be analyzed separately to draw appropriate conclusions for each subsystem. However, the conclusions drawn on one subsystem may not be true for other subsystems. For example, the most significant effect in one subsystem may become insignificant when compared with the significant effects of other subsystems. In order to obtain an overall picture of the entire system, data sets from each subsystem have to be pooled together for analysis. For the combined data, some effects that can explain the difference between original distinctive data sets should be included in the model. Dummy variables are often used to construct such effects. Some information recorded in experiments may serve as an alternative to the use of dummy variables. The following example shows a case of the latter.

Example 10.3: Combining Results of Individual Experiments

A study was performed by the Resistance Welding Task Force of Auto/Steel Partnership to gain an understanding on the influence of some welding factors. Five factors, each with two levels, were chosen for this study. They were button diameter (bd), welding time (wt), heat (ht), force (fo), and machine type (ma). The five factors were applied on nine types of materials, denoted as material A, C, E, G, H, I, K, M, and N, respectively. For each material, a 16-run FFD (minimum aberration 2^{5-1} design) with three replicates was performed. The response of interest is tensile strength (F, peak load). The nine data sets had been analyzed individually and significant effects were found for each data set. The investigators were then tempted to know whether the five factors have similar effects on all nine materials. The nine data sets were combined into one data set for analysis to answer this question.

It was observed that the variations of the nine data sets are different. Box plots of residuals for the nine linear models, each developed for one data set, are drawn in Figure 10.4a. It was found that the difference in variation could be related to the thickness of materials. Figure 10.4b shows that there exists a proportional relationship between the standard deviation of residuals and thickness. It is also observed that some effects have a strong correlation with thickness. For example, as shown in Figure 10.5a, the relationship between constant effects and thickness is almost a straight line, whereas Figure 10.5b reveals a curvature relationship between the main effects of button diameter and thickness.

All these findings suggest that the thickness of materials is a variable that should be included in the whole model to explain the effect caused by materials (note that in the original analysis, the effect of materials was not of concern because analysis was performed on data from the same type of material individually). With thickness (T), in addition to the original five factors included in the analysis, a linear model was developed to describe the dependence of peak load on the factors as follows:

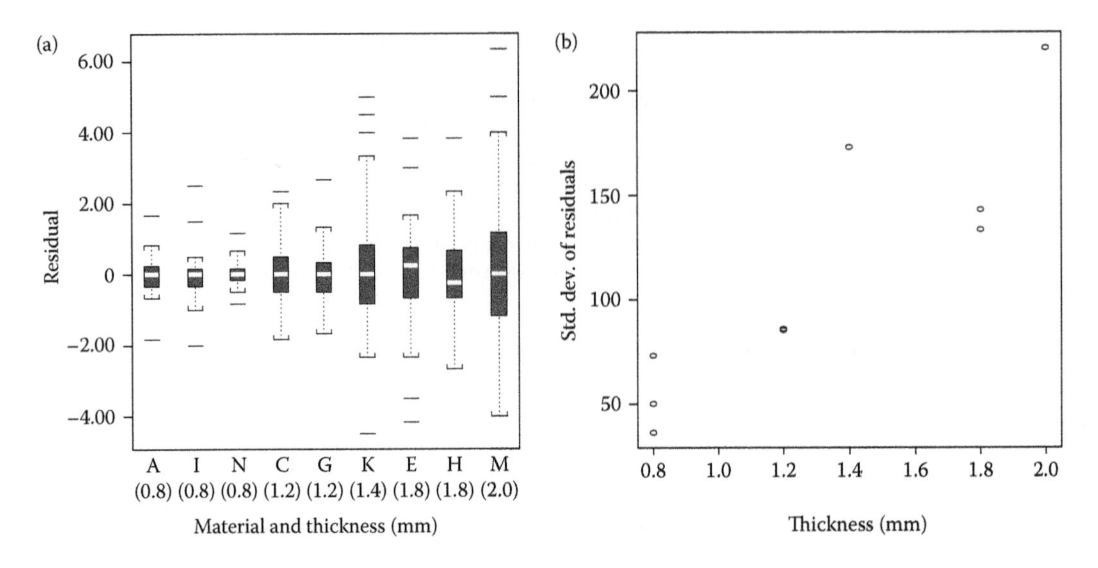

FIGURE 10.4
Analysis of variation in nine models. (a) Material and thickness (mm). (b) Thickness (mm).

$$
\begin{aligned}
F = & (-497.86) + (1174.85)\cdot\text{bd} + (-104.38)\cdot\text{ht} + (-80.11)\cdot\text{fo} \\
& + (-129.18)\cdot\text{ma} + (2330.39)\cdot T + (-94.15)\cdot\text{fo}\cdot\text{ma} \\
& + (-1990.60)\cdot\text{bd}\cdot T + (120.87)\cdot\text{ht}\cdot T + (116.66)\cdot\text{fo}\cdot T \\
& + (128.06)\cdot\text{ma}\cdot T + (76.64)\cdot\text{wt}\cdot\text{ht}\cdot T + (92.59)\cdot\text{fo}\cdot\text{ma}\cdot T \\
& + (966.30)\cdot\text{bd}\cdot T^2
\end{aligned} \tag{10.9}
$$

In this analysis, a dummy variable that describes the difference between the two groups, {A, C, I, K, M} and {E, G, H, N}, of the nine materials was found to be significant.

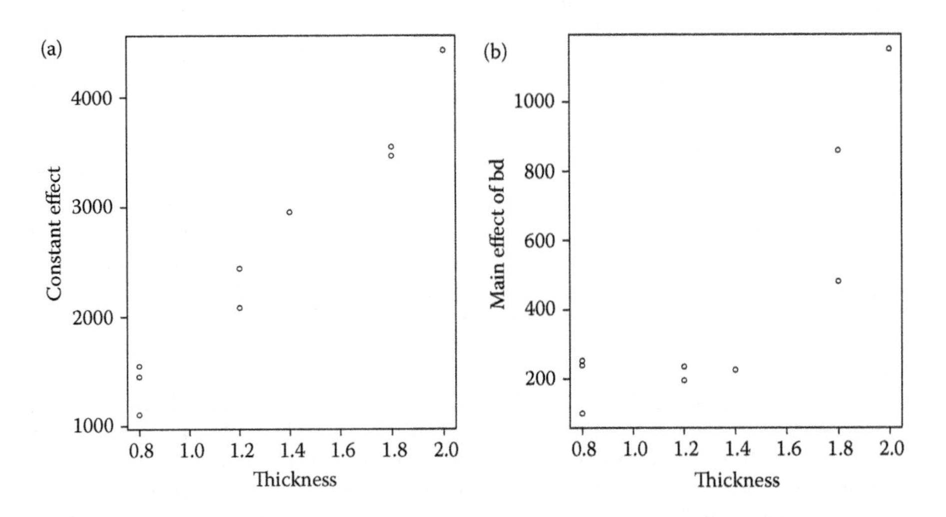

FIGURE 10.5
Relationship between effects and thickness.

However, when the investigators traced the record of experiments, it was difficult to find a physical variable that can explain this observation. In an experimental study, it is suggested that one should record any information that can be gathered as completely as possible for possible later use.

10.3.3.4 Response Surface Exploration

The technique of response surface exploration is an effective tool when the objective of a study of a system is to obtain an accurate approximation function that describes the relationship between the response and the input factors.[1] Response surface exploration is usually conducted when input factors are quantitative. In this case, the true response surface can be approximated locally by a polynomial function of input factors. A rationale of this approach is based on Taylor's expansion, in which a function that satisfies certain conditions can be expressed as a sum of polynomial terms. The coefficient of each polynomial term is an unknown parameter, which is to be estimated using the data. Because it is impossible to use a finite data set to estimate the parameters of an infinite sum of polynomial terms, a model is usually restricted to be of second order in a response surface study. The second-order model can be regarded as a surrogate of the true response function obtained through experimental studies, and the purpose of the investigation, such as maximizing or minimizing the response, achieving a desired value of the response, or understanding the curvature of the response, can be achieved on the surrogate.

Example 10.4: A Study of Dependence of Weld Attributes on Welding Parameters

Three factors, welding current (C), time (*T*), and electrode force (*F*), were chosen for the study. In the experiment, the current had four levels, whereas time and force each had three levels. All 36 combinations of current, time, and force were performed. Several responses (weld attributes), diameter, height, volume, and shape of a nugget were recorded. Only the analysis of nugget diameter is presented here to illustrate the analysis procedure. A commonly used model in the analysis for response surface is a second-order model, which includes all polynomial terms of factors with a power of 2 or less. However, it is expected that the heating rate (=current^2 × time, a third-order polynomial term) could be influential because resistance welding is based on the principle of Joule heating. It was then decided that a third-order model is needed for model fitting of this data so that the influence of heat can be taken into account in the model. Therefore, all estimable effects with powers of at most 3 were considered in the model. In response surface exploration, full models are usually estimated. However, in this study, the cubic effects of time and force, that is, T^3 and F^3, are not estimable. This is because time and force are three-level factors; hence, they are unable to fit a cubic effect. By stepwise model selection, the fitted model was developed as shown below, taking the nugget diameter (ND) as the response:

$$\begin{aligned} \text{ND} &= (6.71) + (1.22)\cdot C + (-0.77)\cdot C^2 + (-0.30)\cdot T\cdot F \\ &\quad + (1.24)\cdot C^2\cdot T + (-0.29)\cdot T^2\cdot F \end{aligned} \tag{10.10}$$

with $R^2 = 0.7754$. This result is further verified by another model selection technique, C_p. From Figure 10.6, it can be observed that the five significant effects identified in stepwise regression appear in most good models identified by C_p. The result provided the investigators sufficient confidence that Equation 10.10 is a good approximation of the true response.

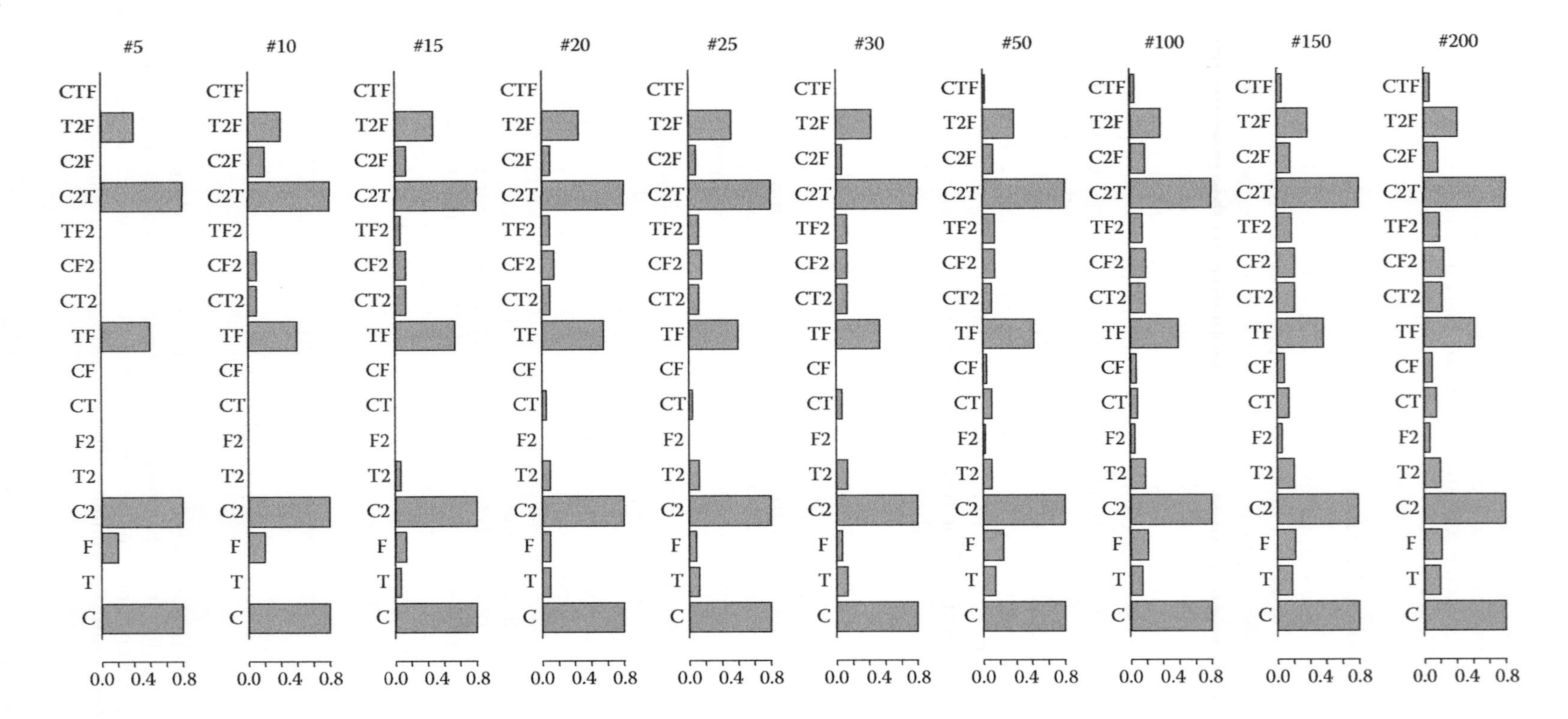

FIGURE 10.6
Appearance frequency of effects in first 5, 10, 15, 20, 25, 30, 50, 100, 150, and 200 best models.

The model can be used for multiple purposes, such as response optimization, that is, maximizing the nugget diameter. To utilize Equation 10.10 for maximizing the response, partial differentiation of ND can be taken on *C*, *T*, and *F*, respectively, and then they are set to zero and solved for solutions. Note that the model is a local polynomial approximation of the true response surface, and the inferences are applied only in the experimental region, using techniques such as interpolation. Any extension of the inference outside the experimental region through extrapolation could be potentially dangerous and the results should be justified and verified. The response surface exploration is only a part of the whole response surface methodology. More details about the response surface methodology can be found in Myers and Montgomery's book.[8]

10.3.3.5 Variance Reduction

Previous sections are devoted to the analysis of average yield (i.e., mean) of the response. However, two treatments that produce similar average yields may cause very different variations. For instance, two schedules are used in welding a material. One has a high welding current and short welding time, and the other has a low current but long welding time. These two schedules may produce similar average button sizes (mean response), but one may have a larger variation than the other. This can be interpreted as one process is less stable than the other, and variation should be considered when selecting a process, as it is an important issue in the quality control of products. The procedure of variation reduction is illustrated in the following example of resistance welding research.

Example 10.5: Dependence of Tensile–Shear Strength on Coupon Size

In a study by Zhou et al.,[9] two factors, width (*W*) and thickness (*T*) of testing coupons, which define the size of a coupon, were chosen for the effect of coupon size. Width was a five-level factor and thickness was a three-level factor. All 15 combinations of width and thickness were performed with 10 replicates for each combination. Two responses, maximum load and displacement, were measured. In this example, only the displacement is presented to demonstrate how variance reduction can be achieved through experimental study. For each combination of width and thickness, 10 replicates were used to calculate the variance. Variance is a numerical index of variation. In the analysis, logarithm of variance was regarded as a response and a statistical model was built to explore the relationship between factors and response. By applying this analysis strategy on the experimental data set, a second-order model was developed as follows:

$$\begin{aligned} \ln D = (-1.17) + (1.62)\cdot T + (-1.71)\cdot T^2 + (-2.82)\cdot W \\ + (1.78)\cdot W^2 + (-0.82)\cdot T\cdot W \end{aligned} \tag{10.11}$$

where *D* is the variance. From this model, it can be seen that choosing a setting with a smaller thickness and larger width can reduce the variation. This result is also supported by the interaction plot in Figure 10.7. For all gauges, variation decreases with width in general, although the reduction in thinner sheets is more significant. This is because the constraint to local deformation around a weld directly depends on the sheet thickness. The size of the localized deformation area is smaller for the thin-gauge specimens than for the thick ones. Therefore, variation on thinner coupons decreases faster than that on thicker ones.

The reduction of variation of displacement can be used to determine optimal testing specimen width, and this example demonstrates how variation can be reduced through experiments. The model developed through the analysis of logarithm of variance is

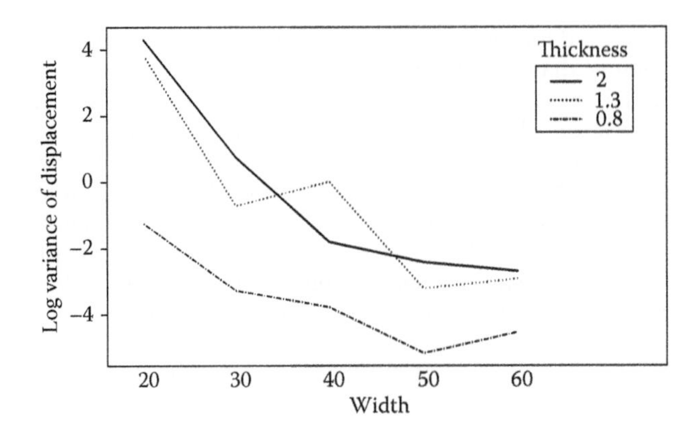

FIGURE 10.7
Interaction plot of width and thickness, with log(variance) of displacement as a response.

usually called a dispersion model. A dispersion model is usually developed, together with a means model (such as the models in Examples 10.1 through 10.4), for choosing a good setting. A means model offers information about the average yield, whereas the dispersion model reveals information about the variation. A good setting based on the two models should be able to generate yields that satisfy the requirements and minimize variation. Sometimes, the two models may have a contradiction on the choice of good settings. A trade-off is inevitable if this happens.

10.3.4 Two-Stage Sliding-Level Experiments

Continuous responses are commonly observed in resistance welding research, as most of the possible responses are of a continuous nature. This section demonstrates the procedure of a special statistical analysis of continuous responses in resistance welding under abnormal conditions.

Normally, a certain number of levels of welding current and time are chosen to cover the range of interest, often in the form of a rectangular matrix. However, such a design is often not efficient in resistance welding study. In practice, the range of acceptable welds is usually limited by two curved boundaries, or a lobe in the welding time–current domain, as shown in Figure 10.8. Therefore, an experiment matrix that covers the entire possible

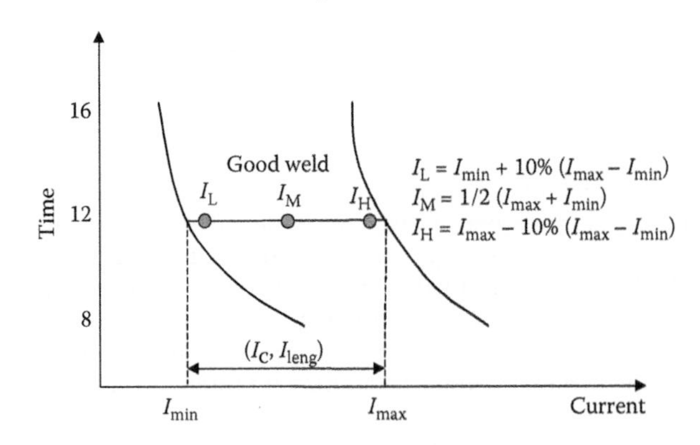

FIGURE 10.8
Determination of current levels.

settings of t and I will leave the settings of low τ and I and high τ and I unused, as in the low end of the matrix, welds are below the minimum acceptable weld size, and the high settings will always render expulsion. Such results can be expected without conducting actual experiments, and can be incorporated into the data for analysis as pseudo-data, as illustrated in Section 10.4.3. A uniform design (or a rectangular matrix) is usually necessary because standard statistical analysis works well on such designs. However, a more efficient way of dealing with such situations is to use a sliding-level design, as developed by Cheng et al.[10] and Li et al.[6]

In their study, the effect of abnormal process conditions was considered by a sliding-level experiment because of the interdependencies among the process variables. Both normal and abnormal process conditions were considered in the experiment. Based on the experimental data, weld size and lobes were analyzed as response variables. Mathematical models were then developed to examine the effects of the process conditions based on predictions. Instead of assigning fixed values to the welding current in the design, it was treated as a slide factor and was determined based on the settings of other process variables.

10.3.4.1 Experiment Design

The abnormal process conditions selected for study in this experiment were axial misalignment, angular misalignment, poor fit-up, and welding parameters, as shown in Table 10.2. Electrode wear was then substituted by electrode size because electrode wear manifests itself through the size change of the electrode face diameter. Three levels of electrode size were used. For convenience, electrode size was denoted as Ed; force, F; weld time, T; and current, I. Because of the interdependency among the process variables, settings of welding current need to be determined to ensure acceptable welds under all the designed conditions of other process variables. This causes the current setting to slide in the weld lobe diagram because the effect of current depends strongly on time. The welding current was chosen to be a slide factor because it has natural boundaries for acceptable welds at given welding times. As such, it can be seen that the experiment needs to be conducted in two stages. First, suitable current settings are found for the designed conditions of other process variables. Then the experiment is conducted using these current settings for making test welds.

The experiment matrix was constructed following the inner and outer array method of Taguchi's robust parameter design.[11] The welding current was arranged in the outer array, and all other variables were arranged in the inner array. For every setting in the inner array, the current limits (I_{min}, I_{max}) were determined first. Then three levels of current (I_L,

TABLE 10.2

Welding Parameters and Levels Used in Experiment

Process Variables	Low	Median	High
Electrode size (Ed)	4.50 mm	6.35 mm	7.87 mm
Force (F)	2.45 kN	3.34 kN	4.23 kN
Weld time (T)	8 cycles	12 cycles	16 cycles
Axial misalignment (Ax)	0	N/A	1.5 mm
Angular misalignment (An)	0	N/A	10°
Fit-up (Ft)	0	N/A	5 mm
Current (I)		Determined accordingly	

I_M, and I_H) were chosen in the outer array for the experiment. This arrangement ensures a balanced mix of the settings of all the variables. Table 10.3 shows a portion of the design matrix.

A weld lobe was developed using standard procedures for determining the current levels. As has been observed in many experiments, the boundaries of a weld lobe of RSW are not deterministic.[12] To increase the probability of making acceptable welds, the low and high current levels (I_L and I_H, respectively) were determined to be 10% of the total current range inside the boundaries of the weld lobes. The middle current level (I_M) is the center point between I_L and I_H. This procedure is illustrated in Figure 10.8.

10.3.4.2 Analysis and Modeling

Because of the complex design of the experiment, statistical methods for the analysis of a conventional design of the experiment cannot be directly applied. A new analysis procedure based on modified stepwise regression has to be developed. First, the coding and model selection need to be addressed.

10.3.4.2.1 Coding System and Model Selection

For regression models, linear effects can be estimated for two-level variables and both linear and quadratic effects can be estimated for three-level variables. Thus, the regression model can be expressed as a summation of the first- and second-order polynomial terms of the variables plus possible interaction terms among them. For example, if x_1 and x_2 are two variables in an experiment, whose levels are 2 and 3, respectively, the regression function can then be represented as

$$f(x_1, x_2) = \beta_0 + \beta_1 x_1 + \beta_2 x_2 + \beta_3 x_1 x_2 + \beta_4 x_2^2 + \beta_5 x_1 x_2^2 \quad (10.12)$$

where the β_i values are coefficients of the factors that need to be estimated from the experimental data. The estimation may not be accurate if collinearity exists among the polynomial terms. Hence, an orthogonal coding system is needed to translate the original polynomial terms x_i into a set of orthogonal vectors. For this, linear and quadratic effects need different transformations.

TABLE 10.3
Two-Stage, Sliding-Level Experiment Design

Inner Array						Current Limits		Outer Array		
								I_L	I_M	I_H
Ft (mm)	Ax (mm)	An (deg)	Ed (mm)	F (kN)	T (cycle)	I_{min} (kA)	I_{max} (kA)	Average Button Diameter (mm)		
0	0	0	6.35	3.3	12	10.56	12.44	4.67	5.71	6.94
5	0	10	6.35	4.2	8	11.25	13.25	3.23	5.14	6.37
0	0	10	6.35	4.2	16	10.40	12.50	1.78	5.19	6.24
...	...	...	...	...	...	...	...	...	...	...

For linear effects:

$$z = \left\{ \begin{array}{ll} \dfrac{x - x_l}{x_m - x_l} - 1, & x_l \le x \le x_m \\ \dfrac{x - x_m}{x_h - x_m}, & x_m \le x \le x_h \end{array} \right\} = \left\{ \begin{array}{ll} -1, & x = x_l \\ 0, & x = x_m \\ 1, & x = x_h \end{array} \right. \tag{10.13}$$

For quadratic effects:

$$z = x \cdot (X^T \cdot X)^{-1} \cdot X^T \cdot y = \left\{ \begin{array}{ll} 1, & x = x_l \\ -2, & x = x_m \\ 1, & x = x_h \end{array} \right. \tag{10.14}$$

where

$$x = \begin{bmatrix} 1 & x & x^2 \end{bmatrix} X = \begin{bmatrix} 1 & x_l & x_l^2 \\ 1 & x_m & x_m^2 \\ 1 & x_h & x_h^2 \end{bmatrix} y = \begin{bmatrix} 1 & -2 & 1 \end{bmatrix}^T$$

x_l, x_m, and x_h are settings for low, middle, and high levels, respectively. This coding system ensures that three-level variables are coded as [−1 0 1] for their linear effects and [1 −2 1] for their quadratic effects, even if their physical settings are not equally spaced. For example, the coding for electrode size is shown in Figure 10.9.

As discussed in previous sections, many model selection methods exist for the conventional design of experiments, such as ANOVA and half-normal plot. Little has been done, as seen in the published literature, for experiments with complex structures such as the one used in this study.

In this two-stage experiment, the model selection has to be handled differently from the all-subset selection procedure because the response variables are different at different

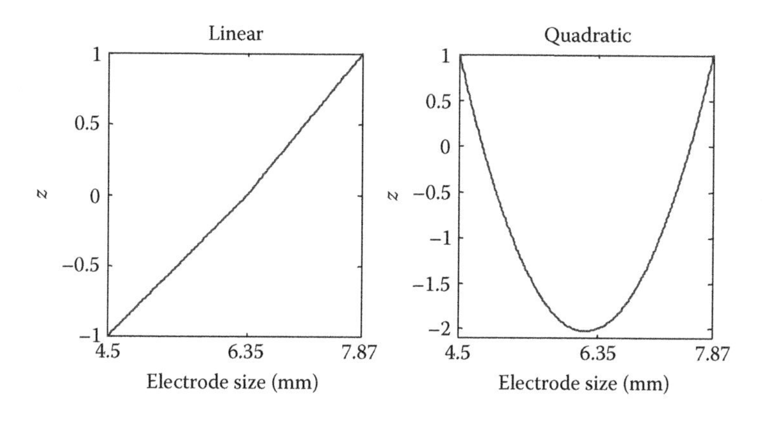

FIGURE 10.9
Linear and quadratic coding of electrode size.

stages. At the first stage, the response is current range, whereas at the second stage, it is button size. In both cases, forward stepwise regression is applied for model selection. However, in the case of button size, the C_p criterion is used as a stopping rule because there are five replicates for each experiment setting and the experimental error can be estimated using the collected data. In the case of current range, the experiment error cannot be estimated because no replicates were conducted for the current range searching. The stopping rule has to be developed based on the engineering estimation of the experiment error. In either case, the principles of effect sparsity and heredity are observed.

10.3.4.3 Analysis of Current Range

As shown in Figure 10.10, the current range can be characterized using two response variables: the center of the current range (I_c) and the length of the current range (I_{leng}), and they are defined as follows for convenience:

$$I_c = \frac{1}{2}(I_{min} + I_{max}) \tag{10.15}$$

$$I_{leng} = \frac{1}{2}(I_{max} - I_{min}) \tag{10.16}$$

I_c represents an average current setting that is determined by the physical process conditions. It can be treated as a normally distributed variable. I_{leng} determines the allowable range of the current for making good welds. By definition, I_{leng} is greater than zero. In regression analysis, it needs to be transformed as follows:

$$I_{lnlen} = \ln(I_{leng}), \text{ such that } I_{lnlen} \in (-\infty, +\infty) \tag{10.17}$$

If I_c and I_{lnlen} have no correlation, they can be analyzed separately as two independent responses.

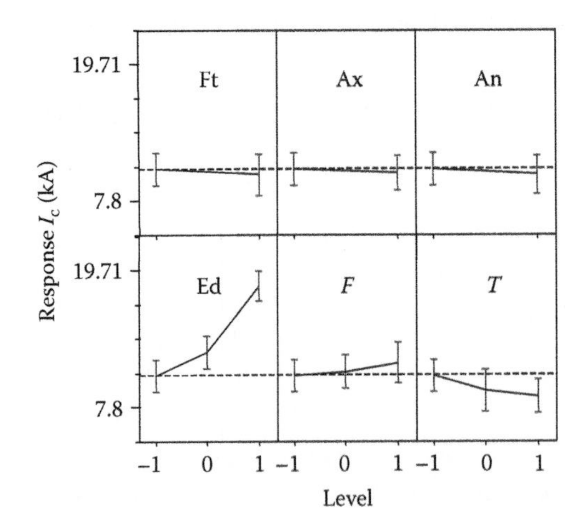

FIGURE 10.10
Average main effects.

The average main effects of the variables are examined before a regression model is developed. The responses vs. level settings for each variable are plotted using box plots in Figure 10.10. It is easy to see that electrode size and time are significant factors, whereas force, fit-up, axial misalignment, and angular misalignment are unlikely to be significant. It is also seen that electrode size, weld time, and force have quadratic effects on the response.

As there are no replicates for current range searching in the experiment, conventional stopping rules for stepwise regression are not applicable. Qualitative methods such as half-normal plot could be used to suggest influential effects for modeling, but the resulting models often underfit and have large root mean square (RMS) errors. From an engineering viewpoint, experiment errors can usually be estimated based on the experimental method. In finding current ranges, for example, there is a certain step that the operator will use to increase or decrease the current to search for the boundaries. By trial and error, he or she can make sure that the boundaries fall in the interval between two steps. This interval can thus be used to estimate the experimental error.

In this study, the step used in boundary searching was 0.1 kA. Thus, the confidence interval is $\Delta I = 0.2$ kA. Assuming a 99.7% ($\pm 3\sigma$) confidence, the standard deviation of the experimental error can be determined as

$$\sigma_e = \frac{\Delta I}{6} = 0.0333 \text{ (kA)} \quad (10.18)$$

From Equation 10.6, it is easy to obtain the error of the center of the current range,

$$\sigma_c = \frac{\sqrt{2}}{2}\sigma_e = 0.0235 \text{ (kA)} \quad (10.19)$$

σ_e is then used as a stopping criterion in the stepwise regression analysis; that is, the regression is stopped when the RMS error of the model reduces to this level. When this stopping criterion is applied, the regression model obtained is

$$\begin{aligned} I_c = {} & 12.80 + 4.10 \cdot \text{Ed}_1 + 0.24 \cdot F_1 \cdot T_1 + 0.62 \cdot \text{Ed}_2 \\ & - 1.13 \cdot T_1 - 0.39 \cdot \text{An}_1 \cdot F_1 + 0.86 \cdot F_1 + 0.39 \cdot T_2 \\ & + 0.29 \cdot \text{An}_1 \cdot \text{Ed}_2 + 0.16 \cdot \text{Ft}_1 - 0.17 \cdot \text{Ed}_2 \cdot T_1 - 0.30 \cdot \text{Ed}_1 \cdot F_1 \\ & - 0.12 \cdot \text{Ed}_1 \cdot T_1 - 0.15 \cdot \text{An}_1 \cdot \text{Ed}_1 \end{aligned} \quad (10.20)$$

with an adjusted R^2 of 0.998 and RMS error of 0.016. In the model, the x_1 and x_2 values stand for linear and quadratic effects, respectively (x = Ft, Ax, An, Ed, F, and T, as defined in Table 10.3).

Similarly, the main effects of the variables on the length of the current range are shown in Figure 10.11. It is clear from the plots that the abnormal process conditions significantly reduce the length of the current range, whereas the force increases the length of the current range. Electrode size and weld time have little effects.

In developing a regression model for the length of the current ranges, the estimation of the experiment error is not as straightforward as that for I_c. This is because the logarithm of the length is used as the response variable. From Equation 10.18,

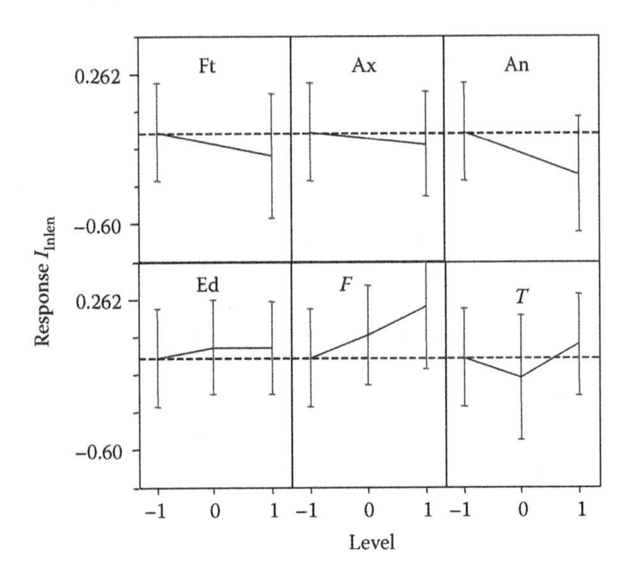

FIGURE 10.11
Main effect plots for length of current ranges.

$$I_{\text{lnlen}} = \ln(I_{\text{leng}} \pm 3\sigma_l) \tag{10.21}$$

where σ_l is the error of the length of the current range. It is easy to see that σ_l takes the same value as σ_c. In order to find the error associated with I_{lnlen}, Taylor's expansion with respect to the mean of I_{leng} ($\bar{I}_{\text{leng}}$) is used to obtain the following approximation:

$$I_{\text{ln len}} \approx \ln(\bar{I}_{\text{leng}}) \pm \frac{1}{\bar{I}_{\text{leng}}} \cdot 3\sigma_l \tag{10.22}$$

From the experimental data, $\bar{I}_{\text{leng}}$ is calculated to be 0.9089. Thus, the error of I_{lnlen} can be estimated as

$$\sigma_{\text{ln len}} = \frac{1}{\bar{I}_{\text{leng}}} \sigma_l = 0.0259 \tag{10.23}$$

The regression model is thus developed as

$$\begin{aligned} I_{\text{lnlen}} = & -0.115 - 0.102 \cdot \text{Ft}_1 + 0.214 \cdot F_1 \\ & - 0.149 \cdot \text{Ft}_1 \cdot \text{Ed}_1 + 0.034 \cdot F_2 - 0.075 \cdot \text{An}_1 \\ & - 0.049 \cdot F_2 \cdot T_1 - 0.091 \cdot \text{Ed}_1 \cdot F_1 + 0.030 \cdot T_2 \\ & + 0.065 \cdot \text{Ft}_1 \cdot T_1 + 0.002 \cdot \text{Ax}_1 + 0.0256 \cdot \text{Ax}_1 \cdot T_1 \end{aligned} \tag{10.24}$$

with an adjusted R^2 of 0.994 and RMS error of 0.020.

10.3.4.4 Analysis of Button Size

The average main effects on the button size are shown in Figure 10.12. Both current and electrode size are seen to have strong influences. On average, the button size increases under poor fit-up and decreases under angular misalignment conditions. Axial misalignment is not seen to have a strong effect. Both electrode force and weld time showed minor quadratic effects.

The model selection for button size is conducted using a forward stepwise regression with C_p as a stopping criterion. Figure 10.13 shows that all the effects before the break-even point should be included in the model. The model is thus obtained as

$$\begin{aligned} D = {} & 5.58 + 1.31 \cdot I_1 + 0.99 \cdot \mathrm{Ed}_1 + 0.38 \cdot \mathrm{Ft}_1 \\ & + 0.50 \cdot F_1 \cdot I_1 - 0.22 \cdot \mathrm{Ax}_1 \cdot I_1 - 0.1 \cdot \mathrm{An}_1 \\ & + 0.22 \cdot T_2 \cdot I_1 - 0.13 \cdot I_2 - 0.20 \cdot T_2 \\ & - 0.21 \cdot \mathrm{Ft}_1 \cdot I_1 - 0.23 \cdot F_1 + 0.14 \cdot \mathrm{Ed}_2 \cdot I_1 \\ & + 0.20 \cdot \mathrm{Ed}_2 \cdot F_1 - 0.13 \cdot F_1 \cdot I_2 - 0.18 \cdot \mathrm{Ed}_1 \cdot F_1 \\ & - 0.06 \cdot T_2 \cdot \mathrm{Ed}_2 - 0.08 \cdot \mathrm{Ed}_1 \cdot I_2 + 0.12 \cdot \mathrm{Ed}_1 \cdot I_1 \\ & - 0.04 \cdot \mathrm{Ed}_2 \cdot I_2 + 0.08 \cdot \mathrm{Ed}_2 \cdot T_2 - 0.08 \cdot \mathrm{An}_1 \cdot \mathrm{Ed}_2 \end{aligned} \tag{10.25}$$

with an adjusted R^2 of 0.9122 and an RMS error of 0.48. I_1 and I_2 are the linear and quadratic effects of the welding current, respectively.

10.3.4.5 Inference and Decision Making

Using the developed models, the effects of the abnormal process conditions can be examined through the predictions of the weld lobes and the nugget growth curves (the change of button size over time). Figure 10.14 shows an example of these predictions under a set

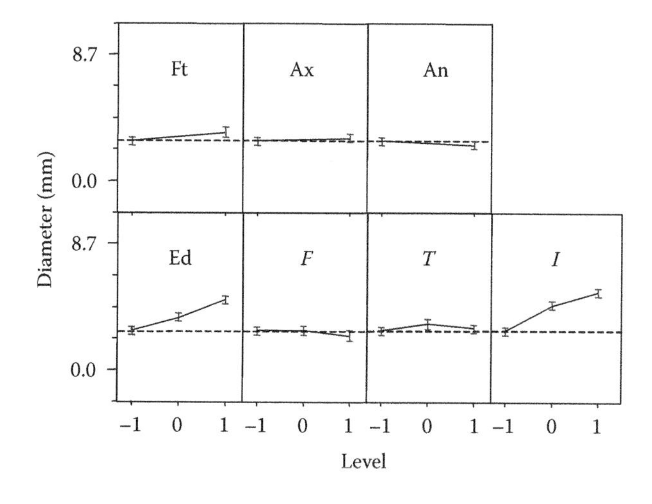

FIGURE 10.12
Average main effects on button size.

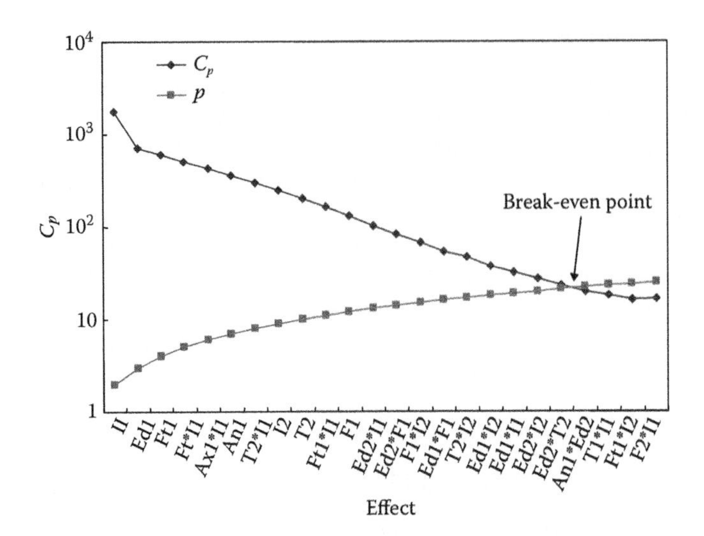

FIGURE 10.13
Model selection for button size.

of process parameters, as indicated in the figure. In general, when poor fit-up and angular misalignment exist, the weld lobe is shifted to the left and becomes narrower. A left shift of the weld lobe implies an early nugget formation and an early expulsion, which may decrease the size of the largest possible welds without expulsion. A narrower weld lobe also indicates that the welding process is less robust under these conditions.

It is also seen that the nuggets grow following different paths under different conditions (Figure 10.15). Whereas they start to form at the same time and eventually reach similar sizes (because of the expulsion limit), the nuggets initially grow faster under poor fit-up and angular misalignment conditions. Therefore, the nugget size variation caused by the abnormal process conditions could be different depending on the welding time used. For example, there could be about 1-mm difference in nugget size at the 7th cycle between

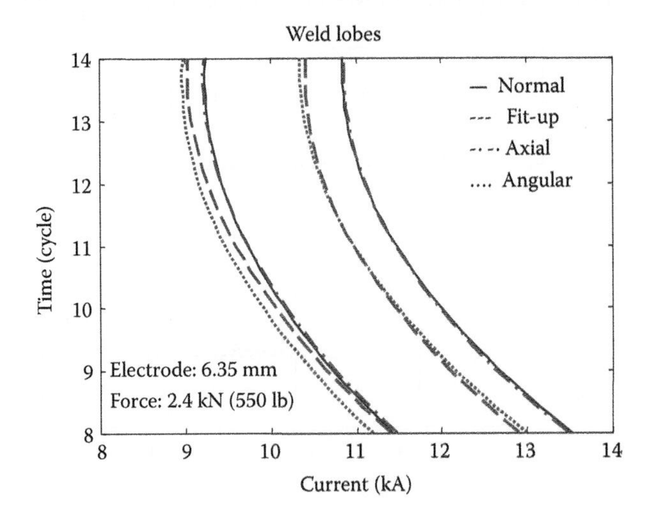

FIGURE 10.14
Weld lobe prediction.

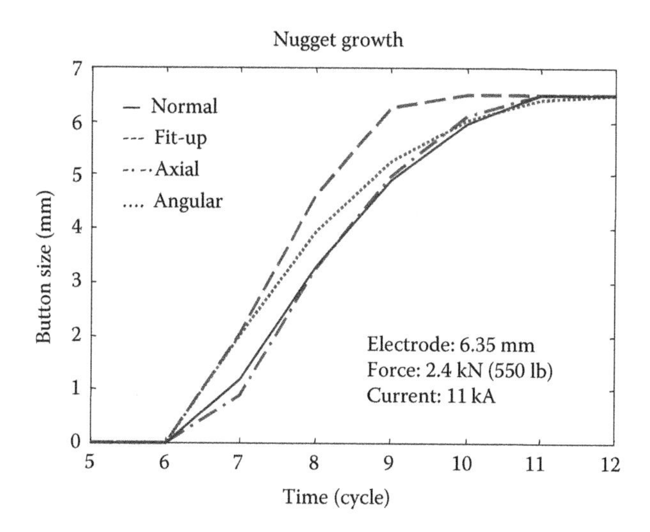

FIGURE 10.15
Nugget growth prediction.

normal and angular misalignment conditions, about 1.5-mm difference at the 9th cycle between normal and fit-up conditions, and almost no difference at the 12th cycle regardless of process conditions.

Axial misalignment did not show a strong effect on either weld lobe or nugget growth. This is because the amount of misalignment (1.5 mm) is too small compared to the electrode size used (24% difference in diameter and 15% in area for 6.35-mm electrodes). Under small axial misalignment, the process behaves similarly to that under normal conditions. Therefore, no significant effects have been observed.

This new two-stage, sliding-level experiment design and its analysis procedure are different from conventional statistical analyses. It can be used to account for the interdependencies among the process variables in RSW. The experiment and analysis revealed that process abnormalities such as axial misalignment, angular misalignment, fit-up, and electrode wear significantly affect the RSW process, and thus cause large variation on weld quality. They generally lead to narrower weld lobes, which implies less robustness of the process. The design and analysis procedures are generic enough to be applied to other processes.

10.4 Experiments with Categorical Responses

A categorical response takes its value over categories. For example, gender is measured as male and female; the size of an automobile may be classified as compact, mid-size, or full-size. Many processes can be characterized by two types of categorical responses: nominal and ordinal. The distinction between them is whether the categories have a natural ordering. Gender in the above example is nominal because the order of male and female is irrelevant in statistical analysis, whereas the size of a car is ordinal because compact, mid-size, and full-size can be regarded as a representative of a truncated continuous measurement of size. In an analysis of experimental data with categorical response, the goal is the same as that in continuous response: building an empirical model for the relationship between response and factors. Such a relationship is usually described through the probabilities of getting an observation from each category. Whereas basic statistical principles as outlined in the previous sections apply, design and analysis of experiments with categorical responses have distinctive characteristics, and they are explained using examples of RSW in this section.

10.4.1 Experiment Design

For experimental design with a categorical response, its main difference from that with a continuous response is its dependence on the true values of parameters, which are usually unknown before experiments. For example, consider an experiment with one two-level factor and binary response. If all the observations at the low level are in the first category and all the observations at the high level are in the second category, then the main effect of the factor has an infinite estimate. This problem usually happens when most levels are chosen on settings whose probabilities are close to 0 or 1. To prevent a situation of this kind, some levels that can generate observations in different categories are required. The choice of such levels needs some understanding about the response curve. To deal with such situations, one approach is to choose a design that optimizes a criterion based on a good guess of parameters. The criterion is usually a function of the information matrix. In a typical statistical model of categorical response, the information matrix depends on the design and the unknown parameters. Therefore, to obtain optimal designs, it is usually assumed that some prior knowledge is available and good initial values of parameters are known. For more information about this approach, see Atkinson and Haines.[13] When the prior knowledge is not available, some two-stage designs,[14,15] which gather information about parameters at the first stage and develop an optimal design based on the information at the second stage, can be used. An alternative to the optimal design approach is to use the FFD or OA. However, in order to appropriately choose the levels of factors, *a priori* knowledge of the response curve is usually necessary.

10.4.2 Analysis and Modeling

In this section, our focus is on categorical responses with two categories, which are also called binary responses. For the cases with more than two categories, refer to Agresti[16] for more details. For a binary response, the commonly used statistical model is the logistics regression model. The difference between logistics regression and multiple regression is reflected in the choice of the parametric model. Once the difference is accounted for, the principles of analysis techniques in multiple regression can be applied to the logistics regression.

Denote the two categories as 0 and 1. The probability that the response y gets 1 on a setting $\mathbf{x}$ is denoted by $P(y = 1|\mathbf{x}) \int p(\mathbf{x})$. Then the logistics regression model can be expressed by the following equation:

$$f(\mathbf{x}) = \beta_0 + \beta_1 x_1 + \cdots + \beta_p x_p \quad (10.26)$$

where $x_1, \ldots, x_p$ are coded effects of $\mathbf{x}$ (see Section 10.3.2 for effect coding), $\beta_0, \beta_1, \ldots, \beta_p$ are parameters, and

$$\log(p(\mathbf{x})/(1 - p(\mathbf{x}))) = f(\mathbf{x}) \tag{10.27}$$

From Equation 10.27, it can be easily derived that

$$p(\mathbf{x}) = e^{f(\mathbf{x})}/(1 + e^{f(\mathbf{x})}) \tag{10.28}$$

The parameters in the model can be estimated by the reweighted least squares algorithm, which performs weighted least squares iteratively. More information about data analysis techniques, such as test of parameters, residuals, and model selection techniques for logistics regression can be found in the work of Hosmer and Lemeshow.[17]

10.4.3 Inference and Decision Making

The inference and decision-making procedure for categorical responses follows the rules outlined in Section 10.2. This, as well as the details of experimental design and statistical modeling, is demonstrated using a study of expulsion limits when welding a low-carbon steel (or drawing steel, DS).

Example 10.6: A Study of Expulsion in RSW

A statistical analysis was performed by Zhang et al.[12] to study the expulsion limits based on experimental results. Unlike previous works on expulsion limits, expulsion was not regarded as an event happening at a particular welding schedule. The occurrence of expulsion was treated as a probability that spans from no expulsion to 100% of welds having expulsion with consideration of random factors. Because the occurrence of expulsion is the concern, the response (yes or no) has categorical characteristics.

One low-carbon bare steel (DS) was used in the experiment. Welding schedules were chosen around potential expulsion boundaries, and adjustment on welding schedules was made during experiments according to previous observations, to effectively cope with the change of expulsion limits. In the experiments, the welding current was varied for fixed electrode force and welding time. The occurrence of expulsion was clearly detected in the signals of dynamic resistance, secondary voltage, and relative displacement between electrodes, as well as visually observed.

Details of experimental design, including ranges of welding parameters, can be found in Section 7.4.2.1.

10.4.3.1 Statistical Analysis

Although statistical analysis has been widely conducted in welding research, commonly used statistical procedures could not be directly used in this study because of the complex nature of expulsion. Certain modifications had to be made in the analysis.

In general, a model is often a function that can explain the relationship between input and output variables. In a statistical analysis of categorical responses, the model is for describing both the deterministic and random phenomena in expulsion experiments. Specifically, for this study, the model must be able to

- Explain and predict the frequency of occurrence of expulsion, using electrode force, current, and time.
- Identify important effects and estimate their magnitudes.
- Describe the randomness of occurrence of expulsion.

The statistical model chosen for this study is the frequently used logistics model, which is ideal for dealing with continuous input and output variables of count data. The main purpose of this study is to understand the relationship between x (welding schedule) and p_x (probability of getting expulsion). In a logistics model, the common link function used to describe the relationship between p_x and x is as described in Equation 10.27. $f(x)$ is a real function of x, and is usually approximated by the sum of polynomial terms of x. In this study, there are three input variables: current (I), time (τ), and force (F), and then $f(I,\tau,F)$ can be approximated as

$$\begin{aligned} f(I,\tau,F) \approx\ & \alpha_{000} + \alpha_{100}I + \alpha_{010}\tau + \alpha_{001}F + \alpha_{200}I^2 \\ & + \alpha_{020}\tau^2 + \alpha_{002}F^2 + \alpha_{110}I\alpha + \alpha_{101}IF + \alpha_{011}\tau F \\ & + \alpha_{300}I^3 + \alpha_{030}\tau^3 + \alpha_{003}F^3 + \alpha_{210}I^2\tau + \alpha_{201}I^2F \\ & + \alpha_{021}\tau^2F + \alpha_{120}I\tau^2 + \alpha_{102}IF^2 + \alpha_{012}\tau F^2 + \alpha_{111}I\tau F \end{aligned} \quad (10.29)$$

where the α_{ijk} values are the coefficients, usually called parameters (not to be confused with welding parameters), to be estimated using information from the data. Equation 10.29 is a third-order polynomial, and more terms can be chosen if more data are available. For details of logistics models, refer to McCullagh and Nelder.[18]

Experimental data usually need to be transformed into an appropriate form before performing statistical analysis. In this case, a coding system and pseudo-data are needed.

10.4.3.2 Coding System and Transformations

In Equation 10.29, $f(I,\tau,F)$ was expressed as the sum of polynomial terms of x. However, the estimation of α_{ijk} may not be accurate due to collinearity between polynomial terms. Hence, an orthogonal coding system is needed to translate polynomial vectors of x into orthonormal vectors by the Gram–Schmidt process. The process is explained as follows.

Let x_I, x_τ, x_F be the vectors that represent the data of current, time, and force, respectively, and let x_I^2, x_τ^2, x_F^2 be the vectors of taking the square of x_I, x_τ, x_F, and x_I^3, x_τ^3, x_F^3 of taking the cube of x_I, x_τ, x_F. Denote

$$\begin{aligned} & u_I = x_I - \left(x_I^T 1\right)1,\ z_I = u_I / \left\|u_I\right\| \\ & u_I^2 = x_I^2 - \left(x_I^{2T} z_I\right)z_I - \left(x_I^{2T} 1\right)1,\ z_I^2 = u_I^2 / \left\|u_I^2\right\| \\ & u_I^3 = x_I^3 - \left(x_I^{3T} z_I^2\right)z_I^2 - \left(x_I^{3T} z_I\right)z_I - \left(x_I^{3T} 1\right)1,\ z_I^3 = u_I^3 / \left\|u_I^3\right\| \end{aligned} \quad (10.30)$$

where 1 is the unit vector, 1 = (1, 1, …, 1), and T indicates a transpose operation. z_I is called the linear effect of current, z_I^2 the quadratic effect of current, and z_I^3 the cubic effect of current. Figure 10.16 shows the relation between $\{z_I, z_I^2, z_I^3\}$ and the original current data, x_I.

It is easy to see that $\{z_I, z_I^2, z_I^3\}$ is orthogonal with unit length. The same transformations are applied to $\{x_\tau, x_\tau^2, x_\tau^3\}$ and $\{x_F, x_F^2, x_F^3\}$. Furthermore, it can be proved that there is a one-to-one transformation between a linear combination of polynomial terms of I and that of I_s which is a standardized form of I, as in the following:

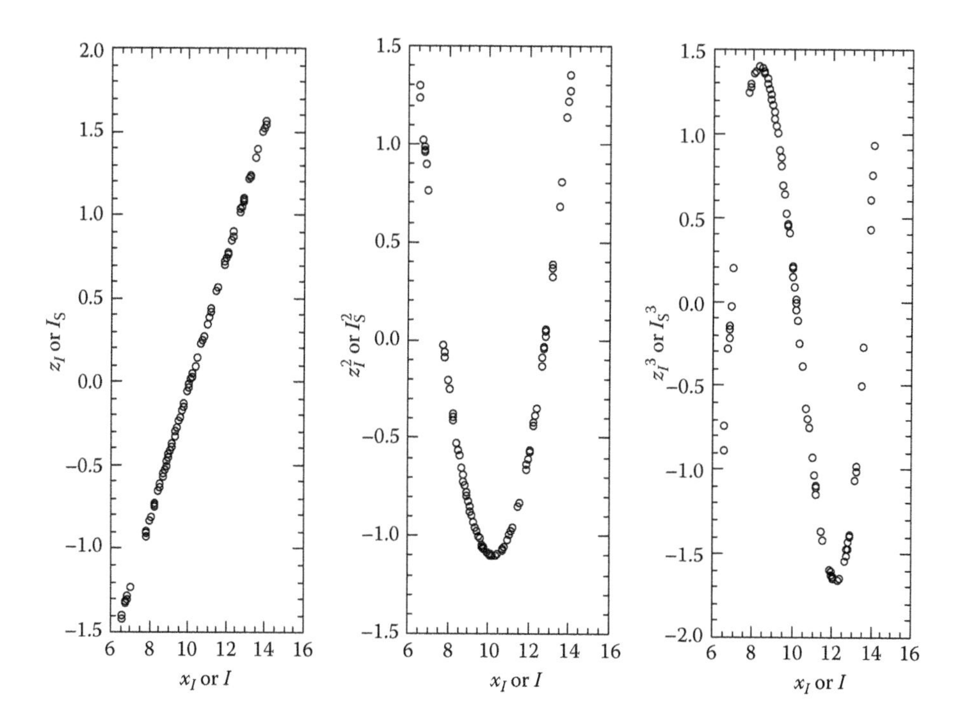

FIGURE 10.16
Relation between original and coded data.

$$I_s = a_{10} + a_{11}I$$
$$I_s^2 = a_{20} + a_{21}I + a_{22}I^2$$
$$I_s^3 = a_{30} + a_{31}I + a_{32}I^2 + a_{33}I^3 \tag{10.31}$$

where the a_{ij} values are the transformation coefficients. Applying similar transformations to time τ and force F, Equation 10.29 can be rewritten as follows:

$$\begin{aligned} f(I,\tau,F) \approx\ & \theta_{000} + \theta_{100}I_s + \theta_{010}\tau_s + \theta_{001}F_s + \theta_{200}I_s^2 \\ & + \theta_{020}\tau_s^2 + \theta_{002}F_s^2 + \theta_{110}I_s\tau_s + \theta_{101}I_sF_s + \theta_{011}\tau_sF_s \\ & + \theta_{300}I_s^3 + \theta_{030}\tau_s^3 + \theta_{003}F_s^3 + \theta_{210}I_s^2\tau_s + \theta_{201}I_s^2F_s \\ & + \theta_{021}\tau_s^2F_s + \theta_{120}I_s\tau_s^2 + \theta_{102}I_sF_s^2 \\ & + \theta_{012}\tau_sF_s^2 + \theta_{111}I_s\tau_sF_s \end{aligned} \tag{10.32}$$

where θ_{000} is the coefficient of constant effect; θ_{100}, θ_{010}, and θ_{001} are coefficients of linear effects; θ_{200}, θ_{020}, and θ_{002} are coefficients of quadratic effects; and θ_{300}, θ_{030}, and θ_{003} are coefficients of cubic effects. The subscripts denote the order of the input variables. Other θ_{ijk} values are coefficients of interaction effects between I_s, τ_s, and F_s.

Because polynomial terms of z in the orthogonal coding system have better orthogonality properties than those of x, the estimators of coefficients in a model formed by polynomial terms of z are more efficient and statistically independent. This makes a model selection procedure accurate, however, at the expense of losing intuitive physical interpretations of the coefficients. The fitted model using the orthogonal coding system can be transformed back to a function of x with more meaningful coefficients. In this example, Equation 10.32 is used to obtain a fitted model, and then it is transformed back to obtain a model of expulsion in the natural scale (Equation 10.29).

10.4.3.3 Use of Pseudo-Data

The settings used in the steel welding experiment are shown in Figure 10.17. It shows that the experiment region of the welding current shifts to the right side when time decreases or force increases. Settings with a low current and short time, as well as those with a high current and long time, were deliberately left out. The reason is that in those regions, expulsion either never happens (low settings) or always happens (high settings), and there is no need to conduct experiments at such settings. Such information that can be obtained without conducting an actual experiment is called prior knowledge.

Although there is no need to conduct experiments, information in such regions is needed to build a statistical model. The Bayesian approach is often used to deal with prior knowledge, which is represented as distributions of coefficients. It is usually difficult to translate

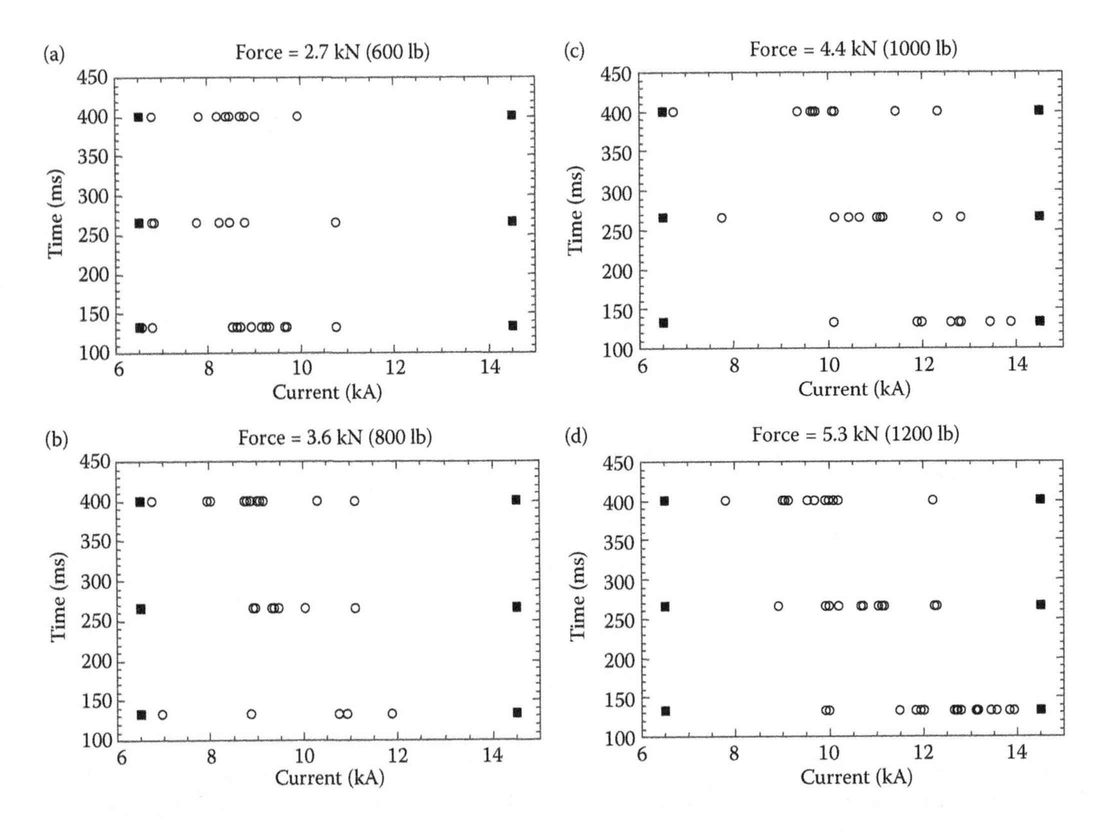

FIGURE 10.17
Treatment combinations including pseudo and actual settings.

the aforementioned prior knowledge of expulsion into distributions of coefficients, so an alternative can be adopted by using pseudo-data to represent the prior knowledge. Twelve pseudo no-expulsion data were created on the low-current side, and 12 expulsion data were created on the high-current side. They are shown as solid dots in Figure 10.17. Using a data set containing the pseudo-data, the fitted model is able to represent the information on expulsion.

10.4.3.4 Analysis and Results

The procedure of statistical analysis of expulsion is presented in detail, using the data of steel welding as an example.

10.4.3.4.1 Statistical Model Selection

Very often only some of the effects are important and have a significant influence on the output (probability of getting expulsion in this case). Other insignificant effects can be screened out by the means of model selection. Besides obtaining a model with only influential effects, another purpose of model selection is to get a balance between goodness of fit and generality.

In this study, a criterion-based method was used for model selection. One of the commonly used criteria for general linear models, the C_p criterion,[4] was applied to each subset of the full model. This criterion is a measurement of both goodness of fit and generality. An appropriate model can be found by comparing the C_p values for each submodel. The procedure of model selection is described as follows:

1. Estimating p_x, the probability of getting expulsion on setting x, by y_x/n_x, the so-called *observed* p_x. n_x is the number of replicates, y_x is the number of expulsions observed on setting x, and y_x/n_x is the portion of the replicates in which expulsion happened. It is an intuitive estimation of p_x when no physical relationship is assumed between p_x and x.
2. Transforming the logistics model (Equation 10.27) and replacing p_x in $\log(p_x/(1-p_x))$ with $y_x n_x$, and then denoting the logistics expression by w_x. To avoid divergence, 0.999 is used for $y_x/n_x = 1$, and 0.001 is used for $y_x/n_x = 0$. The vector of Equation 10.32 can then be expressed as a general linear model:

$$
\begin{aligned}
w_x \approx\ & \theta_{000} + \theta_{100} z_I + \theta_{010} z_\tau + \theta_{001} z_F + \theta_{200} z_I^2 \\
& + \theta_{020} z_\tau^2 + \theta_{002} z_F^2 + \theta_{110} z_I z_\tau + \theta_{101} z_I z_F \\
& + \theta_{011} z_\tau z_F + \theta_{300} z_I^3 + \theta_{030} z_\tau^3 + \theta_{003} z_F^3 \\
& + \theta_{210} z_I^2 z_\tau + \theta_{201} z_I^2 z_F + \theta_{021} z_\tau^2 z_F + \theta_{120} z_I z_\tau^2 \\
& + \theta_{102} z_I z_F^2 + \theta_{012} z_\tau z_F^2 + \theta_{111} z_I z_\tau z_F
\end{aligned}
\tag{10.33}
$$

where w_x is called the dependent variable and $z_I, z_\tau, z_F, z_I^2, z_\tau^2, z_F^2, z_I z_\tau, z_I z_F, z_\tau z_F, \ldots$, are the independent variables in the general linear model.
3. Applying model selection criteria to choose the best statistical model. The C_p criterion is then applied on dependent and independent variables. A small C_p value is preferred. Figure 10.18 shows how C_p values vary with the number of effects for

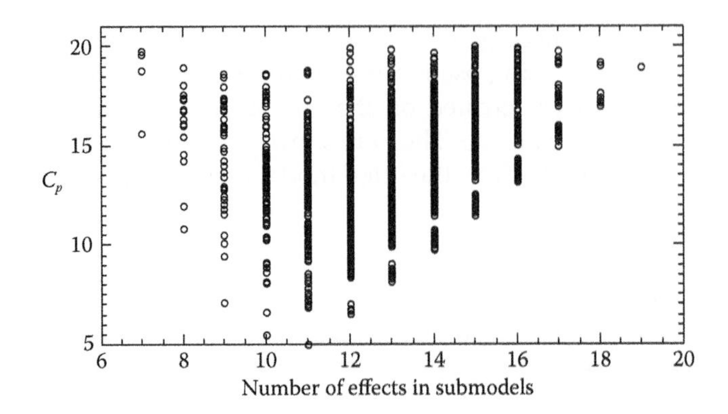

FIGURE 10.18
Relationship between C_p value and number of effects in submodels.

the best submodels. It can be seen that the C_p value first goes down, then up when the number of effects in the submodels increases, which means that a balance between goodness of fit and generality can be reached.

The model selection procedure described above is only based on statistical consideration. Practical knowledge about the process also needs to be used to determine a good model, such as that demonstrated in Section 7.4.2.1.

By applying this model selection method to the data set, a model can be chosen in the orthogonal coding system, which contains linear, quadratic, and cubic effects, and their interactions: z_I, z_τ, z_F, z_τ^2, z_F^2, z_I^3, z_F^3, $z_\tau z_F$, $z_I z_F^2$, $z_I^2 z_\tau$, $z_I^2 z_F$, $z_\tau^2 z_F$.

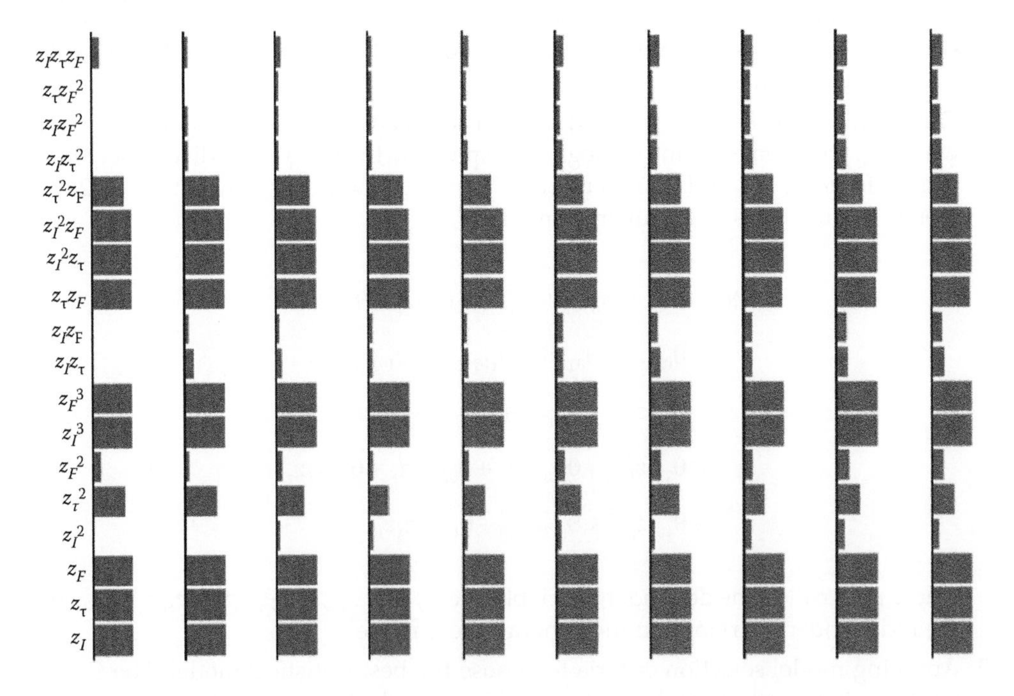

FIGURE 10.19
Appearance frequency of effects in first 5, 10, 15, 20, 25, 30, 50, 100, 150, and 200 best models.

10.4.3.4.2 Identifying Influential Effects

A model selected through previous steps usually contains many effects. Because of the collinearity between effects, less important effects in the chosen model may be replaced by others, and the new model still preserves the same goodness of fit.

The results of model selection can help to identify important effects. Intuitively, if one effect has a strong influence on the response, it should appear in most of the good models. Therefore, the frequency of each effect appearing in most of the best models is calculated, and effects with high appearance frequency are identified as influential effects (Figure 10.19).

Eight effects can be identified as influential for the data set of steel welding: z_I, z_τ, z_F, z_I^3, z_F^3, $z_\tau z_F$, $z_I^2 z_\tau$, and $z_I^2 z_F$. They have frequencies of 1 from the beginning to the end, which means that they appear in all 200 best models. Moreover, two other effects, z_τ^2 and $z_\tau^2 z_F$, also showed high frequency of appearance. It is noteworthy to see that all 10 of these effects were included in the model identified in the previous model selection section.

10.4.3.4.3 Estimating Magnitudes of Effects

After choosing a statistical model by the model selection procedure described above, coefficients θ_{ijk}, the magnitudes of effects in the model, can be estimated. For a logistics model, the estimation proceeds by an iterative weighted least square procedure to get the maximum likelihood estimate of θ_{ijk}.[18] By doing this, the coefficients of the steel welding model are estimated, and the model under the orthogonal coding system can be expressed explicitly as follows:

$$\begin{aligned}\log(p_x/(1-p_x)) \approx{} & (-9.037)+(38.360)I_s+(10.779)\tau_s \\ & +(-16.215)F_s+(-1.816)\tau_s^2+(1.385)F_s^2 \\ & +(3.645)\tau_s F_s+(6.236)I_s^3+(0.677)F_s^3 \\ & +(4.811)I_s^2\tau_s+(-8.253)I_s^2F_s \\ & +(-0.420)\tau_s^2F_s+(5.358)I_sF_s^2\end{aligned} \tag{10.34}$$

It can be transformed back to the coding system of natural scale (with true values of welding current, kA; time, ms; and force, kN), as shown below:

$$\begin{aligned}\log(p_x/(1-p_x)) \approx{} & (-7.6449\times 10^2)+(1.6731824\times 10^2)I \\ & +(7.12636\times 10^{-1})\tau+(9.7174\times 10^1)F \\ & +(-1.54168327\times 10^1)I^2+(-1.49\times 10^{-5})\tau^2 \\ & +(-4.234\times 10^1)F^2+(6.251982\times 10^{-1})I^3 \\ & +(1.4202468)F^3+(-1.540455\times 10^{-1})I\tau \\ & +(8.088965)IF+(6.08688\times 10^{-2})\tau F \\ & +(7.5306\times 10^{-3})I^2\tau+(-1.4449971)I^2F \\ & +(-5.12\times 10^{-5})\tau^2F+(2.6919807)IF^2 \\ \equiv{} & f(I,\tau,F)\end{aligned} \tag{10.35}$$

The fitted probability can be obtained by a simple transformation of the above expression (Equation 10.28) as:

$$p_x = e^{f(I,\tau,F)}/(1 + e^{f(I,\tau,F)}) \tag{10.36}$$

By standardizing the estimated coefficients in Equation 10.25 with respect to their experimental ranges, and then comparing their magnitudes, the influential effects in the natural scale are identified as (in the order of importance): I^2, I^3, I, I^2F, IF^2, $I\tau$, $I^2\tau$, F^2, and IF.

After a statistical model is built, it needs to be judged by its closeness to the original data by using diagnostic methods, such as residual analysis, to see if there is any significant contradiction. Residual analysis on the model presented in Equation 10.35 shows reasonable agreement between the observed and fitted values.

10.4.3.5 Inference and Decision Making

A detailed discussion is shown in Chapter 7, and it shows that the expulsion models created by the statistical procedure outlined in previous sections indeed provide some insights into the expulsion phenomenon. The response surfaces and contour plots generated using models of natural scale, such as the one in Equation 10.35, can be used to obtain a better understanding of the process and to provide a guideline for welding schedule selection. There are many other applications of statistical analysis of categorical responses in resistance welding. For instance, welds can be classified as comformable and uncomformable, instead of being identified by their sizes, which are continuous; tested specimens can be grouped into ones with pull-out weld buttons, interfacial failure, partial thickness failure, etc.

10.5 Computer Simulation Experiments

Computer simulation has been widely used in science and engineering for studying physical phenomena that are too time-consuming, too expensive, too complicated to deal with theoretically, or just impossible for performing physical experiments. In a computer simulation experiment, a computer code is run to simulate the complex physical phenomena and to obtain observations on a response y at various choices of input factors. With the aid of statistical design and computer technology (both hardware and software), computer simulation can provide a quick and accurate solution for a very complex problem. Using such a procedure, the uncertainties associated with physical experiments can be avoided and the cost can be significantly reduced. Besides, more information can be obtained using less number of runs than the conventional design of experiments. (Statistical) computer experiments may serve many purposes, such as optimizing the response, visualizing the influence of factors, or developing a simpler predictor to approximate the complex computer code. The relationship between the response and the factors in a computer experiment is quite different from that in physical experiments. For example, if a computer code is run two times with the same input, the same output is expected, whereas in a physical experiment, the experimental error usually causes different outputs. The lack of experimental error leads to some important distinctions, as follows:

1. In computer experiments, the change of levels of factors is often only a matter of inputting different values into the computer code. It is much easier than in a physical experiment, in which taking more levels often means additional cost or effort for implementation. Therefore, computer experiments may take as many levels of factors as needed. Certain numerical simulations, such as simulating an RSW process need to control the total number of treatment combinations as the calculation cost, often in terms of computer time, can be high.
2. Considerations such as blocking, replication, and randomization in physical experimental design are irrelevant.
3. The difference between a computer model and a fitted model is determined solely by model bias. The usual measures of uncertainty derived from least squares residuals in physical experiments are interpreted as a measurement of model bias in computer experiments.

These differences between computer experiments and physical experiments call for new techniques and different thinking in design and analysis of computer simulation experiments. For a summary, see Sacks et al.[19] and Koehler and Owen.[20]

10.5.1 Experiment Design

There are two main approaches for the design of computer experiments. One is based on the Bayesian modeling and another is based on a technique called space filling. The key issue in the Bayesian approach is to select a design for building an efficient Bayesian model. Because there is no experimental error in the model, the design construction is based on the optimization of some criteria that are related to the model bias. Some optimality criteria, such as entropy, mean squared error, maximin, and minimax, have been proposed. For a detailed discussion of these criteria, see Koehler and Owen.[20] An alternative approach, which is not dependent on the model, is to uniformly choose design points from the experimental region, based on a so-called space-filling property. A class of popular space-filling designs are Latin hypercube designs. An n-run Latin hypercube design for d factors is an $n \times d$ matrix, which consists of d permutations of the vector $(1, 2, \ldots, n)^T$. The one-factor projection of a Latin hypercube design is evenly spaced in the experimental region. One advantage of Latin hypercube designs appears when the response is dominated by only a few factors. No matter which factors turn out to be important, the design ensures that each of those factors is represented in a fully uniform manner. Efforts have been made to choose a good Latin hypercube design. For example, Owen[21] and Tang[22] independently proposed OA-based Latin hypercube designs. Iman and Conover,[23] Owen,[24] Tang,[25] and Ye[26] proposed Latin hypercube designs with small correlations between effects. Park[27] and Morris and Mitchell[28] searched for Latin hypercube designs having good properties for Bayesian prediction.

10.5.2 Analysis and Modeling

A statistical model for computer experiments can be simply expressed as

$$\text{Response} = (\text{linear model}) + (\text{systematic departure}) \tag{10.37}$$

The linear model part is a function of factors that approximate the computer code, and the systematic departure part represents the model bias between the linear model and the computer program. By comparing this model with the model in Equation 10.1, the main difference is that the experimental error in Equation 10.1 is replaced by the departure. There are two main approaches for modeling the departure: Bayesian and frequentist. The Bayesian modeling treats the systematic departure as a realization of a stochastic process Z, in which the covariance structure of Z relates to the smoothness of the response. An advantage of the Bayesian modeling is that it has the exact prediction (i.e., no systematic departure) as the observed response at a design point, and predicts with an increasing departure as the prediction point moves away from the design points. The selection of covariance structure plays a crucial role in constructing Bayesian designs in the analysis. For a detailed discussion of the Bayesian approach and the selection of covariance structure, refer to Sacks et al.[19] and Koehler and Owen.[20] For the user who is familiar with multiple regression modeling, the frequentist approach may be more convenient. In the frequentist approach, all the techniques noted in Section 10.3.2 for the analysis of multiple regression models can be applied to develop a fitted linear model for computer experiment. The only difference is that the residuals are due to model bias, not experimental error. An illustration of Latin hypercube design and analysis based on the frequentist approach is given in Example 10.7.

Example 10.7: Computer Simulation of the Relationship between Quality and Attributes of Spot Welds

In a paper by Zhou et al.,[29] a detailed computer modeling of spot-welded joints was presented and a computer design of experiments was introduced to evaluate the spot weld strength. The effects of weld attributes on weld quality/strength have been evaluated, and the relationship between weld strength and weld attributes has been quantitatively established.

Based on previous studies,[9] a weld's strength can be fully expressed by its peak load, and corresponding energy and displacement at peak load, for a tensile–shear tested specimen. Intuitively, they can be expressed as functions of the joint geometry and material properties, or

$$P_{max} = f_P \text{ (geometry; material properties of base metal, HAZ, and nugget)} \quad (10.38a)$$

$$U_{max} = f_U \text{ (geometry; material properties of base metal, HAZ, and nugget)} \quad (10.38b)$$

$$W_{max} = f_W \text{ (geometry; material properties of base metal, HAZ, and nugget)} \quad (10.38c)$$

where P_{max} is the peak load and U_{max} and W_{max} are corresponding displacement and energy, respectively. In general, all these relationships are unknown. It is also impossible to derive them analytically. Therefore, a computer simulation experiment was used to establish such relationships.

10.5.2.1 Planning of Numerical Experiments

Procedures for the design are detailed in the following sections for quality evaluation of spot-welded specimens.

10.5.2.1.1 Selection of Variables

As in the conventional design of experiments, the first task is to choose experiment variables. Because the numerical computation involves remeshing and refining the mesh to obtain convergence that are time-consuming, it is desirable to have as few number of treatment combinations (runs) as possible. Reduction of variables can be achieved based on prior knowledge on the effects of and relations among the variables. There are two sets of variables needed in this study. One is for geometrical dimensions, which generally include sheet thickness, specimen length, specimen width, sheet overlap, nugget diameter, HAZ size, indentation, sheet separation, and so on. Based on previous studies,[9] the length can be fixed at $L = 150$ mm, and the overlap is equated to the width of specimen. For simplicity, only large-size welds are taken into account, and the nugget diameter is linked to the sheet thickness by $d = 5\sqrt{t}$. The extreme cases of welded joints with a sharp notch around the nugget are considered. Therefore, the number of geometrical variables is reduced and only sheet thickness (t), sheet width (w), HAZ size (h), and indentation (t_i) are treated as variables in the design. The other group of variables includes material properties, which are Young's modulus (E), Poisson's ratio (ν), yield strength (σ_y), ultimate tensile strength (σ_{uts}), and elongation (e). Because the material structures in nugget, HAZ, and base metal are different, different material properties are needed for each part of the weldments. However, they are not independent—material properties of the nugget and HAZ can be approximately linked to those of the base metal by hardness value (H_v), with relations as shown in Chapter 5. By using these equations, five fewer material variables are needed. Furthermore, if only steel is considered, the Young's modulus and Poisson's ratio can be fixed as constants ($E = 210$ GPa and $\nu = 0.3$). Therefore, in the design, only the base metal properties and the hardness ratio (k) between the nugget and base metal are left as material variables.

Therefore, Equation 10.38 can be simplified as

$$P_{max} = f_P(t, w, h, t_i; \sigma_y, \sigma_{uts}, e, k) \quad (10.39a)$$

$$U_{max} = f_U(t, w, h, t_i; \sigma_y, \sigma_{uts}, e, k) \quad (10.39b)$$

$$W_{max} = f_W(t, w, h, t_i; \sigma_y, \sigma_{uts}, e, k) \quad (10.39c)$$

In numerical computation, σ_{uts} is replaced by σ_0, which is the difference between the ultimate tensile strength σ_{uts} and yield strength σ_y. A positive σ_0 ensures that ultimate tensile strength is always greater than the yield strength. Otherwise, σ_{uts} may be less than σ_y in some runs in the design, which is not physically realistic. Table 10.4 lists the ranges of each variable, which are used in the statistical design.

10.5.2.1.2 Latin Hypercube Design

The Latin hypercube method was found to be very useful in conducting computer experiments.[20,26] A class of orthogonal Latin hypercubes that preserve orthogonality among columns

TABLE 10.4

Ranges of Input Variables

T (mm)	h (mm)	w (mm)	t_i	σ_y (MPa)	σ_0 (MPa)	e (%)	k
0.5–2.0	0.1–1.5	30–50	0–20%	205–1725	50–200	2–65	1.0–3.0

is available for this purpose. Applying an orthogonal Latin hypercube design to a computer experiment benefits the data analysis in two ways. First, it retains the orthogonality of traditional experimental designs. The estimates of linear effects of all factors are uncorrelated not only with each other, but also with the estimates of all quadratic effects and bilinear interactions. Second, it facilitates nonparametric fitting procedures, as one can select good space-filling designs within the class of orthogonal Latin hypercubes according to selection criteria.

Table 10.5 is an optimal Latin hypercube design for eight variables based on the maximum distance criterion. Using this criterion, the design points are uniformly distributed in the design space, which eliminates the random effects and ensures that all the points are not too far, nor too close to each other. In this design, there are 33 levels for each variable, ranging from –16 to 16 in coded scale. Figure 10.20 shows the distributions of design

TABLE 10.5

Matrix of Latin Hypercube Design (in Coded Scale)

var1	var2	var3	var4	var5	var6	var7	var8
1	–2	–4	–8	–16	15	13	–9
2	1	–3	–7	–15	–16	14	–10
3	–4	2	–6	–14	13	–15	–11
4	3	1	–5	–13	–14	–16	–12
5	–6	–8	4	–12	11	9	13
6	5	–7	3	–11	–12	10	14
7	–8	6	2	–10	9	–11	15
8	7	5	1	–9	–10	–12	16
9	–10	–12	–16	8	–7	–5	1
10	9	–11	–15	7	8	–6	2
11	–12	10	–14	6	–5	7	3
12	11	9	–13	5	6	8	4
13	–14	–16	12	4	–3	–1	–5
14	13	–15	11	3	4	–2	–6
15	–16	14	10	2	–1	3	–7
16	15	13	9	1	2	4	–8
0	0	0	0	0	0	0	0
–16	–15	–13	–9	–1	–2	–4	8
–15	16	–14	–10	–2	1	–3	7
–14	–13	15	–11	–3	–4	2	6
–13	14	16	–12	–4	3	1	5
–12	–11	–9	13	–5	–6	–8	–4
–11	12	–10	14	–6	5	–7	–3
–10	–9	11	15	–7	–8	6	–2
–9	10	12	16	–8	7	5	–1
–8	–7	–5	–1	9	10	12	–16
–7	8	–6	–2	10	–9	11	–15
–6	–5	7	–3	11	12	–10	–14
–5	6	8	–4	12	–11	–9	–13
–4	–3	–1	5	13	14	16	12
–3	4	–2	6	14	–13	15	11
–2	–1	3	7	15	16	–14	10
–1	2	4	8	16	–15	–13	9

variables projected onto a space of any two variables. As noted above, each variable has a design range. All ranges are evenly divided and distributed to the corresponding levels, which are given in Table 10.6. The comparison results (outputs) are also listed in the table.

In order to effectively conduct the experiment, a generic finite element model was developed so that changes in geometrical variables (width, thickness, nugget size, HAZ size, indentation) and material variables (elastic and plastic properties in base metal, nugget, zones in the HAZ) can be easily implemented. A special code was developed for this purpose, which can automatically update the FEM model and design parameters, using an FEM model of a spot weld as shown in Figure 6.9. To ensure the convergence and accuracy, a large number of nodes and elements are used. A fracture mechanics model is used to cope with the high stress concentration existing around the nugget periphery. Different material properties are used for the nugget, HAZs, and base metal.

10.5.2.2 Results and Inference

Using the results of Table 10.6, models of peak load, maximum displacement, and maximum energy can be derived by the regression method. Figure 10.21 shows the influences of variables on P_{max}, which is traditionally used to describe the quality of spot welds. It

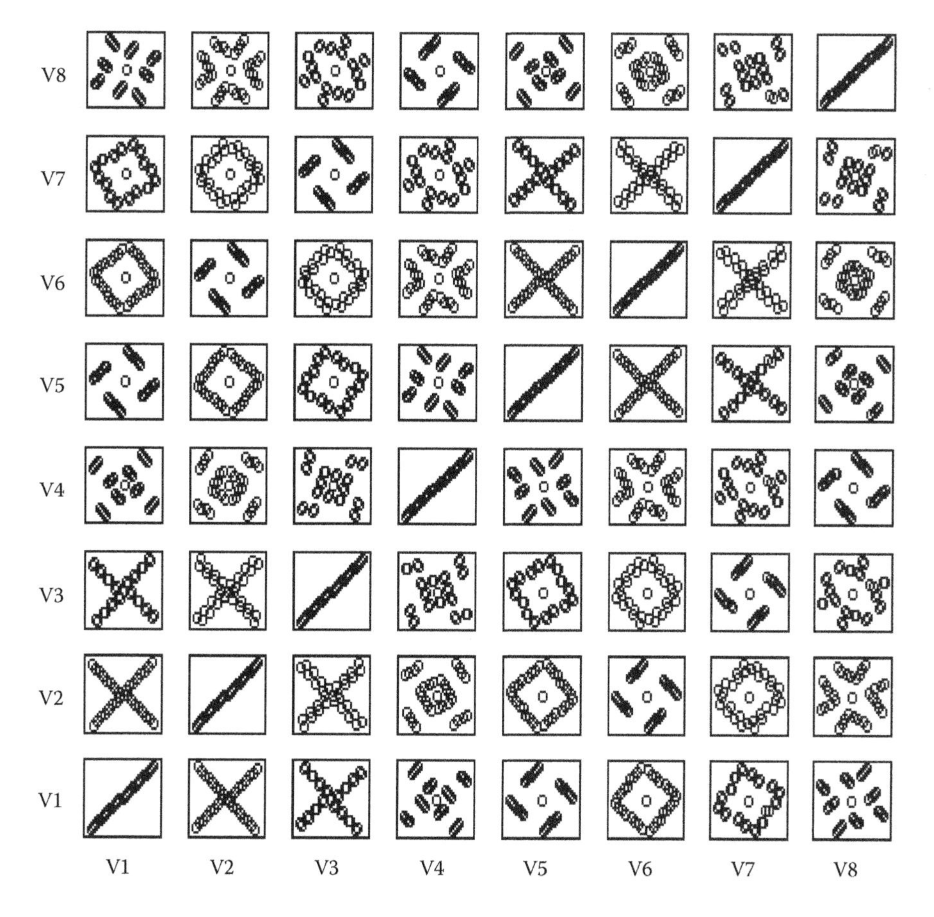

FIGURE 10.20
Distribution of Latin hypercube design.

TABLE 10.6

Matrix of Latin Hypercube Design (in Natural Scale) and Responses

Run	t	h	w	t_i	σ_y	$\sigma_{uts}-\sigma_y$	u	k	U_{max}	P_{max}	W_{max}	J_e
1	1.27	0.74	37.88	0.055	251.06	190.91	0.58	1.48	1.30	4.53	4,683	5.48E-07
2	1.32	0.82	38.48	0.061	297.12	54.55	0.60	1.42	0.93	3.86	2,960	5.16E-07
3	1.36	0.65	40.91	0.067	343.18	181.82	0.10	1.36	1.06	5.32	4,528	4.78E-07
4	1.41	0.91	40.30	0.073	389.24	63.64	0.08	1.30	0.78	4.92	3,003	4.51E-07
5	1.45	0.57	35.45	0.121	435.30	172.73	0.51	2.76	1.38	8.45	8,845	4.57E-07
6	1.50	0.99	36.06	0.115	481.36	72.73	0.52	2.82	1.03	8.43	6,359	4.26E-07
7	1.55	0.48	43.33	0.109	527.42	163.64	0.17	2.88	0.58	8.82	3,120	3.72E-07
8	1.59	1.08	42.73	0.103	573.48	81.82	0.15	2.94	1.00	11.02	7,735	3.54E-07
9	1.64	0.40	33.03	0.006	1,310.45	95.45	0.27	2.03	0.86	15.51	6,989	3.67E-07
10	1.68	1.16	33.64	0.012	1,264.39	159.09	0.26	2.09	1.79	22.78	26,730	3.40E-07
11	1.73	0.31	45.76	0.018	1,218.33	104.55	0.47	2.15	0.44	11.03	2,448	2.75E-07
12	1.77	1.25	45.15	0.024	1,172.27	150.00	0.49	2.21	1.63	24.10	26,740	2.77E-07
13	1.82	0.23	30.61	0.170	1,126.21	113.64	0.35	1.73	0.53	10.30	2,710	3.34E-07
14	1.86	1.33	31.21	0.164	1,080.15	140.91	0.33	1.67	1.44	19.59	18,260	3.12E-07
15	1.91	0.14	48.18	0.158	1,034.09	122.73	0.40	1.61	0.18	4.72	416	2.22E-07
16	1.95	1.42	47.58	0.152	988.03	131.82	0.42	1.55	1.42	20.07	20,540	2.37E-07
17	1.25	0.80	40.00	0.100	965.00	125.00	0.36	2.00	1.05	11.25	7,609	5.58E-07
18	0.55	0.18	32.42	0.048	941.97	118.18	0.29	2.45	0.87	3.46	1,988	2.63E-06
19	0.59	1.46	31.82	0.042	895.91	127.27	0.31	2.39	0.81	3.94	1,976	2.24E-06
20	0.64	0.27	48.79	0.036	849.85	109.09	0.38	2.33	0.80	4.09	2,260	1.81E-06
21	0.68	1.37	49.39	0.030	803.79	136.36	0.36	2.27	0.76	4.52	2,321	1.57E-06
22	0.73	0.35	34.85	0.176	757.73	100.00	0.22	1.79	0.62	3.78	1,476	1.60E-06
23	0.77	1.29	34.24	0.182	711.67	145.45	0.24	1.85	0.72	4.05	1,938	1.41E-06
24	0.82	0.44	46.36	0.188	665.61	90.91	0.45	1.91	0.56	4.08	1,512	1.20E-06
25	0.86	1.20	46.97	0.194	619.55	154.55	0.44	1.97	0.68	4.47	2,141	1.07E-06
26	0.91	0.52	37.27	0.097	1,356.52	168.18	0.56	1.06	1.08	8.22	5,698	1.01E-06
27	0.95	1.12	36.67	0.091	1,402.58	86.36	0.54	1.12	1.12	8.61	6,306	9.23E-07
28	1.00	0.61	43.94	0.085	1,448.64	177.27	0.19	1.18	1.07	10.21	6,978	8.05E-07
29	1.05	1.03	44.55	0.079	1,494.70	77.27	0.20	1.24	1.09	10.57	7,518	7.34E-07
30	1.09	0.69	39.70	0.127	1,540.76	186.36	0.63	2.70	1.41	14.96	13,010	7.18E-07
31	1.14	0.95	39.09	0.133	1,586.82	68.18	0.61	2.64	1.34	14.92	12,360	6.69E-07
32	1.18	0.78	41.52	0.139	1,632.88	195.45	0.11	2.58	1.39	16.67	14,390	6.13E-07
33	1.23	0.86	42.12	0.145	1,678.94	59.09	0.13	2.52	1.25	16.72	12,580	5.60E-07

indicates that the yield strength and sheet thickness have a greater influence than any other variables. The size of HAZ also plays an important role in P_{max}. By selecting most of the effects, P_{max} can be expressed as

$$\begin{aligned} P_{max} = & \, 2.64 - 32.18t + 32.08h - 59.70t_i - 0.0123\sigma_y \\ & + 0.0117\sigma_{uts} + 3.74k + 11.54t \cdot h + 0.0137t \cdot \sigma_y \\ & + 8.022h \cdot \sigma_y + t^2 - 0.00000372\sigma_y^2 \\ & + 0.0000936(\sigma_{uts} - \sigma_y)^2 + 224.94t_i^2 - 28.20h^2 \text{ (kN)} \end{aligned} \quad (10.40)$$

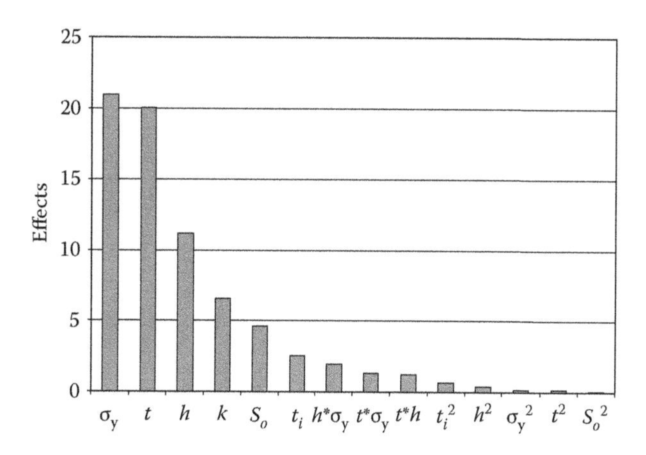

FIGURE 10.21
Variable effects on maximum load P_{max}.

with 99.3% confidence of determination (R = 99.3%).

If only selecting sheet thickness t, yield strength σ_y, and size of HAZ h as variables, P_{max} can be expressed as

$$P_{max} = -6.74 + 2.72t + 0.016\sigma_y - 10.99h + 16.31t \cdot h \text{ (kN)} \tag{10.41}$$

Statistically, it still has a high coefficient of determination (94.5%). Regarding the confidence intervals of the coefficients, if 95% confidence is considered, the intervals are [–15.86, 2.38], [–3.52, 8.97], [0.0128, 0.0197], [–19.90, –2.08], and [9.84, 22.79]. Based on these intervals, the number of significant digits of the coefficients can be determined as shown in Equation 10.41. Although Equation 10.41 has a smaller coefficient of determination, it is preferred to Equation 10.40 for simplicity. In fact, it has a better generality, meaning that it provides better predictions than Equation 10.40.

Following similar procedures, the expressions for W_{max} and U_{max} are obtained as shown in Equations 10.42 and 10.43. They have coefficients of determination of 97.6% and 97.0%, respectively.

$$\begin{aligned} W_{max} = {} & 126{,}966 - 414{,}160t + 325{,}520h - 106.718\sigma_y \\ & + 70.45\sigma_{uts} + 3288k - 6898.8t \cdot h + 22.50t \cdot \sigma_y \\ & + 26.916h \cdot \sigma_y + 164{,}950t^2 - 204{,}840h^2 \end{aligned} \tag{10.42}$$

$$\begin{aligned} U_{max} = {} & 3.41 - 12.49t + 10.26h - 0.012w \\ & - 1.07t_i - 0.0525\sigma_y + 0.0484\sigma_{uts} + 0.347e \\ & + 0.0644k + 5.05t^2 - 6.15h^2 + 0.00000226\sigma_y^2 \\ & - 0.000184(\sigma_{uts} - \sigma_y)^2 \end{aligned} \tag{10.43}$$

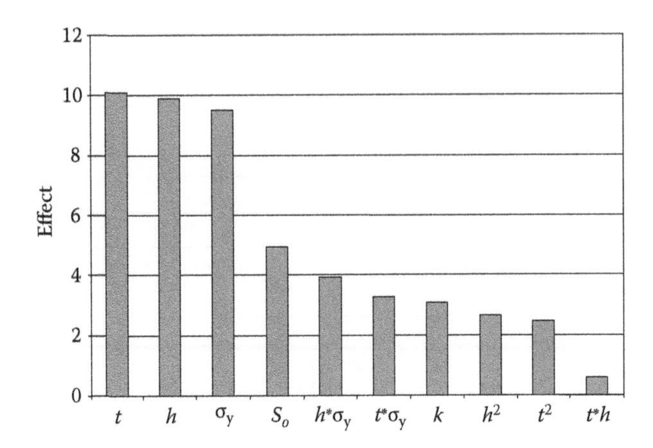

FIGURE 10.22
Variable effects on maximum energy W_{max}.

Variable effects on maximum energy W_{max} and displacement U_{max} are shown in Figures 10.22 and 10.23, respectively. Sheet thickness t, HAZ size h, and yield strength σ_y have the biggest effects for W_{max}. However, for maximum displacement, the most important variables are the quadratic terms of h and t and linear term of h. Therefore, it can be concluded that the most important variable affecting all three responses is the size of HAZ. However, in all cases, some other terms, such as quadratic and interactive terms, cannot be neglected in determining the peak load, maximum energy, and displacement.

This study provides a basic understanding of the dependence of weld quality on both geometric variables and material properties. The findings can be summarized as follows:

- Effects of weld attributes, such as weld diameter, penetration, and indentation, can be analyzed through this integrated numerical analysis.
- The size of a HAZ plays an important role in the analysis of weld strength due to high stress concentration in and around the HAZ.

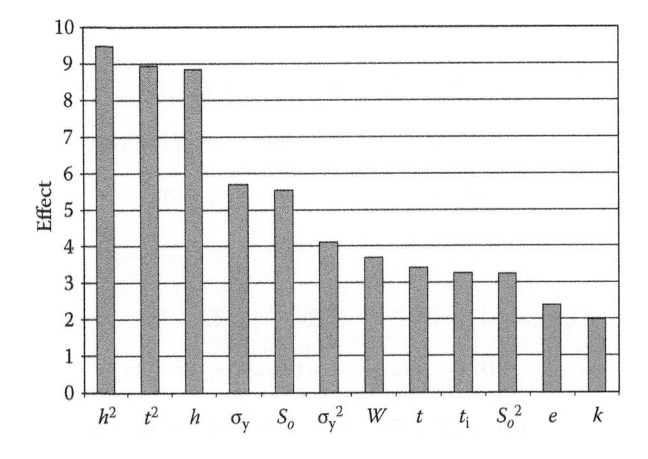

FIGURE 10.23
Variable effects on maximum displacement U_{max}.

- Sheet thickness (and therefore nugget diameter), HAZ, and yield strength of base metal are the critical parameters in determination of spot welding quality.

Although tensile–shear testing was used in the simulation, the method can be extended to other loading modes, such as cross-tension and fatigue.

10.6 Summary

Experimentation can be regarded as a learning process. In the process, questions about a system are formulated, experiments are performed, data are collected for investigation, conclusions are drawn from analyses, and then, based on the results, new questions are formulated for further exploration. This suggests that experiments should be conducted sequentially. A successful experiment requires knowledge of all aspects, such as the importance of factors, appropriate variable ranges, and number of levels. Generally, in the beginning stage of a study, the answers to these questions are not clear. They are learned as more information is gathered from experiments. In a sequential experimentation, some factors may be dropped, others may be added, the experimental region may be changed, and the statistical model may be modified. Consequently, an investigator must keep in mind that in order to obtain a complete understanding of the underlying physical system, a sequence of experiments may be required. It is not recommended that all resources be put on a large, complex, and exhaustive experiment. Instead, several smaller experiments performed in sequence are usually more efficient to reach the target.

References

1. Montgomery, D. C., *Design and Analysis of Experiments*, 6th edition, John Wiley & Sons, New York, 2005.
2. Wu, C. F. J. and Hamada, M., *Planning, Analysis and Parameter Design Optimization*, John Wiley & Sons, New York, 2000.
3. Draper, N. R. and Smith, H., *Applied Regression Analysis*, 3rd edition, John Wiley & Sons, New York, 1998.
4. Sen, A. and Srivastava, M., *Regression Analysis: Theory, Methods, and Applications*, Springer-Verlag, New York, 1990, 234–238.
5. Hamada, M. and Wu, C. F. J., Analysis of designed experiments with complex aliasing, *Journal of Quality Technology*, 24, 130, 1992.
6. Li, W., Cheng, S., Hu, S. J., and Shriver, J., Statistical investigation of resistance spot welding quality using a two-stage, sliding-level experiment, *Transaction of ASME—Journal of Manufacturing Science and Engineering*, 123, 513, 2001.
7. Box, G. E. P. and Meyer, R. D., Finding the active factors in fractionated screening experiments, *Journal of Quality Technology*, 25, 94–105, 1993.
8. Myers, R. H. and Montgomery, D. C., *Response Surface Methodology: Process and Product in Optimization Using Designed Experiments*, John Wiley & Sons, New York, 1995.
9. Zhou, M., Hu, S. J., and Zhang, H., Critical specimen sizes for tensile–shear testing of steel sheets, *Welding Journal*, 78, 305-s, 1999.

10. Cheng, S., Zhang, H., and Hu, S. J., Statistics in welding research design and analysis, in *Proceedings of Sheet Metal Welding Conference IX*, Sterling Heights, MI, Paper No. 5-5, 2000.
11. Taguchi, G. and Konishi, S., *Orthogonal Arrays and Linear Graphs*, ASI Press, Dearborn, MI, 1987.
12. Zhang, H., Hu, J. S., Senkara, J., and Cheng, S., Statistical analysis of expulsion limits in resistance spot welding, *Transaction of ASME—Journal of Manufacturing Science and Engineering*, 122, 501, 2000.
13. Atkinson, A. C. and Haines, L. M., Designs for nonlinear and generalized linear models, *Handbook of Statistics*, 13, 437, 1996.
14. Sitter, R. R. and Forbes, B., Optimal two-stage designs for binary response experiments, *Statistica Sinica*, 7, 941, 1997.
15. Sitter, R. R. and Wu, C. F. J., Two stage design of quantal response studies, *Biometrics*, 55, 396, 1999.
16. Agresti, A., *Categorical Data Analysis*, John Wiley & Sons, New York, 1990.
17. Hosmer, D. W. and Lemeshow, S., *Applied Logistic Regression*, 2nd edition, John Wiley & Sons, New York, 2000.
18. McCullagh, P. and Nelder, J. A., Generalized linear models, 2nd edition, Chapman & Hall, London, UK, 21–135, 1989.
19. Sacks, J., Welch, W. J., Mitchell, T. J., and Wynn, H. P., Design and analysis of computer experiments, *Statistical Science*, 4, 409, 1989.
20. Koehler, J. R. and Owen, A. B., Computer experiments, *Handbook of Statistics*, 13, 261, 1996.
21. Owen, A. B., Orthogonal arrays for computer experiments, integration and visualization, *Statistica Sinica*, 2, 439, 1992.
22. Tang, B., Orthogonal array-based Latin hypercubes, *Journal of the American Statistical Association*, 88, 1392, 1993.
23. Iman, R. L. and Conover, W. J., A distribution-free approach to inducing rank correlation among input variables, *Communications in Statistics, Part B—Simulation and Computation*, 11, 311, 1982.
24. Owen, A. B., Controlling correlations in Latin hypercube samples, *Journal of the American Statistical Association*, 89, 1517, 1994.
25. Tang, B., Selecting Latin hypercubes using correlation criteria, *Statistica Sinica*, 8, 965, 1998.
26. Ye, K. Q., Orthogonal column Latin hypercubes and their application in computer experiments, *Journal of the American Statistical Association*, 93, 1430, 1998.
27. Park, J.-S., Optimal Latin-hypercube designs for computer experiments, *Journal of Statistical Planning and Inference*, 39, 95, 1994.
28. Morris, M. and Mitchell, T., Exploratory design for computer experiments, *Journal of Statistical Planning and Inference*, 43, 381, 1995.
29. Zhou, M., Hu, S. J., and Zhang, H., Relationships between quality and attributes of spot welds, *Welding Journal*, 82, 72, 2003.

Index

N

O

P

X

Y

Z

For Product Safety Concerns and Information please contact our EU
representative GPSR@taylorandfrancis.com Taylor & Francis Verlag GmbH,
Kaufingerstraße 24, 80331 München, Germany

Printed and bound by CPI Group (UK) Ltd, Croydon, CR0 4YY
01/05/2025
01858479-0001